INTRODUCTION TO WIND PRINCIPLES

Thomas E. Kissell
Terra Community College, Fremont, Ohio

Prentice Hall
Boston Columbus Indianapolis New York San Francisco Upper Saddle River Amsterdam
Cape Town Dubai London Madrid Milan Munich Paris Montreal Toronto Delhi
Mexico City Sao Paulo Sydney Hong Kong Seoul Singapore Taipei Tokyo

Editor in Chief: Vernon R. Anthony
Acquisitions Editor: David Ploskonka
Development Editor: Dan Trudden
Editorial Assistant: Nancy Kesterson
Director of Marketing: David Gesell
Marketing Manager: Derril Trakalo
Senior Marketing Coordinator: Alicia Wozniak
Marketing Assistant: Les Roberts
Project Manager: Holly Shufeldt
Senior Art Director: Jayne Conte
Cover Designer: Karen Salzbach
Visual Research Manager: Beth Brenzel
Cover Art: iStockphoto
Full-Service Project Management and Composition: Jogender Taneja/Aptara®, Inc.
Printer/Binder: Edwards Brothers
Cover Printer: Lehigh-Phoenix Color
Text Font: Palatino

Credits and acknowledgments borrowed from other sources and reproduced, with permission, in this textbook appear on the appropriate page within the text.

Figures 6-1, 6-2, 6-3, 12-7 and 12-8 are credited from Kissell, Thomas E., Electricity & Electronics for Industrial Maintenance, 1e ©2005. Electronically reproduced by permission of Pearson Education, Inc. Upper Saddle River, New Jersey.

Figures 6-8, 6-9, 6-10, 6-11, 6-12, 6-14, 6-17, 6-19, 6-21, 12-47 and 12-48 are credited from Kissell, Thomas E., Industrial Electronics: Applications for Programmable Controllers, Instrumentation and Process Control, and Electrical Machines and Motor Controls, 3e ©2003. Electronically reproduced by permission of Pearson Education, Inc. Upper Saddle River, New Jersey.

Figures 6-4, 6-5, 6-6, 6-7, 6-13, 12-1, 12-2, 12-3, 12-4, 12-5, 12-6, 12-9, 12-10, 12-16, 12-17, 12-18, 12-19, 12-20, 12-21, 12-23, 12-24, 12-25, 12-26, 12-27, 12-28, 12-29, 12-30, 12-31, 12-32, 12-33, 12-36, 12-37, 12-38, 12-39, 12-40, 12-41, 12-44, 12-45, 12-56, 12-57, 12-58, 12-59, 12-60, 12-61, 12-62, 12-63, 12-64, 12-65, 12-66, 12-67, 12-68, 12-69, 12-70, 12-71, 12-72, 12-73, 12-74, 12-75, 12-76, 12-77, 12-78, 12-79, 12-80, 12-81, 12-82, 12-83, 12-84 and 12-85 are credited from Kissell, Thomas E., Electricity, Electronics, and Control Systems for HVAC, 4e ©2008. Electronically reproduced by permission of Pearson Education, Inc. Upper Saddle River, New Jersey.

Library of Congress Cataloging-in-Publication Data
Kissell, Thomas E.
Introduction to wind principles / Thomas E. Kissell.
p. cm.
ISBN-13: 978-0-13-212533-8
ISBN-10: 0-13-212533-1
1. Wind turbines. 2. Wind power. I. Title.
TJ828.K57 2011
621.4′5—dc22

2010011731

10 9 8 7 6 5 4 3 2 1

Prentice Hall
is an imprint of

www.pearsonhighered.com

ISBN 10: 0-13-212533-1
ISBN 13: 978-0-13-212533-8

Dedication

I would like to dedicate this book to my wife Kathy,
my brothers Jerry, Bruce, and Rich, and my sister Pat,
who helped shape my early life.

PREFACE

Introduction to Wind Principles covers all aspects of how small, medium, and large wind turbines operate. It is written specifically for students who want to learn enough about wind energy to enter the job market as a wind technician in sales, installation, or repair. It also provides enough content information so that students can understand concepts for installing and troubleshooting wind turbines. This book provides enough detail to give technicians the knowledge they need to handle even the most complex maintenance tasks. It is clearly written and starts out with simple concepts and continues through the most complex issues. Each chapter contains a large variety of pictures and diagrams that help explain basic and complex concepts. The book explains how wind turbine blades harvest wind energy, and how generators convert the energy from the shaft into electricity. The chapters show how electrical and hydraulic systems control the wind turbine blade pitch and yaw position controls. Mechanical systems such as gears, transmissions, and gearboxes are discussed in detail, as are electrical and electronic circuits used to control or convert voltages and control frequency. Many pictures and diagrams are included with each topic, and all mathematical data is provided in tables so the reader does not need calculus or other mathematics to understand the concepts. The chapters in the text can be used individually or in sequential order to meet the needs of all wind energy curricula.

ORGANIZATION OF THE TEXT

The first chapter provides an overview of current wind turbines and the history of wind turbines. This chapter also explains the industrial, residential, and commercial electrical energy demands of the United States and the world, and how energy from wind turbines fits into the supply of green energy. Chapter 1 also explain how to finance wind turbines and discusses return on investment for current wind projects.

Chapter 2 explains the physics of wind energy and how wind turbines are designed to efficiently change wind energy into electricity. This chapter helps students understand all the variables involved in wind energy, how to evaluate a potential site for a wind turbine, and how to predict how much wind energy will be available for conversion.

Chapter 3 explains the operation of vertical-axis and horizontal-axis wind turbines, and pictures and diagrams show all of the basic parts of each type of wind turbine and explain their function. The operation of wind turbine blades is explained in detail so that students can understand the strengths and weakness of one-, two-, three-, and five-bladed wind turbines.

Chapter 4 explains how wind turbines produce electrical power from the wind, the operation of the blades, how wind turbines operate with variable-speed winds, and how to determine wind turbine peak performance.

Chapter 5 provides a very in-depth explanation of how the blades, rotor, low-speed shaft, gearbox, high-speed shaft, and generator work together to convert wind energy to electrical power. Blade pitch control and yaw control are explained in detail so that students can understand how wind turbines are controlled. This chapter explains how programmable logic controllers (PLCs) control the hydraulic and other functions of the wind turbine. An in-depth explanation of hydraulic control, including proportional and servo hydraulic control, are used to control blade pitch and brakes for the wind turbine is given. This chapter also provides detailed diagrams and pictures of PLC and hydraulic components to help students understand these systems.

Chapter 6 explains the theory of AC and DC electricity and how it is used for magnetic fields in AC and DC generators. The operation of motors is explained in detail, including how they are used as synchronous and asynchronous generators. Frequency control of generators is explained, as well as double-fed generators, permanent-magnet generators, and alternators. This chapter helps students learn enough about generators to be able to install, troubleshoot, and repair them.

Chapter 7 provides detailed information about gears, bearings, and drive trains used in wind turbines. Students learn to identify and understand the function of each type of bearing and gear, and how the low-speed rotation of the blades and rotor is converted into higher-speed rotation at the generator shaft.

Chapter 8 provides a complete explanation of the electrical grid in its present form and also provides information about the new "smart grid" and how it will help make electrical transmission more efficient. This chapter provides in-depth detail about the equipment and switchgear used to connect wind turbines to the grid. Connections between wind turbines and substations, and power distribution for underground and overhead service, are also explained.

Chapter 9 explains various tower designs, including monopole and lattice towers. This chapter explains how concrete foundations are designed and poured, and how towers are erected and maintained. Climbing safety and how to work safely above the ground are also discussed. Overspeed safety and lightning safety are explained in detail.

Chapter 10 provides a detailed overview of wind turbines that are currently installed, from single units to large wind turbine farms on land and at sea. Small residential installations, commercial installations, and large wind farm installations are shown so that students can begin to get an idea of how extensive wind turbine installation is worldwide. The top 10 wind turbine farms in the United States are discussed, and students gain an understanding of which states have the largest wind farms and where future growth is predicted.

Chapter 11 explains the steps for installing and testing a wind turbine. It also provides detailed information about the troubleshooting process and how to troubleshoot specific problems on wind turbines. Troubleshooting transmission, generator, mechanical, electrical, PLC, and hydraulic problems are explained in detail so that students can easily master solving problems.

Chapter 12 covers the theory of basic electricity that is needed to understand all of the electrical parts of wind turbines. This chapter is a comprehensive, detailed study of DC and AC electricity, including three-phase and single-phase power, true power, apparent power and power factor, transformers, the National Electric Code, AC and DC motors, and basic electronics used in rectifiers and inverters. This chapter has enough depth to provide students with all the electrical knowledge needed to install, troubleshoot, and repair electrical parts of wind turbines.

ACKNOWLEDGMENTS

I would like to thank my wife, Kathy, for all her help in keeping this project on track, and for her support during this time. I also want to thank Kelly Starkey, my daughter, who provided invaluable help with the glossary.

I want to thank Kevin Walker, John Carpenter, Bruce Meyer, Denny Setzler, Jayne Bowersox, and Kathy Elchert at Terra Community College, who helped me develop the individual topics, obtain pictures, and evaluate the early content. I would also like to thank John Fellhauer, Curtis Stokes, and Paul Schroeder for using portions of this text in their Wind Energy classes at Terra Community College and providing feedback.

I would also like to thank Dan Trudden, Developmental Editor at Pearson Higher Education in Columbus, Ohio, for all his help during this project. He has gone above and beyond what I expected in guiding me through the project. I am especially indebted to his attention to detail and his help in getting many of the permissions and images for this text. I would also like to thank Wyatt Morris at Pearson Higher Education for his help with this project.

I would also like to thank John Fellhauer at SUREnergy and Fellhauer Mechanical Systems Inc., in Port Clinton, Ohio, who generously gave his time and energy to help me get the additional background information and images I needed to complete this book. I also want to thank him for the vast number of pictures he provided to make the book a success.

I would like to thank the following people and their companies for help in providing pictures and diagrams and the permissions to use them in this book:

Joseph R, Andre, Jr., at Nova Lynx

Paul G. Brock, P.E., Assistant Director of Utilities, City of Bowling Green, Ohio

David Brooksbank at Boston Gear

Jackie Catalano, Marketing Communications Manager, Power Transmission Solutions, Emerson Industrial Automation

Steven Crosher and Philippa Rogers at quietrevolution

Bill Currie at Power House Ltd.

John Fellhauer at SUREnergy and Fellhauer Mechanical Systems Inc., Port Clinton, Ohio

Michael French, Graphics Coordinator at Southwest Windpower

Marc Friedman, General Counsel at Harbor Freight Tools

John Gibney at Inventive Solutions, Solar, Wind and Environmental Technologies

Stephanie Havanas, Lead Technical Publishing Systems Specialist at Parker Hannifin Corporation

Janet Mo, Marketing Manager, and Dick Scott at NKE AUSTRIA GmbH

Joe Rand at Kidwind

Charlie Van Winkle, Chris McKay, Christopher Farage, and Jim Stover at Northern Power Systems

Aaron Weida at Bosch Rexroth

Thomas Young at RM Young Company

I would also like to thank the following people who reviewed this text:

Craig Evert, *Iowa Lakes Community College*
Jim Heidenreich, *Cuyahoga Community College*
Isaac Slaven, *Ivy Tech Community College*
David Vrtol, *Highland Community College*
Richard P. Walker, *Texas Tech University*

I would like to offer additional information to students and educators on a wide variety of topics that support wind energy technology. This information is available in the following textbooks I am currently writing and have recently written. These titles include *Introduction to Solar Energy; Industrial Electronics, Applications for Programmable Controllers, Instrumentation and Controls, Electrical Machines and Motor Controls; Electricity, Electronics and Control Systems for HVAC; Motor Control Technology for Industrial Maintenance; Electricity, Fluid Power and Mechanical Systems for Industrial Maintenance;* and *Electricity and Electronics for Industrial Maintenance.*

Thomas Kissell

CONTENTS

Preface iv

Chapter 1 Introduction to Wind Power 1

1.1 Basic Parts of Wind Turbines and How They Generate Electricity 2
1.2 History of Wind Power 4
1.3 Wind Turbine Classifications 6
1.4 Wind Turbine Specifications 9
1.5 Wind Turbine Standards and Certification 11
1.6 The Need for an Uninterrupted Power Supply 15
1.7 Electricity Transmission Limitations 15
1.8 Electrical Demand in the United States 16
1.9 Total Daily Demand for Electricity in the United States 18
1.10 Total Peak Demand 19
1.11 Commercial and Industrial Demand 19
1.12 Residential Demand 20
1.13 Grid-Tied Systems 20
1.14 Stand-Alone Remote Power Sources 22
1.15 Environmental and Ecological Assessments for Wind Power 22
1.16 The Need for Energy from Green Technologies 22
1.17 Future Directions for Wind Power 23
1.18 Political Implications for Wind Power 23
1.19 Financial Implications and Return on Investment 23
1.20 Green Energy Payback Calculations 24
1.21 Tax Considerations for Wind Power 26
1.22 Skills Needed for Green Technology Jobs 28
1.23 Jobs in Wind Power Industries 29

Chapter 2 Wind Resources and Sites for Wind Energy 32

2.1 Basic Physics of Wind Power 32
2.2 The Nature of Wind 35
2.3 Geographic Considerations for Wind Systems 35
2.4 Variations in Wind Speed 36
2.5 Seasonal and Annual Variations in Wind Speed 36
2.6 Day/Night Variations in Wind Speed 36
2.7 Turbulence 36
2.8 Gusting Wind Speeds 37
2.9 Extreme Wind Speeds 37
2.10 Predicting Wind Speed 37
2.11 Turbulence in Wakes on Wind Farms 38
2.12 Turbulence in Complex Terrain 39
2.13 Sizing a Wind Power System for a Particular Application 39
2.14 Determining How Much Wind Power Is Available at a Site 41
2.15 Local Codes That May Affect Wind Power Systems 42
2.16 Site Requirements 42
2.17 Determining the Proper Site for a Wind Turbine 42

Chapter 3 Operation of Vertical- and Horizontal-Axis Wind Turbines 45

3.1 Introduction and Overview of Airfoils 46
3.2 Types of Turbines 48
3.3 Vertical-Axis Wind Turbines 51
3.4 Horizontal-Axis Wind Turbines 53
3.5 Blade Geometry 54
3.6 Number of Blades 55
3.7 Comparison of Blade Types 57
3.8 Advantages and Disadvantages of Single-Bladed, Two-Bladed, and Three-Bladed Turbines 58

Chapter 4 Wind Turbine Performance 60

4.1 Constant-Rotational-Speed Operation 60
4.2 What Makes the Turbine Blade Rotate 61
4.3 Angle of Attack and Blade Pitch 61
4.4 Blade Pitch Control 61
4.5 How Wind Turbines Operate at Variable Wind Speeds 64
4.6 Estimating How Much Energy Is Converted 64
4.7 Testing Wind Turbines 65
4.8 Problems with Wind Generator Turbulence 66
4.9 How to Determine Wind Generator Peak Performance 66

Chapter 5 Basic Parts of Horizontal-Axis Wind Turbines 70

5.1 Overview of the Horizontal-Axis Wind Turbine 71
5.2 The Nacelle and the Nacelle Bedplate 71
5.3 Rotor Hubs and Blade Types 72

5.4 Number of Blades 73
5.5 Rotor Blade Pitch Adjustments and Teetering 74
5.6 Rotational Speed and Rotor Speed Control 75
5.7 Individual Pitch Control of the Blades 76
5.8 Control of Blades for Load Efficiency 76
5.9 Yaw Mechanism 76
5.10 Yaw Drives 77
5.11 Yaw Control 77
5.12 Yaw Drive Brakes 78
5.13 Data Acquisition and Communications 78
5.14 The Anemometer and Wind Vane 79
5.15 Wind Turbine Supervisory Control and Data Acquisition (SCADA) Systems 80
5.16 Wind Turbine Control Systems 80
5.17 Basic Operation of the Programmable Logic Controller 81
5.18 PLC Inputs and Outputs 85
5.19 PLC Analog Control 86
5.20 Controlling Wind Turbines Through Feedback Control to the PLC 87
5.21 Using the PLC for Power Control 88
5.22 Optimizing Blade Tip Speed 89
5.23 Using the PLC to Optimize Blade Torque 89
5.24 PLC Program Storage and Permanent Memory 89
5.25 Controlling the Magnetizing Current in the Generator 89
5.26 Hydraulic Controls 89
5.27 What the Hydraulic System Is Used for on a Wind Turbine 90
5.28 Hydraulic Pumps 93
5.29 Directional Control Devices 96
5.30 Electrical Control Through Hydraulic Solenoid Valves 102
5.31 Hydraulic Cylinders and Hydraulic Motors 102
5.32 Hydraulic Proportional Control Valves 103
5.33 Brakes on Wind Turbines 105

Chapter 6 Generators 109

6.1 Overview of AC and DC Electricity 110
6.2 What Is Alternating Current? 110
6.3 Frequency of AC Voltage 111
6.4 Introduction to Magnetic Theory 111
6.5 DC Generators 114
6.6 AC Motors 118
6.7 Basic AC Alternators (Generators) 122
6.8 Asynchronous AC Generators 125
6.9 Synchronous Generators 126
6.10 Doubly Fed (Double-Excited) AC Induction Generators 127
6.11 Permanent-Magnet Synchronous Generators 127
6.12 Using an Alternator to Produce DC Voltage 128

Chapter 7 Gearboxes and Direct-Drive Systems 129

7.1 Why Gearboxes Are Needed 130
7.2 Advantages of Gear Ratios 131
7.3 Types of Gears 132
7.4 Helical Planetary Gears 136
7.5 Bearings 138
7.6 Gearbox Differential and Spur Gears 143
7.7 Main Gearbox 144
7.8 Drive Trains 145
7.9 Direct-Drive Systems 146

Chapter 8 The Grid and Integration of Wind-Generated Electricity 149

8.1 Understanding the Grid 150
8.2 The Smart Grid 152
8.3 Transformers, Transmission, and Distribution Infrastructures 154
8.4 Grid Code Rules and Regulations 158
8.5 The National Electrical Code and Other Requirements for the Grid 159
8.6 Supplying Power for a Building or Residence 159
8.7 Switches and Connections for Power Distribution 160
8.8 Utility Grid–Tied Net Metering 161
8.9 Overview of Power Quality Issues 162
8.10 Frequency and Voltage Control 162
8.11 Voltage, True Power, and Reactive Power 163
8.12 Low-Voltage Ride-Through 163
8.13 Flicker and Power Quality 164
8.14 System Grounding 165
8.15 Underground Feeder Circuits 165
8.16 Cable Installation 166
8.17 Overhead Feeder Circuits 166
8.18 Wind Farm Substations 167
8.19 Connecting to Residential or Commercial Single-Source Power Systems 167

Chapter 9 Types of Towers, Tower Designs, and Safety 169

9.1 Types of Wind Turbine Towers 170
9.2 Foundations and Concrete Support for Towers 184
9.3 Climbing Towers 186
9.4 Safety Issues When Working with Towers and Climbing Safety 189
9.5 Lightning Safety for Wind Turbines 194
9.6 Overspeed Safety and Overload Controls to Protect Towers 196
9.7 Birds and Bird Safety Around Towers 198
9.8 Tower Maintenance 198

Chapter 10 Wind Turbine Installations and Wind Farms 200

10.1 Project Development 201
10.2 Wind Site Assessment 201
10.3 Site Issues 201
10.4 Visual and Landscape Assessment 202
10.5 Small Residential Wind Turbine Systems 202
10.6 Home-Made Wind Turbine Systems 203
10.7 Commercial Wind Turbine Systems 204
10.8 Wind Farms 206
10.9 Offshore Installations in the United States 211
10.10 Large Offshore Wind Farms in Europe 212

Chapter 11 Installation, Troubleshooting, and Maintenance of Wind Generating Systems 215

11.1 Steps in the Installation of a Wind Turbine 216
11.2 Overview of Troubleshooting 224
11.3 Understanding the Troubleshooting Process 226
11.4 What Is the Difference Between a Symptom and a Problem? 227
11.5 Using Troubleshooting Tables and a Troubleshooting Matrix 227
11.6 Troubleshooting Wind Turbine Generation and Transmission Problems 228
11.7 Troubleshooting Mechanical Problems and Tower Problems 229
11.8 Troubleshooting Electrical Problems 229
11.9 Troubleshooting Hydraulic Problems 232
11.10 Periodic Maintenance of Wind Generation Systems 233
11.11 Major Overhaul of a Wind Turbine 235

Chapter 12 Electrical and Electronic Fundamentals for Wind Generators 237

12.1 Basic Electricity and a Simple Electrical Circuit 238
12.2 Measuring Volts, Amps, and Ohms 242
12.3 Using Ohm's Law to Calculate Volts, Amperes, and Ohms 246
12.4 Fundamentals of Electrical Circuits 248
12.5 Examples of Series Circuits 249
12.6 Parallel Circuits 251
12.7 Series-Parallel Circuits 254
12.8 Capacitors and Capacitive Reactance 255
12.9 Resistance and Inductance in an AC Circuit 256
12.10 Impedance; Calculating the Total Opposition for an Inductive and Resistive Circuit 257
12.11 True Power and Apparent Power in an AC Circuit 257
12.12 Calculating the Power Factor 257
12.13 How to Change Power Factor with Inductors or Capacitors 258
12.14 Volt-Ampere Reactance (VAR) 258
12.15 Three-Phase Transformers 258
12.16 Theory and Operation of a Relay and Contactor 261
12.17 Why Motor Starters Are Used 265
12.18 Fuses 267
12.19 Electronic Components Used in Inverters and Circuits 270
12.20 Using a Diode for Rectification 273
12.21 Light-Emitting Diodes 276
12.22 PNP and NPN Transistors 276
12.23 The Silicon-Controlled Rectifier (SCR) 278
12.24 The Triac 280
12.25 Inverters: Changing DC Voltage to AC Voltage 282

Glossary 287
Acronyms and Abbreviations 295
Index 296

CHAPTER 1

Introduction to Wind Power

OBJECTIVES

After reading this chapter, you will be able to:

- Explain how a wind turbine creates electricity from wind.
- Identify the basic parts of a wind turbine.
- Identify three ways to classify wind turbines.
- Explain the need for an uninterruptible power supply for residential and industrial uses.
- Identify the total daily electrical demand for residential, commercial, and industrial uses of electricity.
- List 10 jobs that will be needed in future wind power industries, and the skills that will be required to do these jobs.

KEY TERMS

American National Standards Institute (ANSI)
American Wind Energy Association (AWEA)
British Wind Energy Association (BWEA)
Brownout
Department of Energy (DOE)
Downwind turbine
Electrical demand
Electrical grid
European Wind Turbine Certificate (EWTC)
Gearbox
Generator
Green technology
High-speed shaft
Horizontal-axis wind turbines
International Electrotechnical Commission (IEC)
International Organization for Standardization (ISO)
Inverter
Kilowatt
Low-speed shaft
Megawatt
Nacelle
National Electrical Code (NEC)
National Renewable Energy Laboratory (NREL)
Peak electrical demand
Power electronic frequency converter
Return on investment (ROI)
Rotor
Turbine blades
Type certification
Type characteristic measurements
Uninterruptible power supply
Upwind turbine
Vertical-axis wind turbine

OVERVIEW

Wind power is becoming a viable source for providing electricity from alternative energy. This chapter provides an overview of wind power, including a history of its first uses through the most modern wind farms in use today. The chapter begins with a discussion of the different types of wind turbines and a description of the standards and certification system used for wind turbines.

The chapter then covers how electricity is used today in the United States in both residential and commercial or industrial applications. Information is provided about the limits of transmission of electricity through the grid and also about the total electrical usage for industry and residential applications in the United States on a daily and an annual basis. Other important data such as the total commercial and industrial demand, and total peak demand, are explained.

Discussion then turns to the types of wind turbines used today in terms of whether they are stand-alone units that provide power directly to a residence or small commercial application, or are tied to the grid by which they supply power to a utility. Some applications that are tied to the grid also provide power to be used at the site.

The last part of the chapter explains environmental and ecological aspects of wind power as well as the need to harvest energy from renewable sources. This part of the chapter discusses future directions, the political implications for wind power, financing and payback, and tax considerations. The types of skills that a technician will need to work in the wind power industry as well as the types of jobs that are available are also discussed.

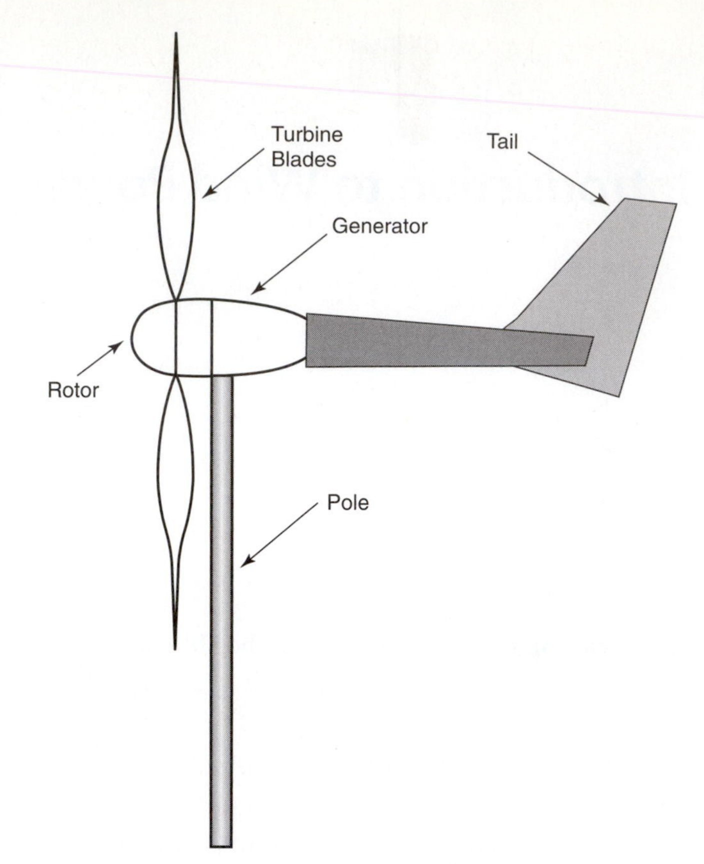

FIGURE 1–1 Typical parts of a small wind turbine.

1.1 BASIC PARTS OF WIND TURBINES AND HOW THEY GENERATE ELECTRICITY

Basic Parts of Wind Turbines

As you begin to learn about wind power and wind turbines, it is important to learn about the basic parts of the wind turbine and to understand how a wind turbine generates electrical power from wind. Figure 1–1 shows a small **horizontal-axis wind turbine**. The basic parts of the horizontal-axis wind turbine include a **turbine blade** connected to a **rotor**, the rotor itself, a **low-speed shaft**, a **gearbox**, a **high-speed shaft**, and a **generator**. Sometimes the turbine blade and rotor together are just called the rotor. The rotor is connected to the generator by a shaft. When the blade and rotor turn the shaft, the shaft turns the generator and the generator produces electricity. Some wind turbines use a generator that produces direct-current (DC) voltage, whereas others use a generator or alternator that produces alternating-current (AC) voltage. The amount of electrical power is rated in units called a watt (W). One watt is produced by 1 volt (V) providing the pressure to move 1 ampere (A) of current. You can find a complete overview of electricity in Chapter 12 of this text. Wind turbine generators are rated by the amount of power they can produce, and the amount of power is rated in watts. One thousand watts is called a **kilowatt** (kW); it is the amount of electrical energy that a 100-W light bulb will use if it is allowed to burn continuously for 10 hours. Utilities bill customers for electrical usage in kilowatt hours (kWh), which indicates the amount of power used over a given time period. To give you a better idea of how large a kilowatt hour of electrical energy is, compare the energy requirements of typical household appliances. For example, a typical automatic washing machine consumes 400 W per hour, an electrical clothes dryer uses 4000 W (4 kW) per hour, a hair dryer consumes 1000 W (1 kW) per hour, and a typical desktop computer uses approximately 120 W per hour. It is important to remember that a watt is the unit of measure for electrical power at any instant of time and is used to determine the amount of voltage and current a load will consume at any instant; a watt hour is a unit of energy in watts used over time in hours, which is used for billing purposes.

Some larger wind turbine generators are rated in **megawatts** (MW). A megawatt is 1 million watts. Figure 1–2 shows an example of several megawatt wind turbines in a rural setting, and Figure 1–3 shows an example of several large-megawatt wind turbines mounted at sea. When a large number of wind turbines are placed together on a wind farm, their output may exceed 1000 megawatts, which is called a gigawatt (GW). One gigawatt of electricity is 1,000,000,000 W (1 billion watts), which is also equivalent to 1000 MW or 1 million kW. A typical U.S.

FIGURE 1–2 Multiple larger megawatt wind turbines in a rural location. (Courtesy of Fotolia, LLC)

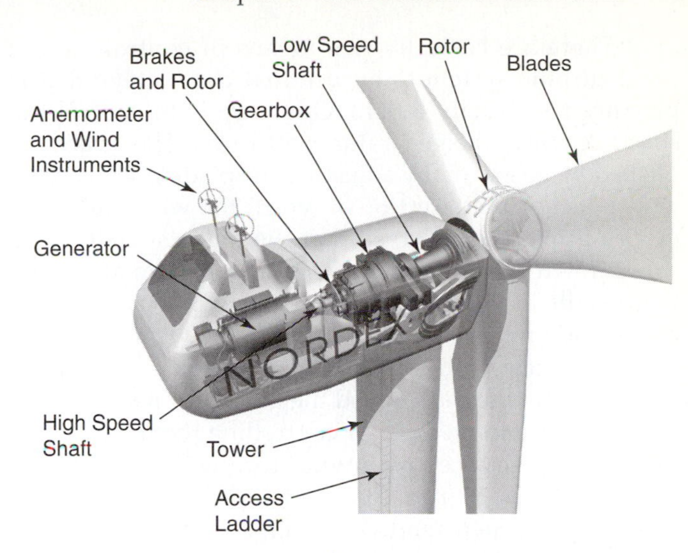

FIGURE 1–4 The basic parts of large horizontal-axis wind turbine. (Courtesy of Nordex.)

household uses around 938 kWh per month or 11,256 kWh per year. More energy efficient homes may use only 8900 kWh per year. Typical industrial and residential demand is covered in Sections 1.11 and 1.12.

The small type of wind turbine shown in Figure 1–1 is guided into the wind by its tail. The generator and rotor assembly are mounted on a tall tower so that it is high above the ground, where it can catch the wind and convert it to movement of the blades and ultimately turn the generator, which produces electricity. The electricity from this type of small wind turbine can be used directly in a home or small application, or it can be stored in batteries for future use. Typically, the small amount of electricity from this type of wind turbine is used to power lights, heat water, or provide electricity for a heating element for a small furnace. Some remote applications may use the electricity generated to pump water.

FIGURE 1–3 Large megawatt wind turbines located at sea. (Courtesy of Vestas Wind Systems A/S.)

Figure 1–4 shows the basic parts of a larger horizontal-axis wind turbine. This figure shows the rotor, low-speed shaft, gearbox, high-speed shaft, and generator, which are the principal parts of the wind turbine and are located inside a nacelle. A **nacelle** is an enclosure or housing for the generator, gearbox, and any other parts of the wind turbine located on top of the tower. The diagram shows how the turbine blades are connected to the rotor, which in turn is connected to the low-speed shaft. This shaft is called the low-speed shaft because the wind causes the turbine blades to turn at somewhere between 10 and 20 revolutions per minute (rpm). The low-speed shaft connects to the gearbox, which has a set of gears that increase the speed of the shaft coming out of it to approximately 1800 rpm. The shaft that comes out of the gearbox and connects directly to the generator is called the high-speed shaft because it turns at a much higher rpm. The low-speed shaft, gearbox, high-speed shaft, and generator are mounted inside the nacelle, which provides a cover that protects all of these parts of the wind turbine from the weather.

The horizontal-axis wind turbine may use a gearbox or set of gears, which changes the slow rotation of the blades into a higher, constant-speed rotation that ensures a generator creates AC voltage at 60 hertz (Hz). The reason this is so important is that all the electricity in the U.S. grid is 60 Hz, and a great many electrical appliances and motors are designed to operate at 60 Hz.

There are basically two means of controlling the wind turbine system to ensure that the voltage that is produced is exactly 60 Hz. One way is to control the speed at which the generator shaft turns. This is accomplished by gears and by adjusting the blade pitch and the yaw, which is the direction at which the wind turbine is pointing into the wind, so that the blade speeds up or slows down to make the high-speed shaft run at an rpm that results in the generator producing 60 Hz. The optimum blade rotation is between 10 and 20 rpm, at which the proper gear ratio can be used to make the high-speed shaft rotate at the exact speed the generator needs.

The second method for controlling the frequency of the voltage produced by a wind turbine is to allow the turbine blades to run freely at any speed the wind provides and then send the voltage that the generator produces to a **power electronic frequency converter**, also called an **inverter**. The inverter accepts single-phase or three-phase AC voltage to its input circuits at any frequency and voltage level. This AC voltage is converted to DC voltage by rectifiers. Capacitors and inductors are used in the DC circuit to remove any remaining image of frequency. The next section of the inverter changes the DC voltage back to single-phase or three-phase AC voltage at exactly 60 Hz. Single-phase and three-phase electricity, frequency, and inverters are explained in detail in Chapter 12. In this type of wind turbine, the speed at which the blades turn is not controlled up to the point where the capacity of the wind turbine is reached, and the generator is allowed to produce voltage at any level and at any frequency. If the wind turbine is producing power in excess of its rating, the blade speed is controlled. When the generator is producing power at any voltage and any frequency, the inverter changes this input voltage to a standard voltage level and frequency so the system can be connected directly to the grid.

If the voltage level coming out of the inverter needs to be increased from 690 to 12,470 V, to match the voltage on the grid, a transformer is used. Some wind turbine applications are designed to use the pure DC voltage of the second circuit of the inverter. In these types of applications, the DC voltage is used directly from this section rather than being sent to the output section of the inverter that produces the AC voltage.

How Wind Turbines Create Electricity

The horizontal-axis wind turbine is designed so that when the wind blows past the turbine blades, they begin to rotate. As they rotate, they convert the power of the wind to rotational force called torque. The torque from the turbine blades is transmitted through the low-speed shaft to the gearbox, and the gear set in the gearbox causes the output shaft to turn at a much higher speed. The generator is connected to the other end of the high-speed shaft and turns at the higher speed to produce AC voltage. In smaller wind turbines, the system is designed so that the wind causes the rudder blade on the tail of the wind turbine to keep the wind turbine blades positioned into the wind. If the wind changes direction, the wind turbine is designed to that it will change its position to keep the blades positioned into the wind.

On larger wind turbines, the nacelle has a control system that allows it to be rotated so that the blades stay positioned so that they are always directed into the wind. This type of control is called yaw control, and it consists of a large circular plate with teeth in it that is mounted directly to the bottom of the nacelle. A high-torque electric or hydraulic motor turns the nacelle gear. The yaw drive motor has a matching gear with fewer teeth, which gives the motor the gear ratio that provides the extra torque needed to make the nacelle rotate. When the yaw drive motor turns its shaft, it causes the big gear that is mounted on the bottom of the nacelle to rotate and turn the entire nacelle so that the blades are directed into the wind. If the wind becomes too strong, the nacelle can also be moved so that the blades are slightly out of the wind, which causes the blades to slow down so the turbine does not overspeed.

Some wind turbines also use blade pitch control to help the turbine be more efficient in low-speed winds and also safe in high-speed winds, by protecting it from overspeed. A computer called a programmable logic controller (PLC) controls the systems in the wind turbine. A number of sensors are connected to the PLC so that it can measure the wind speed and direction. The PLC also has a number of outputs that are used to control the pitch of the turbine blades and the yaw control that positions the nacelle so that the blades are pointed into or out of the wind direction, depending on the conditions. You will learn more about PLCs and these control features of horizontal wind turbines in later chapters. At this point you should be trying to get an idea of the names and functions of all of the parts of the wind turbine so that you can understand some of the history and future of wind turbines.

1.2 HISTORY OF WIND POWER

Wind power has been providing energy since the earliest recorded history. More than 3000 years ago, wind power was used to power sailing vessels on short trips as well as longer trips across large expanses of open water. Records also show that wind power was used in Persia (Iran) in 200 BC to grind grain and pump water. The Romans used wind power similarly. The Dutch began to use wind energy in the fourteenth century to pump water from low-lying areas and swamps to create additional farmland. In the 1800s, wind power was used to provide energy to pump water and grind grain on rural farms throughout the United States. In 1887 Charles Brush built one of the first wind turbines designed specifically to generate electricity and installed it at his home in Cleveland, Ohio. Table 1–1 shows some of the history of the use of wind power. Since 2000, the use of wind power has increased at a record pace.

TABLE 1–1 A Short History of Wind Power

Date	Event	Details
1000 years BC	Large and small sailing ships in various parts of the world (some reports indicate use of wind power as far back as 10,000 BC)	Wind power first used for sailing ships on short trips, and eventually for longer voyages.
200 BC	Wind machines in Persia (Iran)	Used mainly to grind grain and pump water.
250 AD	Roman windmills	Used mainly to pump water.
By the fourteenth century	Dutch use windmills to pump water to drain low-lying areas	Wind power used extensively to help drain swampy areas to provide more farming area.
1850–1950	U.S. farmers use windmills to grind grain, pump water, and saw wood	Rotating shafts of windmills used to power a variety of equipment, usually less than 1 horsepower.
1866	Beloit Wisconsin Eclipse windmill factory opens	Windmills for pumping water on farms and filling water tanks on steam locomotives; hundreds of thousands produced and installed prior to electricity being widely distributed to farms in the 1920s and 1930s.
1887–1888	First automated wind turbine used to produce electricity	Charles Brush built a 12-kW wind turbine with 144 rotor blades that was 50 m in diameter; electricity produced was used to charge 408 batteries in Brush's basement (this turbine ran for 20 years). Some European turbines were installed at or about this time.
1920s	George Darrieus (French) invents vertical-axis wind turbine referred to as "eggbeater windmill"	U.S. Patent granted 1931.
1927	Joe and Marcellus Jacobs open Jacobs Wind Factory	Manufactured generators for farm use, which produced DC voltage to charge batteries.
1931	Yalta, USSR wind turbine	Generated 100 kW from a 30-m (100-ft) tower connected to a 6.3-kV distribution system.
1934	Palmer Cosslett Putman wind turbine to generate electricity	Generated 1250 kW of electricity from a tower over 110 ft (33.5 m) tall with blade diameter over 174 ft (53 m); completed in 1941, abandoned in 1945 due to mechanical failure.
1941	First megawatt wind turbine.	Built in Castleton, Vermont, with 75-ft blades and connected to the power grid.
1954	First wind turbine designed to be connected directly to grid	John Brown Company installed a 100-kW wind turbine, with three blades of 18-m diameter, in the Orkney Islands, UK.
1970	NASA group starts wind turbine research in Ohio	13 experimental wind turbines placed into operation for research.
October 15, 1973 through March 17, 1974	Oil embargo by Organization of Petroleum Exporting Countries (OPEC)	OPEC declared it would no longer ship oil to the United States, as a response to U.S. support of Israel.
1974	Crude oil prices quadrupled, from $3 per barrel to $12 per barrel	Department of Energy created 1977; many efforts to reduce consumption put into effect.
1980	World's first wind farm	20 turbines were installed on a wind farm in New Hampshire; the farm failed to be economical because of turbine breakdown and insufficient wind.

(*Continued*)

TABLE 1–1 (*Continued*)

Date	Event	Details
1981	7.5-MW wind turbinee	NASA installed a 7.5-MW wind turbine in Washington state, which had the largest blades anywhere at the time.
1987	3.2-MW two-bladed wind turbine	NASA installed this wind turbine in Hawaii, which had the largest blades on a wind turbine to that date.
1991	Vindeby, Denmark, first offshore wind farm in Denmark	11 turbines, each 450kW.
1991	Delabole, Cornwall, UK, first offshore wind farm in UK	10 turbines offshore, enough power for 2700 homes.
2001	Enron Energy goes bankrupt	Enron Energy was one of the largest wind energy companies in the United States when it went bankrupt; General Electric purchased the failed wind energy unit.
2003	Wales, UK, first offshore wind farm in Wales	30 turbines, each 2 MW, located 7–8 km offshore.
June 2008	Gasoline reaches $4.00 per gallon ($140 per barrel for crude oil)	Highest oil prices in history caused highest gasoline prices in United States.
July 2008	Oil hits all-time high prices	Oil rises above $140 per barrel, which is an all-time high price.
2009	Wind turbine with the largest blades to date, with 6 MW output	Wind turbine with rotor diameter of 126 m (413 ft) is installed in Emden, Germany.

Some events that helped shape the growth of wind power in recent years were the price and availability of oil, which is used to produce gasoline and other fuels. In 1973 and 1974 the Organization of Petroleum Exporting Countries (OPEC) limited the amount of oil that was put on the market. This caused prices to increase and reduced the amount of oil available to produce gasoline and electricity. The oil embargo caused a flurry of investments in alternative energy sources such as wind energy. As soon as the embargo ended, the price of oil fell and availability increased, so the immediate need for alternative energy fell off quickly. Later, in the 1980s, falling energy availability and rising prices continued to drive more research and investment in wind energy worldwide. In July 2008 oil hit an all-time high price of over $140 per barrel, which was four times higher than the price ($30 per barrel) just three or four years earlier. Each time oil prices spike, investment in wind energy increases, and recent advances in research continue to push technology that makes wind energy and other alternative energy sources more cost-effective.

Another event that changed the way investments in wind energy are made in the United States was the bankruptcy of energy giant Enron, which had financial interests in wind turbine manufacturing and in a number of large wind farms in the United States during the late 1990s and as late as 2001. Enron's large investment in wind farms was later purchased by General Electric and helped that company to become more involved in investing in alternative energy sources and creating a "smart" grid. The switch gear and metering equipment on the smart grid allows information to be both sent and received, whereas information on the traditional grid is sent only one way. You will learn more about the smart grid in Chapter 8.

1.3 WIND TURBINE CLASSIFICATIONS

There are two main types of wind turbines, **vertical-axis wind turbines** (VAWT) and **horizontal-axis wind turbines** (HAWT). The horizontal-axis wind turbine is more efficient than the vertical-axis wind turbine, but the vertical-axis wind turbine can harvest the wind regardless of the wind direction. The horizontal-axis wind turbine must be directed into the wind before its blades can begin to harvest the wind.

Vertical-Axis Wind Turbines

There are many different models of vertical axis wind turbines, but there are two main types, the Darrieus and the Savonius types. Figure 1–5 shows the Darrieus type, which is shaped like an egg beater and uses lift forces on its blades to get them to turn. The egg beater design allows the blade to rotate at speeds faster than the wind. This means that the Darrieus has high speed and low torque, which makes it an excellent design for generating electricity.

FIGURE 1–5 Darrieus-type vertical-axis wind turbine. (Photo courtesy of U.S. Department of Energy.)

One of the problems with a Darrieus-type vertical-axis wind turbine is that it is not self-starting, so it needs a motor to get the blades turning until the wind speed is sufficient to keep it turning. On the positive side, the Darrieus-type wind turbine can harvest the wind and turn it into electricity regardless of the direction in which the wind is blowing. Also, the Darrieus-type wind turbine does not need a tower, which allows the generator to be mounted at its base, at ground level, which makes it easier to perform regular maintenance.

The Darrieus type of wind turbine can be used for a small residential system or a very large system, up to 2.5 MW, such as the one located at Cap-Chat, Quebec, Canada. This very large Darrieus-type wind turbine is 42 m (137 ft) tall and has a diameter of 34 m (111 ft).

Another very large Darrieus-type wind turbines was erected in the 1980s, after the oil embargo, at Amarillo, Texas. It was operated by Sandia Laboratories in conjunction with the U.S. Department of Agriculture (USDA). The wind turbine is 34 m (111 ft) high, but it is no longer used full-time to produce electrical energy.

The Savonius-type vertical-axis wind turbine has blades that have the overall shape of a sail or a cup. Figure 1–6 shows a small Savonius-type vertical-axis wind turbine. One of the advantages of the Savonius-type wind turbine is that it can harvest energy from the wind regardless of the direction the wind blows. The Savonius-type wind turbine is also very quiet, which allows it to be placed close to residential areas and commercial areas.

FIGURE 1–6 Savonius-type vertical-axis wind turbine. (Courtesy of Inventive Solutions.)

Horizontal-Axis Wind Turbines

The most common type of wind turbine is the horizontal-axis wind turbine. Figure 1–7 shows horizontal-axis wind turbines on a wind farm. The horizontal-axis wind turbine is mounted on the top of a pole or tower. It can have one, two, three, or multiple blades. The most common type has three blades. The wind blows through the blades and causes them to rotate, which thereby causes the rotor on the system to rotate. The rotor and blades of the horizontal wind turbine are usually connected to the generator by a shaft. The horizontal wind turbine may have a gearbox or transmission between the blades and the generator. A gearbox may be used because the blades turn at a rather slow speed, between 10 and 20 rpm, and the generator

FIGURE 1–7 Example of horizontal-axis wind turbines on a wind farm. (Courtesy of Vestas Wind Systems A/S.)

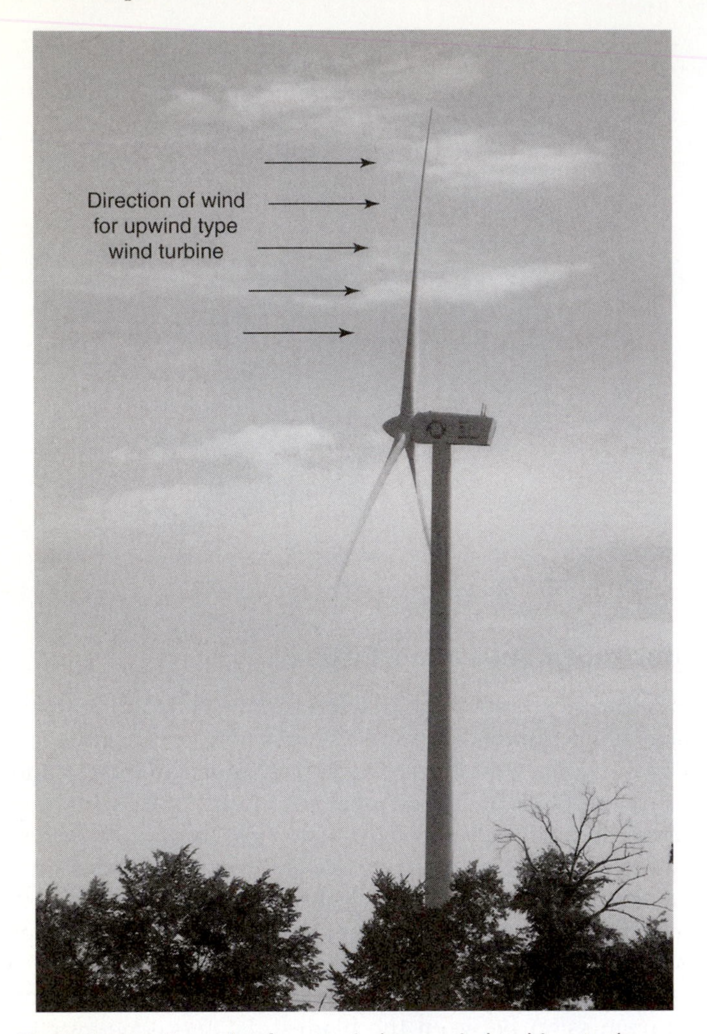

FIGURE 1–8 Example of an upwind-type wind turbine. In the upwind turbine, wind blows over the blades first from the left and then over the generator in the nacelle.

FIGURE 1–9 Example of a downwind-type wind turbine. The wind blows over the nacelle first and then over the turbine blades.

needs to turn at near 1800 rpm. The gearbox or transmission increases the output shaft speed to about 1800 rpm.

The horizontal-axis wind turbines can be further broken down into upwind and downwind turbines. The turbine may be designed as an **upwind turbine**, in which the wind blows over the blades and then over the generator nacelle; or as a **downwind turbine**, in which the wind blows over the generator nacelle and then through the blades. Figure 1–8 shows an example of an upwind-type horizontal-axis wind turbine, and Figure 1–9 shows an example of a downwind-type horizontal-axis wind turbine.

One problem with the upwind designs is that the rotor needs to be made rather rigid and inflexible, and placed at some distance out in front of the tower, because the blades will flex somewhat in stronger winds. If the blade is not mounted far enough in front of the tower, strong winds will cause the blades to flex back into the tower and become damaged when the blades rotate in high winds. A second problem with upwind design is that an upwind machine needs a way to keep the rotor facing the wind. This is accomplished with a yaw mechanism. Yaw control is control of the wind turbine direction. The wind turbine must orient itself into the wind to harvest the energy from the wind most affectively. The yaw control can be designed so the wind turbine moves into the wind automatically, by using a rudder. Other larger horizontal wind turbines may use a yaw motor that has a high-torque electric or hydraulic motor that can move the direction of the nacelle so that the turbine blades are pointed into the wind, or, if the wind becomes too strong, the turbine can be rotated so that it is slightly out of the wind direction, which allows the blades to stay within their design speed even in high winds.

In larger wind turbines, active yaw control is used to control the direction of the turbine blades so they head into or out of the wind. This type of active yaw control typically uses a permanent-magnet-type electric or hydraulic motor to turn the plate on which the turbine is mounted. The yaw motor is controlled by a PLC that receives a signal from an anemometer that measures the speed and direction of the wind and compares it with the speed of the turbine blade. On these larger machines, the turbine blade is connected to a shaft that goes into a multispeed transmission. The generator has a shaft that is connected to the output of the transmission. The speed of the generator on some wind turbines is strictly controlled so the generator output is at 60 Hz, and the yaw control along with pitch control can be used to change the direction the turbine is pointing in relation to the direction of the wind, which controls the speed at which the turbine blade rotates. If the wind speed is too high, the blade pitch control changes the blade pitch so that it harvests less wind energy, and if changing the rotation of the blade does not slow the rotor down enough, the yaw control can rotate the turbine slightly so that it does not face directly into the wind. This will slow the turbine blades further, until the rotor speed is at a safe level. If the wind speed remains too high and has the potential to damage the wind turbine blades, brakes can be set that make the rotor stop moving completely. You will learn more about this in Chapter 5, where turbine control and generator control is explained in more detail. Another type of horizontal-axis wind turbine, called a direct-drive wind turbine, does not have a transmission. Its blade speed does not need to be controlled in order for the turbine generator to produce electrical power at 60 Hz, because it uses an inverter to control the frequency.

In the downwind machine the rotor is placed on the downwind or lee side of the tower. The generator and hub mechanism are also on the downwind side, so any flexing of the blades moves the tips away from the tower. Some smaller downwind turbines are designed so that they orient to the wind direction without a yaw mechanism. This is called passive yaw location, and it makes the system less expensive. Yaw rotation must be controlled or limited so that the generator does not continue to rotate in an unlimited number of (360-degree) turns, because the electrical cables that connect the wind turbine generator to the ground terminals would become tangled. In some larger turbines the yaw control includes a rotation counter, and the number of full rotations is limited to about five full turns. Another way to compensate for the yaw rotating more than 360 degrees is to use slip rings that allow the electrical connection to be continuous even if the turbine rotates multiple times. The slip rings can also become a problem, however, because all the electricity generated by the generator must flow through them to get to the ground terminals. Slip rings wear quickly and require checking and minor maintenance continually throughout the life of the turbine. Because the slip rings are at the top of the tower, where the turbine is mounted, checking them requires climbing the to perform checking and maintenance.

Classification by Size

Another way to classify wind turbines that are used strictly to produce utility power is by their size and their electrical output rated in kilowatts. According to the California Energy Commission, if a windmill produces less than 50 kW, it is classified as a small wind turbine. If the wind turbine produces 50 to 500 kW, it is classified as a medium-size (intermediate) turbine. If the wind turbine produces more than 500 kW, it is classified as a large system. Prior to 1990, the majority of wind turbines installed were small or medium-size systems.

Classification by Number of Blades

Another way to classify wind turbines is by the number of blades that they have. The typical horizontal wind turbine has one, two, or three blades. Some newer wind turbines have five blades. The more blades the wind turbines have, the more wind energy they can convert. If the wind turbine uses one blade, it will need to have a counterbalance to ensure that the blade rotates smoothly. The three-blade turbine is the most popular type among the horizontal-axis wind turbines. The two-blade turbine is less expensive to purchase than the three-blade turbine, but the three-blade turbine will produce more electrical energy than the same size two-blade turbine. If the two-bladd turbine spins faster, it will be able to produce more electrical energy. Some types of wind turbines have four or more blades. Each additional blade allows the wind turbine to produce more electrical power, but each additional blade makes the wind turbine more expensive.

Other Ways to Classify Wind Turbines

Other ways to classify wind turbines relate to the way their output electricity is used or utilized. For example, the electricity that is produced can be sent directly to the grid if the turbine is a grid-tied system. If the turbine is isolated and not connected directly to the grid, it is called a stand-alone or a non–grid-tied system; the electrical energy it produces can be used directly at a residence, small business, or small industry, or it can be stored in a set of batteries. Another type of system is known as a hybrid system, because the electricity that is generated from wind can be connected with wind from other alternative sources, such as solar energy or electricity from hydroelectric sources, and then can be used as an energy source in remote areas.

Still another way to classify wind turbines is by where they are located. For example, they can be located on land or over the water, either on floating platforms or on permanent concrete platforms. Both on land and on water, turbines can be sited either individually or as a group known as a wind farm. Wind farms may have as few as 2 to 10 units or as many as 100 turbines or even more.

Wind turbines can also be classified according to their purpose and who owns them. For example, the wind turbine may be for an individual or small business and is installed for the express purpose of providing part of the electrical supply for the owner. Other, larger wind turbines are owned by investment groups, companies, corporations, cities, or utilities and are installed for the express purpose of providing utility-grade electrical power, just like a coal-fired or nuclear generating station.

1.4 WIND TURBINE SPECIFICATIONS

Wind turbine specifications are provided by the manufacturers of wind turbines. These specifications can be used to compare similar models, and to provide important data that may be needed during planning and installation, such as the capacity of the turbine in megawatts, the blade length in meters and feet, the hub height in meters and feet, and the area swept by the blades in meters and feet. Specifications for the blades include the rpm range, the maximum blade tip speed, and the maximum rated wind speed. Table 1–2 gives specifications for the most widely used models of wind turbines.

Basic specifications for wind turbines include the generator output in kilowatts or megawatts. Other specifications include the cut-in wind speed and the cut-out wind speed. Specifications for the blades include the blades-plus-rotor diameter or individual blade length. Specifications may also indicate the sweep area of the blades, the blade tip speed, and the rotor rpm. Specifications for the tower include the tower height, the height to the rotor, and the power curve for the entire generator. Other specifications may include information about the gearbox or drive train. If an inverter or converter is used, the specifications for that will also be provided. Table 1–3 shows examples of wind turbine specifications categories.

TABLE 1–2 Wind Turbine Sizes by Model

Model	Capacity (MW)	Blade Length[a] (ft)	Hub Height[b] (ft)	Total Height (ft)	Area Swept by Blades (m^2)	Range (rpm)	Max. Blade Tip Speed[c] (mph)	Rated Wind Speed[d] (mph
GE 1.5s	1.5	116	212	328	3904	11.1–22.2	183	27
GE 1.5sl	1.5	126	262	389	4657	10.1–20.4	184	26
Vestas V82	1.65	135	230	364	5281	8.8–14.4	138	29
Vestas V90	1.8	148	262	410	6362	8.8–14.9	157	25
Vestas V100	2.75	164	262	427	7854	7.2–15.3	179	34
Vestas V90	3.0	148	262	410	6362	9–19	200	34
Gamesa G87	2.0	143	256	399	5945	9/19	194	30
Bonus (Siemens)	1.3	102	223	325	3019	13/19	138	31
Bonus (Siemens)	2.0	125	197	322	4536	11/17	151	34
Bonus (Siemens)	2.3	135	262	398	5333	11/17	164	34
Suzlon 950	0.95	105	213	318	3217	13.9/20.8	156	25
Suzlon S64	1.25	105	240	344	3217	13.9/20.8	156	27
Suzlon S88	2.1	144	262	407	6082	xx	xx	31
Clipper Liberty	2.5	146	262	409	6221	9.7–15.5	163	26
Repower MM92	2.0	152	328	480	6720	7.8–15.0	163	25

[a]This figure is actually half the rotor diameter. The blade itself may be about a meter shorter, because it is attached to a large hub.
[b]Hub (tower) heights may vary; the more commonly used sizes are presented.
[c]Rotor diameter (m) $\times \pi \times$ rpm $\div$ 26.82.
[d]The rated, or nominal, wind speed is the speed at which the turbine produces power at its full capacity. For example the GE 1.5s does not generate 1.5 MW of power until the wind is blowing steadily at 27 mph or more. As the wind falls below that, power production falls exponentially.

Source: Data from manufacturers' product brochures and from American Wind Energy Association.

TABLE 1–3 Basic Wind Turbine Specifications

Generator output	Kilowatts or megawatts
Cut-in wind speed	Meters per second or miles per hour
Cut-out wind speed	Meters per second or miles per hour
Blade length	Meters or feet
Rotor diameter	Meters or feet
Rotor speed	Meters per second or miles per hour
Blade tip speed	Meters per second or miles per hour
Direction of rotation	Clockwise or counterclockwise
Rotor height	Meters or feet
Tower height	Meters or feet
Type of blade pitch or yaw control	Hydraulic or electric
Type of braking system	Hydraulic or electric
Power curve for generator	Generator output at specific wind speed
Blade-to-generator connection	Gearbox, transmission, or direct drive
Type of horizontal wind turbine	Upwind or downwind
Type of electronic control	Inverter or converter

Other specifications may include the rated power; rotor diameter; the type of power control, such as electric pitch control or hydraulic pitch control; the speed of the turbine blades; and where the turbine is designed to operate, such as onshore or offshore. If turbine controls such as pitch control or yaw control are used, the specifications for the type of control, such as hydraulic or electrical, will be provided. Any specifications for brakes or braking systems will also be provided.

1.5 WIND TURBINE STANDARDS AND CERTIFICATION

Standards and certification have been needed since the first wind turbines were manufactured, assembled, and operated. The standards are a set of agreed-on conditions that a wind turbine must meet when it is designed, manufactured, assembled, and operated. These standards ensure the safe design and operation of wind turbines. They also indicate how long a turbine should last and how often it should have periodic inspection and maintenance. The biggest problem with standards and certifications is deciding what agency or group should set standards and who inspects and certifies wind turbines. Because wind turbines are designed and manufactured overseas as well as in the United States and Canada, it is sometimes difficult to decide which standards a system must meet. Furthermore, there may be differences between standards for small wind turbines and those for much larger commercial wind turbines that are designed to produce power for electric utilities. Many governmental agencies, including the **Department of Energy (DOE)** in the United States as well as other agencies around the world, also have standards in place covering some parts of wind turbines, such as the electrical system or the towers. In this section you will learn why certification is needed, what agencies provide certification, and what these certifications cover.

Certification is necessary for a number of reasons, including but not limited to, the following.

1. Standards allow buyers to compare different systems in terms of output, size, and expected lifespan.
2. Standards allow agencies that provide financial support or loans for wind turbines to estimate how long the system can be expected to produce energy and how much maintenance will be necessary over its lifespan.
3. Standards and certifications provide a baseline for safety, functionality, and durability to protect consumers from unscrupulous manufacturers or substandard products.
4. Standards for electricity production and electrical output help utilities determine that the electricity produced will not cause problems on the grid, and if the electricity is provided to a residence or a company, that it will be of the same quality as existing electricity on the grid and will not damage any equipment.
5. Standards help to ensure desirable characteristics of wind turbines, such as quality, environmental friendliness, safety, reliability, efficiency, and interchangeability of parts, all at an economical cost.

It is important to understand that no warranty is implied just because a wind turbine meets certification criteria. Certification only indicates that the particular wind turbine model meets the minimum certification standards and that the certification and testing organizations are accredited. It is also important to establish that the manufacturing process and installation are checked for quality by either an internal auditor or someone from the certification agency. Also, follow-up after installation will help to develop a database of issues that may cause problems, conditions that may shorten the lifespan of the wind turbine, or that may suggest health or safety issues. This information can be used to edit certification issues or create new standards. It is also important that all areas of design, manufacturing, and installation of a wind turbine follow sound engineering practices.

National Renewable Energy Laboratory Wind Turbine Certification

The National Wind Technology Center uses the wind turbine certification specifications developed by the **National Renewable Energy Laboratory (NREL)**. This certification is called a **type certification**. Type certification ensures that wind turbines of any particular type, defined by size, form, and use, are designed and manufactured according to specific standards and other technical requirements. This certification also includes additional evaluation completed during installation, operation, and maintenance to ensure that all of these functions are accomplished in accordance with the design standards. Type certification is applied to all turbines of the same type and indicates how they conform to design evaluation, type testing, and manufacturing requirements. An additional, optional certification is called **type characteristic measurements**.

Design evaluation may include evaluation of:

- Control and protection systems
- Loads and load cases
- Structural components
- Mechanical and electrical components
- Design, manufacturing, and installation plans
- Maintenance plans
- Personnel safety

The elements of type testing are

- Safety and functions tests
- Dynamic tests
- Blade tests
- Load tests
- Individual components tests

The elements of the manufacturing conformity evaluation are

- Review and evaluation of the quality system
- Check of implementation of the design documentation to ensure that it explain the correct manufacturing processes and procedures
- Review of auditing to ensure that quality checkpoints are used

The elements of the type characteristic measurements are optional tests whose results must conform to International Electrotechnical Commission (IEC) 1400 standards (described later). These may include the following test elements:

- Power performance tests
- A quality test of output electrical power
- Measurements of acoustic and other noise elements

American Wind Energy Association Standards

The **American Wind Energy Association (AWEA)** is the national trade association for the wind energy industry and includes wind power project developers, equipment suppliers, services providers, parts manufacturers, utilities, researchers, and others involved in the wind industry. It was formed in 1974 in Washington, D.C. The AWEA has established some standards for wind turbine hardware performance. Some of these wind turbine standards are based on IEC standards. These standards include demonstrating the power curve of the wind turbine, the annual energy performance curve, and the sound pressure levels. The strength and safety tests and duration tests are administered as pass/fail.

The AWEA also evaluates other safety and operating parameters, including procedures for turbine operation, systems and/or provisions to prevent dangerous operation in high winds, methods available to slow or stop the turbine in an emergency or for maintenance, procedures and provisions that indicate how the wind turbine will be adequately maintained, procedures for component replacement, susceptibility of the system to any reduction of control function when the system is subjected to low operating temperatures, such as during winter or in locations where the climate is normally cold.

Some basic standards have been established so consumers can compare turbines in terms of annual output versus size. The rating for annual output energy is established at an average wind speed of 5 m/s (11.2 mph). The sound-level standard may be exceeded no more than 5% of the time at wind speeds of 5 m/s (11.2 mph), and the system should produce its rated power when the wind is at 11 m/s (24.6 mph).

The AWEA has been designated as a standards-making organization by the **American National Standards Institute (ANSI)**, which recommends standards for many products in various industries. The AWEA is recognized as the leading wind energy standards organization in the United States. One of the AWEA standards is the Turbine Hardware Performance Standard.

Computer-Aided Design and Modeling for Safety Testing of Wind Turbines

Computer modeling has become so sophisticated in recent years that it can be used to simulate all parts of a wind turbine and every aspect of its operation. Computer tests can show the stresses on all parts of the turbine, such as the blades and drive train. These tests simulate basic operation and are continued until the parts of the turbine fail. Results allow manufactures to predict, for instance, the maximum amount of wind that a blade can withstand before it will fail. Thus, standards that a particular wind turbine can withstand can be determined, or the parts can be redesigned to strengthen the point where it is weakest.

These computer programs can perform finite-element analysis (FEA), which allows the computer to complete a series of mathematical formulas to determine the safe operating range for an individual part or the complete system. FEA can also be used to determine the exact point at which a part will be stressed so much that it breaks down and fails. These simulation programs work so well that they have become an integral part of wind turbine standards and certification.

Small Wind Turbine Certification

Smaller wind turbines are not subject to the same requirements as larger wind turbines that produce power for utility companies. The Small Wind Certification Council (SWCC) is an independent certification body that certifies small wind turbines. The SWCC certification ensures that smaller wind turbines meet or exceed the performance, durability, and safety requirements of similar certifications for small wind turbines. The SWCC small wind turbine certification provides a common North American standard for turbine energy and sound performance.

Another area that is addressed by the SWCC is the need for easy-to-understand labels for wind turbines. These labels indicate annual energy output, rated power, and rated sound level of small wind turbines. The other function of a label is to confirm that the turbine meets durability and safety requirements. The SWCC has submitted its certification standards for small wind turbines to the American Wind Energy Association for its approval.

Standards Organizations in North America and Europe

Several standards organizations have existed in the United States and Canada for many years, and some of their standards apply to wind turbines. These organizations include Underwriters Laboratory (UL), the Canadian

Standards Association (CSA), the International Energy Agency (IEA), the International Standards Organization (ISO), the International Electrotechnical Committee (IEC), the American National Standards Institute (ANSI) and others. These organizations help identify standards for each part of the wind turbine and help to ensure safety, durability, and reliability of systems. Some standards, such as those ensuring safety, have been in place for many years; other standards are newer and were created specifically for wind turbines.

Agencies That Provide Certification in the United States

One agency that provides certification in the United States is the National Renewable Energy Lab (NREL), which is a national laboratory of the U.S. Department of Energy. The NREL has created a wind turbine design code called FAST. FAST stands for Fatigue, Aerodynamics, Structures, and Turbulence, and it is a comprehensive aeroelastic simulator capable of predicting both the extreme and fatigue loads of two- and three-bladed horizontal-axis wind turbines.

A second certification for wind turbines is called ADAMS, which stands for Automatic Dynamic Analysis of Mechanical Systems, and was developed by MSC Software. This program is a general-purpose, multibody-dynamics code with unlimited degrees of freedom that is also used to model robots, satellites, and cars. The modeling software allows for tests of wind turbine prior to assembly and installation. Modeling software allows technicians to test all aspects of the wind turbine components as well as the complete assembled system.

According to the Department of Energy, ADAMS is slower than FAST but more versatile. The DOE indicates that FAST is limited to most standard types of horizontal-axis wind turbines; ADAMS with AeroDyn can model almost any kind of horizontal-axis wind turbine. Both require the AeroDyn subroutine library to model aerodynamics.

The National Electrical Code and Underwriters Laboratory

The **National Electrical Code (NEC)** created by the National Fire Protection Association (NFPA), sets standards for electrical fire prevention and safety. The NFPA was established in 1896 is an international nonprofit association. The NEC is part of NFPA 70, which is updated every three years. In the United States, all wind turbine electrical systems and wiring, including wiring in panels, switches, and generator controls, must conform to the NEC. The NEC also requires proper grounding of all electrical components and circuits to ensure personnel safety and prevent electrical shock hazards.

The NEC is coordinated with the requirements of other agencies such as Underwriters Laboratory (UL) to ensure electrical safety. For example, Article 280 of the NEC and UL 1449 apply to lightning protection and surge. Wind turbines are often the tallest free-standing structures where they are located, which means that a wind turbine stands a good chance of being struck by lightning sometime during its operating life. Underwriters Laboratories has established standards (UL 1449) to help prevent damage or losses due to lightning strikes. UL also provides standards (UL 1741) for electrical inverters. Electrical inverters are used in some wind turbine applications to convert the voltage output from the wind turbine generator to exactly 60 Hz. You will learn more about these standards and codes in later chapters.

The American National Standards Institute

The American National Standards Institute (ANSI) is a private organization that was founded in 1918. Its membership consists of private and nonprofit organizations. Its standards are considered open standards in that all stakeholders help develop them. One goal of any organization that establishes standards is to get its standards accepted by ANSI. ANSI's purpose is to accredit U.S. organizations that develop voluntary consensus standards and to help to ensure the integrity of standards developers. ANSI identifies organizations that use its structure to create and maintain standards, such as the AWEA, which creates performance standards for wind turbines.

Agencies That Provide Certification Outside the United States

Because a large number of wind turbines are manufactured outside the United States, agencies in Europe have taken on the responsibility of creating and administering certifications. Germanischer Lloyd (GL) of Hamburg, Germany, has become the world's foremost certifying body for wind turbines. GL has certified of large ships and other large systems for over a hundred years. GL today certifies wind farms, wind turbines, and their components and is the leading certification body in the wind energy sector, providing both project and type certifications.

The European Wind Turbine Certificate and the British Wind Energy Association

The **European Wind Turbine Certificate (EWTC)** is available for wind turbines manufactured and/or installed in Europe. The standards are specific for the wind turbines and applications used in Europe. Similar certification in the United Kingdom is provided by the **British Wind Energy Association (BWEA)**.

The British certification consists of performance and safety standards that provide a method for evaluating the manufacture of small wind turbines in terms of safety, power, performance, reliability, and acoustic (sound) characteristics. The small wind turbine standards were developed from existing standards for larger wind turbines as specified by the **International Electrotechnical Commission (IEC)**.

The International Organization for Standardization

The **International Organization for Standardization (ISO)** is the world's **largest developer** and publisher of **international standards**, and some ISO standards relate to wind turbines. ISO is a **network** of the national standards institutes in **159 countries**. Basically, the ISO system is a set of procedures that covers all key processes in a business, including designing methods for monitoring processes to ensure that they are effective and keeping records of the processes, checking for defects and corrective actions, and the regular review of processes and the quality system to ensure that they facilitate continual improvement. This means that if a wind turbine manufacturer is ISO-certified, you can be sure the manufacturer is using quality process improvement procedures.

International Electrotechnical Commission Wind Turbine Standards

The International Electrotechnical Commission (IEC) came into being in June 1906 in London, England. The IEC was established to provide global standards to electrotechnical industries throughout the world. Because the main function of wind turbines is to produce electricity, they fall under many of the standards of the IEC. The IEC also works in conjunction with other standards organizations around the world.

Table 1–4 shows examples of the IEC standards for wind turbines, which are also called wind energy standards. This table shows the typical standards and explains what the standards cover and who convenes panels for discussions on the standards together with dates.

TABLE 1–4 IEC Wind Energy Standards

Working Group	Title	Convener	Start	Finish (Plan or Actual)	Purpose	Document Number
WG-1 WG-2 WG-3	Safety Requirements for Large Wind Turbines	R. Sherwin, AWEA, USA	09/1989	12/1993	Principal standard defining design requirements	IEC 1400-1[a]
WG-4	Small Wind Turbine Systems	F. Van Hulle, ECN, the Netherlands	02/1992	01/1994	Principal standard defining design requirements for small turbines	IEC 1400-2[a]
WG-5	Acoustic Emission Measurement Techniques	T. J. DuBois, USA	11/1992	09/1998	Defines acoustic measurements methods	IEC 1400-11[a]
WG-6	Performance Measurement Techniques	T. Pedersen, Riso NL, Denmark	11/1992	01/1998	Defines performance measurement techniques	IEC 1400-12[a]
WG-7	Revision of IEC 1400-1	P. H. Madsen, Riso NL, Denmark	03/1994	01/1999	Edition 2 of IEC 1400-1	1400-1 Ed2
WG-8	Blade Structural Testing	D. van Delft, TU Delft, the Netherlands	03/1994	1999	Defines methods for blade structural testing	1400-23
WG-9	Wind Turbine Certification Requirements	J. McGuire, Lloyds Register, UK	10/1995	1999	Defines certification requirements (harmonized version of several European standards)	1400-22
WG-10	Power Quality Measurements	J. O.Tande, Riso NL, Denmark	02/1996	1999	Defines power quality measurement techniques	1400-21
WG-11	Structural Loads Measurement	F. Van Hulle, ECN, the Netherlands	02/1996	1999	Defines methods for measuring operational loads	1400-13

[a]Published standard.

1.6 THE NEED FOR AN UNINTERRUPTED POWER SUPPLY

One of the problems with electricity generated by the wind is that in most areas the wind is not consistent. The wind may blow at 30 mph on some days but diminish to less than the minimum needed to keep the blades turning at night. Because the wind does not blow at the same speed at all times of the day and every day of the week, electricity generated from wind power must be stored or supplemented by other forms of electricity.

Today, electricity is required for telephones, lighting, heating, and air conditioning, as well as refrigerating food. All of these things have changed from things that are nice to have to things that are needed for health and safety. Our police and security rely on electricity for communications and operations. Food stores and homes require refrigeration to keep foods cold or frozen so they do not spoil. Many homes and small businesses use electricity to operate computers. Stores use electricity to power computers and telephone lines to complete credit card and other financial transactions. Banks and other financial institutions need electricity to keep their networks operating.

Many industries require large amounts of electricity on a continual basis to produce products, process foods and chemicals, and carry out other industrial processes. Many of these processes, such as making steel, chemicals, and pharmaceuticals, are continuous and cannot be shut down because the flow of electricity is interrupted. For this reason, the electrical supply must be continuous and adequate to supply all of these operations.

That electricity generated from wind power must be augmented with traditional sources has been a problem since modern wind turbines were first introduced. Some applications simply tie the wind turbine into the existing grid to offset the use of hydrocarbon fuels such as coal or fuel oil to generate electricity. The problem is not unique to wind turbines, of course, as it is also a problem for solar generation, which becomes minimal at night or on cloudy days, and electricity generated from hydroelectric power plants, which is reduced in dry seasons or when there is insufficient runoff from rain.

Ways to Work Around Varying Amounts of Wind-Generated Electricity

If a small wind turbine is used to produce electricity for a remote homesite, cabin, or farm, a bank of batteries can be used to store electricity generated when the wind is blowing, so that electricity will be available when the wind diminishes. The batteries must be continually checked to make sure they do not overcharge and do not get drained below their minimum rating. Some small wind generators use a small amount of voltage to control the field current for the generator. This type of wind turbine must be connected either to the grid or to a bank of batteries to ensure that this small amount of voltage is always available. If the system uses a bank of batteries, the batteries have a cut-out relay to drop out the battery bank before the load discharges them completely. That way they will have the small amount of voltage necessary for field control when the wind begins to blow at the minimum rate again. If the battery becomes completely discharged and cannot supply field current for the generator, the wind will cause the generator shaft to turn, but the generator will not produce any electricity until the field current is established again. For this reason, some smaller wind turbines that are not tied to the grid and rely on batteries utilize a permanent-magnet generator that does not need an external field current to operate.

Other applications use a small solar generation panel to help keep the batteries at minimum charge so that the small amount of field current is always available. In the worst case, if the wind remains below the minimum needed to turn the blades for an extended period, the batteries can be used to supply the voltage for the loads. If the batteries discharge below the rating needed to supply field current, the solar panel can recharge the battery back above the minimum so that is ready to provide field current when the wind begins to blow again.

If a home or small commercial site uses batteries to store excess power and needs AC voltage, an inverter is used to change the DC voltage back to AC when it is drawn from the storage batteries. Newer batteries that are stronger and can withstand deep discharge are being developed. Some research has also been completed using large capacitors for storage. These new technologies will extend the range of storage capabilities for these applications.

1.7 ELECTRICITY TRANSMISSION LIMITATIONS

Another problems with wind turbines is that the AC voltage that is generated is generally transmitted over a maximum distance of only about 300 to 400 miles. This means that wind turbines must be located relatively close to the sites where the voltage is used, which is not always where the best winds are available.

Typically, AC voltage is produced at a generating site, where the voltage is passed through a set of step-up transformers to get the voltage ready for its trip to the destination where it will be used. Typical voltage levels for transmission depend on how far the voltage needs to be transmitted. For example, for the longest distances, the voltage needs to be stepped up to 765 or 500 kilovolts (kV). For middle distances it is stepped up to 345 or 230 kV, and for the shortest transmission distances the voltage is stepped up to 138 or 69 kV.

Once voltage reaches its destination, substations at the edge of cities or near large industrial users, the voltage is stepped down to 69 or 26 kV for transmission around the local area. If the voltage is routed around a residential area or light commercial area, the voltage is stepped down to 13 or 4 kV so that it reaches the end user

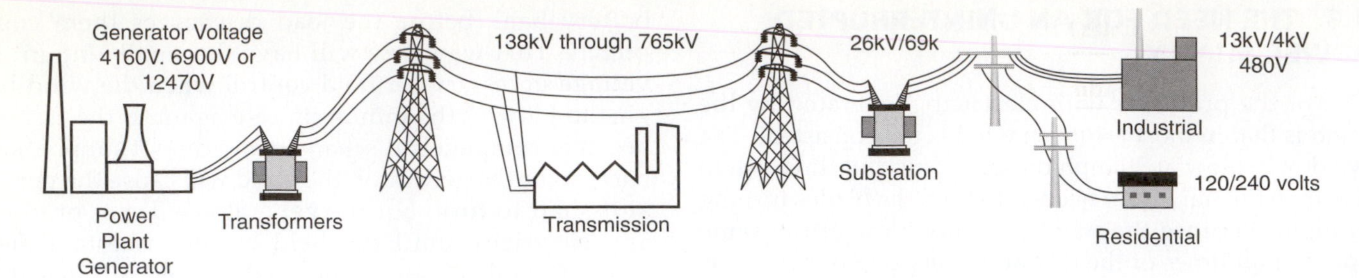

FIGURE 1–10 Typical power distribution system; the amount of voltage at each point is indicated.

at 480 or 240/208 V for industrial users and 240 or 120 V for commercial or residential customers. Figure 1–10 gives an example of these voltages in a diagram that shows typical transmission patterns.

The limit on voltage transmission determines where wind turbines are located and how much voltage is supplied to the transformer. Some wind turbines need larger step-up transformers to get their voltage levels high enough to transmit voltage over the longest distances possible, up to 400 miles. Research is ongoing to design and implement larger wind turbines, up to 6 MW, and put them together in multiple installations for wind farms so that these turbines can use the same step-up transformers and transmission lines.

1.8 ELECTRICAL DEMAND IN THE UNITED STATES

To understand how wind-generated electricity can help provide electricity for an individual application or the grid, you need to have some idea of how large the demand is for industrial, commercial, and residential electricity in the United States. The **electrical demand** is composed of several parts. As a start, the total demand for electricity can be measured by the total amount of electricity needed to supply all consumers over a period of an hour, day, week, month, or year. One of the reasons it is important to understand electrical demand is that it takes several hours to burn or convert the energy source and turn the generator before the electricity can be consumed. This means that electrical utilities need to anticipate usage as far as several hours ahead so that energy to produce the electricity can be put into the system ahead of time. This is not actually as big a problem with electricity generated by wind turbines, because they will produce electricity any time the wind is blowing, and it is put into the grid instantaneously as it is produced.

Another way to calculate the power needed is to break down the users by category or sector, such as industrial, the larger industrial complexes in an area; or commercial, including larger grocery and other stores as well as small and large shopping areas and malls that need large amounts of electrical power for lighting, air conditioning, and refrigeration. Commercial users may also include smaller businesses. The last sector is residential; in some areas, residential demand may be equal to industrial or commercial demand.

Yet another way to measure demand is to determine the peak demand during each day. For example, larger industries may reach peak demand during the early part of the day, when machinery is first started up, and demand may be lower on the second or third shift, when some companies do not operate all of their machinery. This type of demand may also be greater on weekdays and lower on Saturday and Sunday.

Demand may also be measured by the time of year. Demand in southern and southwestern states may be much higher during long hot periods, when air conditioning is used for extended periods. These same areas may not use as much electricity when the weather cools down and air conditioners can be turned off.

What this means to you as you learn about electricity created by wind generation is that demand for electricity fluctuates, and the ability of wind turbines to produce electricity fluctuates with wind conditions. This combination of fluctuating factors makes the problem of supplying electricity from wind quite complex.

Recent demand for electricity in the United States is shown in Table 1–5. This table shows demand for electricity as electricity consumption in billions of kilowatt hours per day for 2008, 2009, and as estimated for 2010. As shown in the table, in 2008 the residential sector consumed 3.51 billion kWh per day, and it was estimated that the residential sector would consume 3.82 billion kWh in 2010. In 2008 the commercial sector consumed 3.64 kWh per day, and it was estimated that the commercial sector would consume 3.70 billion kilowatt hours kWh per day in 2010. In 2008 the industrial sector consumed 2.49 billion kWh per day, and it was estimated that this sector would consume 2.56 billion kWh per day in 2010. In 2008 the transportation sector, which includes electric rail and other electric modes of transportation, consumed 0.02 billion kWh per day, and it was estimated that this figure would not change for 2010.

The row in the table labeled "Direct use" is the amount of electricity that electricity-generating plants use internally. This includes electrical power to pump cooling water and for other applications such as conveyor belts to move coal in coal-fired plants. Direct use in 2008 was

TABLE 1–5 United States Electricity Consumption by Sector

Energy Information Administration Short-Term Energy Outlook—March 2009			
Electricity Consumption (billion kWh per day)	**2008**	**2009**	**2010 est.**
Residential sector	3.51	3.79	3.82
Commercial sector	3.64	3.71	3.70
Industrial sector	2.49	2.74	2.56
Transportation sector	0.02	0.02	0.02
Direct use (power used to generate electricity)	0.43	0.40	0.41
Total electricity consumption	10.06	10.67	10.49
Total electricity consumption retail sales	9.66	10.26	10.10

Source: Energy Information Administration, U.S. Department of Energy.

0.43 billion kWh per day, and it was estimated that this sector would use 0.41 billion kWh per day in 2010.

In summary, the total daily electrical energy consumption in the United States in 2008 was 10.06 billion kWh per day, and in 2009 this value increased to 10.67 billion kWh per day. The total electrical energy consumption for the United States was predicted to be 10.10 billion kWh per day in 2010. The last row in Table 1–5 shows the total amount of electricity sold or billed in the United States. In 2008, 9.66 billion kWh per day were sold or billed, and the estimate for 2010 was 10.10 billion per day.

Total Installed Electrical Generation Capacity

Other information about the amount of electricity used in the United States includes data on the amount of capacity to generate electricity. Table 1–6 shows the cumulative capacity of all of the facilities that can generate electricity, including data for 1980 and then for the consecutive years from 2000 to 2006. These figures include electricity generated from coal, gas, and oil, nuclear power, hydroelectric power, wind, and solar energy. You can see from the table that the capacity in 1980 was 578.6 million kW, and by 2006 that capacity had increased to 1,138.794 million kW. The trend indicates that demand can be expected to increase every year. Capacity must grow at the same rate as demand if we are to keep up with future needs.

U.S. Electricity Generation by Fuel and Sector

The next thing to understand about the amount of electricity generated in the United States is the sources or sectors that generate electricity. Table 1–7 shows sources of electricity generation and the amount that each produces in billions of kilowatt hours per day. Using this table, you can compare the amount of electricity generated by wind energy to the amounts produced by other methods. In 2006, the top-producing sector was coal, with 5.397 billion kWh; nuclear power was second, with 2.157 billion kWh, and natural gas was third, with 2.012 billion kilowatt-hours. The remaining nonrenewable producing sectors included other gases, petroleum, residual fuel oil, distillate fuel oil, petroleum coke, other petroleum, and pumped storage hydroelectric, and their combined production was less than 0.3 billion kWh.

Electrical production from renewable sources is shown separately in Table 1–7, and you can see that in 2006, conventional hydroelectric produced 0.784 billion kWh, wind produced 0.073 billion kWh, geothermal produced 0.04 billion kWh, wood waste produced 0.028 billion kWh and solar produced 0.001 billion.

In 2008, conventional hydroelectric produced 0.715 billion kWh, wind produced 0.127 billion kWh, geothermal produced 0.041 billion kWh, wood waste produced 0.030 billion kWh, and solar produced 0.002 billion. From 2006 to 2008, electricity produced from wind energy increased from 0.073 billion kWh to 0.127 billion kWh, and the projection for 2010 was 0.208 billion kWh, an increase of nearly 3 times.

It is important to understand that the production of electricity from wind energy is increasing at a very rapid rate each year, but the amount produced from wind in 2006 was 0.073 billion kWh, which it was only a small percentage of the total electricity generated, which was 11.136 billion kWh. The projected output for 2010 was 0.208 billion kWh, which amounts to less than 2% of the total of 11.324 billion kWh.

TABLE 1–6 Electrical Generation Capacity for the United States from all Sources, in Billion Kilowatts

1980	2000	2001	2002	2003	2004	2005	2006
0.579	0.942	0.981	1.040	1.089	1.110	1.128	1.139

Source: Energy Information Administration, U.S. Department of Energy.

TABLE 1–7 U.S. Electricity Generation by Fuel and Sector, in Billion Kilowatt Hours per Day

Electric Power Sector	2006	2007	2008	2009 (projected)	2010 (projected)
Nonrenewable					
Coal	5.397	5.475	5.392	*5.273*	*5.255*
Natural gas	2.012	2.232	2.169	*2.195*	*2.236*
Other gases	0.012	0.011	0.012	*0.011*	*0.012*
Petroleum	0.164	0.168	0.116	*0.139*	*0.164*
Residual fuel oil	0.093	0.104	0.061	*0.062*	*0.066*
Distillate fuel oil	0.018	0.022	0.018	*0.018*	*0.019*
Petroleum coke	0.049	0.038	0.034	*0.056*	*0.077*
Other petroleum	0.003	0.004	0.003	*0.002*	*0.002*
Nuclear	2.157	2.209	2.2	*2.209*	*2.218*
Pumped storage hydroelectric	−0.018	−0.019	−0.017	*−0.016*	*−0.016*
Other fuels	0.019	0.019	0.02	*0.022*	*0.023*
Renewable					
Conventional hydroelectric	0.784	0.674	0.715	*0.684*	*0.717*
Geothermal	0.04	0.04	0.041	*0.043*	*0.043*
Solar	0.001	0.002	0.002	*0.003*	*0.004*
Wind	0.073	0.094	0.127	*0.156*	*0.208*
Wood and wood waste	0.028	0.029	0.03	*0.03*	*0.03*
Other renewables	0.038	0.039	0.039	*0.042*	*0.044*
Subtotal electric power sector	10.707	10.974	10.845	*10.79*	*10.937*
Commercial sector					
Coal	0.004	0.004	0.004	*0.003*	*0.004*
Natural gas	0.012	0.012	0.012	*0.012*	*0.012*
Petroleum	0.001	0.001	0	*0.001*	*0.001*
Other fuels	0.002	0.002	0.002	*0.002*	*0.002*
Renewables	0.004	0.004	0.004	*0.004*	*0.004*
Subtotal commercial sector	0.023	0.023	0.023	*0.022*	*0.023*
Industrial sector					
Coal	0.053	0.046	0.047	*0.045*	*0.047*
Natural gas	0.213	0.213	0.2	*0.189*	*0.194*
Other gases	0.027	0.026	0.026	*0.025*	*0.026*
Petroleum	0.012	0.012	0.007	*0.009*	*0.01*
Other fuels	0.014	0.013	0.007	*0.007*	*0.007*
Renewables					
Conventional hydroelectric	0.008	0.004	0.005	*0.005*	*0.005*
Wood and wood waste	0.078	0.077	0.075	*0.072*	*0.074*
Other renewables	0.002	0.002	0.002	*0.002*	*0.002*
Subtotal industrial sector	0.406	0.392	0.371	*0.354*	*0.365*
Total all sectors	11.136	11.388	11.239	*11.166*	*11.324*

Source: Energy Information Administration, U.S. Department of Energy.

1.9 TOTAL DAILY DEMAND FOR ELECTRICITY IN THE UNITED STATES

The daily demand for electricity in the United States fluctuates from month to month and from week to week, depending on the time of year and the location. The total demand on a daily basis also changes by the hour throughout the day. For example, the demand may be greatest during the day and diminish significantly during the late hours of the evening, when people are sleeping. In the South, Southeast, and Southwest, air conditioning is one of the largest residential users of electricity, so the residential demand for electricity is greater in the summer than the winter. The other large part of the daily electrical demand is the industrial demand for large industries. This demand is more consistent on a daily basis, but it does change on Saturday and Sunday, when these plants do not operate at full capacity. The industrial demand also changes slightly in the evening and overnight, when industrial usage drops.

Commercial demand is due mostly to stores and malls, and this demand changes over the hours of each day and is different on different days of the week. For example, commercial stores may open at 9 a.m. and close at 9 p.m., and the demand during those hours will be a maximum. During other hours of the day, this demand is much lower.

The biggest problem with such fluctuating demand is that the amount of electricity generated must change hourly, daily, and weekly to meet this changing demand. This means that some plants, such as coal-burning or nuclear power plants, that use heat to generate electricity, need to adjust the amount of energy input into the generation process hours before the time when the electricity will be needed.

1.10 TOTAL PEAK DEMAND

Total peak demand is different from average demand. Peak demand may fluctuate from a high demand for several hours to a much lesser demand—as much as 30% less—during the remainder of the day. Peak demand is the highest demand for electricity that is needed at any given time. An electrical system must be designed large enough to meet the highest demand that occurs, regardless the day, the day of the week, or the time of the year.

Peak demand in the United States moves across the country with the time zones. For example, peak demand may be reached an hour or two earlier on the East Coast and in the Central Time Zone than it occurs on the West Coast. This is called the *rolling peak,* and it occurs when large cities and industrial users come online first thing in the morning; and the cycling is also evident when cities and industries on the East Coast shut down an hour or two before cities on the West Coast. One of the larger demands occurs at nightfall, when all the lights in a city come on in the evening. This time also changes with the seasons: Nightfall comes much later in the evening in cities in the northern half of the United States in the summer, and it comes much earlier in the evening during winter in this same area. This means that power utilities must continually plan for changes in the time at which the peak will occur from day to day and from month to month.

Peaking Generators

During the past decade, electric utilities have installed a number of systems that can be brought online very quickly. These systems are powered by natural gas or oil and are called *peaking systems.* These systems can be fired up quickly and can add large amounts of electricity to the grid within a very short time during peak load demand periods.

These systems are also typically very clean burning, which means their impact on the environment is minimal. They are also installed very close to the location where the extra demand is coming from, such as areas with a concentration of large industrial sites or where large cities may need extra power for air conditioning on very hot summer days. The peaking generators may not run on other days, but because they are fired with natural gas, they can quickly be brought back up when needed. Peaking generators are also used is when a large coal-fired or nuclear-powered system needs to be shut down for a period of time. These systems can easily produce the extra power that is needed, and they can also be used to make up for fluctuations in wind energy production.

1.11 COMMERCIAL AND INDUSTRIAL DEMAND

In any electrical system, commercial and industrial demand may use up to 60% of the total amount of electricity produced. This demand accounts for a very large part of the total consumption of any system, but it tends to be very predictable. When economic times are good, the commercial and industrial loads will tend to grow slightly; these same loads will become smaller when an economic downturn occurs, such as the worldwide recession of 2009. This reduction in demand occurs because industrial orders become smaller and may be limited or eliminated for short periods when the demand for these products is reduced. **Peak electrical demand** is another important factor that must be considered when discussing electrical energy needs. Peak electrical demand is the largest amount of electrical energy that is needed during any hour at any time over a 24-hour period, on any day of the week or month. Peak electrical demand is different in different parts of each state, as it may be very large in industrial areas and occur mainly on Monday through Friday, whereas the demand in cities, where the load is predominately commercial (malls and stores), the demand may be larger on Friday and Saturday, because the electrical loads for lighting and air conditioning are higher on weekends.

Brownout

When the demand for electricity begins to outpace the supply, the voltage on the entire system becomes lower or droops. This condition of lower voltage is called a **brownout**, and it can damage appliances and other loads that remain connected to the electrical system. There is no defined level for a brownout, but it is generally considered a brownout when voltage falls beyond 8–12% below the typical supplied voltage. This may occur when demand is high, such as on very hot summer days, or if a major fault occurs in the power system, such as an automobile accident that takes out a power pole, or a high wind that takes out one or more power lines when trees blow into them.

Shedding Demand

In some cases, brownout may occur for several days during high-demand periods or when a major storm such as an ice storm or a high-wind storm has damaged part of the electrical transmission system. If the brownout is expected to last for very long, the electrical utility must bring on all of its peaking generators or begin to shed load in some areas. In some parts of California, the utility companies have used rolling brownouts to help keep the voltage up to standard levels. In a rolling brownout, the utility company plans ahead and notifies companies and cities that their power will be shut off completely for a short period of time so that sufficient power will be available for the remaining customers. The company will also ask consumers to turn off as many electrical appliances as possible during these times. Power is shut off to specific areas for a specific length of time, and then it is turned back on and shut off in a different part of the service area for the same length of time. This way, the utility company can stay online even though it does not have enough generated power to supply all of its customers at the same time.

Another technique that is used to help balance the demand and production of electricity is to create contracts with larger users to provide them with a lower utility rate in return for them agreeing to shut down completely during the higher-demand periods. This technique, which is sometimes called *interruptible service,* allows larger users to create a strategy to be able to shut down with a few hours notice if the overall voltage levels in the area are dropping or if a brownout occurs for any reason. These companies receive a very large discount for all of the power they use, in return for shutting down completely on short notice.

Backup Power

Some consumers, such as hospitals, crucial city services, and safety services, may need a backup power supply to support the power from the grid. For example, hospitals may have backup diesel generators that can provide up to 75% of the full load. This power can be used to provide lighting and backup power for essential services such as ongoing surgery. Some large industries have a backup for essential functions such as pouring molten iron or food processing that cannot be interrupted once they are started. It is important to be aware of these systems when connecting a wind generation system to them, because the switching gear that moves between the systems must be able to isolate each system for maintenance or other activities.

An **uninterruptible power supply** is a power supply that is backed up with one or more batteries and an inverter that converts DC voltage to AC voltage. The uninterruptible power supply takes in AC voltage and converts it to DC voltage, which goes into the bank of batteries and then is converted back to AC electricity with a frequency of 60 Hz. Some uninterruptible power supplies allow the AC voltage to pass through to the load and use only a small amount to keep the batteries charged at all times. An electronic switch quickly switches the power over to the backup power in the event of a power loss. This ensures that the load never sees the power outage and always has a source of AC power. The size of the battery in the backup system determines how long the system can run on backup power. In many cases, an uninterruptible power supply is used provide immediate backup after the power failure, and a longer-term backup energy source such as a diesel generator is then brought online to supply electrical power for as long as needed (assuming sufficient fuel is available). Uninterruptible power supplies are used for computer equipment, telephone equipment, alarm systems, and other critical power systems.

1.12 RESIDENTIAL DEMAND

Residential demand in the United States consists of several large loads and several continuous loads. The larger loads in homes include air conditioners, electric heat or heat pumps, electric washing machines and clothes dryers, electric ranges, and microwave ovens. Continuous loads may include lighting and refrigerators or freezers, and these loads tend to run for long periods throughout the day and throughout the year. Air conditioning and electric heating loads are seasonal and occur only at certain times of the year. Electric washing machines and clothes dryers as well as electric ranges tend to be used periodically and on various days of the week. Together these uses create residential demand. When thousands of homes are located together in a city or suburban area, the electrical demand can be as large as or larger than the industrial or commercial demand.

Residential demand also tends to move across the United States with the time zones, with many consumers in an area turning on their lighting at the same time, then turning off the majority of their loads later in the night, when most people go to bed. These loads resume again when everyone starts to wake up in the morning and turn on the loads again. Residential demand also grows when major building projects expand housing in specific areas.

1.13 GRID-TIED SYSTEMS

The **electrical grid** is the name applied to the entire electrical power distribution system in the United States and North America. If a wind turbine is designed to produce power for the grid, it must be connected to an existing section of the grid and its power must meet the specifications for that section of the grid. The grid that distributes electricity across the United States is actually made up of a number of interconnected wires that move electricity from where it is generated to where it is consumed. Electricity may be generated at values of less than 5000 V. This voltage must be stepped up using transformers to higher voltages, 345 kV for long-distance transmission or 500 kV for short-distance transmission, and it may be increased to even higher voltages, up to 765 kV, for

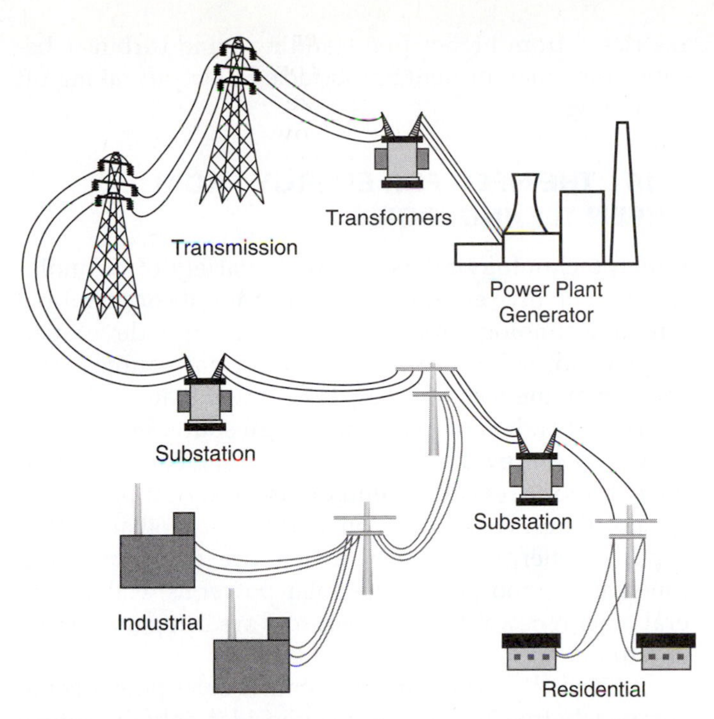

FIGURE 1–11 Electrical power distribution from generating station to residential customer.

transmission over the longest distances. When the electricity reaches the end user, it is transformed back down at a substation to the level at which it will be consumed. Figure 1–11 shows the distribution system from the generating site through transmission lines to a residential customer.

The grid is illustrated by the map in Figure 1–12, broken down by state. The grid is divided into three major sections, the Eastern Interconnect Grid, the Western Interconnect Grid, and the Texas Interconnect Grid. The Eastern Interconnect Grid sections include the Northeast Power Coordinating Council (NPCC), which includes Maine, Vermont, New Hampshire, New York, Massachusetts, and Rhode Island; the Mid-Atlantic Area Council (MAAC), which includes Maryland, Delaware, New Jersey, and parts of Pennsylvania; the East Central Area Reliability Coordination Agreement (ECAR), which includes Michigan, Indiana, Ohio, West Virginia, and Kentucky; the Southeastern Electric Reliability Council (SERC), which includes the southern states; the Florida Reliability Coordinating Council (FRCC), which includes

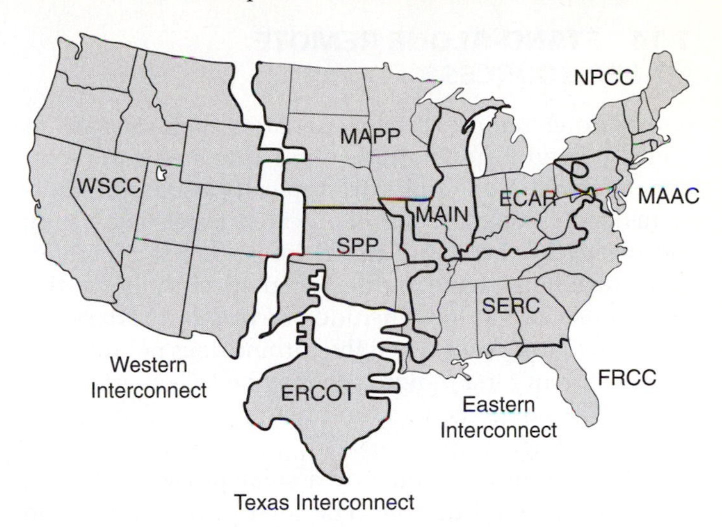

FIGURE 1–12 Electrical grid system for the United States. (Courtesy of NERC.)

all of Florida; the Mid-America Interconnected Network (MAIN), which includes Wisconsin and Illinois; the Mid-Continent Area Power Pool (MAPP), which includes the Dakotas, Nebraska, Minnesota, and Iowa; and the Southwest Power Pool (SPP), which includes Nebraska, Kansas, and Oklahoma.

The Electric Reliability Council of Texas (ERCOT) the majority of Texas, and the Western Systems Coordinating Council (WSCC) includes all of the western states from New Mexico, Colorado, Wyoming, and Montana westward. (ERCOT) is now called the Texas Regional Entity (TRE).

Table 1–8 shows the supply, demand, and capacity margin for the United States for 2009–2010. As shown, the capacity margin for the Eastern Interconnect Grid is 33.5%, for the Texas Interconnect Grid (ERCOT), now called the Texas Regional Entity (TRE), it is 39%, and for the Western Interconnect Grid it is 32.3%. The capacity margin for the entire United States is 30.4%.

The grid provides electricity for each of the areas shown in the map in Figure 1–12, and if you are providing electricity to the grid from a wind turbine, you much identify the part of the grid to which you will be connecting and be familiar with the specifications for that part of the grid. You will learn more about the grid in Chapter 8.

TABLE 1–8 U.S. Net Internal Demand, Actual or Planned Capacity Resources, and Capacity Margins, 2009–2010

	Eastern Grid	Texas Regional Entity (TRE), formerly Electric Reliability Council of Texas (ERCOT)	Western Grid	U.S. Total
Demand (MW)	476,057	46,068	113,504	635,629
Resource supply (MW)	684,015	80,424	170,745	935,184
Capacity margin (%)	33.5	39	32.3	30.4

Source: Energy Information Administration, U.S. Department of Energy.

1.14 STAND-ALONE REMOTE POWER SOURCES

Stand-alone remote power supplies include gas- or diesel-powered generators, wind turbines, and solar panels. Stand-alone units are typically used on remote farms or by other consumers located miles away from the electricity available from the grid. Some smaller wind turbine systems are designed specifically for this type of use in that they include one or more batteries to store power for times when the turbine does not run. The power system for a remote site may be DC- or AC-powered. If AC power is used, an inverter is needed to convert DC voltage to AC.

Some units also include a solar power system to augment the wind turbine. The solar power system can provide battery voltage when the sun is shining, and can back up the wind turbine even if the wind does not blow at sufficient force for several days at a time. The solar power system charges the batteries when the wind is not moving the turbine.

1.15 ENVIRONMENTAL AND ECOLOGICAL ASSESSMENTS FOR WIND POWER

When planning a wind power system, it is important to perform an environmental and ecological assessment. Environmental issues include damage done when wind turbines are installed in an area. In some cases, access roads and infrastructure must be improved before installation can be begun. The environmental impact also includes any changes to the area where the wind turbine is installed.

There have been some recent problems with wind turbine blades causing bird or bat kills. The blades are sometimes so large that they interrupt the flight path of a variety of birds as they rotate. Wind turbines may also produce unacceptable levels of noise. This noise is produced as the turbine blades cut through the air, because the blade tips are moving very rapidly. Newer blades have been designed to be much quieter than previous models, and some types of wind turbines, such as vertical-axis wind turbines, are less noisy than other, horizontal-axis types.

Noise may be an issue when wind turbines are located close to residential areas. For this reason, some larger wind farms are located where winds are stronger and at a distance from urban areas. Also, some wind farms are located in very rural areas, and others are installed offshore, where the noise issue is minimized because fewer people are affected.

Some states and municipalities require noise, environmental, and ecological assessments to be completed before installing a wind turbine. Some communities also restrict the placement of wind turbines, whereas others try to ensure that turbines are located close together in specific areas. Some areas, such as near airports, are also restricted from higher free-standing wind turbines, because they may present an obstacle to aircraft taking off or landing.

1.16 THE NEED FOR ENERGY FROM GREEN TECHNOLOGIES

Green technology refers to a wide variety of technologies that involve emitting minimal amounts of pollution. Green technology today includes energy developed from wind, solar, bio fuels, geothermal sources, and other evolving technologies. The United States has been overly dependent on oil from foreign sources for a large part of its energy. There is also increasing concern about air pollution and carbon emissions from coal-fired electrical utilities. For this reason, and because oil and other types of energy currently in use are not considered renewable, wind power and solar power as well as several other renewable energy sources are being used more and more.

Electricity from wind power and solar power is not completely free. It is free if you consider only the source of energy as compared to coal, nuclear fuel, or gas and oil, but the cost of wind turbines includes the costs of installation and maintenance. Many very large wind turbines are needed to produce the same amount of electricity as current coal-fired plants. When multiple wind turbines are installed, each represents a fairly large investment in periodic inspection and maintenance, which is usually quite expensive because some or all of it must be done well above ground level.

Even though the base energy for the wind turbine is free, currently the cost of producing a kilowatt hour of electricity from the most efficient wind turbines is about 4 to 4.5 times the cost of producing a kilowatt hour of electricity from coal, and about 2.5 times the cost of producing electricity from nuclear power.

The cost of producing energy can be compared by several methods but should include such factors as the cost of transporting coal longer distances to states that do not have coal reserves, and the high cost of maintaining security for nuclear fuel as well as safe long-term storage of spent fuel. And all of these issues may have to be reconsidered if new taxes are levied on traditional energy sources or if fees are imposed for creating pollution.

It is better to look at renewable energy sources such as wind energy and solar energy as good solutions to augment traditional power production for the future and regardless of price, because technical innovations and improvements will be ongoing as long as new installations continue to take place. It is also important that new energy sources that cause less pollution are utilized to the greatest extent possible, while recognizing that energy sources such as wind energy and solar energy at present cannot supply enough energy for our current needs on a full-time basis.

1.17 FUTURE DIRECTIONS FOR WIND POWER

The future of wind power in the United States will include a strong effort to increase capacity as quickly as possible by installing as many large and medium-size turbines as possible. This will be accomplished by government initiatives that include grants and tax relief for current installations and effort to streamline the planning and certification process. Large amounts of stimulus dollars will also be applied to alternative energy sources such as wind power and solar power.

If the price of oil and other types of energy continues to rise, or if these sources are taxed so that their end cost increases, wind power will gradually take a larger role through increased installations and increased production. For the near future, much larger units, up to 6 to 7 MW, are in the final stages of planning and installation. These larger units, when combined with larger wind farms, will make the electrical system equal to small coal-generation system.

Another concern for the future of wind energy is the preparation of a well-trained workforce who have the skills to install, troubleshoot, repair, and complete periodic inspections. Some technicians will be provided through cross training, whereas others will need to be trained from scratch. Other personnel will be required to prepare sites and install equipment as well as additional personnel to manufacture wind turbine blades, towers, and generators.

1.18 POLITICAL IMPLICATIONS FOR WIND POWER

The political implications for wind generation include tax credits and laws that may limit expanded use of some carbon-based fuels and some other energy sources. Federal, state, and local governments will continue to enact laws and regulations that will move toward a more aggressive integration of wind energy with current coal, nuclear, and other methods of generating electricity.

Regulations to improve air quality may limit increased use of coal and other carbon fuels while creating a need for more electricity produced by wind energy. Federal, state, and local governments may also improve the cost factor by creating an artificial pricing structure for electricity produced from wind turbines that will allow it to compete with electricity produced from coal, other forms of carbon-based fuels, or nuclear fuel. These same governments may also create structures to make it more profitable for individuals to produce electricity from wind power and sell it to electrical utilities.

Federal, state, and local governments may also provide tax breaks, grants, and loans for installation of wind turbines and their ongoing maintenance and repair. The federal government is also providing large sums for updating and expanding a smart grid. This research and development of the smart grid and the infusion of millions of dollars of new investments will allow the smart grid to be expanded and make it possible for more wind turbines to connect to the grid and sell back excess generated electricity to the utilities.

1.19 FINANCIAL IMPLICATIONS AND RETURN ON INVESTMENT

There are several factors to consider when trying to determine the return on investment for wind energy. **Return on investment (ROI)** is the amount of money or other value that becomes available after money is invested in a project. For example, if a wind project costs $100,000 to install and start up, the money that the investors receive from the energy the wind turbine produces over a specific period of time is the ROI. The ROI must be large enough to make it profitable for investors to invest in wind turbine projects. Individuals or companies that invest in a project calculate the amount of return and how many years it will take to repay their investment. Factors used in determining ROI for wind energy investment include the ability of the wind turbine to harvest wind energy and turn it into electricity and how often the wind blows strongly enough. The more efficient a wind turbine is and the longer the wind blows at sustained speeds, the better will be its return on investment. Basically, wind turbines with larger turbine blades and larger swept area are more efficient and may yield higher ROI than a smaller unit. Another factor in calculating ROI is the initial cost of the wind turbine system and the cost of installation. Governments can make overt efforts to encourage wind turbine installations by lowering the costs of initial planning and ensuring that installation costs remain low, therby raising the ROI. If government offers grants and guarantees loans, this can reduce the initial cost. If such loans carry low interest rates, ROI again benefits.

Some types of installations provide better output than others. For example, locating wind farms offshore will ensure that a significant amount of wind is available for harvest, and that this wind will blow continuously and strongly so that the wind turbine can produce at near peak levels. When the amount of wind and the duration of the wind increases, the ROI becomes better.

Another thing that can improve ROI is to install large numbers of wind turbines in a cluster called a wind farm. When multiple wind turbines are installed near each other, the installation costs are reduced slightly because improvements to the site infrastructure, such as electrical power substations, grid connections, data-gathering systems, and other technology can be shared.

Other Ways to Obtain Income from Wind Turbines

Companies, farmers, and other land owners can make money from wind turbines in several other ways, such as through contracts with energy companies that install wind turbines on their property and pay them a fee for doing so. The sites are selected because they produce the best winds for the turbines. Because the wind energy companies do not own the land, they must either purchase it outright or lease the land and pay the owner in some way. This may be done by means of fixed payments over a specific period of time and may include established specified increase in the fee at specific dates in the contract. The advantage of this type of escalator clause is that it provides a stable income for the land owner and a consistent known cost to the energy producer.

Another way to receive payment from a wind energy company is to design a contract that pays a royalty or percentage of the income generated. This type of contract needs to include a means by both parties can access all of the records of production and other costs. A problem with this type of contract, however, is that the income may vary from month to month because of variation in the amount of wind. A similar plan uses a fixed amount plus a percentage as a royalty. Yet another type of plan involves an equity partnership, which involves the land owner selling all or a portion of the land to the wind turbine company. If the ownership is percentage-based, then the income is split according to the same percentage.

1.20 GREEN ENERGY PAYBACK CALCULATIONS

One of the most important things about making a decision to install a wind turbine or a series of wind turbines on a wind farm is determining when it will produce enough income to pay off the investment and begin to earn income for the owner. A calculation that helps determine the point at which the wind turbine will produce enough energy to pay back the initial investment is called a payback calculation. Payback calculation involves two parts, all the costs on one side of the equation, and all the income factors on the other side. Table 1–9 lists all of the factors that should be included in calculating costs and income for wind turbines. For example, costs may include the cost of site preparation; the cost of permits and supporting documents for environmental impact assessments and inspections; the cost of shipping the wind turbine to the site, including, perhaps, shipping from an overseas location; the costs of installing the tower and installing the wind turbine and blades; start-up costs,

TABLE 1–9 Factors in Calculating Cost and Income for Wind Turbines

Cost Factors	Income Factors
Site preparation	How much electrical power will be produced daily, weekly, or monthly?
Permits	How much electrical power will be offset or replaced by the wind turbine
Environmental studies	Price of the electrical power the wind turbine is replacing
Inspections	Cost of power that is purchased or bought back by the electric utility
Shipping from manufacturer to site	Efficiency of the wind turbine and size of the blades
Cost of the wind turbine	Location of the wind turbine and how many hours of peak wind it will receive per year
Installation of the tower	Grants or tax credits
Installation of the wind turbine and blades	Accelerated depreciation for tax purposes
Costs to start up turbine and bring online	Renewable energy credits
Electrical gear and connection to the electrical system	
Transformers and substations	
Installation of data-gathering equipment (SCADA)	
Annual inspections	
Periodic maintenance	
Estimated minor maintenance	
Estimated major maintenance	
Salaries of any personnel	

TABLE 1–10 Data Needed to Calculate Wind Turbine Payback Period

Input Data	Notes
Mean power output as a function of mean wind speed and level of wind variability	From manufacturer's data
Estimated cost per kilowatt hour as a function of mean wind speed and turbine lifetime	From a calculation of output of the wind turbine and where it is installed
Estimated payback period as a function of mean wind speed and reference electricity costs	Requires knowing the cost of electricity being replaced or, if it is to be sold to the grid, the buy-back rate
Power output profile showing percentage of time the turbine will produce various levels of power, including the zero-power-output case	From calculation of how long the wind turbine will run at various speeds and durations for each day

including making adjustments at start-up; the cost of connecting to the electrical system and to transformers, substations, and the grid; and the cost of installing the Supervisory Control and Data Acquisition (SCADA) system, which consists of software that continually monitors the wind turbine and the generator output. Other costs include the cost of annual inspections and periodic maintenance as well as a budget for minor and major maintenance that will occur during the life of the electrical and mechanical equipment. In larger systems or on wind farms, you may need to factor in salaries of onsite maintenance personnel.

The income side of the calculation will include estimates of the amount of electrical energy the wind turbine will produce per day, week, and month, so that you can determine how much electrical load you will be replacing with the wind turbine output if the power will be consumed on site as a standalone system. To estimate the output of the wind turbine, you can use the blade sweep area and the estimated amount of wind for the location where the turbine will be installed (there are formulas for this on several online websites). Next you need to calculate the total cost of the replacement power the wind turbine will producing and use the utility billing cost factor to determine the total cost you would pay the utility for this power, including any penalty charges such as exceeding a demand factor or other penalties for excessive usage during peak periods. The electrical rates will be different for residential users compared to commercial or industrial users. Once you have the income and costs calculated, you can determine the length of time it will take to pay back the investment for a wind turbine that will supply power directly to a home, commercial establishment, or industry.

If the system is a direct tied system and the power is being sold directly to the grid, you need to determine the output of the wind turbine over a period of time just as before, but with power being purchased back by a utility, you will need to know the "buyback" price, which is determined by the utility that is purchasing the energy. You need to be aware that the price of power the utility pays for electrical power it "buys back" is substantially less than the price it charges for power it is selling. For this reason, some larger wind turbine farms act as small utility companies and sell the power directly to other end users, instead of selling it to a utility.

Payback Calculators

Online calculator programs or stand-alone computer programs can help you determine the exact values for each of these factors. Some calculators include more or fewer factors, and some weight some factors more than others. Some companies that manufacture and install wind turbines can provide payback calculations that are very accurate, because they have a long history of installing and operating wind turbines.

One online site is www.wind-power-program.com/index.htm, which uses the information listed in Table 1–10 to determine the payback. Figure 1–13 shows a screenshot of this program and the information you will be expected to put into the table. To provide accurate data for this calculation, you will need to know the wind turbine blade diameter; the cut-in speed for the turbine, which is the minimum speed at which the turbine blades begin to move; the cut-out speed, which is the speed at which the turbine blades must be controlled and brought to rest; and the steady power delivered by the turbine based on steady wind speed between the cut-in and cut-out speeds. This information can be entered into the calculator manually, or it can be downloaded by model number for some of the more popular units. The calculation will provide a peak efficiency for the given wind turbine. The peak efficiency is generally between 25% and 35% for small turbines and between 40% and 50% for the largest turbines. These values are based on units of meters per second.

After the efficiency of the wind turbine is established, the amount of power the wind turbine will be capable of producing is determined. For this step you will need to determine how long the wind turbine will run each day and what the average speed of the wind will be during that day for the location where the turbine will be installed. This will help determine the amount of electrical power the wind turbine will produce each day. The information for this part of the calculation must be as accurate as possible to provide an accurate payback. This

Wind Turbine Power Curve Data

Name of Wind Turbine: EZ Wind 2 3.5 meter 2.2 kW

Rotor Diameter in meters: 3.5

Cut-in-Speed (Meters/sec.): 4 Cut-out Speed (Meters/sec.): 28

Initialize Data

Um/s	PkW	U%	Um/s	PkW	U%	Um/s	PkW	U%
1	0	0	11	2.51	25.2	21	2.25	2.8
2	0	0	12	2.55	22.5	22	2.04	2.7
3	0	0	13	2.58	17.9	23	2.02	2.5
4	.12	30.1	14	2.62	15.6	24	2.02	2.3
5	.25	32.2	15	2.68	11.5	25	2.01	2.2
6	.48	33.8	16	2.69	9.5	26	2.02	2.1
7	.74	33.9	17	2.63	6.8	27	0	0.0
8	1.16	34.1	18	2.57	5.3	28	0	0.0
9	1.52	34.6	19	2.48	4.9	29	0	0.0
10	2.14	34.8	20	2.39	3.6	30	0	0.0

Update Calculations

FIGURE 1–13 Screenshot of program that calculates wind turbine power curve data. The wind speed (U) is in meters per second, the amount of power (PkW) is shown in kilowatts (kW), and the efficiency (U) at each speed is shown in percent (%).

information is also specific to the site location and the height of the tower for each given wind turbine. The data may need to be averaged over a day, a week, or a month and adjusted for month-to-month variations depending on the site. Some of this data can be taken from the data provided by the SCADA system from another wind turbine that has been running at this location previously. This is useful when additional turbines are added to a location or to a wind turbine farm.

Once the efficiency of the wind turbine has been determined, the amount of time it will be running and the total amount of energy it will produce in a year can be calculated. The next step in the payback calculation is to determine the cost of the electrical energy the wind turbine is replacing, or the buyback cost of the electricity if its electrical power is produced primarily for a utility on the grid. These amounts will vary from location to location depending on electrical power rates, state legislation, and applicable regulations for producing electrical power from wind turbines. Once these values are known, and the total cost has been found, the payback period can be determined.

1.21 TAX CONSIDERATIONS FOR WIND POWER

Financial incentives are available in the form of tax relief programs that include income tax incentives, corporate tax incentives, property tax incentives, and sales tax incentives. The tax incentives include both federal and state tax incentives as well as loans and other incentives to offset the cost of installing wind turbines. These incentives are designed to help accelerate installation of new systems. Some of the incentives are designed for homeowners and personal-use systems, and others are designed for corporations that are installing and operating larger systems to help add electrical power to the utility supply. Table 1–11 lists some of these programs. This table is reproduced from the database of State Incentives for Renewables & Efficiency (DSIRE), a program that was established in 1995 as an ongoing project of the North Carolina Solar Center and the Interstate Renewable Energy Council (IREC). This table shows only a sampling of federal programs; you should check for the latest programs as they become available and are funded by the

TABLE 1–11 Federal Incentives and Policies for Renewables and Efficiency

Financial Incentive	Name of Program
Corporate deduction	• Energy Efficient Commercial Building Tax Deduction
Corporate depreciation	• Modified Accelerated Cost-Recovery System (MACRS) + Bonus Depreciation (2008–2009)
Corporate exemption	• Residential Energy Conservation Subsidy Exclusion (Corporate)
Corporate tax credit	• Business Energy Investment Tax Credit (ITC) • Renewable Electricity Production Tax Credit (PTC)
Federal grant program	• U.S. Department of the Treasury Renewable Energy Grants • U.S. Department of Agriculture Rural Energy for America Program (REAP) Grants
Federal loan program	• Clean Renewable Energy Bonds (CREBs) • U.S. Department of Energy Loan Guarantee Program • U.S. Department of Agriculture Rural Energy for America Program (REAP) Loan Guarantees
Industrial recruitment support	• Qualifying Advanced Energy Project Investment Tax Credit
Personal exemption	• Residential energy Consecration Subsidy Exclusion (Personal)
Personal tax credit	• Residential Energy Efficiency Tax Credit • Residential Renewable Energy Tax Credit
Production incentive	• Renewable Energy Production Incentives (REPI)

Source: Database of State Incentives for Renewables & Efficiency (DSIRE), an ongoing project of the North Carolina Solar Center and the Interstate Renewable Energy Council (IREC).

federal government. Also, you should be aware that many of these programs have limitations and have deadlines, and the programs may expire after a specific length of time. You can go online to the DSIRE website to check the latest incentives at any time.

At the time of this writing, federal programs include corporate tax deductions, corporate tax depreciation allowances, corporate tax exemptions, corporate tax credits, grant programs, loan programs, industrial recruitment support, personal tax exemptions, personal tax credits, and production incentives.

The current federal tax incentive for residential systems is 30% of the total installed cost of the system, not to exceed $4000. For commercial systems the credit is equal to 30% of expenditures, with no maximum credit for small wind turbines placed in service after December 31, 2008. Eligible small wind property includes wind turbines up to 100 kW in capacity. (In general, the maximum credit is $4000 for eligible property placed in service after October 3, 2008, and before January 1, 2009. The American Recovery and Reinvestment Act of 2009 removed the $4000 maximum credit limit for small wind turbines.)

Additional U.S. Investment Tax Credit

In 2005, Congress created the Federal Energy Policy Act of 2005. This act included a federal tax credit for residential energy property that initially applied to solar-electric systems, solar water heating systems, and fuel cells. In January 2008, H.R. 1424 extended the tax credit to small wind energy systems. This tax credit can be claimed through December 31, 2016. However, it must be applied to the Alternative Minimum Tax, so some low-income homeowners may not be eligible for it.

In general, taxpayers may claim a credit of 30% of qualified expenditures for a system that provides power to their primary residence if that residence is located in the United States. Equipment expenditures can be deducted when the installation is completed, power is being produced and the residence is occupied. Expenditures can include costs for on-site preparation labor, assembly or installation of the wind turbine, and for piping or wiring used to connect the system to the home electrical system. If the federal tax credit exceeds the taxpayer's tax liability, the excess amount may be carried forward to the succeeding taxable year. The maximum allowable credit varies by technology and equipment. Owners of small wind systems with 100 kW of capacity and less can receive a credit for 30% of the total installed cost of the system, not to exceed $4000. The credit is available for equipment installed through December 31, 2016. For turbines used for homes, the credit is limited to the lesser of $4000 or $1000 per kilowatt of capacity.

Another federal program is the Investment Tax Credit (ITC), which is available to help consumers purchase small wind turbines for a home, farm, or business. The owner of a small wind system with a capacity of 100 kW or less can receive a credit for 30% of the total installed cost of the system. **The ITC was originally created as part of the** Emergency Economic Stabilization Act of 2008 and was **modified** by the American Recovery and Reinvestment Act of 2009. The ITC is available for equipment installed from October 3, 2008, through December 31, 2016.

Examples of State Programs

Many states have implemented tax incentives to stimulate new installations of wind turbines. These programs are in addition to the federal tax programs and can be used with federal tax deductions. This means that homeowners, business owners, and commercial energy producers can put together state and federal tax incentives to reduce the financial cost of installing a wind turbine. One example of this type of tax program is that the state of Louisiana provides a tax credit for the purchase and installation of wind energy systems purchased and installed at either a residence or a residential rental apartment complex on or after January 1, 2008. The credit may be applied to personal, corporate, or franchise taxes. The amount of the credit is equal to 50% of the first $25,000 of the cost of the system, including installation.

In Ohio, the Ohio Department of Development's (ODOD) Ohio Energy Office (OEO) has a grant program to implement renewable energy projects that is limited to solar electric, wind electric, and solar thermal systems for commercial, industrial, institutional, and governmental entities in Ohio. The incentive level for this grant is $2.00/kWh. A minimum of 3000 kWh of AC power must be generated annually at the average wind speed for the site. The maximum incentive is 40% of eligible system costs, with the maximum incentive a total $200,000 of eligible system costs. Ohio also has a residential grant that provides up to $25,000 for residential systems through the Residential Wind Energy Incentive (NOFA 09–02). What is unique about the program in Ohio is that this grant opportunity is an Advanced Energy Fund program funded by a rider on the electric bills of customers of investor-owned electric distribution utilities in the state. The utility companies include American Electric Power (AEP), Duke Energy (formerly CINergy), Dayton Power and Light, and First Energy

Checking out the Latest Offers

It is important to check local, state, and federal programs before planning the installation of a wind turbine. These programs change almost monthly as new legislations is approved at the local, state, and federal levels. You can find the most up-to-date information about programs through federal, state, and or local government websites. You can also contact the offices of energy or the offices that control incentives for wind energy.

Some utility companies also provide the latest information about tax incentive and loan programs. Some wind turbine manufacturers and companies that install wind turbines are also aware of all of the tax incentive and loan programs available at any given time.

1.22 SKILLS NEEDED FOR GREEN TECHNOLOGY JOBS

The wind turbine industry will require a large number of electrical technicians and mechanical technicians to assemble, install, troubleshoot, and maintain all sizes and types of wind turbines. The industry will also require a large number of architects and designers as well as planners who will help determine the exact location to place individual or larger number of wind turbines on wind farms.

Another large group of technicians will be needed by manufacturers of metals for the various types of towers and by electronics companies that make generators, programmable logic controllers (PLCs), control boards, yaw motors, and switch gear for the electrical parts of the system. Machinists will be needed to manufacture all the gears in the gearboxes and transmissions, as well as the machining needed for all of the hydraulic pumps, valves, and cylinders.

Another group of jobs will be available in site preparation, which involves everything from road building to preparation of deep footings used to anchor towers. This involves excavating holes for the concrete and mounting any conduits into the concrete for the electrical wires to run underground to the wind turbine.

Other jobs will be found among the people and companies that will be responsible for erecting the towers and placing the nacelle, blades, and generator on the tower. On smaller units this will consist of operating medium-size to large cranes. In some larger systems the nacelle and blades will be lifted into position by helicopters. If the wind turbines are located offshore, another group of technicians will be needed who are capable of working over open water and laying underwater electrical cables and securing them so that the electrical power that is generated offshore can make it to shore.

In some applications, a complete large-scale electrical power distribution system will be needed where large numbers of wind turbines are located on wind farms. The power distribution system will consist of large banks of transformers as well as distribution towers for transferring high-voltage lines over long distances. Another area of job growth will be in companies that manufacture transformers, towers, and high-voltage cables, as well as in companies that locate and install towers and transformers in the field.

Skills Needed to Maintain and Overhaul Wind Turbines

When wind turbines operate for large number of hours, they need routine periodic maintenance. Technicians doing this work must be able to climb the tower to work on the propellers, rotor, gearbox or transmission, shafts and couplings, and the generator. Each of these components must be inspected periodically and lubricated. Sometimes vibration analysis will be required, and adjustments to shafts and couplings must be made.

In some cases the technicians will need mechanical and rigging skills to change out the turbine blades, rotor, gearbox, shafts, or generator. Some of this work may be completed in a large facility on the ground when all of the components are changed out as a system during a major overhaul. In other applications, the parts may be removed and replaced individually, and all of the work is completed at the top of the pole. The overhaul process allows the major parts of a wind turbine to be replaced, which will give the wind turbine an additional 15 to 20 years of operation, this being less expensive than changing out the entire wind turbine. Technicians will be needed for overhaul and periodic maintenance projects, and these technicians may be assigned to a specific site or they may travel from site to site on a contract basis because the work at any one site may be limited to a few weeks per year.

1.23 JOBS IN WIND POWER INDUSTRIES

In the near future, the wind power industry will need a large number of skilled technicians and other professionals to meet the needs of expansion. Some projections put the total number of new jobs at 1 million job openings. The jobs in the wind technology industries are referred to as "green-collar jobs." These jobs include construction manager, director of wind operations and maintenance, wind engineer, senior mechanical and electrical engineer, and environmental engineer.

Planning and Sales Jobs

Other types of jobs that are available in the field of wind energy are planning and sales jobs. Planners may have backgrounds and training in environmental science, and they must understand the requirements of wind availability and other details about site preparation, include connecting to the grid. Sales specialist are needed to meet with clients who express an interest in wind turbine technology.

Educational Requirements for Wind Power Jobs

Some wind power jobs will require only a high school diploma or a vocational school certificate. These types of jobs will be the lowest-level jobs. Many jobs will require degrees from community or technical college programs in electrical or mechanical technology, automation, computer-aided drafting/design, or architecture or project management degrees. Some jobs will require a bachelor's degree some area of engineering or engineering technology. Some jobs will also require either a bachelor's or master's degree in business.

Teachers and Technical Trainers

As the field of wind technology requires more and more trained technicians and trained personnel, the need for teachers and trainers in this area will also expand. Companies that manufacture and maintain wind turbines will need technical trainers to provide proper instruction and training for the maintenance and operation of these large systems. Community and technical colleges as well as universities will need skilled individuals who have the ability to teach these technical programs.

Field Service Technicians

Field service technicians may be employees of the companies that manufacture and install wind turbines, or they may be employed by companies that specialize in wind turbine maintenance. These technicians are usually on call 24 hours a day, seven days a week, so they can respond quickly to problems at operating wind turbine facilities. Field service technicians need electrical skills to work on the electrical panel and the generation equipment, and mechanical skills to work on the mechanical drive train of the wind turbine including the blades, rotor, shafts, and gearbox. They may also need to work on the wind turbine tower. Technicians must be able to climb towers and work at heights above 200 ft. Specialized skills of these technicians may also include the ability to troubleshoot and repair hydraulic systems and their components, PLCs, and electronic systems used in inverters and other wind turbine monitoring systems for.

When a problem occurs in a wind turbine and it stops producing electrical power, the field service technician is called to troubleshoot the entire system and must quickly identify not only the symptoms of the problem but also its root cause. Field service technicians may also complete routine periodic maintenance on wind turbines.

Project Manager and Architectural Jobs

When a wind turbine project is being designed, specified, and installed, skilled personnel are needed to design and implement the system. The tasks include researching specifications, fulfilling permit requirements, and managing the entire installation process. Some of these skills are similar to managing in a construction project, but a thorough understanding of the requirements of wind turbines is necessary to manage turbine installation projects. The needed skills become even more complex when a number of turbines are installed at the same location as part of a larger wind farm.

The project manager must understand all of the requirements for preparing the site, excavating, installing the tower and all the wind turbine components, and then connecting the system to the electrical grid. The project manager relies on specialists in each of these areas, but the project manager's skills keep the project on schedule and within budget. Architectural skills for these types of projects are similar to those for other large construction projects, but additional knowledge about wind turbines is required.

Wind Turbine Assembly Technicians

In the next 10 years the numbers of wind turbines that will be manufactured will be nearly double the number that currently exist. More companies will begin to manufacture and assemble wind turbines throughout the United States and overseas. Jobs in this area will include assembly of the component parts to make an operating wind turbine. Assembly technicians will need to be able to read and interpret blueprints, and follow procedures to assemble the mechanical and electrical components in the wind turbine. This may involve assembling all the components inside the nacelle, including the rotor, low-speed shaft, gearbox, high-speed shaft, and generator. After these components are completely assembled, they are put on a test stand and operated under test conditions so that final adjustments can be made. The final assembly is checked for alignment and vibration of the drive train, as well as the output of the electrical generation system. If an inverter is used, the inverter is connected to the generation system and tested to ensure that its output produces voltage at a set frequency of 60 Hz.

Another job in this process is that of quality control technician, who ensures that proper procedures are followed and specifications are met. The quality technician ensures that every wind turbine that leaves the assembly facility is ready to be installed and will operate without problems.

Turbine Blade Technicians

As the demand for wind turbines increase in both the United States and abroad, a large number of companies are beginning to produce wind turbine blades and other components. Modern wind turbine blades are made of fiber-reinforced epoxy (fiber glass, or other carbon composite and plastic parts. Turning these raw materials into balanced turbine blades requires a set of skills similar to those used in making fiber glass boats or other fiber glass components. Other parts of the system, including the nacelle cover, are made of fiber glass or carbon composite materials to minimize their weight and maximize strength. Another job that has become necessary in the manufacturing of blades is designing and building metal molds. These molds are used to add pressure to the manufacturing process to ensure that turbine blades are strong enough to withstand the strong winds and large torque to which they will be exposed.

Electricians and Electrical Transmission Technicians

When a single wind turbine or a large number of wind turbines are put together on a wind farm, their electrical output must be connected to the grid. This process is similar to connecting any electrical generation systems such as a coal-fired, nuclear power, hydroelectric, or gas-fired generation system to the grid. If a large amount of electricity is being transmitted from the wind farm to the existing grid, additional power poles and transmission towers may be required. This system may also require a number of new transformers and substations to provide the power at the proper voltage level. Electricians and electrical transmission technicians ensure that the power transmission system meets the requirements of the existing transmission utility.

Some of a new installation to transmit high-voltage power from the wind turbine site to the existing grid may be installed underground. Underground cable installation involves burying large conduits (plastic piping) underground, and then pulling electrical cables through it. Modern installation processes allow the conduits to be placed underground with a drilling/boring machine. The drilling/boring machine uses a drill bit that is mounted on the end of the drilling pipe and is mechanically controlled to ensure precise positioning as the drill pushes through the earth at depths of 3 to 10 ft. A very sensitive detection system follows the drilling head as it progresses so that it can be located within inches of its intended path. The drilling pipe is added in 8- to 10-ft sections until the entire run is completed. After the drilling head reaches its final destination in a hole that is dug to the required depth, the plastic pipe that is used for the conduit is connected to the end of the drill. The plastic conduit is stored on large rolls that allow it to be unrolled as the drill pulls the drill pipe sections back to the drilling machine. When the process is completed, the two ends of the conduit remain 3 to 6 ft out of the holes at each end. Later, technicians use a pull cable that is inserted inside the conduit when it is installed underground. The pull cable is then attached to the electrical cables and they are pulled through the underground conduit. The electrical cables are connected to the wind turbine electrical cabinet on one end and to service disconnect equipment where it is installed to the grid on the other end.

Riggers and Tower Installers

Another job connected with installing wind turbines involves erecting the tower and installing the wind turbine nacelle at the top of the tower. This job requires skills that are identified as rigging and installation skills. The rigging skills include attaching cables and lift slings to the tower and/or nacelle so that it can be lifted safely into the final position. The lifting cables and slings are connected to a crane that is used to lift them into position. The crane must be large enough to lift the weight of the tower and tall enough to ensure that it can be erected safely. Rigging also includes the skills needed to move the wind turbine components precisely, even when they are located high above the ground.

Where to Look for Wind Energy Jobs

Probably the best places to look for wind energy jobs is on websites designed specifically to list these jobs. You can also visit the websites of companies that manufacture wind turbines and look for job postings, and you can also find jobs posted by companies that install and erect wind turbines. Some companies that want to hire personnel for their wind energy jobs host job fairs or other types of activities that recruit highly skilled personnel directly.

Questions

1. Name six basic parts of a horizontal-axis wind turbine.
2. Explain what the nacelle is and what is inside it.
3. Explain how a wind turbine that runs its generator at a fixed speed and one that allows the generator to run at any speed can both produce voltage at 60 Hz.
4. Identify the event that occurred in the 1970s that caused a growth in wind power in the United States.
5. Identify the two main classifications of wind turbines.
6. Explain the main difference between Darrieus- and the Savonius-type vertical-axis wind turbines.
7. Explain some of the limitations of using wind power as the sole source of electrical energy.
8. Explain why it is important to understand the total electrical residential, commercial, and industrial demand for electrical power when you study wind energy.
9. Name four federal or state grants that provide funding for wind energy.
10. Identify the factors you would need to consider to determine the return on investment for a wind turbine.

Multiple Choice

1. The difference between an upwind and a downwind horizontal-axis wind turbine is that:
   a. The wind blows over the nacelle and then over the blades in the downwind and over the blades first and then over the nacelle in the upwind horizontal-axis wind turbine.
   b. The wind blows over the nacelle and then over the blades in the upwind and over the blades first and then over the nacelle in the downwind horizontal-axis wind turbine.
   c. Upwind horizontal-axis wind turbines are built upwind of buildings, and downwind horizontal-axis wind turbines are built downwind of buildings.
2. In 2009/2010, the approximate total supply of electricity from all three sectors of the grid was
   a. 669 MW
   b. 935 MW
   c. 15.6 MW
3. A brownout is
   a. A condition in which the entire electrical system is lost.
   b. A condition in which more electricity is produced than can be utilized.
   c. A condition in which the voltage on the electrical grid falls 8–12% lower than its rated level
4. A peaking generator is
   a. A generator that can be brought online very quickly to add large amounts of electrical power to the grid on short notice.
   b. A generator that produces electricity only from hydroelectric sources.
   c. A generator that runs continuously when demand for electricity is low or high.
5. Which of the following are used in basic specifications for wind turbines?
   a. Generator output
   b. Blade length
   c. Tower height
   d. All the above
6. The demand for electrical consumption in the United States in 2008 was
   a. Greater than in 2006
   b. Less than in 2006
   c. About the same as in 2006
7. Which is the correct order of U.S. electrical generation by fuel, from largest to smallest, for 2008?
   a. Wind, solar, coal, nuclear, natural gas
   b. Nuclear, natural gas, wind, solar, coal
   c. Coal, nuclear, natural gas, wind, solar
   d. Coal, natural gas, nuclear, wind, solar
8. For 2009, which is the correct order for electrical consumption from largest user to smallest users?
   a. Residential sector, commercial sector, industrial sector
   b. Industrial sector, commercial sector, residential sector
   c. Industrial sector, residential sector, commercial sector
   d. Residential sector, industrial sector, commercial sector
9. Which of the following are cost factors in calculating return on investment for a wind turbine?
   a. Cost of the wind turbine
   b. Site preparation and permits
   c. Installation cost of tower, turbine, and electrical connections to the grid
   d. All of the above
   e. Only a and b
10. Which of the following jobs will be available in wind energy?
    a. Planning and sales personnel
    b. Field service technicians
    c. Project managers and architects
    d. Electricians and electrical transmission technicians
    e. All of the above

CHAPTER 2

Wind Resources and Sites for Wind Energy

OBJECTIVES

After reading this chapter, you will be able to:

- Explain what causes the wind to blow.
- Explain what causes the direction and speed of the wind to vary.
- Identify factors that cause long-term variations in wind speed.
- Identify factors that influence the selection of a proper site for a wind turbine.

KEY TERMS

Acceleration
Air pressure
Barometric pressure
Boundary-layer wind
Force
Gradient wind
Land breeze
Mass
Pressure gradient force
Prevailing winds
Sea breeze
Wind friction
Wind map
Wind turbulence

OVERVIEW

This chapter will help you better understand the nature and science of wind. You will learn about the nature of the wind and what makes it blow, change direction, and change speed. You will also learn about the geography of certain areas and what makes the wind behave as it does in those areas. You will learn the things that influence variations in wind speeds, such as seasonal and annual variations, day/night variations, and variations due to locations, such as in mountain passes and near seashores.

You also learn about turbulence, gusting winds, and extreme winds, and about what conditions are favorable for wind turbines and what conditions affect wind turbines in other ways. The final part of this chapter will explain how to predict the amount of wind that will be available at a site. You will also learn some things to be aware of when you are selecting a site for a wind turbine.

2.1 BASIC PHYSICS OF WIND POWER

People have been aware of the wind since the beginning of time. In the last several hundred years, wind has been studied scientifically to improve sailing ships, windmills, and finally, in the last 50 years, wind turbines. The physics of the power in the wind has been well studied, and scientists and technicians have learned many things that make today's wind turbines more effective and more efficient.

Originally, most research on wind was used to help predict weather and storms: daily and longer-term weather forecasts, severe thunderstorms, tornadoes, hurricanes, and blizzards. Information about wind was closely tied to variations in the weather. This knowledge eventually helped scientists and technicians who work with wind turbines to understand better the availability of winds and how to predict them.

Scientists have learned to predict where the strongest and most continuous winds will occur, and this has helped them discover where to place wind turbines. They have also come to understand the variations in wind from day to night, season to season, and how the wind reacts to ocean conditions that also vary. To use this knowledge, it is important to understand some basic terms that are used in explaining the nature of wind. The next sections will introduce and explain some of these terms.

Wind Power Terminology

A **force** (F) is defined as a unit of mass being accelerated. **Mass** (m) is the quantity of matter, which is determined by its weight. **Acceleration** (a) is the rate at which an object changes its velocity. Force can be calculated using the equation $F = ma$. Force increases if either the acceleration increases or the mass increases. The unit for measuring force is called a newton, abbreviated as N. Mass is measured in kilograms or pounds, and acceleration is measured in meters per second squared (m/s^2).

When force is applied to an object over a distance, it performs work. For example, if you push or pull a 1-lb block over a distance of 1 ft, you can calculate the amount of work using the formula $W = Fd$, where W is work, F is force, and d is distance, giving 1 ft lb of work. If the force of gravity is pushing directly down on an object, you can include the weight of the object for the force in the calculation. When the force is increased or the distance is increased, the amount of work that is being done also increases.

Pressure is defined as force per unit area and can be calculated by the formula $P = F/A$, where P is pressure, F is force, and A is area. When the force increases, or the area decreases, pressure increases. Common units for pressure include pounds per square inch (psi) or bar, which is approximately 14.5 psi.

Wind direction is identified by the points of a compass and can also be converted to degrees. For example, 00° is north, 090° is east, 180° is south, and 270° is west, and 360° is north again. You will notice that north is defined as 0° and 360°. Directions between the four basic directions are defined as northeast (NE), northwest (NW), southwest (SW), and southeast (SE). The compass directions are also used to identify wind direction. Identifying a wind as a north wind indicates the direction from which the wind is coming. Wind direction always refers to the direction from which the wind is coming.

Other Wind Power Terms

Wind is very closely tied to weather, so you need to know some terms related to weather. **Barometric pressure** is the pressure at which the atmosphere pushes down on the earth. Barometric pressure is expressed in inches of mercury (in Hg) or in pounds per square inch (psi). The pressure at sea level is 29.92 in Hg or 14.7 psi. The inches-of-mercury unit of measure comes from the fact that barometric pressure can be measured with a U-tube manometer, which is a tube that is bent into the shape of the letter U and is open at each end. When this tube is filled with mercury, a column of mercury that is 29.92 in high represents the atmospheric pressure at sea level. As you go higher into the mountains such as Denver, Colorado, which is at approximately 6000 ft, the **air pressure** will be less, 12.1 psi, and the column of mercury will be shorter, a about 24.63 in.

As storms move across the United States, the air pressure changes. In terms of weather forecasting, low or falling pressure indicates that a storm or storm front is coming. If a high-pressure area is developing, weather forecasters can predict better weather, including clear skies. The lowest air pressure reading ever recorded, in the middle of a hurricane, was 25.69 in Hg or about 12.5 psi. The highest air pressure reading ever recorded was about 32 in Hg or approximately 15.6 psi. Extreme high and low pressure measurements are not common, but small fluctuations from day to day are important because they help determine the direction and velocity of the wind.

Wind direction and velocity are related directly to differences of air pressure. As the amount of air pressure changes as it moves across a given distance, it is called a pressure gradient. Weather forecasters use maps that show high- and low-pressure systems and pressure gradients, which indicate the direction and velocity of wind flow. Weather forecasters are interested in the wind velocity and direction at very high altitudes as well as winds at lower altitudes. Technicians and engineers who work with wind turbines are interested mainly in low-altitude winds, less than 200 ft above ground, because these are the winds that are used to turn wind turbine blades. Typically, wind turbines have been mounted at heights of approximately 200 ft, but research is currently being conducted on some wind turbines mounted between 300 and 500 ft high. Because wind turbines in the future may need to be mounted above 300 ft high, the winds at higher elevations will need to be studied for these applications.

High-altitude winds are also subject to changes caused by the rotation of the earth. The rotation of the earth causes a different effect on winds in the Northern Hemisphere than in the Southern Hemisphere. This means that the winds that move across the United States and Canada are different from the winds that move across Australia and much of South America.

Winds that blow near the earth are affected by changes in the surface, such as trees, mountains, and buildings. These things create friction that slows the wind down as well as cause subtle changes in wind direction. Wind measured at different altitudes will have different attributes: Winds near the ground will be affected by friction created by trees, buildings, and mountains, whereas winds several hundred feet above ground will not be affected as much. In some areas in the middle of the United States, such as Nebraska, Iowa, and Minnesota, winds can blow over several hundred miles unabated because the land is very flat. In other places, such as California, mountains obstruct the wind but allow it to flow more predictably through mountain passes.

Another important point to understand is that wind may not blow in a straight line inside a low-pressure or high-pressure area. For example, in the Northern Hemisphere, winds inside a low-pressure area blow in a counterclockwise direction as the front moves through an area. This means that the wind may blow from one direction as the beginning of the front moves through and then slowly change its heading around the compass until it is blowing from nearly the opposite quadrant a few hours later as the front moves out. If the front is a high-pressure area, the wind circulates inside the front in a clockwise motion as the front moves through the area. The wind velocity inside the front varies according to the size of the pressure area directly ahead of the front and that of the next weather front that is coming behind. In the spring of the year in the Northern Hemisphere, several fronts may line up and come through an area very quickly, within 24 hours. What this means in terms of the wind energy available to a wind turbine is that the velocity and wind direction may vary dramatically over a 24-hour period. In other regions, the wind may come from the same direction and at roughly the same velocity from day to day.

Other terms that you will encounter in your study of wind power include boundary-layer wind, gradient wind, and pressure gradient force. **Boundary-layer wind** is wind that is close to the surface of the earth, in the lowest part of the atmosphere. The behavior of boundary-layer wind is influenced by its contact with the earth's surface. Boundary-layer wind usually responds to changes in flow, velocity, temperature, and moisture near the earth's surface in an hour or less. Boundary-layer wind has a large influence on the way a wind turbine will be able to harvest electrical energy from wind blowing over its blades. Boundary-layer winds are affected by the friction caused by the surface irregularities of the earth's surface and by other factors such as temperature changes at the earth's surface, such as temperature differences between summer and winter.

Gradient wind is wind that blows at a constant speed and flows parallel to curved isobars just above the earth's surface, where friction from irregularities such as mountains, trees, and buildings cause changes in the flow. The gradient wind causes the wind speed to change when high- and low-pressure fronts move through an area and as the earth changes temperatures from day to night and from season to season.

The **pressure gradient force** is one of the main forces acting on the air to make it move as wind. Generally, the pressure gradient force is directed from high-pressure toward low-pressure zones, which influences weather patterns to move from high-pressure areas toward low-pressure areas; this movement or pressure differential is one of the things that makes the wind move. The amount of pressure difference influences the speed at which the wind blows—the wind is stronger when the pressure difference is greatest.

Wind friction is also a factor in wind speed, direction, and turbulence. **Wind friction** is the friction between two layers or two currents of air that move in different directions or that move at different speeds. The friction that occurs at the boundary of these two currents is an indication of wind shear. Wind shear is a sudden change of direction. Wind friction may cause winds to become turbulent or cause the direction of the wind to change, perhaps during a spring storm or a thunderstorm, which causes the wind effect on a wind turbine to become unpredictable and unstable. This type of turbulence is also found when temperature differences change quickly as weather fronts move in, especially during the spring months, when the earth begins to warm up from winter temperatures. In some states in the Midwest, for instance, the temperature may change by 20–30°F in less than an hour as cold front air moves quickly over warm air, especially in the spring. These rapid changes can cause severe thunderstorms and tornadoes and are usually accompanied with high winds that change direction frequently. These factors make it difficult for wind turbines to operate during these strong winds and make the efficiency of the wind turbine fall dramatically.

Sea Breeze and Land Breeze

A **sea breeze** is created whenever there is a difference between the water temperature and the land temperature. In the summer when the sun is high in the sky, it shines on the land and the sea and adds heat to both. The land warms more quickly than the water, however, so each day, the land may warm to perhaps 80–85°F while the water remains around 75°F. When this occurs, a sea breeze is generated from the water and moves over the land. At night or evening, when the sun stops heating the earth, the land begins to cool as rapidly as it warms during the day, while the water stays much the same temperature because it is slow to warm and slow to give up temperature. As the land cools, its temperature eventually cools to the temperature of the water, at which time the sea breeze stops blowing or slows to a very gentle breeze. Sea breezes occur more often in the spring and summer, when the land warms more quickly than bodies of water.

If the land cools enough that it is 5–10°F cooler than the water, then the process reverses and a **land breeze** develops, which causes the wind to blow from the land out toward the open water. Farther away from open water, this effect is much less and the wind speed is very light. Land breezes are more common in the fall and winter, as the land begins to cool more than the water bodies.

2.2 THE NATURE OF WIND

It is important to understand and measure wind speed if you are interested in installing a wind turbine and predicting the amount of wind you will have with which to produce electrical energy. One of the natural phenomena of the wind is that it blows at stronger speeds at higher levels. For example, the wind speed at ground level may be 5 mph, but 30 ft up in the air the wind speed may be 20 mph. This is why larger wind turbines today are mounted on taller towers, some as tall as 180 ft. Research is continuing on the feasibility of mounting wind turbines on much taller towers.

The nature of wind is basically a by-product of solar energy. As the sun heats different parts of the world at different rates, it causes winds to blow. The earth rotates so that each side faces the sun for a certain number of hours each day. At the Equator, the amount of time the earth is in the face of the sun is nearly the same day after day. In the Northern Hemisphere, the earth gets more time in the face of the sun in the summer, when the earth is tilted toward the sun and daylight hours may occur from 5 a.m. to 9 p.m. daily. This means that the energy available for wind power is more abundant. In the winter, the Northern Hemisphere tilts away from the sun and the amount of daylight is less. The rapid changes of weather caused by winter storms also provide adequate energy for wind power. The nature of the wind depends on the cycles of the earth's rotation and the seasons. You will learn more about other natural cycles that cause stronger or weaker winds in the next sections of this chapter.

What Makes The Wind Blow?

When the wind blows, it is moving from one point of high air pressure to a point of lower air pressure. The larger the air pressure difference between the two points, the stronger the wind will blow. Today it is possible to forecast or predict wind speed from day to day by measuring the pressure difference between weather fronts. A weather front is a band of air pressure moving over a specific area.

The direction of the wind depends on the location of two air pressure systems. These pressure systems are constantly changing from day to day and from season to season. Several factors contribute to establishing the air pressure at two points; one is the temperature, and another is the presence of a weather front.

For example, in the spring of the year, the surface temperature of the earth goes through drastic changes on a daily basis. One day an area may have 3 or 4 in of snow on the ground and a temperature of 25–30°F. The next day, a warm front may bring warmer air, perhaps 50–60°F, from the south. This drastic change of temperature will also bring strong winds.

Another way to think of how the wind blows is to open a window or door when it is cold outside. You will notice that you will feel the cold air move in at your feet, and that the warm air moves out the door or window at eye level. The warm air moves up and out to the cold area, and cold air replaces the warm air and immediately moves lower toward the floor. In nature, winds blow the same way, caused by high- and low-pressure areas as well as areas of warm air and areas of cold air. The important thing to remember is that there is nothing people can do to make the wind blow or make it blow stronger. All we can do is learn where a fairly steady wind normally occurs and then use these sites to locate wind turbines to harness as much energy as possible.

Prevailing Winds

In the United States the **prevailing winds** blow from the West Coast to the East Coast. Weather fronts also move generally in this direction. In addition, other weather fronts move up from the Gulf of Mexico across the southern states as they move in a northeasterly direction. Prevailing winds are important to wind turbines because in some areas of the country the winds blow more strongly and more continuously. Wind turbines may also be located in areas that do not have strong winds, because the wind in these areas blows fairly continuously and therefore yields enough electricity to make wind turbines worthwhile.

2.3 GEOGRAPHIC CONSIDERATIONS FOR WIND SYSTEMS

The best winds for wind turbines are usually found in specific geographic locations. For example, in the United States, wind farms are found in several geographic locations. One area is in central Texas, where the winds blow off the plains. Others locations where multiple wind turbines are located because of geographic conditions include north central Colorado; on a mountain top in Massachusetts; at the San Goronio Mountain pass in San Jacinto, California; in Indiana, Kansas, Iowa, Nebraska, and Minnesota, where steady winds blow nearly continuously off the flat farming lands; and also in the mountains of New Mexico.

These locations include mountain areas where winds blow nearly continuously, and flat plains where obstructions are minimal and winds blow unabated. Other good locations include shorelines along the Atlantic and Pacific Oceans as well as along the Great Lakes, such as the western shore of Lake Superior just west of Sault Ste. Marie, Ontario, Canada. Some other excellent geographic locations include offshore installations such as are found in Sweden, Great Britain, and other countries where strong winds occur just offshore.

Another consideration in placing wind turbines, in addition to geography, is that wind turbines must be located where they can be serviced easily and not too far from where they can be connected to the grid. This means that some otherwise ideal locations, where the wind blows constantly, may be too far or too expensive to provide the necessary electrical cables and towers to get the electrical power to where it will be consumed.

2.4 VARIATIONS IN WIND SPEED

Sometimes there are changes in the climate of a particular area that causes a change in winds that may last for several decades. For example, in the 1930s, a long drought took over areas of southwest Kansas, southeastern Colorado, and the panhandles of Texas and Oklahoma. When vegetation in crops begin to dry up in this region, the winds began to blow stronger and stronger and more continuously, until large clouds of dust were created. The strong winds lasted for several years, until the drought was broken and vegetation was restored to the areas. Many other places across the country have also experienced major changes to the land that cause changes in wind speed and duration. Scientist can document these changes, but because they take a long time to occur, they are generally not a concern for the placement of wind turbines. However, engineers and technicians who study site locations for wind turbines should be aware of the phenomenon, and should try to estimate how long the winds will continue at that location if they are caused by a variation of geography.

2.5 SEASONAL AND ANNUAL VARIATIONS IN WIND SPEED

Engineers and technicians who select site locations for wind turbines must also be aware of seasonal and annual wind variations. These variations are caused by weather changes from summer to winter. For example some locations may have strong continuous winds through the winter months, and then, when the area heats up through the summer, the winds may be lighter and variable. In other locations, the winds may be stronger and more continuous during the summer than in the winter. This is normal in some southern locations. When a potential wind turbine location is being investigated, wind data from all of the months must be analyzed to ensure that the wind turbine will be able to produce adequate electricity throughout the year. In some cases the variation from season to season is so dramatic that it would not be effective to place a wind turbine at that site, because it would be able to produce adequate voltage for several months when the winds were at their highest, but at other months the wind would not be strong enough to produce the amount of electricity required for the investment.

Seasonal and annual wind variations are also important in areas where wind turbines are located offshore. Some offshore sites also are not suitable because hurricanes may occur during several months through the summer and early fall. The sites may have sufficient winds during some months, but the winds inside tropical storms and hurricanes are so strong that they would be likely to damage the wind turbines and their platforms. Other offshore locations may have suitable winds, but other problems may arise in the winter when ice forms around the platforms. Ice is a strong adversary of anything in its way, because it will move when water levels change, and it also moves in strong winds.

When engineers and technicians review sites for wind turbines, they are looking for sites that do not have strong variations in their winds from season to season. An ideal site has the same amount of strong wind energy all year long, so that the output of the wind turbine generator will be a dependable power source. If the amount of wind is low throughout the year, the site will not be suitable. Sometimes sites that have strong winds for several months will be selected because the amount of energy produced by the available winds will make more energy than is needed for the minimum return on investment.

2.6 DAY/NIGHT VARIATIONS IN WIND SPEED

Another problem that must be anticipated when a site for a wind turbine is being investigated is to determine the variations in wind speeds from day to night. At some locations, the wind tends to die down during the evening, as the earth cools. If the wind does not produce the same amount of energy during the night as it does in the daytime, the site may not be useful to produce energy for an electric utility. You will remember that demand for electrical energy is reduced slightly at night, but it is still a strong need. If the wind turbine is to be used as a backup energy source for a coal-fired or nuclear power plant, then the problem is not as significant. In some small residential applications, the wind turbine can produce enough energy that some can be stored in batteries during the daytime, or other times when the wind is blowing, so that the home can use this power from the batteries during the nighttime or whenever it is needed.

2.7 TURBULENCE

When a liquid flows in a straight line, it is called laminar flow. When a liquid becomes disturbed and the flow is no longer in a straight line, it is called turbulent flow. Wind is considered a liquid in this sense and when its flow is not in a straight line it is also considered turbulent flow. **Wind turbulence** is a condition in which the wind does not blow in a straight line and does not blow continually at the same speed. Sometimes these winds are called swirling winds. They do not produce the same energy as strong straight-line winds. Other turbulence is caused by storms that move through an area. The turbulence in a storm is caused by sudden changes in the wind velocity and wind direction, which will cause the wind forces on the turbine blades to change rapidly.

Another type of wind turbulence occurs when the wind passes the blades of a wind turbine and swirling winds are created in the wake. This is not a problem if only one wind turbine is located on a site, but it may be one if multiple wind turbines are located on the same site and the swirling winds from one turbine get in the air stream of the next wind turbine. This problem occurs randomly and may not occur at all wind speeds. It may be a problem at some sites, however, so the choice of a location for multiple wind turbines may become more complex.

2.8 GUSTING WIND SPEEDS

Gusting wind speeds may occur at any time because of fluctuations in the wind. This is especially problematic when winds are light and variable. When this occurs, the wind direction may change quickly and gusts may build up, causing the wind to increase temporarily by 20–30%. Such sudden increases in wind do not automatically convert into increases in wind energy that can be used by a generator to produce more electrical energy. If the wind turbine is using technology to keep the generator shaft speed constant, wind gusts will cause the blades, rotor, and shafts to turn faster as the wind speed increases, and return quickly as the gust moves through. If the wind speed does not continue, the stress on the shafts will increase and decrease sharply, which causes more stress on the blades, shafts, bearings, and any gears.

If the wind turbine is designed to run freely regardless of shaft speed, the generator will simply produce more electricity as the gust comes through, and then return to a lower output as the gust goes by. If the wind direction changes abruptly when the gust occurs, the wind's angle of attack on the turbine blades will also change, and the efficiency of the blades to harvest the wind energy will be reduced.

2.9 EXTREME WIND SPEEDS

At times a wind turbine will experience extremely high wind speeds. When this occurs, the wind speed may exceed the design wind speed level. If the extreme high winds continue, the blades must be rotated through pitch control so that the wind does not strike the blade directly, thus slowing the blades during the high winds. Another way to control blade rotational speed during high winds is through yaw control, changing the direction at which the nacelle is pointed into the wind so that the blades do not collect all of the energy of the high winds. Only the most complex wind turbines have a yaw control system fitted with wind speed and wind direction sensors together with computer controls that can determine if the wind is exceeding the design speed. Smaller, less expensive wind turbines may use passive yaw control, in which a tail fin is used to let the wind itself move the nacelle to the proper direction.

All wind turbines have a maximum wind speed at which they can operate before their blades and rotor are damaged. To keep them from being damaged, most wind turbines have a protection mode they go into where the blades are stopped. On some wind turbines this is done with mechanical or hydraulic brakes. When this occurs, the controller on the wind turbine senses the high-speed winds through its anemometer and sends a command to change the pitch or change the yaw direction to limit the torque on the blades. If the highest winds exceed the maximum safe speed, the controller sends a signal to apply the brakes and lock the rotor until the wind speed is lower. You will learn more about yaw control and brakes in Chapter 5, where hydraulic controls are discussed.

Extreme wind speeds as well as thunderstorms, hurricanes, and blizzards are also a concern when engineers and technicians are determining locations for wind turbines. For example, the land throughout the Midwest provides strong sustained winds during most of the year, but the region is susceptible to strong spring thunderstorms that can spawn tornadoes or blizzards, which can cause extremely high winds that can damage wind turbines. A number of new technologies are being developed that will move or adjust the wind turbine blades to a safe condition when extremely high winds occur. When these technologies are perfected, they will allow wind turbines to be located in places that may not be possible today.

High wind speeds are very problematic for upwind horizontal wind turbines because the high wind strength tends to bend the tips of the turbine blades into the tower. To withstand these high winds, the turbine blades must be designed to be stronger and more durable so that the wind does not spend as much at the tips. This may make the blades less efficient, but it is necessary to keep them from bending into the tower. In some upwind horizontal wind turbines, the nacelle is designed so that it sits on the tower in such a way that the blades extend farther out from the tower, so that even if they flex during high winds, the blade tips cannot touch the tower. Another way around this problem is to use a downwind horizontal wind turbine, with which the wind blows past the tower first and then through the blades. If a high wind hits the blades unexpectedly, they can bend outward away from the tower so they cannot cause any damage.

2.10 PREDICTING WIND SPEED

The need to project wind speeds and forecast the strength and longevity of winds is very important information for selecting the proper site for a wind turbine. This information is also necessary to determine the return on investment (ROI) before deciding to purchase a wind turbine. This science is getting better as more and more wind turbines are being installed across the United States and other parts of the world. The data that is collected through the supervisory control and data acquisitions (SCADA) system is put into a data bank where it can be analyzed.

Wind predictions have become more and more important over the past 20 years for applications such as seagoing vessels, open-water racing vessels, and airplanes. These technologies rely on such information to plan individual trips that can take advantage of winds that will aid the trip or avoid winds that may cause problems. This technology is now used to help predict the winds for sites where wind turbines may be erected. The weather models used data from local geography and previous years' wind strength and direction to help create a model specifically for that site. The model can be used in addition to local weather reports and satellite maps to determine weather fronts and predict the strength and duration of winds for a given month or year. The data is used in conjunction with other known information about the wind turbine, so that a fairly accurate prediction can be made to determine the amount of electricity a wind turbine can produce at that location.

If one or more wind turbines already exist at a location, the accuracy of these reports can be improved, and the wind model can be made more accurate because of the wind data that has been taken at that location over the period of time the wind turbines have been installed and running. Another useful piece of data comes from records of local weather conditions that have been recorded for a specific area. In the last 50 years, weather stations, television stations, and other government installations have recorded weather information on an hourly basis over long periods of time. This information includes temperature, wind speed, wind direction, and humidity. This information can be put into highly sophisticated models to predict the total amount of wind energy that will be available for a given area over a year's time. Some companies have created large database of this information and sell it to companies that have an interest in setting wind turbines in an area. The accuracy of this data has become better over the years, and the predictions and forecast have also become more accurate. This information helps identify the best locations for wind turbines.

When this type of information is used in addition to the data collected from SCADA systems from an operating wind turbine, it is especially useful for helping create more accurate databases for predicting and forecasting winds. The data collected from SCADA systems includes not only wind speeds and wind direction but also the amount of electrical power that the generators produce when the wind is blowing. This is especially useful because it indicates the current wind speed and whether the blades are receiving winds at the best attack angle or if the blades need to have their pitch changed when the wind is too strong or too weak. As more and more wind turbines are installed at a given site, and this data is collected and shared, the accuracy of the prediction and forecast for adding more wind turbines at that site will become better.

2.11 TURBULENCE IN WAKES ON WIND FARMS

Engineers and technicians who work with multiple wind turbines on wind farms have known for years about the amount of turbulence that each wind turbine puts back into the wind stream. This phenomenon was studied comprehensively by NASA in the 1970s and 1980s. From 1974 through 1981, the NASA Glenn Research Center in Cleveland, Ohio, using funding from the National Science Foundation (NSF) and the Energy Research and Development Administration (ERDA), constructed and operated an experimental 100-kW wind turbine at the NASA site at Plum Brook in Sandusky, Ohio. As part of this research, NASA completed smoke tests with wind turbines to determine the effects of wake turbulence. Figure 2–1 shows a picture of this research, where smoke was used to show the turbulence patterns with a two-bladed wind turbine. From this research, important data was gathered to help determine the distance needed between wind turbines on a wind turbine farm.

FIGURE 2–1 NASA uses smoke tests with a wind turbine to study wind turbulence. (Courtesy of U.S. Department of Energy.)

When large numbers of wind turbines are located in one area such as on a wind turbine farm, the direction the wind blows will change constantly and at some point one or more turbines will be in the wake of other turbines. No matter how you place or locate the wind turbines, at some point, some of the wind turbines will be in the wake of other turbines. When this occurs, turbulence may make the turbines in the wake less efficient at harvesting energy from the wind turbine. Figure 2–2 shows the effects of wind turbulence on a number of wind turbines on a wind farm. When multiple wind turbines are located near each other, the wind that passes through the blades of one

FIGURE 2–2 When wind turbines are place in rows, turbulence may occur. (Courtesy of EWT International.)

wind turbine may create turbulence for the turbines that are in line with the first one. If this occurs, the wind energy that is available from the turbulent wind is less than the wind energy that was harvested by the first set of blades. If additional turbines are located in this same line, they will also suffer the same inefficiency. On many wind farms, the wind turbines are located specifically to reduce the effects of turbulent winds.

One way to combat the problem of wake turbulence is to place the wind turbines a specific distance apart, so that the turbulence from the wake of the first turbine does not bother the second and additional turbines on the site. Software programs and data from each turbine type can help to determine the distance each turbine should be placed from the others.

Figure 2–3 shows another example of multiple wind turbines on a wind farm, in which you can see how far apart the wind turbines are spaced so they will not cause turbulence problems with each other.

FIGURE 2–3 Example of multiple wind turbines in a wind farm in Minnesota. (Courtesy of John Fellhauer, jfellhauer@surenergy.us.)

2.12 TURBULENCE IN COMPLEX TERRAIN

Another problem that affects wind turbines is turbulence caused by complex terrain. For example, when the wind blows over the top of a hill or mountaintop, or any other obstruction, wind turbulence is created on the back side of the obstruction. The turbulence may be so strong as to make the location unsuitable for a wind turbine.

This type of turbulence also occurs near some wooded or forested areas. If the wind turbine is placed on a site near a wood lot or woods, it may get "clean" air that is not obstructed some of the time, depending on the direction of the wind, but when the wind changes and comes through the woods and trees first, this creates such strong turbulence that the area may not be a good site for a wind turbine.

Turbulence may also be a problem when wind turbines are located at the edge of a city. When the wind blows toward the city, there are no problems with wind turbulence. If the wind comes through the buildings in the city first, there may be a lot of turbulence before it reaches the turbines.

This problem may even occur when the wind blows across a building or small trees before it gets to the wind turbine. One way to combat this type of turbulence is to set the wind turbine as high as possible. The extra height not only combats wind turbulence, it also harvests a stronger wind. This is why some of the largest wind turbines are now being installed on towers that place them over 300 ft in the air.

2.13 SIZING A WIND POWER SYSTEM FOR A PARTICULAR APPLICATION

When you set out to determine the size of a wind turbine, you need to know what the wind turbine will be used for. For example, if you want the wind turbine to supply electricity for a small home or business, you will need to know the usage and demand that the wind turbine is providing electricity for. The annual electrical consumption of a household of two adults and two children in the United States is considered to be approximately 4500 kWh. This means that the wind turbine needs to produce 1.45 kWh per day.

Because the production of electricity by any wind turbine varies throughout the day and night and also from day to day and from season to season, you must consider the monthly or yearly production of electricity to be sure that it is sufficient. This presumes that either batteries for storage or backup power from the grid will be available so that continuous power can be provided to the home at all times.

If you use the average value of 4500 kWh per year as the amount of power consumed, you can determine the size of wind turbine needed to provide that much annual power. Another good rule of thumb is to figure approximately 1 m of blade diameter for each kilowatt hour needed. This means that a wind turbine with a blade

FIGURE 2–4 Wind turbines mounted on existing lighting poles. (Courtesy of Southwest Windpower.)

diameter of approximately 5 m is needed to provide the annual power of 4500 kWh for a residential application.

Sizing a Wind Power System for a Commercial Application

Another typical application for wind turbines is commercial. For example, some automobile dealerships have very extensive costs because of the amount of lighting they need to provide for their lots. This type of application is well suited for a single wind turbine or several small wind turbines that can be mounted on existing lighting poles or on larger towers designed specifically for the wind turbine (lights can also be mounted on these towers). Figure 2–4 shows smaller wind turbines mounted directly on existing light poles at an automobile dealership.

Again, it is important to identify the specific electrical usage and demand electrical power that the wind turbine will be replacing. In some cases, electrical power will be used to offset some, rather than all, of the electrical load. In this type of application a 50-kW wind turbine will provide about 87,000 kWh annual output with average winds of 12 mph. The rotor diameter for an average 50-kW wind turbine is approximately 50 ft, so the centerline for the hub and rotor will need to be over 80 ft tall. This may create a problem with the local zoning board, because many commercial applications are located inside city limits. Therefore, some commercial applications may use multiple smaller 10-kW wind turbines rather than one large 100-kW, because these smaller wind turbines may fit on towers that are the size of conventional lighting posts and therefore not be limited by zoning restrictions. In this application, the wind turbines charge batteries and the lights use the stored electricity until the batteries are discharged, then they switch over to conventional electrical power from the utility company.

Another application for wind turbines is municipal water and wastewater treatment plants that use large quantities of electricity at a nearly constant rate to run their pumps. Because this type of application uses electricity at an almost constant rate, the size of the wind turbine can be designed to carry most of the load. Wind turbines for this type of application may be sized from 100 to 250 kW. The size of the wind turbine will depend on factors such as the amount of electrical power the plant uses daily and any restrictions on height and noise for wind turbines if the location is inside city limits or near residential areas. In some of these applications, the wind turbine produces all the power the water treatment plant needs and sells any extra back to the utility through a grid connection. The grid connection is also needed in case the wind is not strong enough to produce sufficient power on some days.

Another application for wind turbines is to supply power to remote locations where it is needed. Figure 2–5 shows a wind turbine installed on a high-voltage transmission line. In this application the wind turbine uses the transmission tower that is already in place and provides power to transmit data along the power line.

Other small industrial applications for wind turbines may include any application where the electrical demand is rather constant and backup power from the

FIGURE 2–5 Wind turbine located on a power transmission line. (Courtesy of Southwest Windpower.)

FIGURE 2–6 A wind turbine located at a railroad switching yard.

FIGURE 2–7 A wind turbine located at a school. (Courtesy of John Fellhauer, jfellhauer@surenergy.us.)

grid is also available. The wind turbine is usually designed to produce approximately 80% of the demand, and then the remainder is taken from the grid. Figure 2–6 shows a wind turbine in a railroad switching yard. The electrical demand typically used for lighting for the switching yard is fairly constant, and the wind turbine can provide electrical power for the majority of time, with the remainder of the power coming from the grid.

Wind Turbines for Schools and Colleges

Another good application for medium-sized wind turbines is school and college buildings. There are two basic reasons that these sites are excellent for wind turbines. First, their demand for electricity is very large during daylight hours and smaller during nighttime hours, which matches with the wind availability in many locations. This means that a wind turbine can be sized for the electrical load of the building, and the production from the wind turbine will typically meet this demand. At night, when the wind typically diminishes, the load for the buildings reverts to lighting loads for a small percentage of the building and parking areas. Any electrical power that is produced will also help to keep the utility costs for the school building more constant over time, whereas typical electrical power from a utility increase in price every year. This makes the wind turbine a great match for these applications.

Another reason to install wind turbines at educational facilities and campuses is that it provides a learning experience for students and faculty to get a better understanding of wind energy. Having a wind turbine on site is a reminder to all the students and faculty of the possibilities that wind energy possesses to aid in providing renewable energy sources. It also is useful if the educational facility is teaching courses involving renewable energy or wind energy specifically. Other classes from first grade through college can utilize the data from the wind turbine in their observations and measurements of efficiency. The data will also provide a real-life application by which students can study how much wind is available and how much electricity can be produced by the wind turbine at that site. Figure 2–7 shows a wind turbine mounted at a school.

2.14 DETERMINING HOW MUCH WIND POWER IS AVAILABLE AT A SITE

The most important part of determining how much wind energy is available at a site is to determine the wind speed and duration. The wind must be measured at the height at which the wind turbine will be mounted. For example, wind measured at 30 ft above the ground may be minimal at a site, but if the wind turbine is to be mounted at 180–300 ft above the ground, there may be more than enough wind.

If other wind turbines are mounted nearby, you may be able to gather SCADA data from one or more of these sites. If you are planning to install a wind turbine at a site that has not previously had wind turbines, you may need to take measurements over a period of time to determine how much wind is available. Measurements of the continuous wind speed and direction can be accomplished automatically using metrological equipment or an anemometer that records the wind speed over a period of time. This can be accomplished by taking the average wind speed once a minute over an hour and converting the readings into an average wind speed per hour. It is important that the instrument used to make the measurements be set at the correct height and calibrated before any data is captured.

The instruments used to measure the wind speed and direction can also be connected to a data storage unit or to transponders that send the information to a computer or a network site where the data can be logged and stored for later analysis. You can hire a company to provide this service for larger wind turbines, but it may be too expensive for small residential wind turbines.

Another way to obtain the needed information about average wind speed is to consult the wind resource data published by the U.S. Department of Energy. If data is available for your area, it is usually detailed enough to allow a wind power technician or engineer to predict the amount of wind energy that will be available at that site.

2.15 LOCAL CODES THAT MAY AFFECT WIND POWER SYSTEMS

When wind turbines were first being installed in larger numbers in the 1970s and 1980s, there were not many codes that affected them. As more and more wind turbines were installed, local governments began to create codes governing size, height, safety concerns, noise, and locations where wind turbines could be installed. State codes have tend to be more uniform and to encourage the development and installation of wind power wherever possible. Local codes are driven by local conditions, which either promote wind power or try to contain the growth of wind turbines. One provision that has been included in many zoning codes is a height limitation or restriction that makes it impossible to place wind turbines at the proper height.

For example, installing a wind turbine on on private land in an agricultural area is very seldom prohibited, because the site is usually on a farm far away from neighbors or anyone who might object. A wind turbine in a residential neighborhood or inside city limits, however, will affect neighbors and others nearby. There may be safety concerns about the size, or people may feel the wind turbine looks too industrial or mechanical to fit into a residential neighborhood. Others may be concerned about placing wind turbines in locations that have scenic beauty value, such as lakeshores and seashores or mountain areas where scenic vistas draw tourists. The noise that wind turbines create when wind passes over them and their blades rotate may also be a problem in some residential areas.

Codes may also address whether the wind turbine is considered a permanent or temporary structure. Because a wind turbine has a limited life of 20–40 years, it may be considered a temporary structure, and a plan for disassembly and removal may also have to be provided at the time the permit to erect it is requested.

Some wind energy associations are trying to get a common set of topics included in all residential codes so that municipalities cannot create additional problems to impede installation of wind generating equipment. As time goes on, other concerns become more important to address and must be included in some codes.

Since the wind turbine's electrical output may be connected to existing wiring in a building or may be connected with the electrical grid, all of the wiring and interconnections must meet existing electrical codes that have been established by city, county, or state governments. Also, any codes that govern the erection of towers apply to wind turbine towers.

2.16 SITE REQUIREMENTS

When selecting a site to place a wind turbine, you must understand some of the site requirements that will ensure the site is the best for your application. In some cases, the site will be determined by ownership of the land or where a land lease is possible. In these cases, the best you can do is to determine where the turbine should be placed on that parcel of land.

If a large wind turbine is being placed, the conditions will help narrow down the location. For example, if the system is a grid-tied system, it must be located somewhere close to where the grid crosses the property. If the wind turbine is being used to produce electricity for a building, it will need to be located somewhere near the building so that the amount of electrical cable is not too long and it will not have large amounts of voltage drops due to long wires.

It is also important to identify and avoid natural or man-made wind obstacles. It is not uncommon to find one or more wind obstacles that will divert some of the wind energy. When you select the site for the wind turbine, you can check it out and visually inspect for any obstacles.

The wind turbine will need to be the tallest structure around, so that its blades have a clear path to harvest as much wind at its height as it can. This is why many wind turbines are located in agricultural areas, where the land is usually flat for many miles around. The wind will be availabel across the flat land without being disrupted in any way. It is also important to consider road and highway access, because large cranes and other transport equipment may be needed at the site to deliver the wind turbine, construct it, or perform maintenance.

2.17 DETERMINING THE PROPER SITE FOR A WIND TURBINE

A useful tool when you are selecting a site for locating a wind turbine is a **wind map**. Figure 2–8 shows a wind resource map of the United States. From this map you can see that the areas along the Atlantic Coast, the Pacific Coast and the Gulf of Mexico have the highest wind resources. Additional maps are availabel through the weather services that show the actual direction the wind is blowing and the wind speed. The wind speed is shown on these maps in miles per hour and meters per second. Wind maps are currently provided by the major weather websites, and they are updated every 5–6 hours.

When you are deciding on a site for a large wind turbine, you need to remember that a wind speed of 10–14 mph at a height of approximately 70 m (230 ft) is necessary to get minimum energy. If the site does not allow taller towers or if the wind speed is minimal on many days, the turbine will not produce enough energy to make it pay back on its investment.

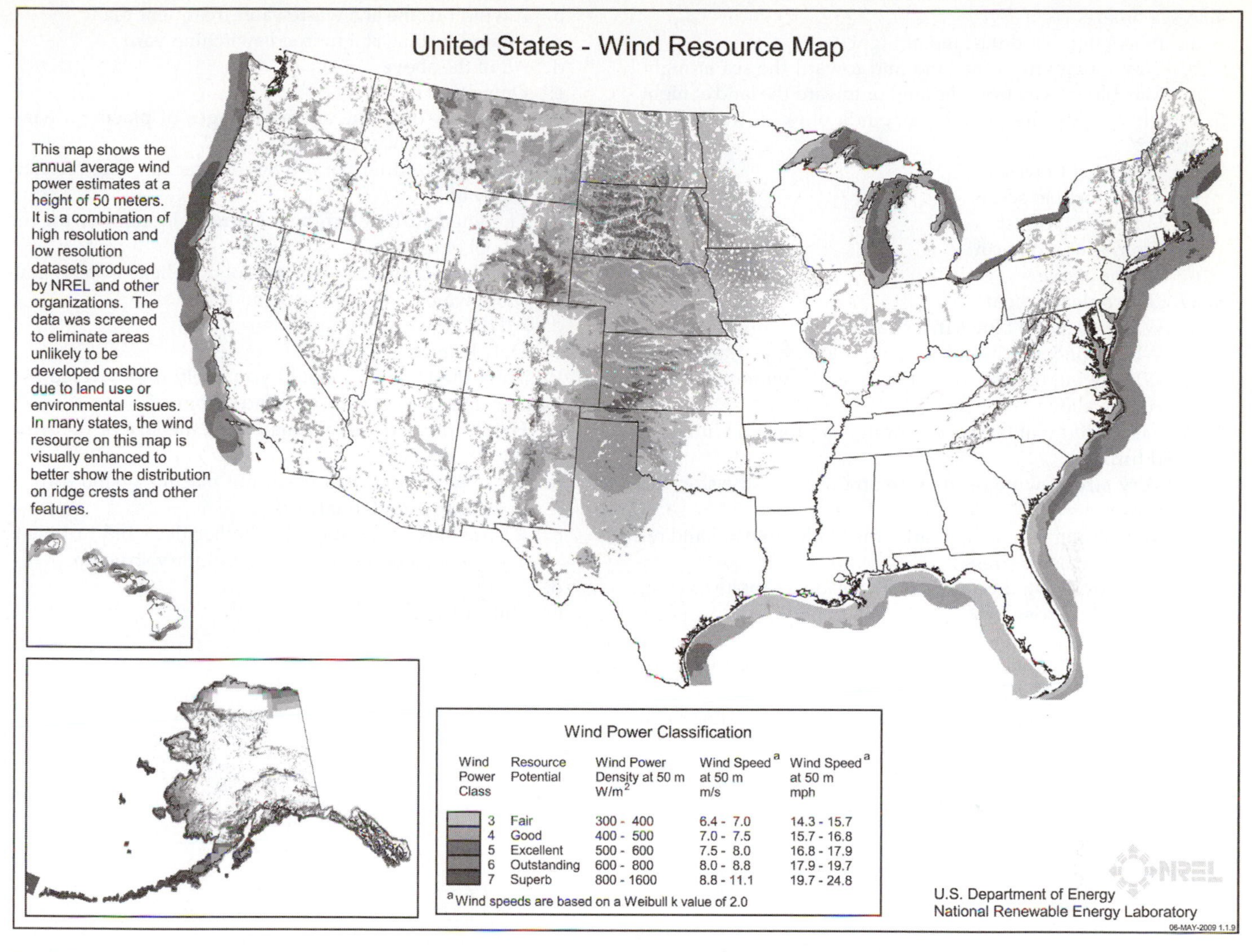

FIGURE 2–8 A map of U.S. wind resources. (Courtesy of U.S. Department of Energy, National Renewable Energy Laboratory.)

Questions

1. Define atmospheric pressure.
2. Explain what makes the wind blow.
3. Explain how sea breeze is created.
4. Explain what causes a land breeze.
5. Explain what causes the difference in wind speed from summer to winter.
6. Explain why there are differences in wind speed between day and night.
7. Explain why there may be turbulence between wind turbines on a wind farm.
8. Give three examples of special types of commercial wind turbine applications.
9. Identify two things you should be aware of when you are selecting a site for a wind turbine.
10. Explain how you would be able to determine how much wind power is availabel at any site you are evaluating for the placement of a wind turbine.

Multiple Choice

1. What is a wind map?
   a. A wind map shows the average temperature where the wind starts to blow.
   b. A wind map shows the average amount of time the wind blows each day.
   c. A wind map shows the direction and the speed of wind for a given amount of time for that location.
2. Atmospheric pressure is
   a. Approximately 100 psi at sea level
   b. Approximately 14.7 psi at sea level
   c. Approximately 20 psi at sea level
3. Typically, the wind blows
   a. More strongly during the day and less at night
   b. More strongly during the night and less during the day
   c. The same during the day and night

4. A sea breeze
   a. Blows in toward the land at night
   b. Blows away from the land and toward the sea at night
   c. May blow away from the land or toward the land at night
5. Which way do the prevailing winds blow in the United States?
   a. From east to west
   b. From north to south
   c. From west to east
   d. From south to north
   e. All the above
6. Wind turbulence occurs
   a. When the wind blows in a straight line
   b. When wind swirls and does not blow in a straight line
   c. When wind blows more strongly in the winter
   d. All the above
7. Why is a wind that blows too strongly a problem with for a wind turbine?
   a. Very strong winds put large amounts of stress on the tower.
   b. Very strong winds may cause the blades to bend and receive excessive stress.
   c. Very strong winds may cause the blades to rotate too fast.
   d. All of the above
   e. Only a and c
8. Which of the following is an example of a commercial or industrial application of a wind turbine?
   a. A wind turbine located on a lamp pole at an automobile dealership
   b. A wind turbine at a wastewater treatment plant
   c. A wind turbine at a railroad switching yard
   d. All of the above
   e. Only a and b
9. Which of the following are advantages of placing a wind turbine on a school or college campus?
   a. The wind turbine can provide power to offset the electricity bill.
   b. The wind turbine can be a learning laboratory for the school or college.
   c. The electrical load for the school or college is fairly constant, which is good for a wind turbine.
   d. All the above
   e. Only b and c
10. Which of the following would you likely find covered in a local or state code governing wind turbines?
   a. Height restrictions that may limit the size of the wind turbine tower
   b. Noise restrictions that may limit placing a wind turbine in or near a residential area
   c. Restrictions depending on whether the wind turbine is considered a permanent or a temporary structure
   d. All the above
   e. Only a and c

CHAPTER 3

Operation of Vertical- and Horizontal-Axis Wind Turbines

OBJECTIVES

After reading this chapter, you will be able to:

- Explain the advantages and disadvantages of vertical-axis wind turbines.
- Explain the advantages and disadvantages of horizontal-axis wind turbines.
- Explain how wind moves past a turbine blade and causes it to rotate.
- Identify how lift helps wind turbine blades convert wind power to electrical energy.
- Explain how the design of the wind turbine blade assists in creating lift for the blade.
- Explain the advantages and disadvantages of one-blade, two-blade, and three-blade horizontal axis wind turbines.

KEY TERMS

Airfoil
Angle of attack
Downwind horizontal-axis wind turbine
Drag
Furling
Horizontal-axis machine
Leading edge
Lift
Lift-to-drag ratio
Nacelle
Stall
Teetering
Tip speed
Trailing edge
Torque
Upwind horizontal-axis wind turbine
Vertical-axis wind turbine

OVERVIEW

This chapter provides an overview of the operation of vertical- and horizontal-axis wind turbines. It introduces the terms and vocabulary you will encounter while learning about these wind turbines. The chapter also introduces the hardware associated with vertical- and horizontal-axis wind turbines. You will learn more in-depth details of the theory of operation, installation, troubleshooting, and repair in later chapters of this book.

The chapter includes a variety of pictures and diagrams that show the basic operation of the turbine blade airfoil, the types of turbine blades used, and one-bladed, two-bladed, and three-bladed horizontal-axis wind turbines. The chapter concludes with a comparison of the systems that summarizes the advantages and disadvantages of each. This will help you understand the vast variety of designs and controls used in wind turbines today. You also learn why certain designs are required with certain types of wind turbines.

3.1 INTRODUCTION AND OVERVIEW OF AIRFOILS

The turbine blade on a wind generator is called an **airfoil**. The design of the wind turbine blade (airfoil) may be slightly different for each type of wind turbine. To understand airfoils, you must understand a number of terms such as lift, drag, lift-to-drag ratio, stall, angle of attack, and torque. This section will help you understand how wind turbine blades are similar to aircraft propeller and wings, and see how their design has evolved to take advantages of technology related to converting wind power to electrical energy. You will need to become familiar with these terms to understand the technology that goes into the design and application of wind turbine blades.

Lift

Lift is defined as a condition when air moves past the airfoil as it does on an airplane wing, creating a low-pressure area on the top side of the blade and allowing it to rise. The low-pressure area occurs because the distance the air travels over the top of the airfoil is longer than the distance the air moves under the bottom of the airfoil. Because the same air is moving over and under the airfoil, a pressure difference is created. Figure 3-1 diagrams how wind passing over an airfoil creates lift. The pressure at the top of the airfoil is less than the pressure at the bottom of the airfoil, which creates lift on the airfoil. This lift is used in an aircraft wing to cause the aircraft to rise off the ground and fly. The faster the air travels past the airfoil, the more lift is created.

When the wind moves past a wind turbine blade, it causes the same effect as lift. This effect makes the turbine blade turn. In the simplest terms, when air passes over the rounded side (the top) of the turbine blade and a low-pressure area is created, the turbine blade moves in the direction of the low-pressure area. The difference in pressure causes the wind to push the flat side of the blade and pull the rounded side of the blade, so the blade begins to rotate in the direction of the low-pressure area. Another way to think of this is that when wind blows across the turbine blade, the blade rotates so that the rounded front leading edge is always moving in the direction in which the blade is rotating. Figure 3-2 shows the **leading edge** and **trailing edge** of a wind turbine blade.

The stronger the wind blows, the more pressure differential (lift) is created, the faster the turbine blade turns, and the more kinetic energy the turbine blade can harvest from the wind. This energy is converted to rotational energy through the shaft on which the blade is mounted. In some wind turbines, the electrical generator is mounted to the other end of this shaft. In other wind turbines, the shaft on which the turbine blade is mounted may be connected to a transmission and a second shaft coming out of the transmission powers the electrical generator. As lift increases, it is easier for the turbine blade to move through the air.

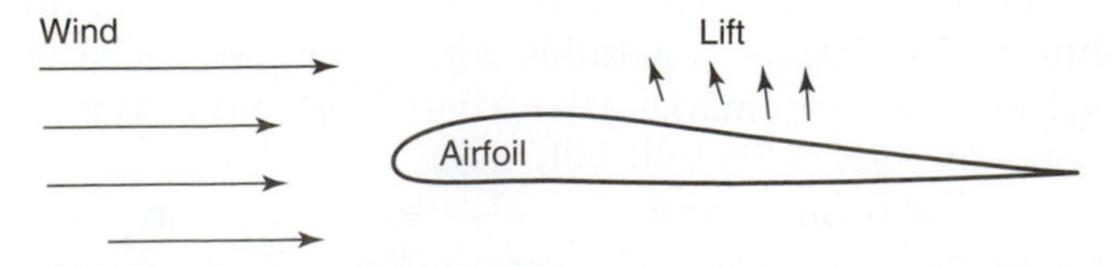

FIGURE 3-1 Air moving past an airfoil, creating lift on the top of the blade.

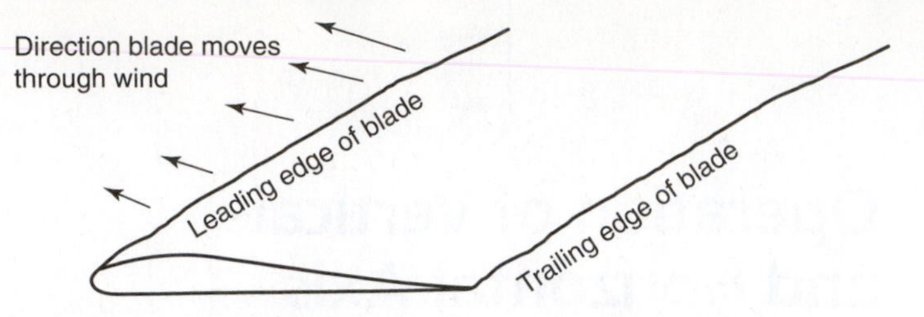

FIGURE 3-2 A typical wind turbine blade, showing the direction the blade moves through the air. The leading edge and trailing edge are identified.

Angle of Attack

The angle at which the wind strikes the turbine blade is called the **angle of attack**. When the wind blows directly over a blade as shown in Figure 3-3a, the angle of attack is zero. The angle of attack increases as the front of the blade rotates upward, as shown in Figure 3-3b, and the amount of lift increases. When the turbine blade has rotated up to where it is producing the maximum lift, as shown in Figure 3-3c, the amount of angle of attack is at a maximum;this angle is called the critical angle of attack. When the angle of attack increases to the critical angle of attack, the turbine blade begins to lose its ability to convert energy from the wind.

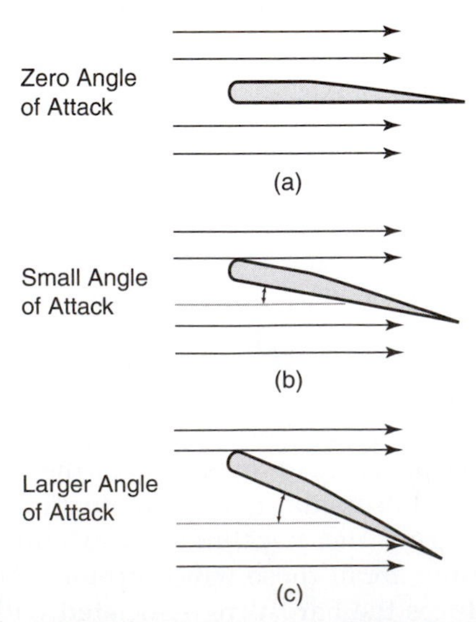

FIGURE 3-3 A blade changing its pitch angle with respect to the direction of wind flow: (a) the blade is horizontal to the wind and the angle of attack is zero; (b) the blade is tilted slightly upward and the angle of attack is fairly small; (c) the blade is tilted upward to its maximum position and the angle of attack is largest.

FIGURE 3-4 Wind turbine blades positioned into the wind so they can harvest the most wind and turn as fast as possible.

Figure 3-4 shows a wind turbine blade harvesting the maximum amount of wind, which allows the blades and rotor to turn as fast as possible. The pitch of the blade can be adjusted, and Figure 3-5 shows the blade pitch angle changed so that the leading edge of the blade is directly into the wind, and the trailing edge is directly away from the wind. In this position, the wind turbine blade harvests the minimal amount of wind energy, so the blade will not rotate. The wind turbine blades in this picture are positioned this way during maintenance, and the brakes are applied so that the blades will not rotate.

Pitch Control Versus Stall Control

Wind turbine designers use the concept of stall to control the ultimate speed of the turbine and help prevent the generator from overloading. **Stall** occurs when the turbine blade is nearly flat and no longer has lift. A wind turbine is subjected to the highest and lowest winds that flow at its location. Sometimes, such as during storms and other severe weather conditions, high winds may occur. When these high winds occur, the turbine blades increase their speed, and the output of the generator may increase to the point at which the generator becomes overheated and damaged. Also, high winds may damage the turbine blades and the tower if the generator is allowed to increase its output higher and higher uncontrolled. There are two basic ways to control the speed of the wind turbine blades: pitch control and stall control. When pitch control is used to control the turbine blade speed during periods of high wind, the control system on the wind turbine uses wind speed instruments such as anemometers to measure the wind speed and the controller uses the wind speed data to adjust the blade pitch to a point at which the blade begins to stall. When the blade's pitch is changed to a point at which it begins to stall, it produces less torque to the shaft and the turbine blades begin to slow down, just as if the wind speed had lowered. This is sometimes called **furling** the blades. If a variable-pitch blade is used, the wind turbine blade can be slowed significantly by moving the blade pitch until the narrow profile of the blade is faced into the wind and the blade can no longer harvest wind energy.

FIGURE 3-5 Wind turbine blades pitched so that the leading edge is directly into the wind and the trailing edge is directly out of the wind. In this position the wind turbine blades will not rotate, and they are locked by a brake for maintenance.

On some turbines, the pitch control is passive, so the blade will move to a point at which the blade begins to stall whenever its speed becomes too great. If passive pitch control is used, it is designed to protect the wind turbine from overspeeding. You will learn more about blade pitch control in Chapter 5.

Other wind turbines use stall-regulated concepts. On the stall-regulated wind turbine, the blades do not adjust during operation; instead, they are locked in place, a situation called fixed pitch. The stall is accomplished through the basic design of the blades, which cause the blade to stall naturally as the wind increases beyond a safe speed. This design allows the blades to harvest the maximum amount of wind when the wind speed is below the design speed, and allows the wind to make the blades stall when the wind speed becomes too high.

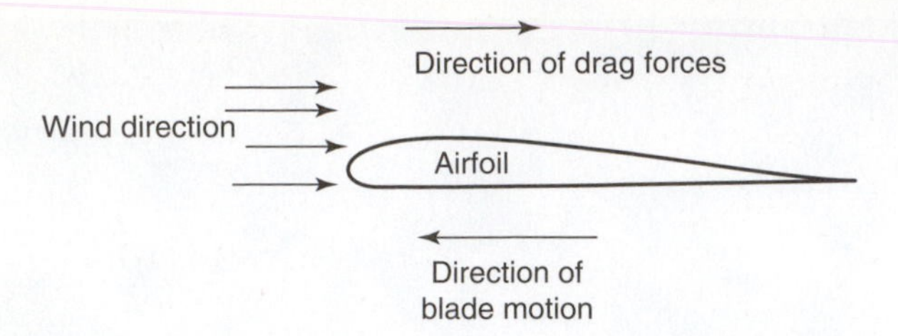

FIGURE 3-6 A turbine blade (airfoil) moving to the left through the wind. The drag forces hit the leading edge of the blade and are present in the same direction as the wind.

Drag

Drag is the force that opposes the motion of the airfoil as it moves through the air. Figure 3-6 shows an airfoil moving through the air and the drag forces that oppose the airfoil and try to stop it from moving through the wind. Remember that when the turbine blade rotates, its leading edge moves into the wind.

Drag is expressed in terms of the drag coefficient, which is a dimensionless number (it has no units). Typically, the only area of a wind turbine blade that is used in the calculation of drag is the front area (leading edge) of the blade, which moves directly into the wind. Design engineers try to design a wind turbine blade that has the smallest amount of drag. If the design engineers can keep the amount of drag small, the wind turbine blade will function better in converting wind energy to electrical energy. As drag increases, the efficiency of the turbine blade decreases.

A typical drag coefficient for wind turbine blades is 0.04; compare that to a typical automobile, which has a drag coefficient of approximately 0.30. Even though the drag coefficient for a blade is fairly constant, when the wind speed increases, the amount of drag force pushing against the blade also increases. The lower the drag coefficient number, the better the aerodynamic efficiency is.

Lift-to-Drag Ratio

The **lift-to-drag ratio** is a ratio of the value of lift to the value of drag. A higher lift value and a lower drag value provides a higher lift-to-drag ratio. The higher the lift-to-drag ratio, the more efficient the turbine blade is at converting wind energy into shaft torque, which produces more electricity at the generator. Most turbine blades are designed with the highest lift-to-drag ratio near the tip of the blade. The blade is designed so that it has more material with very high strength near the point where it connects to the hub, and the amount of material near the tip of the blade is reduced so that lift is increased and drag is reduced.

The drag caused by the blade cutting through the air is not constant. When the wind speed increases, both lift and drag increase with the square of the wind speed. When the air becomes denser, the amount of drag forces it creates increases. This means that when air gets colder it becomes more dense and the amount of drag it creates increases. When a wind turbine is located at higher altitudes, the air is less dense and the amount of drag caused by the air is less than at lower altitudes. Drag also increases if there is higher humidity or if it is raining or snowing, which means that the amount of drag is continually changing, even though the drag losses caused by the blade design are fairly constant.

Turbine Blade Tip Speeds

The tip of the turbine blade travels at the highest speed of any part of the turbine blade when it is rotating. Conversely, the part of the turbine blade that is connected to the hub, near the shaft, is traveling at the slowest speed when the blade is rotating. **Tip speed** is defined as the measured speed at the blade tip as it rotates through the air. Because the tip is traveling at the highest speed, it comes under considerable stress caused by centrifugal force when it is rotating. Designers and planners must be aware of the maximum tip speed that a wind turbine can handle for each blade and the maximum amount of wind speed it sees.

High tip speed is defined as speeds between 65 and 85 m/s, which is about 145 to 190 mph. High tip speeds are needed to make the turbine blade more efficient. At very high speeds the turbine blade receives maximum stress, however, which increases the rate of deterioration.

The turbine blade is designed to withstand very high winds and large loads on the generator. It is made of special materials that make it durable over years of exposure to the sun and weather and able to withstand the high speeds and torque from wind turbulence. Another specification that is important to blade speed is called the tip speed ratio. The tip speed ratio is the ratio between the speed of the blade at the tip and the speed of the wind. A tip speed ratio of 6–7 is optimal; and as the ratio becomes lower, the blades become less efficient.

It is important to remember that the fewer blades a wind turbine has, the faster these blades need to turn to harvest the same amount of energy as a wind turbine with more blades. For example, a three-bladed wind turbine does not have to turn as fast as a two-bladed wind turbine to harvest the same amount of energy. Therefore the tip speed ratios of a two-bladed wind turbine and a three-bladed wind turbine will be different.

Tip speed is generally discussed during the blade design phase, when the size of the wind turbine and the size of its blades are being considered. Once the wind turbine is installed, the speed at which the blades turn, measured in revolutions per minute (rpm), is used to specify the cut-in speed, which is the slowest speed at which the blades can turn and still produce electrical power, and the cut-out speed, which is the fastest speed at which the blades will be allowed to turn so as not to damage the wind turbine.

3.2 TYPES OF TURBINES

Two types of wind turbines are in use today: vertical-axis wind turbines and horizontal-axis wind turbines.

FIGURE 3-7 A Darrieus-type wind turbine blade. (Photo courtesy of U.S. Department of Energy.)

Vertical-Axis Wind Turbines

Among **vertical-axis wind turbines**, there are two basic types: the Darrieus-type wind turbine, which looks like an eggbeater; and the Savonius-type turbine. Figure 3-7 shows an example of a large Darrieus-type wind turbine. You can see the size of the wind turbine in relation to the worker standing on the base. One advantage of the Darrieus type of wind turbine is that it can harvest the wind from any direction.

Figure 3-8a shows a picture and Figure 3-8b shows a diagram of a Savonius-type vertical axis wind turbine, which has its design enhanced by adding a wing to it to make it more efficient than the traditional Savonius wind turbine. The diagram shows a top-down view of the wing that is added to the Savonius wind turbine. The most important aspects of this Savonius design are the wing and the turbine scoops. First, the wing that has been added to the front and rear of the Savonius wind turbine keeps the unit always pointing into the wind and guides the wind into the turbine. Second, the turbine scoops are designed as for an airplane wing, creating positive force in the scoop and negative force on the back side of the scoop. This creates lift, pushing the turbine faster than the velocity of the wind hitting it. Third, the diverter portion of the leading wing directs the wind to the outer edge of the scoop on the right. In a typical Savonius design, with the turbine cylinder without the wing, the wind is coming in the front side of the cylinder and rotating the cylinder. However, wind is also coming in on the back side of the cylinder, and this is counteractive and actually slows the rotation. The arrows in Figure 3-8b show how the diverter on the leading wing diverts the wind to the left and right of the scoops. The wind that is diverted to the left side of the scoops (back side) prevents the counteraction on the back side. The combination of the wind hitting the outside edge of the scoop on the right side and blowing past the back side of the scoop on the left side enhances the performance of the wind turbine.

(a)

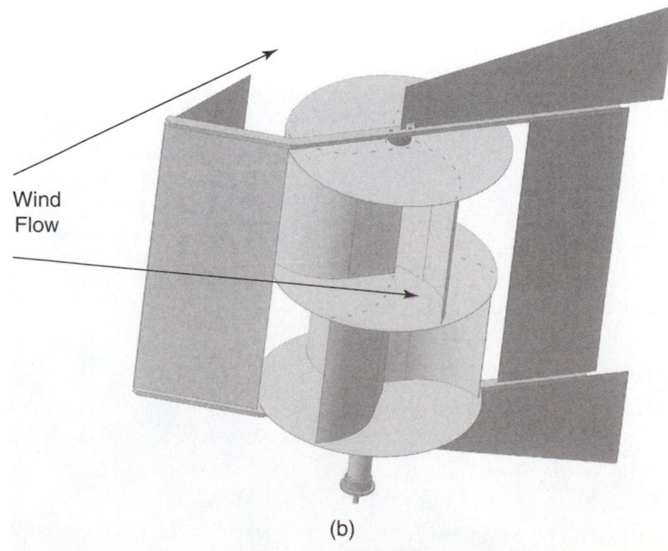

(b)

FIGURE 3-8 (a) A vertical Savonius wind turbine with a wing added to it. (b) Diagram of the Savonius wind turbine with the wing added. (Photo and diagram courtesy of Inventive Solutions, www.insol.us.)

FIGURE 3-9 A horizontal-axis wind turbine.

Horizontal-Axis Wind Turbines

Several vertical-axis wind turbines installed in the 1980s were large enough to produce utility-scale power, but they are no longer used. Vertical-axis wind turbines are not as efficient on a large scale as **horizontal-axis wind turbines**. Vertical-axis wind turbines today are mainly smaller versions used for residential use, or where noise is a problem or height restrictions exist. Larger wind turbines are usually of the horizontal-axis type. Figure 3-9 shows an example of a typical horizontal-axis wind turbine.

Horizontal-axis wind turbines can be divided into upwind and downwind turbines. With an **upwind horizontal-axis wind turbine**, the wind blows over the blades first and then over the generator. With a **downwind horizontal-wind turbine**, the wind blows over the generator and nacelle and then through the blades. Figure 3-10 shows an example of an upwind-type horizontal-axis wind turbine, and Figure 3-11 shows an example of a downwind-type horizontal-axis wind turbine.

Because the upwind machines have the rotor facing the wind, they have the advantage of avoiding the wind shade behind the tower. Wind shade is anything that blocks the direct flow of wind to the turbine blades. The majority of horizontal wind turbines use the upwind design.

One problem with the upwind designs is that the rotor needs to be made rather rigid and inflexible, and must be placed at some distance out in front of the tower, because the blades will flex somewhat in stronger winds. If the blades are not mounted far enough in front of the tower, strong winds will make the blades flex back into the tower and possibly be damaged. Some earlier upwind machines did not account for this, and the blades flexed into the tower and were severely damaged. Many of the blade strikes occurred when the wind direction changed suddenly and the blades flexed into the tower. A second problem with the upwind design is that an upwind machine needs a way to keep the rotor facing the wind. The direction the wind turbine blades are facing is called yaw, and the direction the blades on the wind turbine are pointed is constantly changed to keep the wind turbine producing electrical power at the most efficient rate. This is accomplished with yaw control on the wind turbine. The wind turbine must orient itself into the wind to harvest the energy from the wind most effectively.

The downwind machine has the rotor placed on the downwind or lee side of the tower. Because the gen-

FIGURE 3-10 An upwind-type horizontal-axis wind turbine.

FIGURE 3-11 A downwind-type horizontal-axis wind turbine.

erator and hub mechanism is on the downwind side, any flexing of the blades moves the tips away from the tower, which protects the blade tips from ever striking the tower. Some smaller downwind turbines are designed so that they orient into the wind without a yaw mechanism. This is called passive yaw location, and it makes the system less expensive. Yaw rotation must be controlled or limited so the generator does not continue to rotate in an unlimited number of 360-degree turns, because the electrical cables that connect the wind turbine generator to the ground terminals would become tangled. In some larger turbines the yaw control includes a rotation counter, and the number of full rotations is limited to about five full turns. Another way to compensate for the yaw rotating more than 360 degrees is to use slip rings instead of connecting the electrical cables between the point where the nacelle rotates and where it is mounted on the tower. Slip rings allow the electrical connection to be continuous between the nacelle and the point where the electrical cables are connected at the top of the tower even when the turbine yaw rotates the nacelle multiple times. The slip rings themselves can become a problem, however, because all the electricity generated by the generator must flow through them to get to the cables that carry the electrical power to the electrical terminals mounted in the electrical panel on the ground. Slip rings also wear quickly, and they require checking and minor maintenance continually throughout the life of the turbine. Because the slip rings are at the top of the tower, where the turbine nacelle is mounted, checking them requires climbing the tower to do the maintenance.

In larger wind turbines, active yaw control is used to control the direction that the turbine blades head into or out of the wind. This type of active yaw control typically uses a permanent-magnet-type high-torque motor or a hydraulic motor to turn the plate on which the turbine is mounted, The yaw motor is controlled by a programmable logic controller (PLC) that receives a signal from an anemometer that measures the speed and direction of the wind and compares it to the speed of the turbine blade. On these larger machines, the turbine blade is connected to a shaft that goes into a multispeed transmission. The generator has a shaft that is connected to the output of the transmission. The speed of the generator is strictly controlled so that the generator output is always at 60 Hz, and the yaw control is used to change the direction the turbine blades are pointing in relation to the direction of the wind, which helps control the speed of the turbine blade rotation. If the wind speed increases too much, the yaw control can rotate the turbine slightly so that it does not face directly into the wind, and the turbine blades will slow down until the generator is running so that it produces 60 Hz. The wind turbine also uses blade pitch control to help control the speed at which the blades turn as the wind speed increases and decreases. You will learn more about this in Chapter 5, where turbine control and generator control are explained in more detail.

3.3 VERTICAL-AXIS WIND TURBINES

Vertical-axis wind turbines are used where wind speeds are lower and the wind direction is not constant. They are typically not used in production of commercial electrical power; rather, they are also used in residential areas and urban areas where noise reduction is necessary or height restrictions do not allow a larger horizontal-axis wind turbine. Vertical-axis machines also operate with lower levels of vibration. They will naturally produce power regardless of the direction of wind flows and regardless of the amount of wind velocity.

One of the advantages of the vertical-axis machines is that you do not have to worry about the direction the machine is facing. Because they are round, they will rotate and produce electricity regardless of which way the wind is blowing or whether the wind is gusty and changes direction frequently. Vertical-axis machines are not as widely used as horizontal-axis wind turbines. In fact, vertical-axis wind turbines represent only a small percentage of the wind turbines in use today. The choice between a horizontal-axis wind turbine and a vertical-axis wind turbine should not be considered a competition between the two; rather, the type of wind turbine that is specified should be based on which one provides the best results for the application and location where the wind turbine is to be installed.

Figure 3-12 shows a typical Darrieus design and Figure 3-13 shows a Savonius-type vertical-axis wind turbine. The Savonius-type wind turbine creates rotational energy by use of drag forces, and the Darrieus-type wind turbine turns its shaft using lift forces. You can see the smaller generator mounted directly below the blades on the Savonius-type wind turbine in Figure 3-13.

Vertical-axis wind turbines can produce electrical power at lower speeds and at a variety of changing speeds, so it is not necessary to keep the generator speed

FIGURE 3-12 Darrieus-type vertical-axis wind turbine. (Photo courtesy of U.S. Department of Energy.)

FIGURE 3-13 Typical Savonius-type vertical-axis wind turbines.

FIGURE 3-14 Quietrevolution QR5 vertical-axis wind turbines. (Courtesy of quietrevolution.)

constant. And, because the generator speed is not kept constant, the frequency of the generated voltage is not always at 60 Hz. Instead, this type of turbine uses an inverter that accepts three-phase voltage produced by the generator and changes it to produce single-phase AC voltage at 60 Hz for residential use, or the inverter can produce three-phase AC voltage at 60 Hz for commercial and industrial use. Other electrical components can be used to change the AC voltage to DC voltage; alternatively, if DC voltage is required, a DC generator can be used.

Figure 3-14 shows an example of another type of vertical-axis wind turbine called the Quietrevolution QR5. This unit looks like a piece of working art, yet it quietly produces enough energy to supply a home or small office even at lower or variable wind speeds. The unit requires a three-phase electrical connection, which may not be available at some smaller commercial or residential sites. This type of smaller vertical-axis wind turbine can be purchased for approximately $50,000; at the present time, it is sold only in the United Kingdom. The rotor is approximately 5 m (16.4 ft) tall and 3.1 m (10.17 ft) in diameter, and it weighs approximately 450 kg (990 lb). The blade sits on a 3.4-m (11.15-ft) shaft where height restrictions are a concern, or on a shaft at approximately 6 m (19.69 ft) or 15 m (49.21 ft) on a tilt-down mast.

This type of turbine uses a direct-drive, permanent-magnet synchronous generator integrated into the base of the rotor. The peak power with wind speeds of 14 m/s is 7.4 kW aerodynamic and 6.2 kW DC. The British Wind Energy Association (BWEA) rated power at 11 m/s is 4.2 kW aerodynamic and 3 kW DC. This output can provide between 5000 and 11,000 kWh per year, depending on the amount of wind at the site. The aerodynamic helical blade design results in very smooth, quiet operation.

Because of the vertical-axis geometry of the rotor, the blade tips are close to the axis of rotation and travel a shorter distance per revolution. This results in a lower blade tip speed, further reducing noise.

Smaller vertical-axis wind turbines operate well in urban environments, where they provide quiet, vibration-free operation. In urban areas, wind speed and directions are frequently changing, and wind speeds tend to be lower because of buildings and other objects that create wind shadows. Vertical-axis wind turbines are able to generate voltage at low wind speeds, and they do not have to change direction to catch usable wind. Another advantage of using vertical-axis wind turbines in urban areas is that their design tends to be pleasing to the eye. The blades can also be painted or coated to match the surroundings, such as the color of buildings or homes.

Vertical-axis wind turbines also have several other advantages. Because they harvest the wind regardless of the direction it is blowing, they do not need any type of yaw control. The gearbox and generator may be placed on the ground, which means they can be serviced without climbing up a pole or tower. In fact, they may not need to be put on a tower at all, or the tower or base can be significantly shorter than for a horizontal-axis wind turbine.

Some disadvantages of vertical-axis wind turbines include the fact that some are not self-starting, so they may need a motor to get them turning initially, which means that a battery backup is required or the system must be tied to the grid to get power for the starter motor. Vertical-axis wind turbines are also not as efficient as horizontal-axis wind turbines, and the wind at lower levels is not as strong and consistent as it is at higher levels, where horizontal-axis wind turbines are mounted. As a result, the cost of a kilowatt hour of power is higher than for horizontal-axis wind turbines. Also, most vertical-axis wind turbines require guy wires to stabilize them, which may be a problem if they are mounted on a smaller residential plot or in a farm field, where people could trip over them or farm equipment could get tangled in them.

3.4 HORIZONTAL-AXIS WIND TURBINES

Horizontal-axis wind turbines have the main shaft and electrical generator at or near the top of a tower in the nacelle. The turbine blades may be pointed into the wind or out of the wind, depending on the design. Figure 3-15 shows a typical small residential-type vertical-axis wind turbine. Figure 3-16 shows two large units that are used to generate upwards of 2 MW each; their output is sent to electric utilities. Figure 3-17 shows a large commercial-type horizontal-axis wind turbine that produces about 80–100 kW.

FIGURE 3-15 Small residential-type horizontal-axis wind turbine. (Courtesy Southwest Windpower.)

FIGURE 3-16 Large horizontal-axis wind turbines that provide electricity to a utility. (Courtesy of Vestas Wind Systems A/S.)

Figure 3-18 shows the basic parts of a large horizontal-axis wind turbine. The turbine blade (1) has three blades, and you can see from the arrow indicating the wind direction that this is a upwind-type wind turbine. The blade is mounted on the rotor (2), and it has variable pitch control (3) that allows the turbine blade pitch to be adjusted so the turbine converts the most wind energy into electricity. This type of turbine has a braking mechanism (4) that is used to stop the blade from rotating when it encounters unsafe conditions or when maintenance must be done. The blades must be pitched (rotated) to the stall position when the brakes are applied so that wind will not cause them to rotate.

The turbine blade and rotor are mounted on the front end of the low-speed shaft (5), and the other and of

FIGURE 3-17 Horizontal-axis wind turbine for commercial application.

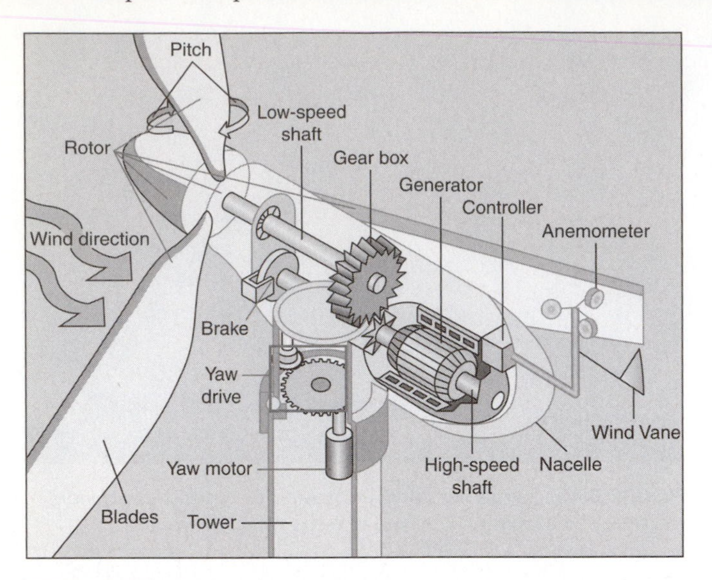

FIGURE 3-18 Basic parts of a large horizontal-axis wind turbine. (Courtesy of U.S. Department of Energy.)

the shaft has the main gearbox gears (6) mounted to it. The generator (7) is mounted to the high-speed shaft (12) that comes out of the gearbox. This means that when the wind makes the turbine blade rotate, that energy is transmitted through the low-speed shaft to the gearbox and then out of the gearbox through the high-speed shaft to the generator. A computer or PLC controls the pitch of the blade, the yaw position, and the generator loading. An anemometer (9) measures wind speed, and a wind vane (10) measures the wind direction; this information is sent to the controller to determine the direction the wind turbine should be pointed and the necessary pitch of the turbine blades so that the wind turbine harvests the maximum amount of wind energy. All of these parts are housed inside a compartment called a **nacelle**. The nacelle provides an enclosure that protects the generator and gearbox from the weather.

The nacelle sits on top of the yaw control assembly, which consists of the yaw drive (13), which is a large gear, and the yaw motor (14), which has a gear mounted to its shaft that engages the larger yaw gear. The gear ratio between the smaller gear on the yaw motor and the larger gear on the yaw drive allows the motor to have sufficient power and torque to rotate the nacelle so that the blades face into the wind. All of these parts are securely mounted to the tower (15). The tower may be 40–100 m (approximately 130–328 ft) high so that it is tall enough to position the turbine blade into the strongest wind flow. Today, most towers for larger wind turbines used to produce electrical power for utilities are in the range of 65–100 m tall. The Encore E126, recently installed in Germany, has a tower that is 138 m (453 ft) high. The taller the wind turbine tower, the more wind it is likely to harvest.

Horizontal-axis wind turbines may use a gearbox or set of gears, which changes the slow rotation of the blades into a faster, constant-speed rotation that ensures a generator creates AC voltage at 60 Hz, the standard for the U.S. electrical grid.

There are several ways to control the generator to ensure that the voltage produced is at exactly 60 Hz. One way is to control the speed at which the generator shaft turns, so that the generator rotates at the exact speed needed to produce voltage at 60 Hz. This is accomplished by gearing and by adjusting the blade pitch so that the blade speeds up or slows down to keep the high-speed shaft at an rpm that ensures the generator produces electricity at 60 Hz. For some larger wind turbines, the optimum blade rotation is between 10 and 22 rpm, so an appropriate gear ratio is used to make the high-speed shaft rotate at 1800 rpm, the speed the generator needs to produce at 60 Hz.

Another way to control the output frequency of the generator is to use a double-feed inductive-type generator in which the AC field current is tightly controlled to 60 Hz by feeding the current through an electronic inverter that produces exactly 60 Hz. The field current for this type of generator can also come from the grid, where the frequency is 60Hz. When the field current is controlled at 60 Hz, the generator will output power at 60 Hz.

Still another method of providing power to the grid at exactly 60 Hz is to allow the generator to run freely at any speed the rotation of the main blade causes. When the generator turns at any speed, the frequency of the voltage it produces varies with the speed. Because the frequency of the output voltage varies, it is sent to an electronic control called an inverter. When an inverter is used with a wind turbine, it is sometimes called a power electronic frequency converter (PEFC). The inverter accepts single-phase or three-phase AC voltage to its input circuits at any frequency. This AC voltage is converted to DC voltage by rectifiers. Capacitors and inductors are used in the DC circuit to remove any remaining image of frequency. The next section of the inverter changes the DC voltage back to single-phase or three-phase AC voltage at exactly 60 Hz. If the voltage needs to be increased, a transformer is used. Some applications are designed to use the pure DC voltage of the second circuit of the inverter, and the DC voltage is used at this point rather than sent to the section of the inverter that produces AC voltage.

3.5 BLADE GEOMETRY

Turbine blades must be designed so that the turbine can produce electricity at below rated wind speed operation, at rated wind speed operation, and when the wind is above rated wind speed operation. The geometry of the blades is designed to maximize the amount of electricity the wind turbine can produce at every wind speed. The design of turbine blades may be different for different applications. It is important to understand that the amount of wind, the duration of the wind, and the speed of the wind will be different depending on the location of the wind turbine. For example, along coastlines, winds tend to be stronger and available for longer periods of time. If the turbines are located offshore, they will be subjected to different types of winds compared to a turbine on a farm in the Midwest. Other areas may have only limited low-speed winds available, so the blade geometry may be slightly different to take advantage of the prevailing conditions. Another aspect of blade geometry design involves the number of blades the turbine has. The number and size of blades and their geometry depend on the amount of electrical energy the turbine will need to produce given the prevailing winds at the turbine's location.

Energy Unlimited, Inc., has developed variable-length wind turbine blades. The main feature of these blades is that they are designed with a tip that automatically extend outward in response to light winds and retracts in stronger winds. This design allows the blade to change its length automatically in response to wind speed. The design uses centrifugal force and a mechanism that allows the blade to be extended, which results in higher energy capture in low-wind conditions. This design also moves the blade tip back in and reduces its length to minimize mechanical load in high-wind conditions. The increase in efficiency associated with this innovative development can increase production by as much as 25%.

3.6 NUMBER OF BLADES

Wind turbines may be designed with one, two, three, or multiple blades. Smaller, residential-size units are designed for cost efficiency and the size of the electrical load of the home. Turbines used for commercial production of electric power may be two-bladed or three-bladed and designed for much larger energy loads. The vast majority of horizontal-axis wind turbines used in the commercial production of power for utility companies are three-bladed turbines.

The number of blades may require a larger horizontal-axis wind turbine to change the pitch of the blades and the direction the wind turbine faces as it moves into or out of the wind to ensure that it harvests the largest amount of energy from the existing wind. The blade pitch and the direction the turbine faces is generally controlled by a computer or PLC. An anamometer measures the wind direction and the speed of the wind and constantly sends this information to the computer, which determines the best direction to maximize generator output. This information is also used to determine minimum and maximum wind speed, which is used to allow the wind turbine blades to start turning or to slow down or stop when the wind exceeds the maximum designed standard to prevent damage to the turbine blades or the generator. The computer also measures the generator output as well as the turbine blade speed and moves the blades farther into or out of the wind depending on the conditions. The turbine is rotated through 360 degrees by a yaw motor that is mounted on a large gear mechanism. Some larger turbines use a series of yaw gears so that the yaw motor can move the wind turbine nacelle in the strongest winds. A protection system is designed into the yaw control to ensure that it limits the number of full rotations the turbine makes to five full rotations or less in one direction or the other. A rotation counter is designed to limit the total rotations and cause the turbine to change its rotation back to the other direction so that the cables do not become tangled.

Three-Bladed Wind Turbines

The majority of modern large wind turbines use three blades with the rotor position maintained upwind. These wind turbines use yaw motors to ensure that the three blades are always facing into the wind.

Figure 3-19 shows an example of a three-bladed wind turbine. The three blades provide the most energy conversion while limiting noise and vibration. The three blades also provide more blade surface for converting wind energy into electrical energy than a two-bladed or single-bladed wind turbine.

FIGURE 3-19 Three-bladed wind turbine. (Courtesy of Vestas Wind Systems A/S.)

FIGURE 3-20 A large wind turbine blade in transit. (Courtesy of John Fellhauer, jfellhauer@surenergy.us.)

The main disadvantage of the three-bladed turbine is that it weighs more than a two-bladed or single-bladed turbine, making the turbine more difficult to transport and lift into place on a tower. Figure 3-20 shows the large size of a turbine blade in transit. You can see that a large truck and trailer is needed to move a single turbine blade. This also means that one or more very large cranes are needed to set the tower and turbine in place. The tower to hold the larger three-bladed turbines must also be larger and reinforced to support the weight and to withstand the large wind that is harvested to turn the generator to produce its maximum output.

Two-Bladed Wind Turbines

Two-bladed wind turbine designs have the advantage of saving the cost and the weight of one rotor blade in comparison to a three-bladed turbine, but it has the disadvantage that it requires higher rotational speed to yield the same energy output. This is a disadvantage in terms of both noise and wear of critical bearings, shafts, and gearboxes. Two-bladed turbines have experienced high fatigue failures of the blade and other mechanical parts. Figure 3-21 shows an example of a two-bladed wind turbine.

One way to limit the wear and fatigue problems in the blade is to use **teetering**. A teeter hub allows the

FIGURE 3-21 A typical two-bladed wind turbine. (Courtesy of Southwest Windpower.)

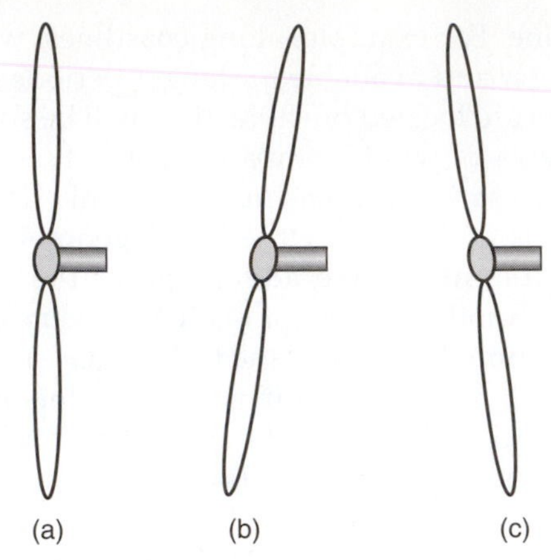

FIGURE 3-22 Two-bladed turbine blades mounted on a teeter hub: (a) blades positioned perpendicular to the mounting shaft; (b) wind flexes the top tip of the blades back toward the shaft; (c) wind flexes the bottom tip of the blade back toward the shaft.

blade to flex inward or outward from the shaft on which it is mounted. The teeter movement is much like a child's teeter-totter or seesaw that moves back and forth, and this motion allows the blade to flex rather than remain rigid, thereby reducing wear and fatigued, which may cause cracks to develop. Figure 3-22 shows the operation of a teeter hub. In Figure 3-22a, the blade is at 90 degrees to the shaft. In Figure 3-22b, the hub has teetered (shifted and changed) so that the top of the blade is flexed inward to the shaft. In Figure 3-22c, the hub is teetered so that the top of the blade is flexed outward from the shaft. In addition to allowing the blade to flex in higher winds and thus reduce fatigue and wear, teetering also protects the two-bladed turbine from heavy shocks to the turbine when a rotor blades passes the tower. The teetering is limited to a small range (approximately plus or minus 2 degrees), but it is sufficient to limit the wear and tear caused by the higher tip speed of the two blades.

Another way to improve the efficiency of the two-bladed turbine is to make the two blades thicker and wider than traditional turbine blades, so that the two blades can convert more wind energy. The thicker blades also mean that the blades are stronger and better able to resist the fatigue problems that plagued early two-bladed turbines. New composite materials allow the increased size without adding substantial weight to the blade. These materials also allow the blade to be produced at a lower cost. Even with these more efficient blades, however, the two-bladed turbine is still slightly less efficient than the three-bladed turbine.

On the other hand, the two-bladed turbine is easier to install than the three-bladed turbine. The two-bladed turbine can be lifted into position after the turbine blades have been mounted to the nacelle on the ground. This makes installation faster and safer.

Single-Bladed Wind Turbines

Single-bladed wind turbines are used in some limited applications, but they are the least used of all the horizontal-axis wind turbines. To rotate smoothly, single-bladed turbines

FIGURE 3-23 Single-bladed horizontal-axis wind turbine with one counterbalance.

must have one or two counter-balances. Figure 3-23 shows a single-bladed wind turbine with one counterbalance, and Figure 3-24 shows a single-bladed wind turbine with two counterbalances. The major advantage of this type of wind turbine is that it saves the cost of additional turbine blades, but single-bladed turbines must run at much higher speeds to convert the same amount of energy from the wind as two-bladed or three-bladed turbines with the same size blades. Because the single-bladed turbine must run at higher speeds, there is more wear and fatigue on the blade and bearings in the mounting mechanism, which in turn means higher maintenance costs over the life of the turbine. Single-bladed turbines also require extensive setup procedures to ensure that the blade is mounted perfectly and balanced to limit oscillation and vibration. A teeter hub is usually used with them.

Single-bladed turbines are used in some applications because they weigh much less than two-bladed or three-bladed wind turbines, so a smaller crane can be used to hoist the turbine into position.

Five-Bladed Wind Turbines

Some wind turbines have five blades to produce more electrical energy from the wind. Figure 3-25 shows an example of a five-bladed wind turbine. In this figure you can

FIGURE 3-24 Single-bladed horizontal-axis wind turbine with two counterbalances. (Courtesy of Powerhouse Wind Ltd., www.powerhousewind.co.nz.)

FIGURE 3-25 A five-bladed wind turbine.

see that the wind turbine blades are large, and because there are five blades, the wind generator will produce more electrical energy from the same amount of wind than a three-bladed wind turbine with the same size blade. Notice the thickness of the tower. The tower and base are mounted into the roof of the building, which is a concrete-reinforced building. This type of five-bladed wind turbine needs a very strong base and tower to hold the wind turbine in the wind. Also notice the cowling around the blades, which helps direct wind directly into the blades.

3.7 COMPARISON OF BLADE TYPES

Wind turbine blades can be compared in a number of ways, such as by size, weight, material, and the way they are manufactured. We have discussed some of the advantages and disadvantages of one-bladed, two-bladed, and three-bladed wind turbines. Wind turbine blades can be made from a variety of materials, from wood for smaller blades to aluminum and other metals for small and medium-sized blades, to composite materials (plastic and fiberglass) for the smallest through the largest turbine blades. Some smaller blades are made of fiberglass-reinforced polyester or wood-epoxy.

Turbine blades must be designed and manufactured so that they are stiff enough to prevent the blade tips from being pushed into the tower by high winds, yet agile enough to efficiently convert wind power into electricity. The blades are located on the main shaft on a rotor at a considerable distance in front of the tower, so they are far enough out to prevent them from accidentally touching any part of the tower when the blades are rotating.

Most commercial wind turbine blades are made from fiberglass with a hollow core, or other materials maybe used in the core to make it more rigid. The core

FIGURE 3-26 Small wind turbine blades on a residential-type wind turbine. (Courtesy of Southwest Windpower.)

may be filled with plastic foam or other lightweight substance. A typical fiberglass blade that is about 50 ft in length weighs about 2500 lb. Other types of turbine blades are constructed of fiberglass composite using a vacuum infusion method, which is a leading-edge process that ensures a consistently lightweight and strong blade.

FIGURE 3-27 A large wind turbine blade for a very large wind turbine. (Courtesy of John Fellhauer, jfellhauer@surenergy.us.)

Turbine blades range in size from very small blades, as shown on the residential-type turbine in Figure 3-26, which has blades that are less than 24 in long, to vary large blades such as the one shown in Figure 3-27, which must be transported on a large truck or railroad car.

3.8 ADVANTAGES AND DISADVANTAGES OF SINGLE-BLADED, TWO-BLADED, AND THREE-BLADED TURBINES

The best way to compare the advantages and disadvantages of the single-bladed, two-bladed, and three-bladed turbines is to show them in a table, as presented in Table 3-1.

TABLE 3-1 Advantages and Disadvantages of Three-Bladed, Two-Bladed, and Single-Bladed Wind Turbines

Type of Wind Turbine	Advantages	Disadvantages
Three-bladed turbine	1. Quietest of the three types of turbines. 2. Least amount of vibration. 3. Available blade pitch control allows blade to change to catch maximum amount of wind. 4. Taller towers allow turbine blades to catch more wind than shorter towers. 5. Aesthetics. 6. Lowest energy cost.	1. Heavier than the other two types of turbines. 2. Most capital-expensive of the three types. 3. Requires active yaw control to make blades face into the wind. 4. Requires the largest cranes to erect. 5. Requires the largest and heaviest tower. 6. Larger blades are more difficult to transport.
Two-bladed turbine	1. Yaws automatically when wind direction changes. 2. Produces more energy than single-bladed turbine.	1. Noisier than the three-bladed turbine. 2. Produces less energy than the three-bladed turbine.
Single-bladed turbine	1. Least expensive. 2. Lightest weight of the three types of turbine. 3. Easiest to erect because of light weight and because the blade can be mounted while it is on the ground. 4. Requires the smallest and lightest tower.	1. Noisier than the three-bladed turbine. 2. Must run at highest speed to produce the same amount of electrical power. 3. Produces the most vibration at the blade. 4. Aesthetics.

Questions

1. Explain how lift is created when wind blows across the blade of a wind turbine.
2. Explain what the lift-to-drag ratio is.
3. Explain why teetering may be needed for a two-bladed horizontal-axis wind turbine.
4. Explain what blade tip speed is and why it is important.
5. Discuss the advantages and disadvantages of a three-bladed horizontal-axis wind turbine.
6. Discuss the advantages and disadvantages of a two-bladed horizontal-axis wind turbine.
7. Discuss the advantages and disadvantages of a single-bladed horizontal-axis wind turbine.
8. Explain how the wind blowing past the blades of a horizontal-axis wind turbine makes the blades begin to rotate.
9. Explain how the wind blowing past a vertical-axis wind turbine makes the blades begin to rotate.
10. Discuss the advantages and disadvantages of the vertical-axis wind turbine.

Multiple Choice

1. The wind passes over the nacelle first and then over the blades in
   a. A downwind horizontal-axis wind turbine
   b. An upwind horizontal-axis wind turbine
   c. A vertical-axis wind turbine
   d. All wind turbines
2. When wind is blowing past a turbine blade,
   a. It moves past the trailing edge first.
   b. It moves past the leading edge first.
   c. It can move past the leading edge or the trailing edge, depending on the direction the wind is blowing.
   d. It causes the blade to stall immediately.
3. A _____ can produce the most electricity when the wind is blowing at 20 mph.
   a. Single-bladed wind turbine
   b. Two-bladed wind turbine
   c. Three-bladed wind turbine
   d. A vertical-axis wind turbine that is the same size
4. A ______ is the most expensive when compared to other wind turbines its size.
   a. Single-bladed wind turbine
   b. Two-bladed wind turbine
   c. Three bladed wind turbine
   d. It depends on whether it is an upwind or a downwind machine.
5. A(n) ______ does not need a yaw motor to continually harvest wind, regardless of the direction.
   a. Upwind horizontal-axis wind turbine
   b. Downwind horizontal-axis wind turbine
   c. Vertical-axis wind turbine
   d. all of the above
   e. b and c
6. In a comparison of wind turbines that are all the same size, which one will begin to turn its blades at the lightest wind speed?
   a. An upwind horizontal-axis wind turbine
   b. A downwind horizontal-axis wind turbine
   c. A vertical-axis wind turbine
   d. The blades of all wind turbines will begin to turn at the same wind speed.
7. The advantages of vertical-axis wind turbines are that
   a. They typically cost less than horizontal-axis wind turbines of the same size.
   b. They can produce energy regardless of the direction of the wind flow.
   c. They are typically quieter than horizontal-axis wind turbines.
   d. All the above
8. Which type of wind turbine blades need counterbalances?
   a. Three-bladed wind turbines
   b. Two-bladed wind turbines
   c. Single-bladed wind turbines
   d. None of the horizontal-axis wind turbines
9. Lift is produced on the wind turbine blade
   a. When the wind blows across the blade from the leading edge to the trailing edge
   b. When the wind blows across the blade from the trailing edge to the leading edge
   c. When the blade is set to the furled condition
   d. Any time the wind blows, regardless of the direction from which it blows against the turbine blade
10. When the wind speed increases across a wind turbine blade,
    a. The lift and drag forces increase.
    b. The lift and drag forces decrease.
    c. The lift increases and the drag decreases.
    d. The lift decreases and the drag increases.

CHAPTER 4

Wind Turbine Performance

OBJECTIVES

After reading this chapter, you will be able to:

- Explain why the constant rotational speed of the generator shaft is important for a wind turbine.
- Explain how wind makes a turbine blade rotate.
- Explain the meaning of angle of attack as it applies to wind turbine blades.
- Explain how to estimate how much energy can be converted from the wind.
- Use a wind power curve to determine the peak performance for a wind turbine.

KEY TERMS

Active stall control
Angle of attack
Air density
Airfoil
Amplifier
Betz's law
Blade pitch
Closed-loop control
Error
Feedback sensor
Finite-element analysis
Lift
Output
Passive pitch control
Passive stall control
Peak performance
Pitch control
Power curve
SCADA (Supervisory Control and Data Acquisition)
Setpoint
Summing junction
Turbulence
Wind turbine peak performance
Yaw control

OVERVIEW

This chapter investigates wind turbine performance and how it affects the operation of a wind turbine. You will learn how the wind makes a wind turbine blade rotate. You will also learn how the blade works as an airfoil, and how changing the angle of attack can make the blade more or less effective. You will also learn how wind turbines can be controlled to optimize their operation, producing the maximum amount of electricity for any given wind speed. This chapter also shows how wind turbines are tested and how you can determine the wind generator peak performance.

4.1 CONSTANT-ROTATIONAL-SPEED OPERATION

Operation at a constant rotational speed is necessary in larger wind turbines to ensure that the frequency of the voltage produced by its generator is as close to 60 Hz as possible. The speed at which the generator shaft turns determines the frequency the voltage will have. Typically, the speed of the rotor and blades is between 10 and 20 rpm. In these types of wind turbines, a gearbox is used to increase the speed of the **output** shaft that sends power to the generator. The output shaft speed from the gearbox is controlled to about 1800 rpm, which ensures that the frequency of the output voltage of the generator is as close to 60 Hz as possible.

When the wind speed changes, the speed of the rotor and blades also changes. The gearbox continually changes gears and keeps the output shaft speed close to 1800 rpm regardless of the input speed from the turbine blades and rotor. It is important for the rpm to remain constant, to keep the output voltage of the generator at 60 Hz.

On smaller wind turbines that are not connected to the grid or that do not use the AC voltage directly, the frequency of the output voltage is not a problem. In these types of wind turbines, the output voltage may be converted to DC voltage, which can be used directly or stored in batteries, or it can be connected to an electronic inverter, which takes in the voltage as DC voltage and produces an output AC voltage that is at exactly 60 Hz. These types of wind turbine systems can allow the turbine blades and rotor to turn at any speed and still produce electrical energy. In later chapters you will learn more about inverters and storage devices.

4.2 WHAT MAKES THE TURBINE BLADE ROTATE

In Chapter 3 you learned about airfoils and the design of wind turbine blades. The wind turbine blade is designed as an **airfoil**, which allows it to produce **lift** as wind blows across the face of a turbine blade. Because the turbine blade is anchored at one end to the rotor at its base, the tip of the blade begins to move with the wind, which causes the blade to swing in a controlled arc at the rotor. The faster the wind blows, the faster the blade moves, and its rotation causes the tip of the blade to move faster and faster.

4.3 ANGLE OF ATTACK AND BLADE PITCH

The angle at which the wind strikes the turbine blade is called the **angle of attack**. The position or location of the blade with respect to the wind turbine is called the **blade pitch**. The position or pitch of the blade can be adjusted on some wind turbines from zero to 90 degrees. When the pitch of the blade is adjusted so that the wind blows directly over a flat blade, the angle of attack is zero. It is easier to visualize the angle of attack if you think of the blade lying horizontal, like an airplane wing. When the pitch of the blade is changed so that of the front edge of the turbine blade rotates upward, the amount of lift increases and the angle of attack increases. When the pitch is increased so that the turbine blade is rotated up to where it is producing the maximum lift, the amount of angle of attack is at its maximum, which is called the critical angle of attack. When the angle of attack increases to the critical angle of attack and beyond, the turbine blade stalls and loses its ability to convert energy from the wind. The angle of attack can be explained the same way when the pitch of wind turbine blade is changed so that the wind causes more or less lift as it moves over the blade.

The pitch of the wind turbine blade can be controlled to change the angle of attack for the blade. When the angle of attack is changed, the speed at which the wind turbine blade rotates varies even with the same wind speed, and the amount of electricity that the wind turbine can produce can be controlled. The angle of attack can be changed so that the wind turbine can begin to produce electricity at the lowest wind speeds, and the efficiency of the wind turbine can be increased throughout the range of wind speeds.

4.4 BLADE PITCH CONTROL

Wind turbines must operate in a large variety of wind conditions, from very light winds to very heavy winds. It is important on larger systems that the blades have a **pitch control** that allows them to have their pitch adjusted so that the system operates in an optimal condition. The main function of the pitch control is to ensure that the rotor turns at a specified speed that ensures the generator produces voltage at the proper frequency. The speed at which the generator shaft turns determines the frequency of output AC voltage. In the United States, the frequency for AC voltage is 60 Hz. The turbine blades are attached to the rotor, which is connected by a shaft to the gearbox. The output shaft of the gearbox is connected to the generator. When the wind blows through the wind turbine blades, it causes the blades to move and make the rotor turn. The speed of the turbine blades at the rotor is usually between 10 and 20 rpm, and the output shaft of the gearbox turns the generator at approximately 1800 rpm, which allows the generator to produce voltage at 60 Hz. If the wind speed changes, the pitch of the blade can be adjusted to speed up or slow down the rotor to ensure that the generator turns at a speed as close to 1800 rpm as possible so that the output frequency of the generator is controlled to as close to 60 Hz as possible.

If the wind is too strong, the blade pitch control can also adjust the pitch of the blades so that they are not damaged by the high wind and consequent overspeed rotation. Pitch control is also necessary to reduce the load on the blades and the tower during the highest winds. This section explains how the pitch control operates to change the pitch of the blades. You will learn about active and passive pitch control as well as about the latest type of control.

Blade pitch control is also useful to adjust the blades to gather as much wind as possible when wind speed is at its lowest, so that the blades can turn even when the wind is light. Blade pitch control is also used to adjust the blades whenever the wind is blowing so that the wind turbine is able to harvest the maximum amount of energy at any wind speed.

Optimum Wind Speed for a Wind Turbine

As you begin to learn about blade pitch control, you will see that it depends on the type of wind turbine and its size. Blade pitch control is needed because the wind may sometimes become so strong that it can damage the blades or the rotor, gear train, and generator. Adjusting the pitch of the blade can slow a turbine down. For most wind turbines, the optimum wind speed for maximum power output is about 15 m/s (33 mph). When the wind speed exceeds its maximum rating, the blades and rotor will spin exceedingly fast and can damage the mechanical assembly.

How Pitch Control Adjusts the Blades

When the blade pitch changes, the profile that is shown to the wind is altered. Figure 4-1 shows the blade pitch in three different positions. In Figure 4-1a, the blade is shown at full pitch and the profile of the blade is full. When the blade is at full pitch, it can harvest the maximum amount of wind that is available. In Figure 4-1b, the blade is at half-pitch, only part of the blade profile is shown to the wind, and the blade can harvest only approximately 50% of the available wind. The blade in Figure 4–1c is at 25% pitch, and the blade is able to harvest only a small portion of the wind. As the blade pitch changes and the profile becomes smaller, the blade harvests less of the available wind, and this in turn causes the blades to slow their rotational speed so that the wind turbine slows to a safe speed.

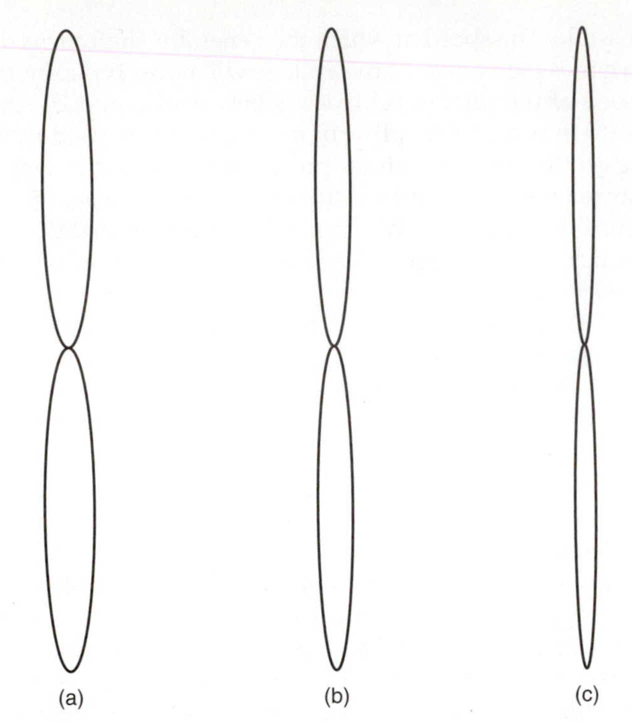

FIGURE 4-1 Blade pitch: (a) turbine blade at full pitch; (b) turbine blade at half-pitch; (c) turbine blade at quarter-pitch.

Closed-Loop Pitch Control

Control of wind turbine blade pitch is accomplished using a **closed-loop control** whose output changes the amount of hydraulic pressure and flow to the blade control apparatus. Figure 4-2 shows a basic closed-loop control circuit. The **setpoint** is the maximum speed at which the blades can rotate. A **feedback sensor** measures the rpm of the blades and rotor. The setpoint and feedback signal are compared in the **summing junction**, and the difference is called **error**. The amount of error is sent to the controller, which is typically a programmable logic controller (PLC). The PLC sends an output signal to a hydraulic amplifier, which changes the pressure and flow that controls the turbine blade pitch. An **amplifier** takes a small electrical signal and converts it into a larger electrical signal that can power the hydraulic valve. The PLC controller also determines how quickly to change the amount of output signal through a proportional, integral, and derivative (PID) algorithm. The PID control can make the response fast enough to change several times in each second, or it can be slowed to make the changes only once or twice during each rotation of the rotor. The PID tuning is specific to the type of wind turbine and its application. If the speed of the rotor starts to exceed the

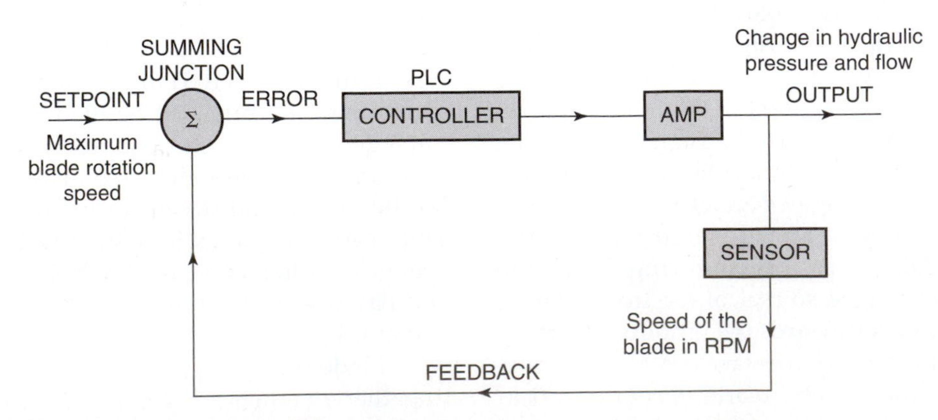

FIGURE 4-2 Closed-loop system to control blade pitch.

setpoint value, the feedback sensor senses this increase in speed. When the speed is above the setpoint, the error becomes negative and the PLC controller sends the output to the hydraulic amplifier and makes changes to the blade pitch control, which makes the blade slow down. As the blade changes speed, the speed of the rotor is slowed to a point at which it is safe to operate. If the wind speed slows down, the rotor will slow more, and the feedback signal will indicate that the speed is reduced. The summing junction then produces a positive signal and the PLC adjusts the hydraulic amplifier to adjust the blade pitch in the opposite direction, which allows the turbine blades to speed up again and harvest the maximum amount of wind. This oscillation of blade pitch continues until the wind speed begins to cause the blades and rotor to overspeed again. This closed-loop control keeps the turbine blade pitched to its maximum level to harvest the maximum amount of wind energy until the rotor begins to rotate above the safe rpm level. At the point where it begins to exceed the safe rpm, the loop control changes the hydraulic amplifier to change the blade pitch and bring the rotor speed down below the unsafe speed. This type of control can be set to change the pitch 0.5 to 1 degree at a time. It can also be set so that the amount of change is larger, 2–3 degrees at a time. The PID control can be adjusted to determine how much change is sent to the hydraulic amplifier as the amount of error changes.

Other sensors can be used in the loop controller to indicate when the maximum torque on the blades or on the tower is occurring, and the pitch of the blades can be adjusted to lower the forces on the blades and tower. On some very large wind turbines, the blade pitch loop control can also be combined with a yaw loop controller, which can adjust the position of the turbine into and out of the wind to keep the stress on the tower and blades below the maximum, safe conditions as well as protect the drive train and generator from unsafe conditions. The yaw for a wind turbine is the direction the nacelle is pointed with respect to where the blades are compared to the direction the wind is blowing. **Yaw control** changes the direction the wind turbine points and can point the wind turbine blades into the wind when wind speed is moderate or low, and move the nacelle so the wind turbine blades are not pointed into the wind when the wind speed is too high.

Energy to Turn the Blades

The energy source used to change the pitch of the wind turbine blade may be either hydraulic fluid or a mechanical system. On the largest wind turbines, a hydraulic system is used to adjust the blade pitch. In one type of system control, a hydraulic cylinder is connected to the base of the blade; when the cylinder extends, it moves the pitch of the blade to its minimum position, and when the cylinder is retracted, and it moves the pitch of the blade to its maximum position. The PLC control system uses a hydraulic amplifier to control the position of the hydraulic cylinder from full extension to full retraction smoothly with proportional control. Proportional control allows the PLC to change the amplifier output in very small increments, which converts the movement of the cylinder to degrees of pitch control. In the best systems, the amount of change in the movement of the cylinder is as small as 0.1 degree of pitch.

In other smaller wind turbine applications, a mechanical system changes the pitch of the blades using a strong spring. When the wind is at its strongest, the wind makes the blade pitch change. The spring tension then changes the blade pitch to slow down the wind turbine. In other types of systems, the blades are designed so that strong winds cause a condition called stall, reducing the amount of energy the blade can get from the wind. This type of control will be discussed in the next section, on passive and active stall control.

Passive Stall Control for Wind Turbines

Passive pitch control is also called **passive stall control**. This type of control depends on the design of the turbine blades to help slow the turbine down in the event of winds that become too strong. Wind turbines that use passive stall controlled employ a design in which the rotor blades are bolted onto the hub at a fixed angle. When the blade is connected at this angle, the blade assumes a profile in the face of the wind that causes the blade to stop harvesting the wind energy if the wind becomes too strong. When the wind blows at an unsafe speed, the angle of the blade causes the blades to stop producing torque, so the rotor speed is reduced to a safe level. The blade is designed so that higher wind speeds create turbulence that reduces the lift on the blade and slows it down. Wind turbulence occurs whenever the wind does not blow or move in a straight line. The turbulence builds slowly as the wind speed increases, and this causes the blade to correct the rotor speed continually as the wind speed increases.

This type of control is known as passive control because there are no moving parts to fail and no expensive controls are needed. However, it makes the overall performance of the wind turbine less efficient.

Active Stall Control for Wind Turbine Blade Pitch

Another way to control the wind turbine blade pitch is called **active stall control**. In this type of control the pitch of the turbine blade is changed to a point at which airfoil activity on the blade is stalled. When the blade is in the stall position, it does not produce any energy to rotate the rotor. Active stall control adjusts the blade pitch, but the goal in this type of control is to take control of the blade and rotor and bring them to a complete stop if necessary. This type of stall control makes it easier for the wind turbine to the brought under control in emergency stop condition. The advantage of this active control is that it can also be used to pitch the blades during start-up or when wind speed is low, and it makes the wind turbine more efficient. This may also allow the wind turbine to begin turning at a much lower wind speed than it normally would without pitch control.

These types of control systems are discussed in more depth in Chapter 5, where you will learn more about rotor speed control, yaw control, pitch control, and other hydraulic control systems. You will also learn more about PLC controls in Chapter 5.

Individual Blade Pitch Control

The latest type of blade pitch control is called individual blade pitch control. In larger wind turbines with multiple blades, it is now possible to adjust the pitch of each blade independently. The PLC control and the sensors used in this system allow each blade to be adjusted multiple times during each revolution of the rotor. This individual blade pitch control system allows the wind turbine to maximize efficiency at all wind levels, while protecting the blades and the wind turbine from damage that may be caused high winds.

This type of blade pitch control allows the pitch of each blade to be adjusted independently in small increments while the blades are rotating. This means that the pitch of one blade may be adjusted 2 or 3 degrees more or less than the other blades. It also means that during any one rotation, each blade may be adjusted separately to provide the best overall efficiency of the blades for the given wind speed.

This type of control is especially useful when wind speeds are very gusty and change frequently. It also works well when the direction of the land changes slightly, because the blades adjust with these changes to make the turbine remain at its most efficient settings. The newest types of sensors, PLC controls, and proportional hydraulic systems provide the means for this most advanced type of control. This type of control system is very expensive, but on the largest wind turbines a change of 1% or 2% in efficiency helps pay for the system very quickly.

4.5 HOW WIND TURBINES OPERATE AT VARIABLE WIND SPEEDS

When engineers begin to design a wind turbine installation, they must take into account what the electrical power generated will be used for. If the electrical power is to be used on site in a small residential or commercial application, a smaller wind turbine that produces electrical power without regard to a controlled frequency can be specified. The electrical power can be used as DC power for electrical heating, water heating, or lighting that does not require a specific frequency. If the loads require AC voltage at 60 Hz, an inverter can be used to convert the DC voltage or AC voltage that is not at 60 Hz when it is produced by the turbine generator to exactly 60 Hz. These smaller units can be placed where the wind speed is variable and the system can produce the required electrical power as the wind changes speed from very slow speeds to speeds near 35 mph without worrying about the effects of the changing wind speeds.

When a mid-sized wind turbine is designed to provide power for a larger electrical load such as a commercial establishment, an inverter can be used to convert the electrical power that is produced at any wind speed to exactly 60 Hz. The electrical inverter must be sized slightly larger than the highest demand for the electrical load it is supplying.

Larger wind turbines designed to supply electrical power to the grid at exactly 60 Hz must have a control system such as a gearbox or transmission to convert the low speed of the turbine blades and rotor to the 1800 rpm needed at the input of the generator shaft.

The largest wind turbines may use a variable pitch and yaw controller that uses hydraulic controls to adjust the blade pitch and yaw to ensure that the turbine blade speed and the gearbox produce an output shaft speed of 1800 rpm to the generator. These larger systems allow the turbine to harvest the wind at a wide variety of wind speeds, from a low of 7–10 mph up to 35 mph. The control systems change the pitch and yaw to ensure that the blade and rotor rotational speed are within the specifications that allow the generator to produce power at 60 Hz.

Some very small wind turbines are designed with a tail that makes the wind turbine nacelle move with the wind direction and wind speed. These smaller units begin producing electrical power when wind speeds of approximately 7–10 mph occur. When the wind turbine blades first begin to rotate, at the lowest wind speeds the amount of electrical power is minimal. As the wind speeds reach between 15 and 35 mph, the wind turbine begins to produce a more constant amount of electrical power. When the wind speed reaches 35 mph, the control system for the wind turbine slows the blades or stops them completely to ensure that they are not damaged.

4.6 ESTIMATING HOW MUCH ENERGY IS CONVERTED

When you are trying to calculate how much energy is converted from the wind, you must keep in mind the basic laws of physics. One of these laws is that energy cannot be created or destroyed; it can only be converted from one form into another. Another basic law is that no machine is perfectly efficient, which means that wind turbine blades can convert only a percentage of the wind energy into electrical energy. The energy in the wind is called kinetic energy, which means that it is energy in action, and it is available to be converted. When the wind turbine blades begin to move and cause the generator shaft to move, this is mechanical energy, and it produces electricity. The electrical energy that is produced can be manipulated through transformers and transmitted along power transmission lines, and then it can be used by turning switches on or off to control electrical loads such as heating, lighting, and turning motor shafts for fans, air conditioning, and refrigeration. The turbine is not able to convert all the energy from the wind, in part because some of the wind energy passes through the blades.

To calculate how much power is available at any given wind speed, we must start with the definition of power. Power is a unit of energy divided by time. It is also important to understand that power can be expressed in many different units, such as watts for electrical energy or horsepower for mechanical energy. We can also convert back and forth between electrical and mechanical power. For example, 1 horsepower of mechanical energy is equivalent to 33,000 foot pounds per minute, and is also equivalent to about 746 W of electrical energy.

How much electrical power is available in a given amount of wind power depends on the velocity of the wind, the size of the turbine blade, and the density of the air. The formula is

$$\text{Wind power} = (\text{turbine blade diameter})^2 \times (\text{wind velocity})^3 \times (\text{density of the air in the wind}) \times (\text{a constant that includes the number of blades})$$

If you have a multiple-bladed wind turbine, you will need to add one or more constants to determine the turbine blade sweep area and any variations in unit conversions. In symbols, the above formula is

$$P_w = D^2 \times V_w^3 \times d_a \times K$$

where
P_w = wind power
D^2 = diameter of turbine blades
V_w = velocity of the wind
D_a = density of the air in the wind
K = some constant that includes the number of blades

In this formula the value of the wind velocity is cubed, which means that if the wind speed doubles, the amount of energy available will increase eight times. Because the diameter of the turbine blades in this formula is squared, it also means that the power in the wind will increase by four times if the blade length is doubled. This is why larger and larger wind turbines are designed and located in places where the wind speeds are constantly among the highest values, around 33 mph. This higher wind constantly blowing against the tower and blades means that these large wind turbines can produce megawatts of electrical energy and also withstand the wear and tear caused by the high winds.

On the other hand, smaller wind turbines are less efficient because their blade size is smaller and their towers are smaller. They are also located where lighter winds or fluctuating wind speeds occur. Design engineers can oversize the turbine slightly to ensure that it produces the amount of electrical energy needed, even though the turbine is not as efficient.

Betz's Law and the Efficiency of Energy Conversion for Wind Turbines

In discussing the conversion of wind energy for a wind turbine, you need to know about the work of Albert Betz, a German physicist. In 1919 Betz determined that the best efficiency a wind turbine can achieve is approximately 59%. He found that if it were possible for the efficiency to increase beyond 59%, more of the energy in the wind would be removed, and basically the wind would not have enough energy to continue to blow. If this occurred, the wind would stop blowing in the area around the wind turbine blade, which would create a calm area. In simplest terms, this means the wind turbine blades can remove the maximum amount of energy when the wind turbine is around 59% efficient and still leave enough energy in the wind to allow it to continue to blow. Betz's conclusion is now known as **Betz's law**.

Efficiency of the Gearbox and Generator

Another part of determining the efficiency of a wind turbine is understanding the efficiency of the mechanical system that includes the rotor, gearbox, shafts, and generator. Each of these components experiences mechanical losses that affect its efficiency. When these are combined with the limits of the wind turbine blades to harvest wind and convert it into mechanical energy, the overall efficiency of the wind turbine decreases. Thus the overall efficiency of the wind turbine can be improved by making the individual components more efficient. On larger systems that generate 2 MW and up, a small increase in efficiency will produce a fairly large increase in the amount of energy produced.

4.7 TESTING WIND TURBINES

When a wind turbine is designed, it must be tested to verify the amount of electrical power it can produce at every given wind speed. Its cut-in wind speed must also be determined and verified so that it can be published in the turbine data sheets. Some components of the wind turbine, such as the blades, rotor, gearbox, and generator, can be tested separately in a laboratory. This allows the engineers and technicians to determine the efficiency of the individual components before they are used as a complete system.

A wind turbine generator can be tested by connecting it to a dynamometer or a load bank, which allows the output power from the generator to be used. When the output load is varied as in real conditions, the generator will show the efficiency it will have in converting mechanical to electrical energy. The load on a dynamometer can be designed to run at a specific constant value over a specific period of time, or it can increase and decrease its load to show the affects of changing load conditions.

Using Finite-Element Analysis to Test Blade Designs

Today it is possible to put the complex design of the turbine blade into a computer program that can provide an analysis of the efficiency of the blade design in harvesting wind energy. This type of program can also determine the weaknesses in the turbine blade, or simulate a failure that might destroy the blade. The computer program can put specified forces and torques on the turbine blade in increasing and decreasing loads to show the effects of changing wind speeds and how the blades will respond to them. It is also possible to make changes to the design of the blade and compare the efficiency of the blade and failure points.

This type of computer program uses a technique called **finite-element analysis** (FEA), and it employs complex mathematical algorithms to ensure its accuracy. As computers have become more complex, with bigger memories and faster calculating speeds, it has become possible to get very accurate results and to provide multiple tests before building blades for the turbine. Over time, FEA programs have become more and more accurate and have made the testing process more reliable. When combined with actual data from the field, FEA makes it easy to predict the efficiency of additional wind turbines of similar size.

Using SCADA to Verify Actual Operating Data Against Predictions

Today it is possible to collect a wide variety of data on any operating wind turbine with a **SCADA (Supervisory Control and Data Acquisition)** system. The data from the SCADA system provides information such as wind velocity, wind direction, and the amount of electrical power that is being produced. When this data is gathered over a long period of time and stored in a computer databank, it can be analyzed so that very accurate projections can be developed.

This method of projecting wind turbine efficiency is very accurate, especially when multiple wind turbines are erected on a similar site, such as on a wind farm. SCADA has allowed projections and calculations on wind turbine efficiencies to become more and more accurate.

4.8 PROBLEMS WITH WIND GENERATOR TURBULENCE

Turbulence is loosely defined as wind that does not blow in a straight line. It is also known as swirling wind, and when it moves across the turbine blade, it does not provide the same energy as straight-line winds. In some locations, turbulence from the wind occurs a small percentage of time. This is why some locations are not suitable for the installation of wind turbines. Wind turbulence can be caused by buildings, trees, and other obstructions along the direction the wind blows. This type of turbulence may be natural or it may be caused by man-made objects and buildings.

The amount of turbulence in these locations may change from season to season, such as when trees have leaves and when the leaves have fallen from the trees. Other changes in turbulence may occur from season to season as wind direction changes with the weather. It is possible to identify all of these types of turbulence for a given location, and a decision can be made whether a wind turbine should be installed in this location.

Another type of wind turbulence may be caused by the wind turbine itself as air moves over its blades and the blades begin to rotate. If a wind turbine is located near a house or building, the turbulence may create noise and vibrations on surrounding facilities. If the noise is excessive, the effect on humans may become intolerable, and changes may have to be made to the wind turbine or the wind turbine removed altogether.

When multiple wind turbines are located on the same site, this phenomenon may become amplified and cause greater problems. The turbulence caused by one wind turbine may create a condition such that the wind moves on to the blades of the next turbine and affects the efficiency of the second turbine. This is especially problematic when wind farms have many wind turbines, sometimes over a hundred turbines, installed and the wind passes from one to the next.

New designs and testing of wind turbine blades can identify these turbulence problems and limit their effects. Design changes have been made to turbine blades to minimize the amount of wind turbulence that they cause. These changes may make the wind turbine blade less efficient, but the trade-off is to provide an environment that is not affected by the turbulence or noise of the wind turbine.

4.9 HOW TO DETERMINE WIND GENERATOR PEAK PERFORMANCE

There are several ways to measure the peak performance of a wind generator. **Wind turbine peak performance** occurs when the output of the wind turbine generator is at or above its rated output. One way to measure peak performance is to use a graph of a power curve. A **power curve** is a graph that shows the wind speed and the output power of the wind turbine over a range of wind speeds from zero to the maximum wind speed for which the wind turbine is designed. Figure 4-3 shows a typical table and Figure 4-4 shows a graph of a power curve for a wind turbine. On this graph the wind speed is shown across the bottom of the graph on the horizontal axis from 2 to 21 m/s. (Note that the chart shows the wind speed in increments of 3 m/s.) The output of the generator is shown on the vertical axis on the left side, and it indicates power in kilowatts from 0 to 70 kW. According to this power curve, the wind turbine produces its maximum output of 63–65 kW when the wind is between 15 and 19 m/s.

A second way to measure peak performance includes consideration of the air density. Density is a measure

Generator Output vs. Wind Speed

Wind Speed (m/s)	Wind Speed (mph)	Power (kW)
2	4.5	0
3	6.7	0
4	8.8	0.7
5	11.2	2.2
6	13.4	6.5
7	15.6	13.8
8	17.9	22.7
9	20.1	32.1
10	22.4	41.3
11	24.6	46.5
12	26.8	52.8
13	29.1	58.7
14	31.3	61.7
15	33.5	62.9
16	35.8	63.8
17	38	64.7
18	40.2	63.5
19	42.5	62.6
20	44.7	61.7
21	46.9	61.2

FIGURE 4-3 Wind speed and generator output data for calculating a power curve for a wind turbine.

of how much mass is contained in a given unit volume (density = mass/volume). **Air density** is a measure of how much mass is contained in 1 ft^3 of air. The amount of moisture (humidity) in the air, the air pressure, and the temperature all affect the air density. Air becomes more dense when it is cooled, when its pressure increases, and when it is higher in humidity.

When the air is dense, the output of the wind turbine increases. Air density is not included in some performance measurements, and in other measurements the test is conducted in the most dense air, so the output is slightly higher than if it were measured in less dense air.

It is important to understand that the wind curve is different fo each wind turbine. Variables that will affect the power output include the diameter of the turbine blades and the height at which the turbine is mounted above the ground. Once the power curve has been established, it is easy to determine the total output of the wind turbine as the wind varies. If you have the data table for the power curve, you can enter the values into a spreadsheet and use the spreadsheet to create a graph.

Use of Wind Power Curves in Advertising

The wind curve indicates how much power a wind turbine should produce at any given wind speed. The maximum value from the wind power curve may be used in marketing wind turbines, and for comparisons between competing models, so the values are sometimes higher than the actual output. If you are using power curves as part of a purchasing decision, you may want to request actual SCADA data from several installed wind turbines. Some manufacturers use projections and calculations to determine the values on the wind power curve.

It is also important to remember that rated power is not necessarily peak power. Peak power is the amount of electrical power the wind turbine can produce at the highest rated wind speed. This number is not particularly valuable, however, because wind speeds that high do not occur on a continuing basis, and the wind turbine generator and gearbox may not be rated for that exceptional load. If you are using a wind power curve to help determine a payback for a particular wind turbine, you must remember that you will need to use an average wind speed rather than the maximum for ongoing operation.

Testing by Independent Laboratories

Power curve testing should be done by independent testing laboratories as well as by manufacturers. If the data is

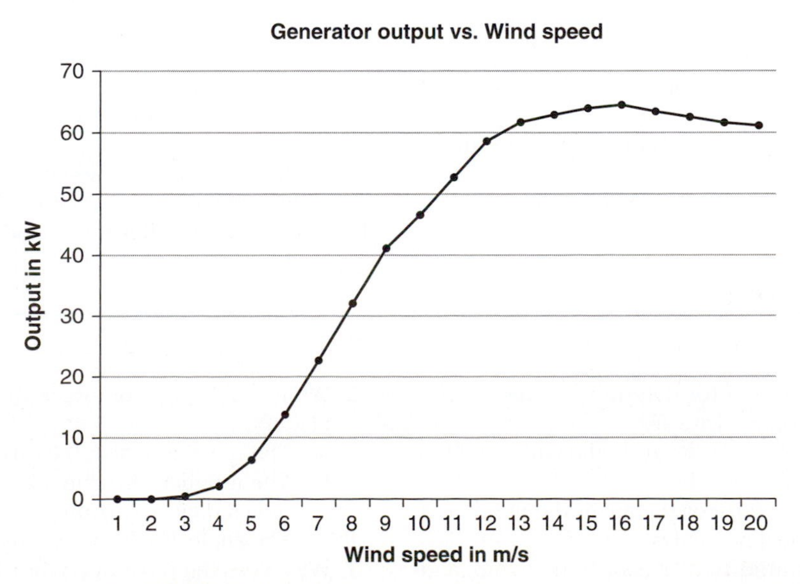

FIGURE 4-4 Graph for a power curve of a 100-kW wind turbine.

to be used to compare different wind turbines, the same set of testing criteria must be used for all the models. One agency that performs testing is the National Renewable Energy Laboratory (NREL) of the U.S. Department of Energy (DOE).

The NREL provides tests in conjunction with other agencies such as the National Aeronautics and Space Administration (NASA). NASA has a large wind tunnel located at Moffet Field in Silicon Valley, California. A wind tunnel is a testing laboratory designed specifically to create large wind flows under controlled conditions. Originally, wind tunnels were used to test aircraft wings and aircraft stability under operating conditions. The wind tunnel is a large dome that is filled with compressed air, and then the air is slowly released to pass through a tunnel where the wind turbine blades and rotor are positioned. The release of the air into the wind tunnel creates a realistic simulation of wind blowing at the height and level where a wind turbine would normally be located on top of its tower. Because the wind turbine is mounted near the ground in the tunnel, it can easily have a large number of sensors connected to it to provide vast amount of data when it is under load. The wind tunnel at Moffett Field is 80 ft × 120 ft in area, and it can produce low- and medium-wind speeds to test wind turbines. The data from these tests is very important because it is collected by a scientific laboratory that is entirely independent for any manufacturer.

Another independent testing center is the National Wind Technology Center near Boulder, Colorado, which is operated by the U.S. Department of Energy. This testing center provides a unique setting because naturally occurring winds of up to 70 mph (31 m/s) are used for research and testing. These high winds not only allow the turbines to be tested for maximum output under realistic conditions, they also allow testing of safety systems used to stop the blades during extremely high wind conditions.

In 2008 the NREL begin testing small wind turbines with outputs of less than 100 kW. Because mostly small companies manufacture these wind turbines, comprehensive testing is not cost-effective and it had been difficult to get data from independent sources. The NREL has established test parameters so that data for different wind turbines can be compared.

Measuring Generator Output

Another major part of determining the performance of a wind turbine is to measure the efficiency of the generator. The output of the electrical generator changes with the speed at which the generator shaft rotates. You will learn more about different types of generators and their individual characteristics in Chapter 6. You will find that different types of generators have different efficiency ratings. The field current of some generators can be adjusted so that their efficiency remains fairly constant over a range of shaft speeds.

Because the generator efficiency varies, it is possible that the blades of the wind turbine are harvesting the maximum amount of wind energy and converting it to mechanical energy, but the speed of the rotation of the rotor is not at the generator's most efficient point. This will make the overall efficiency of the wind turbine less than its rating. For this reason, the actual peak output may occur at several different wind speeds when the turbine blades reach their maximum and again when the generator reaches its maximum.

Questions

1. What is a wind curve, and what is it used for?
2. Why is constant rotational speed of the generator shaft important for a wind turbine?
3. Explain why pitch control for a wind turbine blade is useful on larger wind turbines.
4. What is an inverter, and how does it allow a wind turbine to rotate at any speed?
5. Explain what setpoint, feedback, and error are for a closed-loop system.
6. Explain the difference between active stall control and passive stall control.
7. Explain why some smaller wind turbines do not need rotational speed control.
8. What is the effect on a wind turbine when the air becomes more dense?
9. Explain why Betz's law states that the best efficiency a wind turbine can achieve is about 59%.
10. Explain what stall is for a wind turbine blade.

Multiple Choice

1. The three parts of the formula for determining the amount of power a wind turbine can create are
   a. Wind velocity, the density of the air in the wind, and the diameter of the turbine blades
   b. Wind direction, the density of the air in the wind, and the diameter of the turbine blades
   c. Wind velocity, the humidity of the air in the wind, and the diameter of the turbine blades
2. What is the angle of attack in reference to a wind turbine blade?
   a. The speed at which the turbine blades are turning
   b. The number of turbine blades divided by the speed at which they are turning
   c. The angle at which the wind strikes the blade
3. Why does the pitch of a wind turbine blade need to be adjusted on some wind turbines?

   a. Changing the pitch can make the turbine blades slow down when the wind is very strong.
   b. Changing the pitch can make the blades catch more wind when the wind speed is slower.
   c. Changing the pitch can allow the wind turbine to run at its optimum speed at all times.
   d. All of the above
4. What does SCADA stand for?
   a. Supervisory Control and Data Acquisition
   b. Service Center and Data Acquisition
   c. Supervisory Control and Data Applications
   d. All the above
5. What factors affect the amount of energy in the wind?
   a. Wind speed
   b. Wind density
   c. Wind direction
   d. All of the above
   e. Only a and b
6. Betz's law and the efficiency of energy conversion for a wind turbine indicates that the maximum amount of wind a wind turbine can harvest is
   a. Approximately 50%
   b. Approximately 59%
   c. Approximately 75%
   d. Approximately 99%
7. What is the difference between active and passive stall control of wind turbines?
   a. Active stall control uses the design of the blades to create the stall condition, and passive stall control uses a hydraulic or mechanical system to change the pitch to cause the stall condition.
   b. Active stall control uses a hydraulic or mechanical system to change the pitch, and passive stall control uses the design conditions of the blade to create the stall condition.
   c. Both active and passive stall control use hydraulic or mechanical systems to cause the stall condition.
8. What is a wind curve?
   a. A wind curve shows the wind speed in miles per hour or meters per second and the amount of electrical power that is produced at that speed.
   b. A wind curve shows the speed of the blades and the frequency for the output at any speed.
   c. A wind curve shows the torque on the blades at any given wind speed.
9. Why is it important for the generator shaft for some wind turbines to run at a specific speed?
   a. The specified speed is necessary so that the voltage generated by the wind turbine is at exactly 60 Hz.
   b. The specified speed is important to ensure that the turbine runs at near-optimum specifications.
   c. The specified speed is needed to keep the stress on the blades constant.
10. What do NREL and DOE stand for?
    a. National Electrical Lab and U.S. Department of Energy
    b. National Renewable Energy Laboratory and U.S. Department of Electricity
    c. National Renewable Energy Laboratory and U.S. Department of Energy

CHAPTER 5

Basic Parts of Horizontal-Axis Wind Turbines

OBJECTIVES

After reading this chapter, you will be able to:

- Identify the basic parts of a horizontal-axis wind turbine.
- Describe the four basic parts of a programmable logic controller (PLC).
- Explain the function of a PLC input module.
- Explain the function of a PLC output module.
- Identify the basic parts of a hydraulic system.
- Explain the operation of a directional control device.
- Explain the operation of a flow-control device.
- Explain the operation of a pressure-control device.
- Identify the different types of brakes that are found on wind turbines.
- Explain how the brakes in a wind turbine are activated.
- Identify the parts of closed-loop system diagram.

KEY TERMS

Analog control
Blade pitch
Blade tip speed
Cartridge valve
Check valve
Closed loop
Directional control
EPROM
Error
Feedback
Flow control
Hydraulic control
Hydraulic pump
Hydraulic reservoir
Hydraulic servo control valve
Image register
Input module
Ladder logic
Load efficiency
Open loop
Output module
Pilot pressure
Pressure control
Process variable (PV)
Program mode
Programmable logic controller (PLC)
Proportional valve
Rotor
Rotor hub
Run mode
Sensor
Setpoint
Summing junction
Supervisory control and data acquisition (SCADA)
Valve spool
Vane pump
Yaw drive

OVERVIEW

In this chapter you will learn more about the basic parts of the horizontal-axis wind turbine. You will learn more about the blades, the road or hub, and the nacelle, which bring wind energy into the wind turbine. You will also learn about the yaw mechanism that positions the nacelle into or out of the wind direction so the blades can harvest the maximum amount of wind energy.

You will also learn about the programmable logic controller (PLC) and see how it controls on/off inputs and outputs, as well as variable inputs and outputs called analog control. Analog control is a signal that can be varied between 0 and 100% and causes the system to respond between 0 and 100% or, if the signal is from a **sensor**, it can measure any variable value from 0 to 100%. You will also see how the PLC is used for data monitoring and data acquisition for the wind turbine. The PLC is also used to optimize the blades and the rotor speed to ensure that the wind turbine runs as efficiently as possible, while protecting the blades, rotor hub, low-speed shaft, gearbox, high-speed shaft, and generator.

The last part of the chapter introduces hydraulic systems and components that are used to control various parts of the horizontal-axis wind turbine. You also learn about hydraulic proportional control and hydraulic servo control, which is used to provide variable control for adjusting blade pitch, yaw direction, and brakes.

5.1 OVERVIEW OF THE HORIZONTAL-AXIS WIND TURBINE

Figure 5-1 shows a two bladed horizontal-axis wind turbine with all of its covers removed. The blades are not attached. This wind turbine is an 80-kW unit that is approximately 10 ft long. In this figure you can see that the blades will be connected with bolts to the blade **rotor** that extend out parallel on the left and right sides of the front of the wind turbine. The hub is connected to the gearbox through a low-speed shaft. The gearbox output shaft is connected directly to the generator input. All of these components are mounted on a nacelle bedplate that is then topped with a cover. A hydraulic system is mounted directly to the rear section of the wind turbine, right behind the generator. The hydraulic system provides control of the rotor brakes and the pitch of the blades. On larger wind turbines, a cooling system may also be mounted in this area to cool the generator and gearbox if needed. Sensors such as an anemometer and wind direction instruments are mounted on the top of the nacelle at the rear to collect data about the wind direction and wind speed. This information is fed into the controller and is used to optimize the wind turbine's performance.

The nacelle bedplate, with all the components mounted on it, is mounted to the top of the tower. A yaw mechanism is bolted to the bottom of the nacelle bedplate, which provides a means to adjust the direction of the entire nacelle as it sets on top of the pole. Changing the direction of the nacelle is called yaw control and is generally accomplished by a large electric or hydraulic motor.

Blade pitch hydraulic cylinder
Blade rotor and connection for blades
Hydraulic control system
Generator
Gearbox

FIGURE 5-1 Basic parts of a horizontal-axis wind turbine.

5.2 THE NACELLE AND THE NACELLE BEDPLATE

Figure 5-2 shows a nacelle mounted on a pole, with the top of the nacelle open so workers can inspect the basic parts. In this figure you can see the rotor on the front of the nacelle, which is connected through the low-speed shaft to the gearbox. The generator, which is partly concealed, is connected to the output of the gearbox by a high-speed shaft. All of these parts are mounted on the nacelle bedplate, and they are all completely covered by the nacelle cover. The yaw mechanism is mounted to the bottom of the nacelle bedplate.

FIGURE 5-2 The top of the nacelle is open so a technician can check the gearbox and generator. (Courtesy of Siemens Wind Energy.)

FIGURE 5-3 A completely assembled nacelle, ready to be installed on the top of its tower. (Courtesy of John Fellhauer jfellhauer@surenergy.us.)

Figure 5-3 shows a completely assembled nacelle that is ready to be lifted to the top of its tower. The yaw mechanism at the bottom of the nacelle bedplate provides a means to bolt the nacelle to the tower. Once the nacelle is secured to the top of the tower on this unit, its blades can be individually attached to its rotor. If the wind turbine ever needs to be completely overhauled, the nacelle cover can be opened or removed and the parts on the nacelle bedplate can be removed individually, or the entire nacelle bedplate can be removed and replaced with all the parts already mounted in place on the bedplate. Figure 5-4 shows a nacelle being hoisted to the top of its tower with a large crane. When the nacelle is mounted on the top of its tower, its blades can be attached.

FIGURE 5-4 Nacelle being hoisted with a crane to the top of its tower. (Courtesy of Northern Power Systems Inc.)

FIGURE 5-5 The rotor hub. The blades are bolted to this hub, which is then bolted to the shaft protruding through the front of the nacelle. (Courtesy of Northern Power Systems Inc.)

5.3 ROTOR HUBS AND BLADE TYPES

The **rotor hub** of a horizontal-axis wind turbine is actually part of a rotor assembly. The rotor hub is shown in Figure 5-5. Figure 5-6 shows the blades after they have been bolted to the rotor hub. This completed blade assembly with the blades connected to the hub will be hoisted into the air as an assembly, where it will be bolted as a complete unit to the rotor plate that is part of the low-speed shaft. Figure 5-7 shows a nacelle being lifted into place; you can see the rotor plate mounted to the low-speed shaft that protrudes out the front of the nacelle. Figure 5-8 shows a crane that has started to hoist the blade assembly into place at the front of the nacelle. Figure 5-9 shows the workers ready to bolt the blades and rotor hub onto the

FIGURE 5-6 The blades have been bolted to the rotor hub, and the assembly is ready to be lifted to the nacelle. (Courtesy of Northern Power Systems Inc.)

FIGURE 5-7 The mounting plate for the rotor is shown connected to the low-speed shaft and protruding out the front of the nacelle. The rotor is bolted to this plate. (Courtesy of Northern Power Systems Inc.)

FIGURE 5-8 The complete blade assembly is hoisted to the nacelle that is already mounted on the top of the tower. After it is positioned correctly, the blade assembly is then bolted to the rotor plate. (Courtesy of Northern Power Systems Inc.)

FIGURE 5-9 The blade assembly is shown bolted to the rotor plate. (Courtesy of Northern Power Systems Inc.)

plate that is on the front of the low-speed shaft that extends out the front of the nacelle. The workers align the holes in the blade assembly with the holes in the rotor plate and place large bolts into the holes. Washers are placed on the bolts and then nuts are screwed onto the bolts to secure the blade assembly to the rotor.

For mid-sized wind turbines, the blade assembly may be bolted onto the nacelle while the turbine is on the ground, and then the entire nacelle and blades are lifted into place with large cranes. For some smaller systems, the rotor hub is mounted to the nacelle while it is on the ground. In these applications, the nacelle is attached to the tower while it is on the ground. Next the nacelle is positioned so the blades can be attached to the rotor hub one at a time while the nacelle and the tower are still on the ground. After the blades have been attached to the rotor, the nacelle, blades, and tower are all lifted into place as a unit with a large crane.

On some of the largest wind turbines, the blades are so large they have to be bolted to the rotor hub one at a time after the nacelle is placed on top of the pole. In these types of applications the nacelle is lifted to the top of the tower and bolted into place. The blades are so heavy that they have to be lifted one at a time to the rotor hub, where technicians bolt the blades to the rotor. This type of installation is very complex because of the weight of each blade and the fact that each blade must be connected to the hub after the nacelle is place on the pole high above the ground.

5.4 NUMBER OF BLADES

Horizontal-axis wind turbines may have one, two, three, four, or five blades. The more blades the wind turbine has, the more wind energy it can harvest when the blades are turning at the same speed. If a one- or two-bladed wind turbine is used, it must turn at a faster rpm than a three-bladed wind turbine to harvest the same amount of wind energy. Figure 5-10 shows a one-bladed wind

FIGURE 5-10 A single-bladed wind turbine. The smaller arms are counterweights. (Courtesy of Powerhouse Wind Ltd.)

FIGURE 5-11 Two-bladed wind turbines. (Photo courtesy of U.S. Department of Energy.)

turbine. It looks as though it has three blades, but the smaller arms are actually counterweights and do not harvest wind as a blade does. The counterweights are needed to balance the rotor as it rotates. Figure 5-11 shows a number of two-bladed wind turbines. In this type of wind turbine, the two blades are used to balance each other, and both blades harvest the wind.

Figure 5-12 shows a three-bladed wind turbine. This type of wind turbine can harvest more wind energy than a two- or one-bladed wind turbine when spinning at the same speed. If a one-bladed or two-bladed wind turbine is to harvest the same amount of energy as a three-bladed wind turbine, the two-bladed wind turbine must spin much faster than the three-bladed wind turbine, and the one-bladed wind turbine must spin even faster than the two-bladed wind turbine. These higher speeds mean that the bearings on the rotors of one-bladed and two-bladed wind turbines tend to wear out more quickly than the bearings on three-bladed wind turbines.

FIGURE 5-12 Three-bladed wind turbines. (Courtesy of Vestas Wind Systems A/S.)

FIGURE 5-13 A five-bladed wind turbine.

Figure 5-13 shows a five-bladed wind turbine that has a special cowling mounted around it to funnel the wind directly through the wind turbine blades. This type of wind turbine can harvest more wind than units with fewer blades. The major drawbacks of having more turbine blades is that each extra blade adds to the cost, and the towers for wind turbines with more blades must be much stronger to hold the wind turbine steady in the strongest winds.

5.5 ROTOR BLADE PITCH ADJUSTMENTS AND TEETERING

Several types of adjustments can be made to the rotor blades of a horizontal-axis wind turbine. The pitch of the blade can be adjusted so that the blade can harvest more wind, and on some smaller two-bladed wind turbines, the blades can teeter, which relieves stress on the blades in higher winds. **Blade pitch** is the rotation of the wind turbine blade from 0 to 90 degrees on its axis where it is attached at its base to the rotor.

The pitch of the rotor blades allows the blade to be adjusted (rotated at its base) so that the angle of attack between the blade and the wind direction is maximized, so the blade can harvest the maximum amount of wind energy; or the blades can be adjusted so that the blades are turned to the zero point, where the blades do not capture any wind. Blades in this position are called *feathered*, where the front edge, rather than the side, of the blade is pointed directly into the wind. Figure 5-14 shows the blades of a wind turbine feathered so that it will not turn during a maintenance procedure. Figure 5-15 shows a wind turbine with its blades rotated nearly flat to the

FIGURE 5-14 Wind turbine blades feathered so the blades will not turn during maintenance. Notice that the edge of the blade is facing into the wind.

FIGURE 5-15 Wind turbine blades rotated flat so that they harvest the maximum amount of wind energy.

wind, so they will harvest the maximum amount of wind. On larger wind turbines, the blades can be pitched to any position between completely flat and fully feathered. The hydraulic system adjusts the blade pitch for the wind conditions to ensure that the blades harvest the maximum amount of wind energy. The blade pitch can also be adjusted to protect the wind turbine blades from rotating too fast if the wind speed becomes too high.

Blades on two- and three-bladed wind turbines can all be adjusted by the same amount at the same time, or, on other, more complex wind turbines, each blade can be adjusted individually to make the wind turbine more efficient. Blade pitch can be adjusted to make the wind turbine harvest more wind at very low speeds, which allows the wind turbine to begin turning its blades at lower wind speeds than when the blades are fixed. You will learn later how hydraulic and mechanical systems are used to adjust the turbine blades.

Another adjustment to blades that can be made to ensure they can rotate safely is called teetering. When the wind blows at its strongest, the tips of the blades of a wind turbine may bend into the nacelle to the extent that they nearly touch the tower. If the wind turbine has just two blades, this action may be amplified so much that the blades come into contact with the tower and therefore may be damaged during the highest winds. A teeter hub allows the blades to flex when the wind blows at this strong level. The teeter hub lets the wind flex (teeter) the top of the blade back toward the direction of the tower, which draws the bottom of the blade away from the tower so it has plenty of room to clear the tower. The stronger the wind, the more the blade teeters, which ensures plenty of space between the tip of the blade and the tower. Teetering is actually limited to a small range (approximately plus or minus 2 degrees), but that is sufficient to ensure that the blade tips never come close to the tower as they pass the pole during the bottom part of their sweep when the wind is blowing at its strongest.

5.6 ROTATIONAL SPEED AND ROTOR SPEED CONTROL

The rotational speed of the blades of all horizontal wind turbines must be controlled at both the high-speed end and the low-speed end. Control is needed at the high-speed end to keep the blades from becoming damaged by rotating too fast. This can occur when the blade is as long as 125 ft and is turning at near-maximum speed of 18 to 22 rpm: The tip speed may exceed 260 ft/s, which is equal to more than 175 mph. If the blade is longer or the rpm is faster, the tip speed may exceed 200 mph. At this speed, the blade can be damaged by prolonged stress from the torque that the wind applies when it makes the blade twist, and by the centrifugal forces at the extreme tip of the blade at high speeds. On larger wind turbines, the maximum speed is controlled by feathering the blades to begin to make them stall at higher speeds, which unloads them and slows their rotational speed.

The maximum speed can also be controlled by changing the yaw position to move the nacelle slightly out of the wind. When the yaw control has the blades pointed directly into the wind, the blades will rotate at their fastest rate; when the yaw control moves the nacelle so that the blades are not pointed directly into the wind, blade speed is not able to reach its maximum.

Some larger wind turbines have a program in their controller that allows the blades to accept higher wind speeds during wind gusts. The high wind speed during gusts can be converted to energy by allowing the blades to overspeed for a short period of time. When this occurs, the excess wind energy is converted to electrical energy and is removed from the generator for a short period of time, without damaging the blades, shafts, or gearbox.

Even smaller wind turbines can suffer blade or rotor damage if the blades are allowed to rotate too fast. Each turbine has a maximum rated design speed that should never be exceeded. Wind turbines have a number of ways to control the maximum speed. On smaller wind turbines, speed control is managed by pitch control or by stall control. Stall control occurs when the angle of the blades is fixed at a point where the blade no longer harvests the wind energy, because the angle of attack of the wind causes the blade to allow wind to move past the blade but not convert the wind energy into rotational energy. On smaller wind turbines with mechanical systems, the blade position control has several fixed positions to which it can move the blade. The blade moves between the fixed positions in small increments until the speed of the turbine slows or speeds up enough, depending on the need. When the turbine is trying to start at very low wind speeds, the blades increment to a position where the maximum amount of energy can be harvested at the slow wind speed. If the wind speed increases to an unsafe level, the blade position indexes back to where the blade begins to cause a stall condition, and the turbine blades begin to slow down.

On the smallest wind turbines, a governor-type device is used to control the maximum speed. When the speed increases, the governor weights swing outward and put more pressure on the blade rotor. If the speed becomes too high, the weight of the governor increases the braking action and slows the turbine blades to a safer speed.

5.7 INDIVIDUAL PITCH CONTROL OF THE BLADES

On larger wind turbines, the control system is so sophisticated that it can adjust the pitch of each blade individually to provide optimal load and efficiency to the wind turbine. A hydraulic system uses several valves and a closed-loop control algorithm to ensure that the pitch of the blade is always optimized to harvest the maximum amount wind. This system has individual sensors on each blade, as well as sensors that can measure the efficiency of electricity generation and also the direction and velocity of the wind. As the wind conditions, direction, and velocity constantly changes, the sensors detect the slight changes and make modifying changes to the pitch of the blades. Each blade may be pitched at a slightly different angle to ensure that the blades are supplying the optimum amount of energy. Previously the technology was not as complex, and all the blades were pitched the same amount at the same time.

5.8 CONTROL OF BLADES FOR LOAD EFFICIENCY

The blades of a wind turbine can control the load efficiency of the wind energy that is converted to electricity. **Load efficiency** is a measure of the rate at which the wind turbine blade can harvest wind energy and of the ability of the generator to convert the torque to electricity. Load efficiency is measured in percent, from 0 to 100%. Load efficiency is changed by adjusting the pitch of the blades to ensure that the maximum amount of wind energy is harvested and converted to electricity through the generator. When the wind is light, the blades can be adjusted to ensure that they are turning fast enough to generate electricity continually. If the wind speed is too slow, the blades turn slowly and the controller de-energizes the generator, thus allowing the blades to continue to rotate slowly. The controller can adjust the blade pitch so that the blades rotate even in light winds, so the generator can re-energize when the blade speed increases enough to produce electrical power.

When the wind is strongest, the controller can adjust the blade pitch to ensure that the turbine is harvesting the maximum amount of wind energy, while slowing the blade rotation enough to keep it under control. The controller continually samples the output of the generator and compares it to the wind speed and the blade rotation speed, then continually adjusts the blade pitch to ensure that the wind turbine is running at its optimum conditions. This type of control allows newer wind turbines to be more efficient than older models that do not have blade pitch control.

5.9 YAW MECHANISM

The yaw mechanism on a wind turbine rotates the nacelle to control the direction in which the nacelle and blades are pointing. The **yaw drive** is generally moved by high-torque electric motors, but hydraulic motors may be used. The nacelle can be rotated so that the blades are directly into the wind when the turbine is trying to harvest the largest amounts of wind, and it can move the blades out of the wind when the winds are too strong and could damage the wind turbine. The yaw mechanism is a large ring that is mounted on the bottom of the nacelle. The yaw ring has gear teeth that mesh with the yaw drive motor. When the drive motor shaft rotates, the yaw ring rotates and makes the nacelle swivel or rotate. Figure 5-16 shows a typical yaw mechanism on the bottom of a nacelle. The yaw control can use up to seven yaw motors

FIGURE 5-16 Yaw ring gear and drive motor on the bottom of the nacelle. The gear on the drive motor engages the yaw ring and makes the nacelle rotate on the pole. (Courtesy of Northern Power Systems Inc.)

to move the yaw ring on the largest wind turbines, because the nacelle and blades on these turbines are so large. The wind turbine in Figure 5-16 is smaller, so it needs only one yaw motor to rotate the nacelle.

On some smaller downwind vertical-axis wind turbines, the wind itself adjusts the yaw so that the blades remain directed into the wind. This type of wind turbine has a large fin on the nacelle that causes the wind to move the nacelle so that it is always oriented into the wind. This type of wind turbine uses a yaw counter to make sure the yaw makes only four to six complete revolutions in any direction, before it is forced back in the opposite direction. If the yaw was allowed to rotate continually in one direction, it would cause the large electrical cables to twist to the point where they would break. Because the electrical cables carry power from the generator, which is mounted on the yaw bedplate, which rotates, to the bottom of the tower, which does not rotate, the cables could become twisted to the point that they would break off or at least become stressed. The yaw counter ensures that the yaw is only allowed to rotate in one direction for four to six turns, after which the yaw mechanism drives the yaw back in the opposite direction to unwind the cables.

5.10 YAW DRIVES

The yaw drive motor shown in Figure 5-16 can be powered by an electric or a hydraulic motor. If the yaw drive motor is electric, it will be geared down to a very low speed that can produce a very high torque. Permanent-magnet motors work well as yaw drives, because they produce very high torque at low speed. The yaw ring also has a gear ratio that allows the yaw motor to move the large load, which includes all the parts in the nacelle that are connected to the nacelle bedplate. Because the yaw motor runs at high torque and low speed, it can only move the nacelle a small amount at a time.

5.11 YAW CONTROL

The yaw drive motors are controlled by the PLC or computer controller for the wind turbine. The controller takes in data from the anemometer, the sensor that measures the wind speed and direction. This information is fed into a closed-loop control system, which determines how much the drive should adjust the nacelle and in which direction to ensure the generator is receiving as much power as possible from the blades.

Figure 5-17 shows two large wind turbines with their blades pointing in slightly different directions. This difference in direction is due to the variation of wind direction and the way the control system in each turbine is adjusting the pitch of the blades and the direction of the blades through the yaw control to maximize the wind energy conversion to the generator. These two wind turbines are located within a thousand feet of each other, and the wind direction and speed is varying enough at this distance to cause the yaw mechanism control on each wind turbine to move it to a slightly different position. On systems in which a PLC or computer is controlling the yaw position, it also checks for the number of rotations the

FIGURE 5-17 These two wind turbines are located close to each other, but their yaw mechanisms are pointing the blades of the two turbines in slightly different directions.

nacelle has made, and ensures that the nacelle does not wind up the cables from the generator too tight. If the nacelle has rotated too many turns in one direction, the computer control will reverse the rotation several times in order to put slack back into the cable from the generator.

If the wind turbine is a smaller type, it may adjust automatically to variations in the wind direction, but on larger machines the nacelle position may be adjusted to keep the blades pointing in the direction the wind is coming from so that the blades can harvest the maximum amount of energy from the wind. If the wind direction is variable, the nacelle continually adjusts so that the blades are always facing the optimum direction.

5.12 YAW DRIVE BRAKES

There are times when the position of the nacelle needs to be locked, so it cannot move when the wind changes direction. At these times a yaw drive brake is applied on the yaw ring. The brake is usually a disc-type brake that is similar to the disc brakes on an automobile and consists of two pads mounted on either side of the yaw ring. Figure 5-18 shows a diagram of this type of brake. When the brakes are not set, the disc brake pads are not touching the yaw ring, and it is allowed to rotate freely. When the brakes are set, the hydraulic brake system applies hydraulic fluid to the cylinder, which causes the brake pads to move against the yaw ring and apply pressure. The hydraulic pressure to the brake cylinder is increased until the pads apply enough pressure to cause sufficient friction between the brake pads and the yaw ring to hold the yaw ring in position. The brake pads and the hydraulic braking system are sized large enough to hold the yaw ring in position in the strongest winds to which the wind turbine will be subjected. On some systems, usually smaller turbines, the hydraulic system keeps the brakes open and large springs are used to set the brakes. On most systems the nacelle brakes and the blade rotor brakes are both designed to be fail-safe, which means that if there is a loss of hydraulic pressure or some other problem, heavy springs cause the brakes to set. You will learn more about the hydraulic system later in this chapter.

5.13 DATA ACQUISITION AND COMMUNICATIONS

Wind turbines have a wide variety of needs for acquiring data, such as wind speed, direction, generator voltage and current, and other important information about power production and maintenance issues such as the number of hours the blades have been rotating. The PLC or computer controller for the wind turbine gets information from numerous sensors that provide wind direction, wind speed, generator voltage and current, nacelle direction, blade pitch, and other information that is used for control of the wind turbine. This same information can be archived and sent to a data acquisition system through a network link. Figure 5-19 shows an example of a data access point link that takes data from the wind turbine and sends it over a telephone, Internet, or Ethernet connection. All of this data is sent to a central location and archived so that it can be retrieved and analyzed at a later date. Over time, this data provides a continuous representation of how

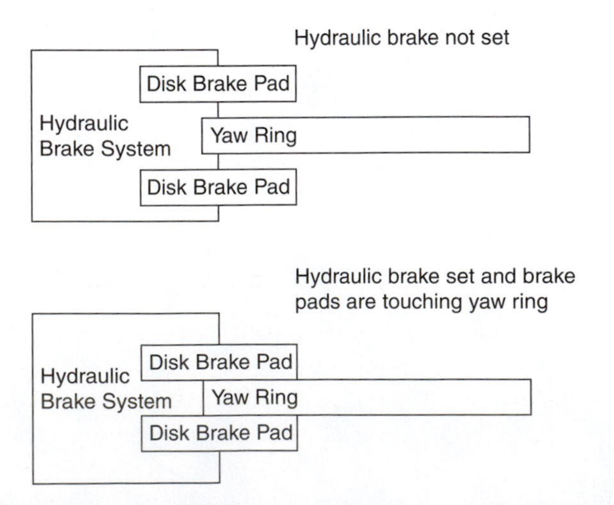

FIGURE 5-18 Yaw brakes shown in the not-set position (top) and in the set position (bottom), where they are applying pressure to the yaw ring. In some systems, the hydraulic system keeps the brakes open and large springs cause the brakes to set.

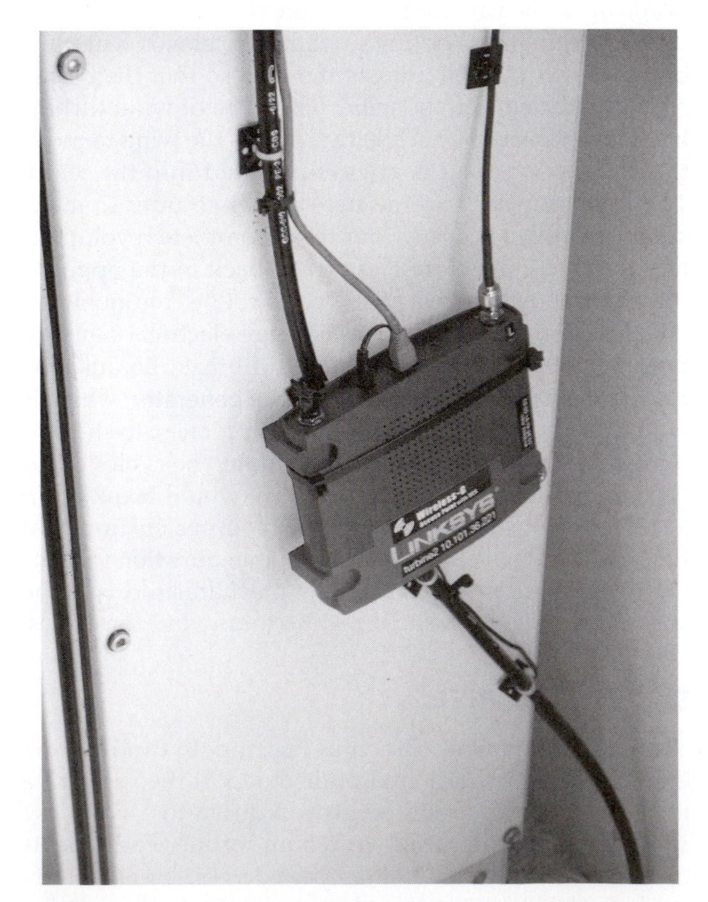

FIGURE 5-19 Data access point that sends data from the wind turbine through a telephone, Internet, or Ethernet connection to a data center.

well the wind turbine is operating and how much wind is available at various hours of the day on different days of the month. This information also shows which months produced the best winds and whether other wind turbines are located nearby.

Much of this data is also used to identify maintenance issues. For example, the total number of hours a generator produces electrical power, and the amount of voltage and current that is produced, can be tracked. This information is useful when it is combined with actual maintenance processes to determine the life of bearings, gears, shafts, and other mechanical parts of the wind turbine. The data that is collected from the wind turbine can be checked live as it occurs, or it can be pulled from the archive and used in a spreadsheet or graph to help make projections. Technicians can use this data to determine if there are problems with the system, by comparing the wind speed and direction to the amount of electrical power produced. If the amount of electrical power produced in a 20-mph wind is substantially less today than it was last month, it may be an indication that the wind generator has a problem.

5.14 THE ANEMOMETER AND WIND VANE

The anemometer is mounted on the top of the nacelle, usually near the back, where it can gather wind speed information. Because the anemometer is located at the same height as the blades, it is a useful instrument to indicate the amount of wind that is available to the blades at any given time. When the anemometer is connected to the PLC or computer controller for the wind turbine, it can sample wind speeds several times a second and provide an average once a minute, or more frequently. One type of inexpensive anemometer is a simple permanent-magnet DC generator, which uses spinning cups mounted on a shaft as a propeller. When the wind blows through the cups, they rotate, which causes the shaft of the DC generator to turn. The faster the wind blows, the faster the cups spin and turn the shaft of the small generator. When the shaft of the DC generator turns, it produces a small amount of DC voltage. For example, the voltage can range from 0 to 12 V.

Another type of anemometer uses spinning cups to catch and sample the wind, but instead of turning a small generator, this type of anemometer uses small magnetic sensors called a reed switch, which create a small electrical pulse as they move past a metal tooth gear mounted near the reed switch. Each time the magnetic field from the reed switch passes over the metal tooth, a small electrical pulse is produced. A frequency counter circuit accepts the pulsed signal and converts it to wind speed data. The larger the number of pulses in the signal, the higher is the wind speed. The anemometer is calibrated to be accurate within 1–2% or better so that it can provide accurate wind speed data. Figure 5-20 shows a cup-type anemometer and wind direction instrument and its electronic circuit that converts the wind speed signal to a 4- to 20-mA loop. Some anemometers send their information as a 0- to 10-V or 0- to 12-V signal.

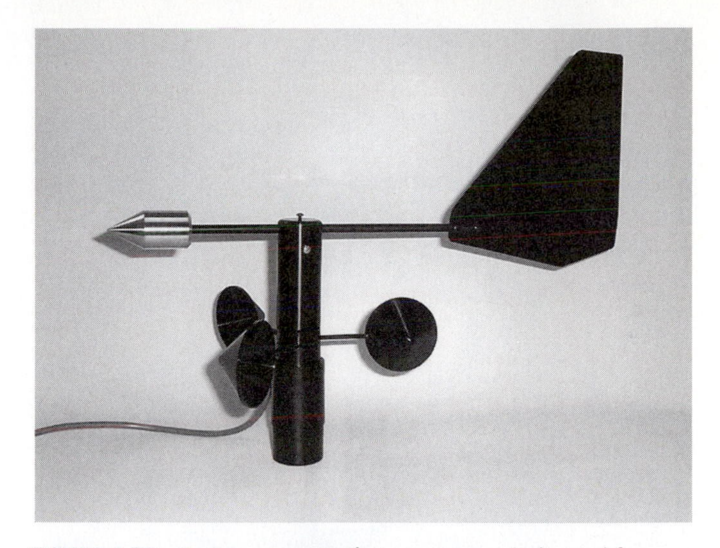

FIGURE 5-20 An anemometer that measures wind speed for a wind turbine. This sensor uses an electronic circuit that produces separate 4- to 20-mA signals for wind speed and direction. (Courtesy of Nova Lynx Corp.)

Wind Direction Indicator

The wind direction indicator shown in Figure 5-20 is a vane blade that is moved by the wind until the vane is point in the same direction as the wind is blowing. The shaft of the wind vane is connected to a 20-kΩ potentiometer, which is basically a variable resistor that is adjusted by a rotary shaft. The movement of the shaft can change the resistance from 0 to 20 kΩ within one revolution of the shaft (360 degrees). The shaft movement is not limited to one revolution, but each direction of the compass will produce the same amount of resistance change from the potentiometer. For example, 0 degrees may align with 0 Ω resistance, 90 degrees with 5 kΩ, 180 degrees with 10 kΩ, 270 degrees with 15 kΩ, and 360 degrees with 20 kΩ. The 20-kΩ potentiometer signal is sent to a circuit board (also shown in Figure 5-20) and is part of a 4- to 20-mA loop circuit. The signal for wind direction, a value from 4 to 20 mA, is sent to the PLC analog input module, where it can be used to control the pitch of the blades, the direction of the nacelle, and also be sent as archived data for wind speed and direction.

Ultrasonic Wind Measurement Instruments

Another type of wind measurement instrument is called an ultrasonic wind measurement instrument. This type of sensor measures wind speed, wind direction, and air temperature. Figure 5-21 shows an example of an ultrasonic wind measurement instrument. Other brands of ultrasonic wind instruments can also measure air density. The air density and temperature measurements are important because the amount of wind energy available to the turbine blades is greater in more dense, warm air. The

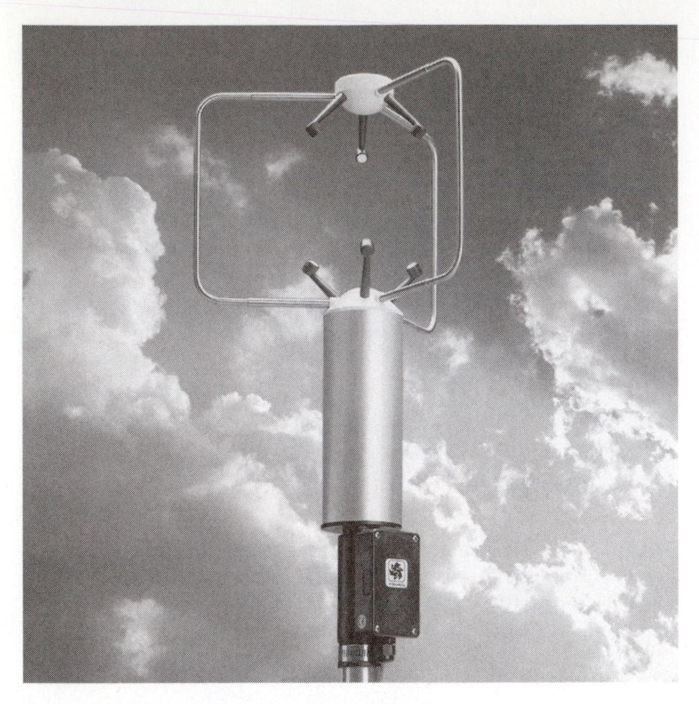

FIGURE 5-21 An ultrasonic wind-measuring instrument. (Courtesy of R. M. Young Company, www.youngusa.com.)

ultrasonic wind instrument also has an integrated heating circuit that melts snow and ice during the winter months to ensure that it operates accurately.

The ultrasonic-type sensor has multiple sensor tips, and an ultrasound signal is sent between the tips. The speed of the ultrasonic signal is constant when there is no wind; when the wind begins to blow, it causes the ultrasonic pulse to be changed, and this information is converted to wind direction and wind speed.

5.15 WIND TURBINE SUPERVISORY CONTROL AND DATA ACQUISITION (SCADA) SYSTEMS

The **supervisory control and data acquisition (SCADA)** system has been used in industry for many years. Industrial SCADA systems monitor important production information. The SCADA system for a wind turbine collects data on a single wind turbine or on a large number of wind turbines for a wind farm.

The SCADA system can be used to start, stop, or reset wind turbine generators remotely, either individually or for groups of wind turbines in a wind farm setting. Other information the SCADA system can collect includes vibration information, diagnostic information about the generator, information from the low- and high-speed shafts, and information from the gearbox. The data from the SCADA system can also track maintenance problems for comparison with similar wind turbines as well as track maintenance that has been performed on each wind turbine.

After a number of years of data gathering, the data can be used for production predictions and load planning. When a large number of wind turbines on a wind farm are connected to the grid, the SCADA system can help with grid loading and integration with coal-fired and nuclear-powered electrical generation plants. SCADA systems have been used for coal-fired and nuclear-powered electrical generation stations over the years, and much of the electrical production and generator data are similar to the data collected from wind turbines. Security information about the wind turbine is also provided through the SCADA system. This security information may include video camera or other security data such as fire protection system information for each wind turbine.

The SCADA data is usually captured through the PLC or computer controller as well as individual sensors around the system. Generally, sensor data is also used in controlling the wind generator, so this data serves two purposes, as one set of data goes to the PLC or computer to operate and control the wind turbine, and the other set of data is sent to be stored as archive data.

The SCADA system can use telephone, Internet, Ethernet, RS232 serial system, Profibus, wireless, and other network protocols to send data between wind turbines, their PLCs or computers, and the network between wind turbines on a wind farm installation. Because some of the data is proprietary, basic network protection such as firewall software is used to protect the information from outside intrusion. The SCADA system also has a SCADA server, which is a dedicated computer on the network that gathers and stores the data.

5.16 WIND TURBINE CONTROL SYSTEMS

The largest wind turbines use computer control systems to control the pitch of the blades, the direction in which the nacelle is pointed (yaw), all of the system brakes, and the load on the generator. On smaller wind turbines, the computer control may only control the blade pitch, yaw position, and the generator; on others, the computer control system may only control the generator. The computer control system may be a dedicated controller or a PLC. On larger wind turbines, the controller is generally mounted in the nacelle. On mid-size and smaller wind turbines, the controller can be mounted in the nacelle or in the electrical cabinet on the ground with the switchgear for the system. In the next sections of this chapter you will learn how PLCs and dedicated controllers use input signals from sensors along with computerized logic to energize/de-energize output components. Figure 5-22 shows a typical electrical control cabinet for a wind turbine.

The wind turbine computer controller is also used to oversee or control generator functions and the connection to the grid if the wind turbine is tied to the grid. The control for the electrical interconnect is accomplished with relays and contractors, while other parts of the control are accomplished by the computer.

FIGURE 5-22 Computer control system for a mid-sized wind turbine.

FIGURE 5-23 A PLC-type dedicated controller used to control a mid-sized wind turbine.

5.17 BASIC OPERATION OF THE PROGRAMMABLE LOGIC CONTROLLER

The **programmable logic controller (PLC)** is a dedicated computer controller that is designed to solve logic (AND, OR, NOT logic) that specifically controls on/off-type devices such as relays, contactors, and hydraulic solenoids and switches. The PLC also allows other control devices of varied voltages to be easily interfaced to provide simple control. The PLC is different from other computers in that it is designed to have industrial-type switches, such as 110-V pushbuttons, 110-V limit switches, proximity switches, and photoelectric switches, connected to it through an optically isolated interface called an **input module**. The PLC uses a similar interface called an **output module** to energize and de-energize the coils of lower-voltage contactors and motor starters that control high-voltage three-phase motors. Figure 5-23 shows a typical dedicated controller in an electrical control panel.

The computer part of the dedicated controller or PLC is called a central processing unit (CPU), and it allows a program to be entered into its memory that will represent the logic functions. The program in the PLC is not written in a normal computer programming language such as BASIC, Fortran, or C++. Instead, the PLC program uses contact and coil symbols to create ladder logic to indicate which switches should control which outputs. These symbols look similar to a typical electrical-relay ladder diagram. In most systems the program can be displayed on a computer screen, where the actions of switches and outputs are animated as they energize hydraulic solenoids and turn other outputs on or off. The animation includes highlighting the input and output symbols in the program when they are energized. Other systems use light-emitting diodes (LEDs) on input and output modules that illuminate when an input or output is energized to show its status. These features make it easy to troubleshoot any electrical components in the wind turbine to determine if they are on or off and if they are working correctly. Another feature that makes a PLC-type controller valuable in controlling wind turbine functions is that the computer is designed to work under severe conditions, and will not lock up or reset as a desktop computer may.

PLCs are very useful in controlling wind turbines because the logic program can be easily edited during the design phase, and then the memory can be locked down and stored in an electronic memory module called an **EPROM** (erasable programmable read-only memory). The EPROM stores the program in a memory chip that does not require battery backup, which ensures that it cannot be erased by power loss, and it will always be available to the controller once the program is stored in it. Another feature of the EPROM is that it cannot be changed by a technician or engineer who does not have the special EPROM writer. This is an important feature in the control of the wind turbine to ensure that only qualified engineers and technicians can make changes in the controller program. Some wind turbine manufacturers use a stock-type PLC, whereas others modify the systems and manufacture a dedicated computer controller. Regardless of whether the system is a PLC or a dedicated computer controller, the operation is similar.

Figure 5-24 shows an example of a PLC program. You may notice that the program looks like an electrical

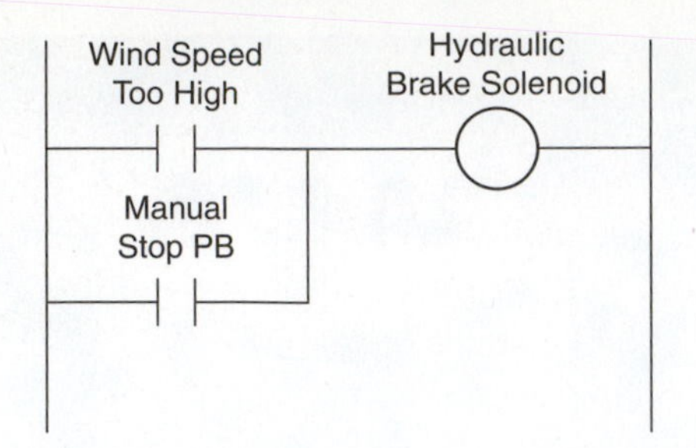

FIGURE 5-24 Typical ladder logic program in a PLC to control the hydraulic brakes on a wind turbine.

relay diagram. As the program (diagram) gets larger, additional lines are added. This gives the program the appearance of a wooden ladder that has multiple rungs. For this reason the program is called a ladder diagram, and it may be referred to as **ladder logic**.

The first PLCs were designed in the early 1960s for use in factory automation. The original function of the PLC was to provide a substitute for the large number of electromechanical relays that were then used in industrial control circuits. Early factory automation used large numbers of relays, which were difficult to troubleshoot. The operation of the relays was slow, and they were expensive to rewire when changes were needed in the control circuit. In 1969 the automotive industry designed a specification for a reprogrammable electronic controller that could replace the relays. This original PLC was actually a sequencer-type device that executed each line of the ladder diagram in a precise sequence. In the 1970s the PLC evolved into the microprocessor-type controller used today. These early PLCs allowed a program to be written and stored in memory, and when changes were required to the control circuit, changes were made to the program rather than to the electrical wiring. For the first time, it also became feasible to use the controller to troubleshoot each component in the system and determine quickly if the component was working or if it was causing a problem. This feature allows very complex circuits to be controlled and troubleshot in a very simple manner.

Understanding a Simple PLC Program

The ladder logic program show in Figure 5-24 shows two input switches that control an output solenoid. The switches are input signals from a wind overspeed sensor and a manual-stop pushbutton, and the output solenoid controls the hydraulic fluid to the parking brake for the wind turbine. The way the program works is that the computer part of the PLC is asking an "if–then" question. Because the two switches are in parallel, *if* either condition becomes true, *then* the controller will energize the output solenoid, which will set the brake. The PLC controller is constantly executing the program and asking the question. If the wind switch senses that the wind speed is too high and dangerous, or if the manual-stop pushbutton has been depressed, then the output solenoid is energized. The PLC scans its program and asks this logic question over and over at a time interval of about 20 ms (0.020 s). This means that if either switch changes state to indicate that its condition is true, the PLC will sense this change, energize the hydraulic solenoid, and apply the brakes within 20 ms.

Another important feature of the PLC is that a technician can check the program either locally, by attaching a laptop computer, or remotely, through the SCADA system or other network interface, and thereby determine whether either switch or the output is energized. If there is a problem and the wind turbine has been stopped by the brake when these conditions are not true, a technician can quickly determine which switch is causing the problem. This feature alone makes the PLC very valuable for troubleshooting wind turbines, either locally or from a distance, when a fault occurs.

The ladder logic program is the simplest type of program used in the PLC. In larger wind turbines, the PLC can also execute very complex loop control programs to control the yaw position, the turbine blade pitch, and other more complex variable systems that allow the system to be controlled between 0 and 100%. These types of control systems use analog control, which controls the output to a number of positions between 0 and 100%, depending on the amount of resolution. You will learn more about analog control later in this chapter.

Basic Parts of a Simple Programmable Controller

All PLCs have four basic major parts: power supply, processor, input modules, and output modules. A fifth part, a programming device, such as a laptop computer, is not considered a basic part, because the PLC does need one if its program is loaded from an EPROM chip. A typical PLC has input devices such as manualstop pushbutton switches that are wired directly to the input module, and the hydraulic solenoids that turn the brakes on and off and are connected directly to the output module. Before the PLC was invented, each switch was connected directly to the solenoid it was controlling. In a PLC the switches and output devices are connected to the modules,

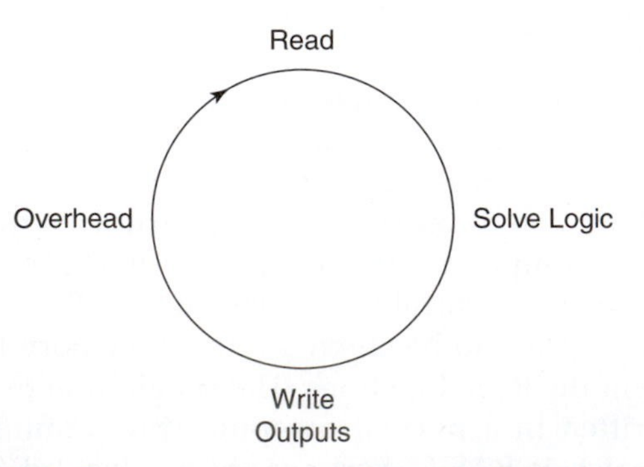

FIGURE 5-25 The scan cycle for a PLC starts with the Read cycle, then progresses to the Solve Logic cycle, Write Outputs, and finally Overhead cycle.

and the program in the PLC determines which switch controls which output using its if–then ladder logic program. In this way, the physical electrical wiring needs to be connected only once during the installation process, and the control circuitry can be changed unlimited times through simple changes to the ladder logic program.

Using a Laptop or Desktop Computer to Program the PLC

The PLC is programmed with software that is provided by the PLC manufacturer. This software can be used on a laptop or desktop computer where the PLC program is created and stored so it can be downloaded to the PLC. Once the program has been downloaded to the PLC, it can be stored on an EPROM and secured with a password so that it cannot be changed by anyone who does not have authorized access. Once the program is stored on the EPROM, it is automatically reloaded into the PLC whenever the power has been turned off and then back on. This ensures that the PLC never powers up without the program being loaded and verified, which makes it a very reliable controller. The software also allows the programmer to add comments to each rung of the logic to indicate to a troubleshooting technician what that section of the program is doing.

Another feature of the PLC software that resides on the laptop computer is that it can animate the PLC program with colors and text and indicate which inputs and outputs are turned on or off and also the time at which the transition took place. This feature makes it very easy for a technician to log onto the controller from a network or Internet node and troubleshoot the wind turbine from anywhere in the world.

The PLC program has features that allow the program to become more complex if necessary, such as on-delay and off-delay timers, counters, and a wide variety of math functions. The advantage to using timer and counter functions in the PLC is that they do not involve any electronic parts that could fail, and their preset times and counts are easily altered through changes in the program. Additional PLC instructions provide a complete set of mathematical functions as well as manipulation of variables in memory files. Modern PLCs and dedicated wind turbine computer controllers have the computing power of many small mainframe computers as well as the ability to control several hundred inputs and outputs.

Scanning a PLC Program When It Is in Run Mode

The PLC processor examines its program line by line, which is the way it solves its logic. The processor in the PLC actually performs several additional functions when it is in **run mode**. These functions include reading the status of all inputs, solving logic, and writing the results of the logic to the outputs. When the processor is performing all of these functions, it is said to be scanning its program. Figure 5-25 shows an example of the program scan cycle.

The PLC provides a means to determine when the switches that are connected to its input module are energized or de-energized. When the contacts of a switch are closed, they allow voltage to be sent to the input module circuit. The electronic circuit in the input module uses a light-emitting diode and a phototransistor to take the 110-V AC signal and reduce it to the small-voltage signal used by the processor.

The contacts in the program that represent the switch change state in the program when the module receives power. If you use the programming panel to examine the contacts in the PLC program, they will highlight when the switch they represent is energized. The PLC will transition the state of the programmed contacts at this time. For example, if the contacts are programmed normally open, when the real switch closes and the module is energized, the PLC will cause the programmed contact to transition from open to closed. This means that the programmed contacts will appear to pass power through them to the next set of contacts. If both sets of contacts are passing power in line 1 of the program, the output will become energized.

Image Registers

One of the least understood parts of the PLC is the image register. The **image register**, also called the *image table,* is a memory section of the PLC that is designed specifically to keep track of the status (on or off) of each input and output. The image register is 1 bit wide. Figure 5-26 shows an image register. In this diagram you can see that

Inputs	0	1
	1	1
	2	0
	3	0
	4	1
	5	1
	6	0
	7	0
	8	1
	9	1
	10	0
Outputs	11	1
	12	1
	13	1
	14	0
	15	0

FIGURE 5-26 Image register for a PLC. Inputs and outputs that are energized have a 1 in their image table location, and they have a 0 if they are turned off.

each of the inputs and outputs for the PLC has one memory location where the processor places a 1 or 0 to indicate whether the input or output connected to that address is energized or de-energized. The processor determines the status of all inputs during each scan cycle by checking each one to determine if the switch connected to it is energized. If the switch is energized and the input module sends a signal to the processor, the processor places a binary 1 in the memory location in the image register that represents that input. If the switch is off, the processor places a binary 0 in the memory location in the image register. This means that the image register contains an exact copy of the status of all its inputs during the Read part of the scan.

In the second part of the PLC scan, the processor uses the values in the image register to solve the logic for that particular rung of logic. For example, if input 1 is ANDed (in series) with input 2 to turn on output 12, the processor looks at the image register address for input 1 and input 2 and solves the AND logic. If both switches are (on) high, a 1 is stored in each of the image register addresses, and the processor solves the logic and determines that output 12 should be energized.

After the processor solves the logic and determines the output that should be energized, the processor moves to the third part of the scan cycle and writes a 1 in the image register address for output 12. This is called the Write cycle of the processor scan. When the image register address for output 12 receives the 1, it immediately sends a signal to that output module address. This energizes the electronic circuit in the output module, which sends power to the device, such as the motor starter coil or solenoid coil, that is connected to it.

It is important to understand that the PLC processor updates all of the addresses in the image register during the Read cycle even if only one or two switch addresses are used in the program. When the processor is in the Solve logic part of the scan, it solves the logic of each rung of the program from left to right. This means that if a line of logic has several contacts that are ANDed (connected in series) with several more contacts that are connected in an OR (parallel) condition, each of these logic functions is solved as the processor looks at the logic in the line from left to right. The processor also solves each line of logic in the program from top to bottom. This means that the results of logic in the first rung are determined before the processor moves down to solve the logic on the second rung. Because the processor carries out the logic scan on a continual basis approximately every 20–40 ms, each line of logic appears to be solved at the same time.

The only time the order of solving the logic becomes important is when the contacts from a control relay in line 1 are used to energize an output in line 2. In this case the processor actually takes two complete scan cycles to read the switch in line 1 and then write the output in line 2. In the first scan, the processor reads the image register and determines that the start pushbutton has closed. When the logic for rung 1 is solved, the control relay is energized. The processor will not detect the change in the control relay contacts in the second line until the second scan, when it reads the image register again. The output in the second line is finally energized during the Write cycle of the second scan, and power is sent to the motor starter coil that is connected to the output address. It is also important to understand that the processor writes the condition of every output in the processor during the Write part of the scan, even if only one or two outputs are used in the program.

The image register is also useful to the troubleshooting technician, because the contents of each address can be displayed on a screen. The technician can compare the image register with the input and output modules and the actual switches and solenoids that are hard-wired to the modules.

A separate image register is also maintained in the processor for the control relays. A *control relay* is a relay that exists only inside the PLC. In some systems these relays are called *memory coils* or *internal coils*. A control relay has one coil and one or more sets of contacts that can be programmed either normally open or normally closed. When the coil of a control relay is energized, all of the contacts for the relay change state. Because control relays reside only in the program of the PLC and do not use any hardware, they cannot have problems such as dirty contacts or an open circuit in their coil. This means that if the contacts of a control relay do not change from open to closed, you would not suspect the coil of having a malfunction. Instead, you would look at the contacts that control power to the control relay coil in the program: One or more of them will be open, which is what causes the coil to remain de-energized.

When the processor is in run mode, it updates its input, output, and control relay image registers during the write-input/output part of every scan cycle. Because the image registers are continually updated approximately every 20–40 ms, it is possible to send copies of them to printers or color graphic terminals at any time to indicate the status of critical switches for the machine process. This allows the technician to do minor troubleshooting when a problem occurs, before a troubleshooter is called. For example, if a technician who is checking the operation of the wind turbine or doing preventive maintenance discovers a problem by observing the PLC program, the technician can make the repair rather than call a more expensive troubleshooter. The conditions that can keep a wind turbine from starting or running correctly can be inserted into the PLC program as a fault, and the fault can be monitored on a color graphic display that shows the status of the fault. For example, on a large system, if the hatch or door on the nacelle is ajar, a technician can see this on the fault monitor. The copy of the input image register from the PLC will indicate that the door switch is open. The technician can then take the appropriate action instead of calling a troubleshooter. This way, the PLC not only controls the machine operation, it also helps in troubleshooting problems.

Run Mode and Program Mode

When the PLC is in run mode, it is continually executing its scan cycle. This means that it monitors its inputs, solves its logic, and updates its outputs. When the PLC is in **program mode**, it does not execute its scan cycle. The term *program mode* was originally used because programming in older PLCs could not be changed while they were executing their scan cycle in run mode.

The programming software for a PLC may allow you to write the ladder logic program in PLC software on a personal computer and later download the program from the personal computer to the PLC. If you are connected directly to a PLC and you are writing the ladder logic program in the PLC memory, it is called *online programming*. If you are writing the ladder logic program on a personal computer or programming panel that is not connected to a PLC, it is called *offline programming*. Most modern PLC programming software allows the program to be written on a personal computer without being connected to the PLC. This allows new programs or program changes to be written at a location away from the PLC. If the PLC does not have a microprocessor chip, all programming must be completed offline, and then you must download the program changes to the PLC memory or directly to an EPROM so the PLC can read the memory from the EPROM when power is first turned on.

5.18 PLC INPUTS AND OUTPUTS

Inputs and outputs modules are connected to the PLC processor by electronic circuits in the rack that holds the modules. The location that the module has in the rack is called a *physical address*, and this location is used by the troubleshooting technician to physically connect the wires from the input switch or output device. This physical location also allows the troubleshooter to go to that terminal and module and check the wire for voltage if there is a problem.

Up to this point in our discussion, we have identified inputs and outputs in generic terms. In reality, the processor must keep track of each input and output by an address or number. Each brand-name PLC has its own numbering and addressing schemes, which generally fall into one of two categories: sequential numbering of each input and output, or numbering according to the locations of input and output modules in the rack. The sequential numbering system is used in most small-sized PLCs and in about half of the medium-sized systems.

Input and Output Instructions

When individual electrical switches are wired to an input module, the switch contacts can be normally open (NO) or normally closed (NC). For example, an emergency stop button is wired as a normally closed switch, so that if the switch fails or if a wire falls off the switch, it will be in the fail-safe condition that will stop the system. Other switches are wired as normally open switches. When an input instruction is used in the PLC program, it can be used as a programmed set of normally open or normally closed contacts in the program. The symbol for the NO contacts looks like this, –] [–, and the NC contacts look like this, –]\[–, with a slash between the contacts to indicate that they are closed. The PLC checks the input module and image register to determine whether the physical switch is closed, and then it changes or transitions the programmed switch contacts to their opposite condition.

The processor keep tracks of the status of the circuit by placing a 1 in the image register address if it is energized and a 0 if it is de-energized. Because the input/output module uses a small LED as a status indicator, a troubleshooter can check the front of the module and also determine whether the input circuit is energized or de-energized.

The output instruction includes the traditional *coil*, which is shown by the symbol -()-. This instruction is called *output*. Specialized outputs called the *latch*, –(L)–, and the *unlatch*, –(U)–, are also provided. The latch coil output maintains its energized state after it is energized by an input even if the input condition becomes de-energized. The unlatch coil output must be energized to toggle the output back to its reset (off) state. The latch coil and unlatch coil must have the same address to operate as a pair.

Latch coils are used in wind turbine applications when a condition should be maintained even after the input that energized the coil returns to its de-energized state. It is important to understand that the latch coil bit will remain HI (on) even if power to the system goes off and comes back on. This feature is important in several applications such as fault detection. Many times a system has a fault such as an overtemperature or overcurrent condition. When the fault is detected, the machine is automatically shut down by the logic, and by the time the technician or operator gets to the machine, the condition (e.g., an overtemperature) is back to near normal as the system cools down over time, and it becomes difficult to determine why the system shut down. If a latch coil is used in the detection logic, the fault bit will set to the ON condition when the overtemperature occurs, and the bit will remain HI (ON) until the fault reset switch energizes the unlatch instruction after the troubleshooting technician has acknowledged the fault.

Another application for the latch coil is for cooling-water pumps or fans that are included in the nacelle. A latch coil is used in these circuits because the motors must return to its energized condition automatically after a power loss. If a regular coil is used in the program, a technician would have to go to each motor or pump and depress the start pushbutton to start the motor again. If a latch coil is used, the coil will remain in the state it was in when the power went off. This means that if the latch coil was energized and the motor was running when the power went off, the latch coil will remain energized when power is returned and the motor will begin to run again automatically.

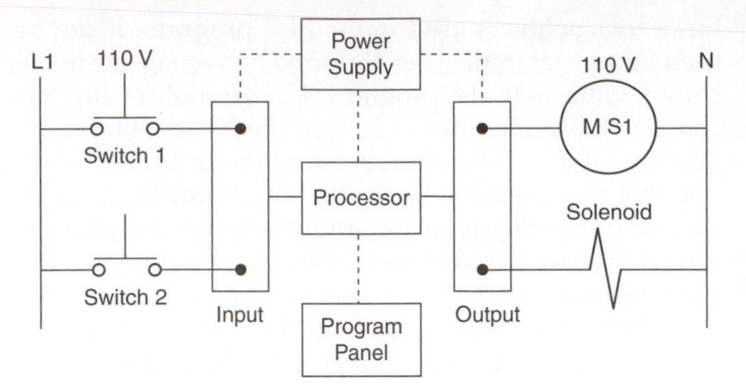

FIGURE 5-27 Electrical diagram of inputs and outputs connected to input and output modules of the PLC.

Wiring Input Switches and Output Devices to the PLC

All the input switches and output solenoids used in winds turbines with a PLC system must be connected to the input or output module hardware. Figure 5-27 shows an example of a normally open and a normally closed pushbutton switch wired to two input addresses and a coil for a motor starter (MS1) that controls a **hydraulic pump** and a hydraulic solenoid coil connected to two output addresses. You can see from this diagram that 110 V AC is used to supply the voltage to the switches to use to send their signals to the PLC input module, and 110 V AC is also used to provide power to the coil of the motor starter and the coil of the hydraulic solenoid that are connected to the output module. This allows standard electrical switches and other components to be used in the wind turbine system.

5.19 PLC ANALOG CONTROL

PLCs can also use **analog control** to transmit information between sensors and output devices such as wind-speed sensors and yaw-positioning motors. The term *analog* means that the process value, such as the position of a motor, can be any value between 0 and 360 degrees. A speed sensor for the low-speed shaft can measure values between 0 and 100 rpm. Another sensor can measure the speed of the high-speed shaft. For example, suppose that a shaft's top speed is 3600 rpm. The analog value that represents the shaft's speed at any time will have to be a value between 0 and 3600 rpm. These values are called the minimum (min) and maximum (max) analog values. This is also called 0–100% control. In essence, many of the processes in the wind turbine can be represented by minimum and maximum values. As an everyday example, consider a light switch. If you turn the switch to the ON position, the light is fully on, and if you switch the light to the OFF position, the light is fully off. This is an example of on/off control. If the light has a dimmer switch on it, it is an example of analog control. You can turn the light one-fourth on, which is 25% on. You can also adjust the dimmer switch so that the light is halfway on, which is 50%. You can set the dimmer control to virtually any value between the minimum value (off) and the maximum value, which is 100% on. Another example of analog control is the water valve that is on a kitchen sink. You can open the valve a little bit, halfway, or completely, depending on how much water you want to come out of the faucet.

In analog systems we can describe the action in terms of percent, or we can use actual values, such as rpm or amount of hydraulic pressure. We generally use voltage or current to control these parameters in a PLC, so we need to talk about the voltage and current systems that are commonly used for analog control. Table 5-1 shows the most common voltage and current signals used in wind turbine systems. The table also shows the PLC values for a 12-bit chip, which are 0–4095. You can see that the voltage can be 0–10 V, and the current can be 4–20 mA. Of these, 0–10 V is easy to calculate percentages mentally with a maximum value of 10. You should notice the 4- to 20-mA system does not start at 0, but we still refer to it as a minimum-to-maximum (min-max) analog system. The 4- to 20-mA system is the most common system used for instrumentation. The system starts at 4 mA instead of 0 so that it can indicate when a broken wire or open circuit has occurred in the system. If 0 were used, there could be a zero value when the system was at its minimum point as well as when the circuit is open. This would make it difficult to determine which condition is causing the zero value. With 4 mA as the minimum value, a technician can quickly make a current measurement and determine whether the system is indicating 4 mA because it is at its minimum value or it is indicating 0 mA because there is a broken wire somewhere in the electrical system. Because broken wires and electrical opens are the most common problems in instrument systems, the 4- to 20-mA system

TABLE 5-1 Analog Values for Milliamps, Voltage, and Motor rpm

Analog Value (%)	PLC Value (0–4095)	Milliamps (4–20 mA)	Voltage (0–10 V)	Motor rpm (0–3600)
100%	4095	10 mA	10 V	3600 rpm
75%	3073	16 mA	7.5 V	2700 rpm
50%	2047	12 mA	5.0 V	1800 rpm
25%	1024	8 mA	2.5 V	900 rpm
0%	0	4 mA	0 V	0 rpm

The digital values for a 12-bit chip are shown in the PLC value column because the values for the digital chip are 0–4095.

is frequently used for analog systems. You will see examples of analog signals used in hydraulic control later in this chapter.

PLCs can also have analog input and output modules. An analog input module allows analog values to be used as alarm inputs, or signals from sensors that report wind speed, wind direction, and amount of voltage and current the generator is producing. These analog signals also vary voltage or current from a minimum value to a maximum value. The analog voltage signal usually is a 0–10 V, and the current signal is usually 4–20 mA. The analog input module for the PLC can have two, four, or eight separate channels to read the signals. This means that the voltage signal can come from a wind speed sensor or the wind direction sensor and be connected to separate channels. The wind sensor is calibrated for the maximum wind speed it can measure. For example, the wind speed sensor can be calibrated so that the sensor sends 10 V when the wind speed is 100 mph, 5 V when the wind speed is 50 mph, and 2.5 V when the wind speed is 25 mph. The main feature of the analog signal is that every voltage between 0 and 10 V is available for the sensor to send to indicate the level of the wind speed for any value of wind speed from 0 to 100 mph. The analog system generally has a resolution of approximately 0.1 V.

The analog input signal from the wind speed sensor can be used to adjust the pitch on the blades at various points, and it can also be used to determine when to apply brakes to the low-speed or high-speed shaft if the blades are turning too fast. In addition, the sensor signal can also be used to indicate an alarm condition if the wind speed is higher than 75 or 80 mph. One nice thing about the analog input signal is that once it is measured and brought into the PLC analog input module, it can be used for multiple purposes.

Because the analog value is a number, it can be used in comparison statements to compare against known values for greater-than, less-than, or equal-to comparisons, and the output of these instructions can be used to set multiple alarms. Other math functions such as averaging can also be used to make the values more accurate over a longer period of time.

The analog signal can also be used in a closed-loop calculation whose answer is an analog output value. For example, if the analog output signal is used to apply hydraulic pressure to the brakes for the rotor or low-speed shaft, the analog value can be made larger so that more hydraulic pressure is applied to the brake pads if the shaft is not slowing down fast enough. The analog output signal can be sent out of the PLC as a variable voltage (0–10 V) or a variable current (4–20 mA).

5.20 CONTROLLING WIND TURBINES THROUGH FEEDBACK CONTROL TO THE PLC

Many of the control systems in a wind turbine use analog control and feedback sensors to control their range of control. **Feedback** control is also called **closed loop** control. Figure 5-28 shows a closed-loop diagram with all of the parts identified. It will be easier to understand these parts if they are applied to a specific wind turbine application such as the yaw position control. From the block diagram you can see that the system starts with a **setpoint** (SP) signal. The SP is the desired value, which is the desired direction. For example, if we wanted the desired yaw direction to be set at 90 degrees (east), the SP would be adjusted to 90 degrees.

The next part of the control system that we will examine is the **process variable (PV)**, which is the signal that comes from the sensor that indicates the directional heading of the nacelle. For example, in this system that controls the nacelle direction, the PV is the signal from a directional sensor that is the sensor for this closed-loop system. The PV is also called the *feedback signal*, and it is the present value or actual value of the position at the instant the sensor reading takes place. Because the sensor reading is continuous, the PV changes continually to indicate the changing direction of the nacelle.

The **summing junction** is the place in the control system where the setpoint (SP) is compared to the sensor (process variable) PV. Mathematically, this means that the PV is subtracted from the SP (SP – PV). For example, if the SP is 90 degrees and the PV signal indicates that the actual direction is 80 degrees, the difference is 10 degrees.

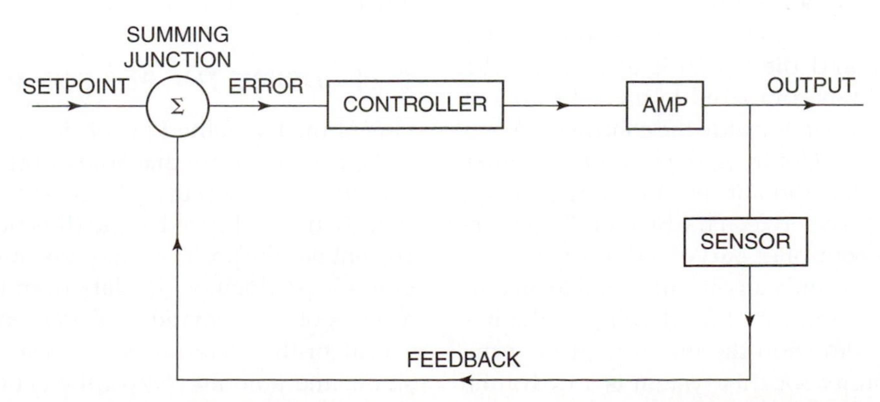

FIGURE 5-28 Closed-loop (feedback) diagram showing the setpoint, process variable, error, controller, and output.

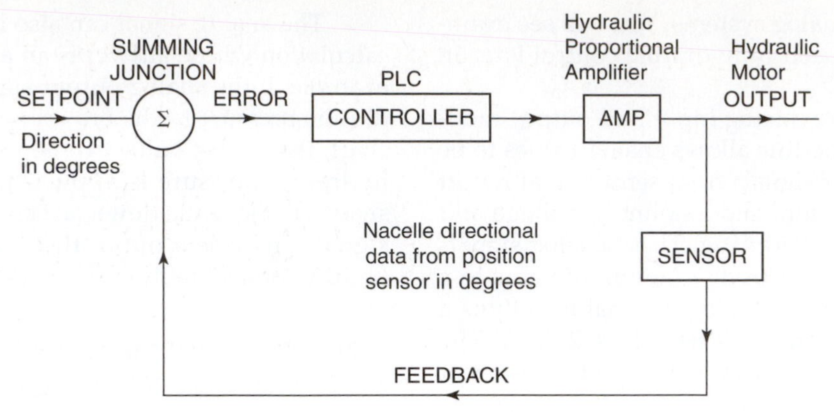

FIGURE 5-29 Nacelle positioning loop, which uses a hydraulic valve to control the fluid flow to a hydraulic motor that changes its speed and direction to move the nacelle ring so the blades point in the desired direction.

The summing junction is identified by the Greek letter sigma (Σ).

The difference between the SP and the PV is called **error**. The error is a comparison of the value of the setpoint (SP), where you want the nacelle to be, and the process variable (PV) signal that indicates the nacelle's actual location. Both the error signal and the process variable signal are measured in values of 0–100%, and the difference between them is found by subtracting the value of the process variable (PV) from the value of the setpoint (SP). The error can be a positive value if the SP is larger than the PV, or it can be a negative value if the SP is smaller than the PV. The controller then changes the output by some value to make the nacelle change direction. For example, the output for this system is a hydraulic motor that rotates a shaft to make the nacelle ring rotate and change direction so the blades are pointing into the wind. The important thing to remember about the value or amount of error is that the controller is always trying to make adjustments to the output to make the value of the error smaller, so that the process variable is as close to the setpoint as possible.

The nacelle **directional control** system in this application is illustrated in Figure 5-29. In this control diagram the components of the basic control system are identified. The controller is a loop controller. When this loop is programmed into the PLC, the SP is the new position to which the nacelle needs to move. The new direction can be determined from a calculation that indicates what the best direction is. The sensor for this system is a directional sensor much like a compass, and the signal is also called the process variable. The SP is compared to the PV signal at the summing junction, which is inside the controller. When the loop control begins its operation, the PLC or computer does a mathematical calculation at the summing junction to determine the difference between the SP and PV; this error signal is sent to the controller part of the system.

The controller then sends an output signal to an amplifier that controls the amount of fluid going to the hydraulic motor and the direction the motor is turning the nacelle ring. The amplifier for this system is a hydraulic proportional amplifier that adjusts the amperage to the hydraulic solenoid valve from 0 to 100% with a 0- to 10-V DC signal. The controller sends the voltage signal as an output signal to the amplifier, and the amplifier changes the position of the hydraulic valve that controls the direction and speed of the hydraulic motor shaft.

The block diagram for the control system shown in Figure 5-29 is called a *loop* because a sensor samples the direction the nacelle is pointing as a compass position in degrees and send its PV signal back to the summing junction where it can be compared to the SP. The sensor signal is also called the *feedback* signal because it gets fed back to the summing junction. When this happens, it is called *closing the loop*, and the system is called a *closed-loop system*.

If the system is put into manual mode, the feedback signal is not used, and then the system is called an **open-loop** system. In most systems it is possible to use an auto/manual switch to control the PV (sensor) signal that is used for feedback. If the switch is closed, in auto mode, the sensor signal is used as feedback, and if the switch is open, in manual mode, the sensor signal is not used. The auto/manual switch determines whether the loop is operated in open-loop or closed-loop mode. When the system is in open-loop mode, the output is adjusted manually by the operator. For example, if a maintenance crew needs the nacelle to move to a specific direction so that maintenance can be performed, the positioning loop can be put in manual mode and the nacelle moved manually to the direction needed.

5.21 USING THE PLC FOR POWER CONTROL

One of the loops the PLC or dedicated computer controls is the power control that adjusts the blades and the nacelle position. This loop uses the sensors that provide information about wind speed, wind direction, current blade pitch, current nacelle position, and how much electricity the generator is producing. The data from these sensors goes into a series of mathematical calculations called an algorithm. The algorithm determines the best blade pitch, nacelle position, and amount of load the generator should have for the wind turbine to operate at optimum performance.

5.22 OPTIMIZING BLADE TIP SPEED

Another loop that the PLC can control on larger wind turbines is blade rotation speed, which in turn also controls blade tip speed. The **blade tip speed** is the speed at which the blade moves through the air as it rotates, and this speed is measured at the very tip of the blade. The tip speed of the blade is much faster than the speed at the point where the blade is attached to the rotor. If the tip speed is too fast, it will damage the blades over time. If the PLC loop controls the blade pitch and the nacelle yaw position, it can control how fast the blade is rotating in all different wind speeds and wind directions. The PLC loop control provides an optimum speed and a maximum speed for the blades to send to the setpoint (SP) of the rotor speed loop. If the rotor speed loop is less than the maximum, the loop changes the nacelle yaw position and the pitch on the blades to ensure that the wind turbine generator is producing the maximum amount of electrical energy for the conditions.

5.23 USING THE PLC TO OPTIMIZE BLADE TORQUE

The blade torque is another condition that is important to control. When the wind is blowing, the blades turn the rotor, which in turn rotates the low-speed shaft, the gearbox, and the high-speed shaft. The high-speed shaft turns the generator shaft, which produces electrical power. As the electrical power begins to increase, the electrical magnetic forces on the generator try to turn the generator shaft in the opposite direction. Without going into too much detail at this point, the amount of force the magnetic fields in the generator produces as a force trying to turn the generator shaft in the opposite direction will be nearly equal to the force from the blades that make the shaft turn in the first direction. These opposing forces cause a large amount of torque on the blades. When the wind is blowing strongly and the generator is producing maximum amount of electrical energy, the torque can become greater than a safe design amount and cause permanent damage to the one or more of the blades.

The PLC may have a loop to control the maximum amount of torque the blades endure. This loop provides information to trim the blade pitch, rotate the yaw position of the nacelle, or begin to unload the generator until the torque returns to a lower, safe condition.

5.24 PLC PROGRAM STORAGE AND PERMANENT MEMORY

Once the program has been written and tested for the PLC to control a wind turbine, the program must be saved and stored so it can be used as a backup if the PLC loses its memory, or if a new PLC replaces the original one. If the PLC needs to be changed out for any reason, all of the wiring to the modules and the input/output modules remains in the rack and is not changed out. This makes it very easy to change out the PLC in the event there is a problem. The main issue when a PLC or dedicated computer is changed out is to ensure that the original program together with any subsequent changes is loaded into the new PLC. For this reason, the original PLC program must be stored in a secure medium to ensure that it is always ready to be loaded. One way to save the program is on an EPROM (erasable programmable read-only memory) that is stored in a socket on the main processor board for the PLC. Any time power is cycled to the PLC, the program in the EPROM is automatically uploaded into the PLC memory. The EPROM ensures that the original program is always reloaded into the PLC even if someonehas made unauthorized changes to the program. The EPROM is a rugged memory chip that does not need any electrical power to maintain its backed-up state. Additional EPROM chips can be made and stored on site or, if a major change is made to the PLC program, the EPROM can be burned with the new updates after they have been checked out.

If the wind turbine is connected to a network or SCADA system, a copy of the program with documented comments can be backed up over the network and stored on a central server for the SCADA system. A technician who needs a copy of the PLC program can use a laptop computer to log into the SCADA system and download the program to the PLC, or can use the program to burn a new EPROM. It is important that printed copies of the program with documentation are available as well as the copy of the program that is stored electronically in memory somewhere.

5.25 CONTROLLING THE MAGNETIZING CURRENT IN THE GENERATOR

Another part of the wind turbine that can be controlled by the PLC or dedicated computer is the amount of magnetizing current that is sent to the generator. You will learn more about generators in Chapter 6, where you will find that the amount of magnetizing current sent to the generator fields, along with the speed at which the generator shaft is turning, determine the amount of electrical power the generator creates. The amount of magnetizing current that is provided to the generator ultimately determines how much electrical power it produces when its shaft is turning. The PLC can increase or decrease the amount of magnetizing current to increase or decrease the generator's load, which in turn increases or decreases the torque on the blades, rotor, shafts, and gearbox.

5.26 HYDRAULIC CONTROLS

Hydraulic control systems use hydraulic oil under pressure to extend and retract cylinders and cause hydraulic motor shafts to rotate. Hydraulic actuators are used to adjust the pitch of the turbine blades by rotating them at the point where they are attached to the rotor hub, and they rotate the complete nacelle to ensure that the blades are positioned into or out of the wind, depending on need. To work on a wind turbine, you will need to be able to identify the basic parts of any hydraulic system and understand its operation. This information will also help you to follow a basic hydraulic diagram and analyze the operation of a system during troubleshooting or installation.

5.27 WHAT THE HYDRAULIC SYSTEM IS USED FOR ON A WIND TURBINE

The hydraulic system on a wind turbine consists of a hydraulic power supply, actuators, control valves, and hoses. The basic parts of this system include the reservoir, which is a tank that holds the hydraulic fluid; the pump, which moves the fluid; the pressure-control device, which sets the pressure for the system; the directional control device, which switches the fluid to the proper cavity on the cylinder to cause it to open and close; the cylinder, which has a rod that extends and retracts; and the hydraulic motor, which rotates a shaft and all the tubing to route fluid throughout the system. The system may also have a flow-control device, which controls the speed at which the cylinder extends or retracts by setting the amount of fluid that is flowing. The more fluid is flowing, the faster the cylinder rod moves. The shaft of the hydraulic pump is turned by an electric motor. The hydraulic pump, electric motor, reservoir, and control valves are mounted together as a hydraulic power pack, an example of which is shown in Figure 5-30.

One of the simplest hydraulic systems is a hydraulic cylinder that extends and retracts. This type of cylinder may be used to adjust the pitch of the wind turbine blade. Figure 5-31 shows a typical hydraulic system for a wind turbine, with all of its parts. The actuator is a hydraulic cylinder. When the cylinder is extended, the rod moves out of the cylinder, and when the cylinder is retracted, the rod moves back into the cylinder. If the cylinder is used to control the pitch of the wind turbine blade, the rod is moved to the extended position to move the blade in one direction, and when the cylinder is retracted, it moves the blade back to its original position. The diagram also shows that fluid starts from the reservoir and is pumped into the system. An electric motor turns the hydraulic pump. A pressure regulator adjusts the pressure in the system by bypassing some of the fluid through the pressure regulator and back to the reservoir. Fluid to the cylinder is controlled by the directional control valve. In this circuit, the directional control valve is a three-position valve that is actuated by a manual handle. When the handle on the left side is activated, the directional valve shifts to the left side, which directs fluid flow to the cap end of the cylinder and makes the rod extend. When the handle on the right side of the directional valve is activated, the directional control valve shifts to the right side, directing fluid to the rod end of the cylinder and making it retract. The variable flow control and check

FIGURE 5-30 Hydraulic power pack consisting of a hydraulic pump and motor mounted on a hydraulic reservoir. (Courtesy of Parker-Hannifin Corporation.)

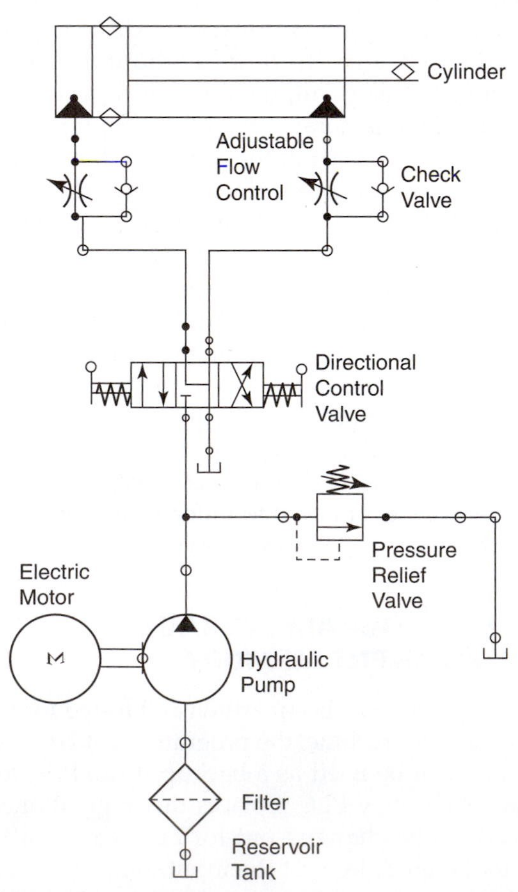

FIGURE 5-31 Diagram of a typical hydraulic system used to control a hydraulic cylinder for a wind turbine.

valve controls the amount of fluid that flows to the cylinder, and this in turn controls the speed at which the cylinder extends and retracts. The **check valve** in the flow control circuit on the right side of the diagram blocks oil when it comes from the rod end of the cylinder, which forces the fluid through the flow control and controls the speed at which the cylinder rod extends. The check valve on the left side of the circuit blocks fluid flow when the cylinder rod is retracting, and the variable flow control valve controls the speed at which the rod retracts.

Figure 5-32a shows a hydraulic cylinder designed specifically to change the pitch of wind turbine blades, and Figure 5-32b shows the blades of a horizontal wind turbine with arrows indicating the directions the blades rotate and change pitch. In some wind turbines, the pitch of all of the blades is connected and the blades move simultaneously. On other wind turbines, the pitch of each blade can be adjusted separately by a hydraulic cylinder connected directly to each blade pitch control system. If the blade pitch is controlled by a hydraulic system, a hydraulic cylinder extends or retracts to adjust the pitch of the blade. On two-bladed wind turbines, the pitch control for both blades is connected with a bar, so only a single hydraulic cylinder is needed to change the position of the bar, which changes the pitch of both blades at the same time.

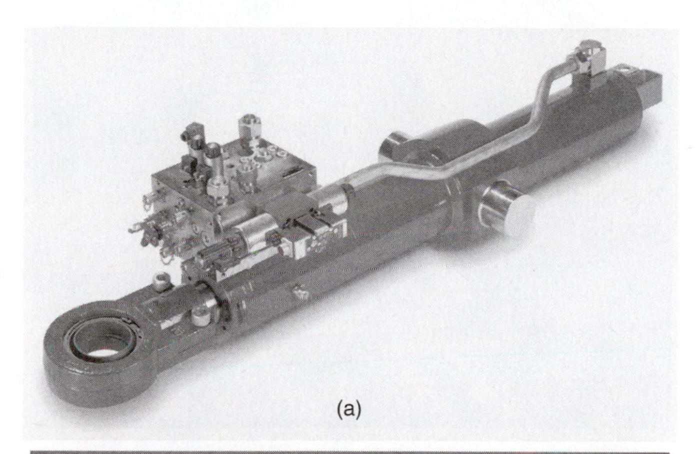

(a)

(b)

FIGURE 5-32 (a) A hydraulic cylinder can be used to adjust the blade pitch. (Courtesy of Bosch Rexroth Group.) (b) The arrows show the direction in which the blade pitch rotates the blades.

Hydraulic System Terminology

To fully understand the operation of a hydraulic system, you must also understand all the terms associated with it. The main purpose of any hydraulic system is to do work. *Work* is defined as exertion of a force—such as a push or a pull—through a distance. This can be expressed as a formula: Work = force × distance, or $W = f \times d$; the answer is expressed in foot-pounds (ft-lb). For example, if a hydraulic cylinder pushes 100 lb of weight a distance of 2 ft, the hydraulic system exerts 200 ft-lb of work.

Force is the pressure exerted per unit area, often expressed as pounds per square inch (psi). Pressure is then the force per square inch.

Power is defined as the rate of doing work. This means that you can determine the power a hydraulic system produces if you can measure the amount of work it produces over a measured amount of time. The formula for power is expressed as: Power = work ÷ time or $P = W/t$. The usual units for power are foot-pounds per time (ft-lb/time).

Remember, work is force × distance. The common unit for power in hydraulic systems is the horsepower (hp). One horsepower is equal to 33,000 ft-lb/min, or 550 ft-lb/s. For example, if the hydraulic system produces 99,000 ft-lb in 1 min, you can calculate the horsepower in the following manner:

$$\frac{99{,}000\,\text{ft-lb/min}}{33{,}000} = 3\,\text{hp}$$

To calculate the problem using 550 ft-lb/s, you must first express 99,000 ft-lb/min in terms of seconds. To do this, divide 99,000 by 60 s/min, which gives 1650 ft-lb/s. Then divide this value 55 ft-lb/s—the answer will still be 3 hp. In most cases, when you are working around hydraulic systems, the work the system does is defined by the number of gallons of fluid that are pumped in 1 min (gpm) and the amount of pounds of force that is exerted per square inch (psi). The formula for this type of calculation is

$$\text{hp} = \frac{\text{gpm} \times \text{psi}}{1714}$$

This formula can also be expressed as

$$\text{hp} = \text{gpm} \times \text{psi} \times 0.0007$$

It is important to understand that horsepower can also be expressed in electrical units as 1 hp = 746 W, and it can be expressed in units of heat as 1 hp = 42.4 Btu (British thermal units).

When you first learn about hydraulic systems and encounter formulas, you may become discouraged, but when you are working with hydraulic systems you will generally use printed charts or look-up tables printed on sliding charts to determine force, flow, and horsepower. You may also be able to obtain software for a computer to perform essential calculations and provide information about the system on which you are working. Another good resource is the technical staff at most hydraulic equipment sales offices, who have a number of ways to help you determine necessary elements of the system. The main thing is to remember where the formulas are printed and not to worry about memorizing them.

Fluid Dynamics

Fluid dynamics is the science of the flow and pressure of hydraulic fluids. Pressure, as we stated earlier, is defined as force per unit area. An example of pressure is 100 lb/in^2, which we usually write as 100 psi. This means that 100 lb of force is exerted on every square inch of surface, which is like placing a 100-lb weight on every square inch.

Flow is the movement of fluid through piping, controls, and actuators. Flow can be described in two ways. The first way is by its velocity, which is how fast the fluid particles move past a given point. The second way is to measure the amount of flow, which is called the volume. Volume can be described as the number of gallons of fluid or the number of cubic inches. The conversion of gallons per minute (gpm) to cubic inches per minute (in^3/min, or cu in/min) is shown next. Note that in the term *cubic inches per minute*, the word *per* is replaced with a slash (/). We see that gpm = 231 in^3/min. This means that if you know the flow is 693 in^3/min and you are trying to find the number of gallons per minute, you can use the formula

$$\text{gpm} = \frac{\text{in}^3/\text{min}}{231}$$

If you know the number of gallons per minute, you can calculate the number of cubic inches per minute by multiplying the number of gallons per minute by 231. For example, if you have 4 gpm, the formula isnn

$$\text{in}^3/\text{min} = 4\text{ gpm} \times 231 = 924\text{ in}^3/\text{min}$$

$$\text{gpm} = \frac{693\text{ in}^3/\text{min}}{231} = 3\text{gpm}$$

You may need to calculate cubic inches per minute or gallons per minute when you are trying to figure what size of pump you need or the size of tubing you need.

The Reservoir

The reservoir, also called the tank, serves several functions in a hydraulic system. Figure 5-33 shows a typical **hydraulic reservoir**. The first and most obvious function of the reservoir is to provide a place to store sufficient fluid for all applications of the system. For example, if the system has multiple cylinders, they may not all be moving at the same time, so only a small portion of the fluid is used. If all the cylinders are moving at the same time, the amount of fluid being used is greater. The reservoir holds the extra fluid the system needs during peak demand times.

FIGURE 5-33 A reservoir with pump and valves on top.

The second function of the reservoir is to provide a means of helping to dissipate to the surrounding air all the heat that the fluid picks up as it moves through the system. If the fluid picks up too much heat and the reservoir cannot pass it all to the air, a heat exchanger may be added to the system, which uses water or air passing over the unit to remove additional heat. The reservoir has a large surface area to help allow heat to transfer from the fluid to the outside air. The tank may have internal baffles to help keep air bubbles that occur when oil is returned to the reservoir from reaching the oil-pickup area of the reservoir.

Another function of the reservoir is to provide a platform for mounting the pump and motor. Because the reservoir is made of heavy steel, it provides an excellent place to mount the pump and motor, which also saves space by keeping these components off the floor. The reservoir has one or more drains that are used to empty the tank if the fluid becomes contaminated or needs to be changed. Access plates are mounted on the top or sides of the tank. These plates are bolted onto the tank so that they can be removed if personnel need to look into the tank or get

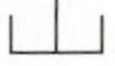

FIGURE 5-34 Hydraulic symbol for return to tank or reservoir.

inside it for any reason. The tank may also have an oil-level gauge called a sight gauge. The sight gauge is mounted on the side of the tank, where it is very visible and allows the level of the oil to be checked without opening the tank. A filler access hole is provided to allow a place to add oil. The filler also has a breather that allows clean air to enter the tank. Typically the tank is not pressurized, so the breather allows outside air to enter the tank at all times. A drain plug is provided in the bottom of the tank to allow it to be drained for maintenance or to change the oil.

Figure 5-34 shows the hydraulic symbol for the tank. This symbol may be used in several locations in a hydraulic diagram to show where a line originates that returns to the tank, rather than having to draw the line all the way back to the tank. Most drawings have multiple lines through which fluid returns to the tank, so the symbol makes the diagram look cleaner and easier to read. This symbol is sometimes referred to as the *return to tank* symbol.

FIGURE 5-35 Example of a hydraulic pump. (Courtesy of Parker-Hannifin Corporation.)

5.28 HYDRAULIC PUMPS

A hydraulic pump converts electrical energy from a motor to fluid energy when its shaft is rotated. The output of the pump is defined by its rate of flow, which is called the pump's *displacement*. Pumps are rated by the amount of output flow and maximum operating pressure when the shaft is turning at its rated speed.

One class of pump is called a positive-displacement pump. A positive-displacement pump delivers a specific amount of fluid for each revolution of its shaft or for each cycle of the pump. The positive-displacement pump can have a fixed or variable displacement. A second class of hydraulic pump is the non–positive-displacement pump. The output of a non–positive-displacement pump varies with the size of the input and output ports, the type of impeller, and the back pressure of the system. Figure 5-35 shows a positive-displacement pump. Other pictures and diagrams in this section will show the internal workings of different types of positive-displacement pumps, such as a piston pump, a gear pump, and a vane pump.

A hydraulic pump moves fluid, and the fluid cannot be compressed. When the pump causes flow, the fluid creates pressure when it fills all of the hoses and valves completely and a restriction is created. When the amount of restriction is controlled, the pressure in the system can be controlled. Non–positive-displacement pumps do not pump the same amount of fluid with each rotation of the shaft. This means that some amount of fluid may bypass the impeller or propeller on the shaft when the flow of fluid is restricted. This characteristic is very useful in some systems, such as transfer pumps and sump pumps, because the pump will not become damaged if unwanted restrictions such as dirty filters or foreign material create blockages in the system. This characteristic is typically not useful in most hydraulic systems used in wind turbines, however, because the flow may be unpredictable. Most hydraulic systems in wind turbine applications use positive-displacement pumps, because they will continue to try to pump the same amount of fluid under all kinds of changing conditions within the system. The main way the volume of fluid pumped through the system is changed with the positive-displacement type of pump is that one or more bypass routes back to the reservoir are provided. The flow through the bypass route, such as a pressure-relief valve, can be adjusted, which basically changes the volume of fluid moving through the system.

Figure 5-36 shows a large electrical motor on the left and a large hydraulic pump on the right, which is the

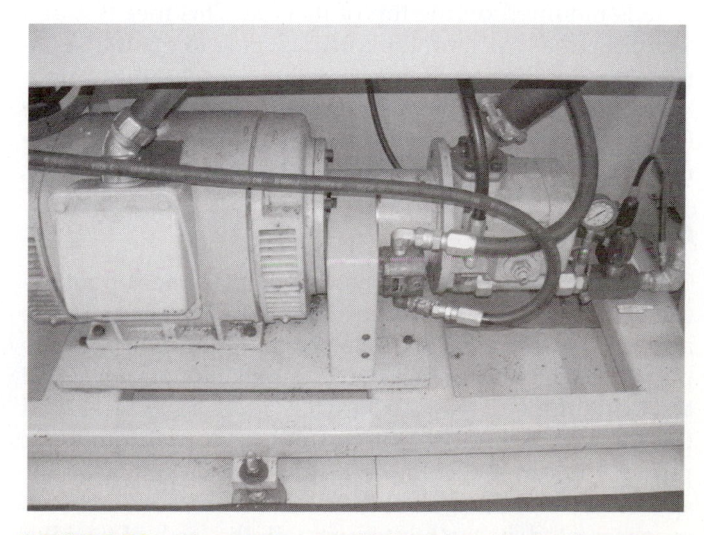

FIGURE 5-36 A large AC motor driving a large hydraulic pump and a smaller pilot pump.

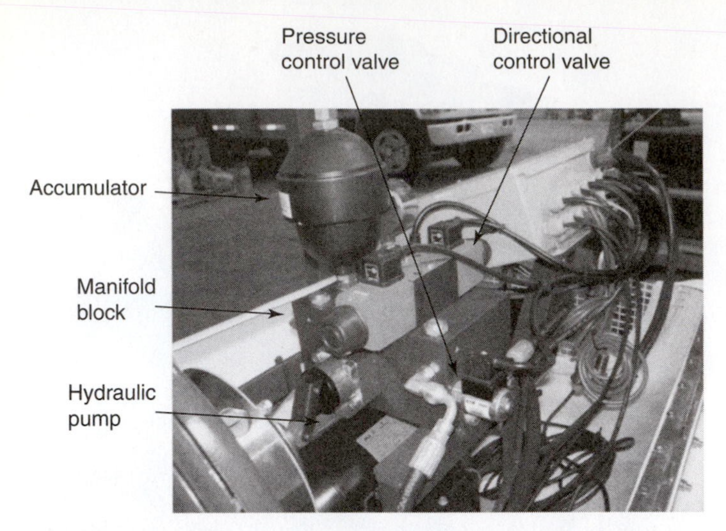

FIGURE 5-37 Hydraulic system for a wind turbine, including a manifold block, accumulator, pressure-control valve, and directional control valve. The hydraulic pump and the electric motor that drives it are under the manifold block.

FIGURE 5-38 Example of a double-gear positive-displacement pump.

main hydraulic pump for the system. A smaller pilot pump is shown in front of the larger main pump, which provides a lower volume and pressure of hydraulic fluid for the pilot circuit. You will learn more about the pilot circuit and pilot control of valves in later sections.

When the hydraulic system is designed for smaller wind turbines that are around 100 kW in output, the system must be more compact and the components such as the electric motor that drives the hydraulic pump and the controls are all much smaller. Figure 5-37 shows the accumulator, manifold block, pressure-control valve, and directional control valve for a wind turbine hydraulic system. The hydraulic pump and the electric motor that drives it are located under the manifold block and are not visible in the picture. The shaft of the hydraulic pump is keyed or has a spline so it can be driven by an electric motor through a coupling. The hydraulic system for this type of wind turbine is very compact, and it must be inspected while the nacelle is mounted on the top of its pole. This means that a technician has to climb the tower or pole to get to the nacelle, remove the nacelle cover, and then inspect the components. Most wind turbines of this size have a small platform for the technician to stand on while inspecting or making repairs to the wind turbine while it is in the air.

Gear Pumps

One of the basic types of positive-displacement pump is called a gear pump. This type of pump has two gears inside a pump housing. Figure 5-38 shows a double-gear-type pump, in which the motor drives the shaft of one gear, and the teeth of the drive gear move the second gear—called the idler gear—which causes fluid to flow through the pump. You can see from the picture that oil can fill in the spaces between the gear teeth. When the shaft of the driven gear is rotated, its teeth intersect with the teeth of the idler gear, and the oil that was located in between the gear teeth is displaced and caused to flow through the pump. Each rotation of the drive gear causes a specific amount of oil to move through the pump. That is why this type of pump is called a positive-displacement pump.

Vane Pumps

Figure 5-39 shows another type of hydraulic pump called a **vane pump**. The vanes are metal rectangles that slide back and forth in the slots in the rotor. The rotor in this figure has 12 slots, but only 4 vanes are in the slots so you can see the way they slide in the slot and what the slots look like when they are empty.

Figure 5-40 shows the vanes and rotor inside the pump housing. In this figure you can see that the pump housing is an oval rather than a circle. The oval shape allows space for the vanes to slide out of their slots so they are fully extended against the side of the pump housing.

FIGURE 5-39 The vanes of a vane pump can slide freely in the slot when the pump is turned. Only 4 of the 12 vanes are shown in this figure.

FIGURE 5-40 The vanes of a vane-type pump in a vane cartridge insert. The stationary part of the pump, called the pump housing, is oval in shape. When the rotating parts on the inside move, the centrifugal force causes the vanes to slide out of the slot at the wide part of the oval to gather oil and then slide back into the slot when vanes move to the narrow side, which causes oil to be pumped. Only 4 vanes are shown; the pump would normally have vanes in all 12 slots.

When the vanes are fully extended, there is space for hydraulic fluid to fill in the void. As the rotor rotates, the vanes are moved to a part of the pump housing that causes the vanes to slide back into the rotor, and this action causes the space where the oil is to be reduced significantly. This action forces the hydraulic oil out of the pump. Because any oil that is trapped between the vanes is forced out of the pump, the pump is considered to be a positive-displacement pump.

Now that you have a basic understanding of how this type of pump works, we can begin a more detailed description of its operation. The pump shaft is inserted into the middle of the pump rotor section, and the spline on the pump matches the spline in the rotor. This allows an electric motor to be coupled to the pump shaft, and the rotation of the electric motor causes the pump rotor to spin continually. Because the pump housing is oval, the vanes are forced outward by the centrifugal force caused by the rotor spinning. When the vanes are extended, fluid is trapped between them at the inlet. When the rotor section rotates, the vanes with oil in them are rotated to the other side of the pump housing, which does not have extra space. This action forces the vanes back into their slots, reducing the space where the oil is trapped, which forces the fluid to flow out the pump outlet. As long as the rotor section can rotate and the vanes can slide freely in and out of their slots, the pump will pump fluid. The vanes and housing are called a *cartridge* and is easily replaced during pump maintenance if there is wear or if the vanes begin to stick. Over time, the vanes do become worn, or they can begin to stick in their slots, which keeps them from sliding completely out and gathering sufficient oil, at which point the pump begins to lose the amount of volume it originally pumped.

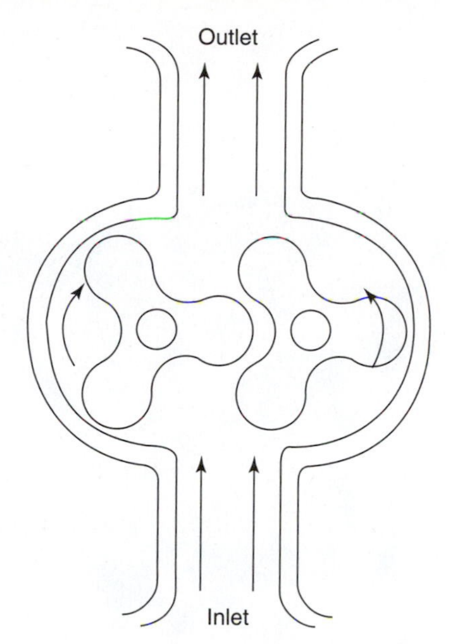

FIGURE 5-41 A lobed hydraulic pump.

Lobe Pumps

Another type of positive-displacement hydraulic pump is called a lobe pump. Figure 5-41 shows a diagram of a lobe pump, which has two lobes that intersect each other. One of the lobes is driven by the pump shaft and the other one becomes driven as the lobes intersect. The basic theory of operation of this pump is that fluid flows into the vacant space where the lobes are not touching, which is called the pump inlet, at the bottom of the pump. As the two lobes rotate, fluid is pumped when the lobes intersect and the area between them becomes reduced. This causes the lobes to push fluid to the outlet as they rotate. As long as the input pump shaft is rotating, the lobes will rotate and continually bring fluid into the inlet section of the pump and move it through the pump and on to the outlet. This type of pump operates similar to the gear-type pump, but the lobes do not tend to wear out as fast.

Variable-Volume Pumps

Positive-displacement pumps that have a fixed volume move the same volume of fluid with each rotation of their shafts. If the volume of oil being pumped needs to be varied with a fixed-volume pump, the main way to change the volume is to change the speed of the electric motor. Because the speed of an electric motor cannot be changed easily while maintaining its horsepower rating at slow speeds, variable-volume hydraulic pumps were developed. Figure 5-42 shows a type of variable-volume pump called a radial piston pump. The pump housing is at the bottom, with the pistons and swash plate at the top. These pumps change the volume of fluid that is pumped with each rotation of their shafts by changing the length of the piston stroke. Figure 5-43 shows the pistons removed from their cylinders. The pump housing on the left shows the cylinders into which the pistons fit. The

FIGURE 5-42 A variable-volum piston pump, showing the pump housing, pistons, and swash plate.

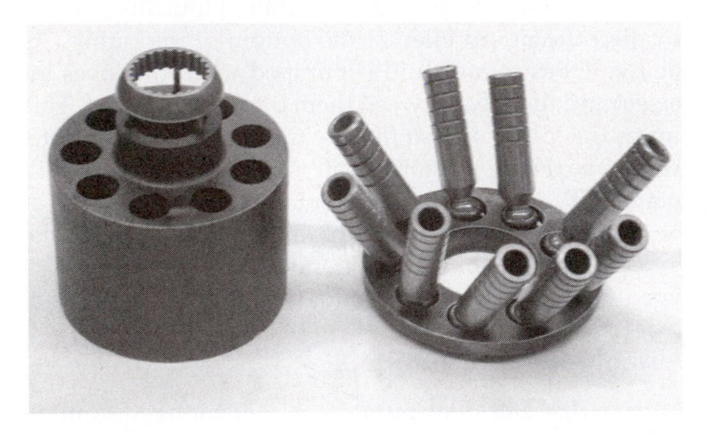

FIGURE 5-43 A variable axial piston pump, showing the cylinders in the pump housing on the left and the pistons mounted in the swash plate on the right.

pistons are mounted through a swash plate that can be adjusted to change the stroke of each piston. When the swash plate is moved in one direction, the stroke of each piston is at its maximum point and the volume of oil each piston pumps is a maximum. When the swash plate is moved in the opposite direction, the stroke of the pistons is moved to the minimum position and each piston pumps the minimum amount of fluid each time the piston strokes.

Figure 5-44 shows the same pump, with the swash plate tilted so it allows the pistons to move to their

FIGURE 5-44 Variable axial piston pump showing the swash plate and pistons. When the swash plate is at the maximum angle as on the left, the pistons on the far left side of the pump move through their maximum stroke and pump the maximum volume. When the swash plate is at its minimum angle as on the right, the pistons on the far left side of the pump pump their minimum stroke and pump their minimum volume.

maximum stroke on the left, and with the swash plate tilted in the opposite direction to move the pistons to their minimum stroke on the right. In the variable piston pump the volume can be adjusted manually by turning an adjustment rod that moves the swash plate, or the swash plate can be continually adjusted by changing the position of a cylinder rod that is controlled by a variable analog proportional valve or servo valve. The analog proportional valve or servo valve allows the swash plate to be moved in very small increments from minimum to maximum to vary pump output flow between 0 and 100%. Computer control can be used to make an analog proportional valve or servo valve continually change the output volume of the pump to meet varying requirements.

The application for the hydraulic system for the wind turbine is the determining factor in whether a fixed or a variable positive-displacement pump is used. A variable-volume pump can be used in conjunction with variable flow control valves to adjust the amount of fluid flowing, which will control the speed of the actuators that are used to change the yaw position for the nacelle or to change the pitch of the blades. The variable pumps are used to provide fluid at precisely controlled speed and pressure. The variable pumps are generally controlled by feedback sensors, such as pressure and flow sensors, to keep the system operation as accurate as possible.

5.29 DIRECTIONAL CONTROL DEVICES

All hydraulic systems need some type of directional control devices to force fluid to flow in the directions required by the application. The simplest form of directional control is a check valve; more complex forms of direction control include two-way, three-way, or four-way directional control valves. Directional control valves can be actuated manually, hydraulically by a pilot system, or electrically by solenoids.

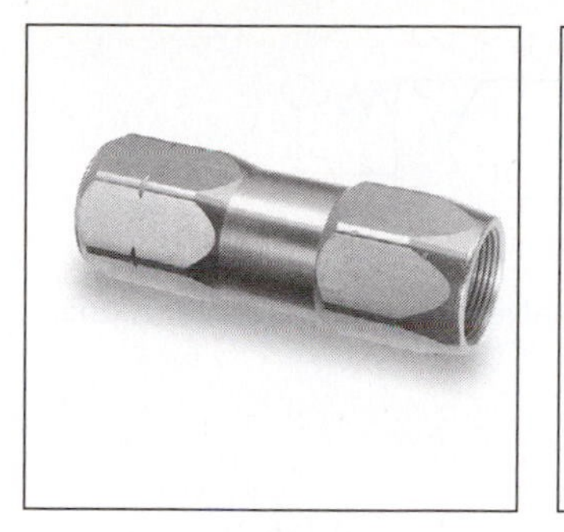

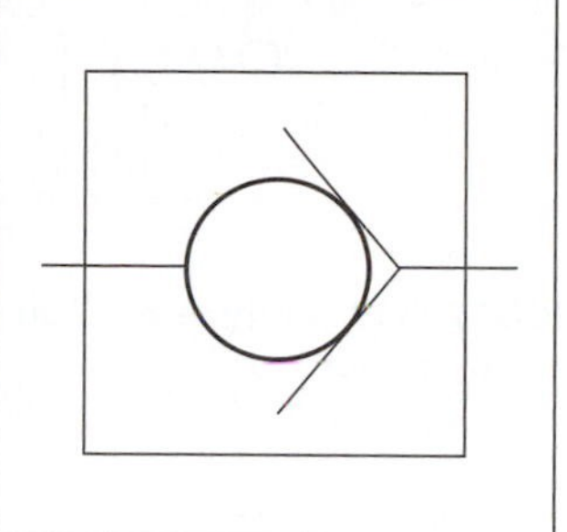

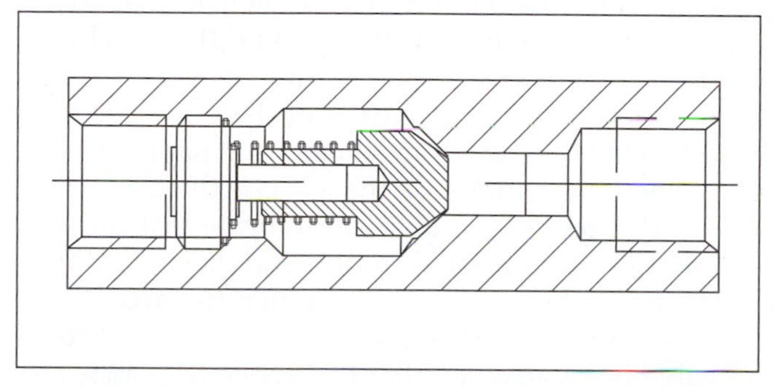

FIGURE 5-45 The check valve is shown with the ball (poppet) unseated by the flow from left to right. When the flow is moving from right to left, the ball moves against its seat and causes the flow to stop. A symbol for a check valve is also shown at the upper right of the figure. (Courtesy of Parker-Hannifin Corporation.)

Check Valves

Check valves are used in hydraulic systems to ensure that fluid flow is always in the proper direction. Figure 5-45 shows an example of a check valve that consists of a valve body and a ball. The ball is also called a poppet. When fluid flows through the check valve in the correct direction, the ball moves away from its seat to allow maximum flow. When the flow is in the wrong direction, the ball is forced against its seat, and all flow is stopped. Some check valves also have a spring to hold the poppet (ball) in the proper location.

Directional, Pressure-, and Flow-Control Valves

Hydraulic systems are controlled by directional, pressure-, and flow-control valves. This section will explain how these valves operate and how they are used in wind turbine hydraulic systems. Pictures and diagrams are provided to explain where the valves are and how they operate. This information will help you when you must adjust or troubleshoot these hydraulic components.

Directional Control Valves

Directional control valves are used to control the direction in which fluid flows through a hydraulic system. When the directional control valve is used to extend and retract a hydraulic cylinder, the directional valve changes the direction of flow from the rear of the cylinder (cap end), to make it extend, to the front of the cylinder (rod end), to make it retract. Directional control valves are categorized by the number of positions they have (two or three) and the number of ways they can act (two-way, three-way, or four-way).

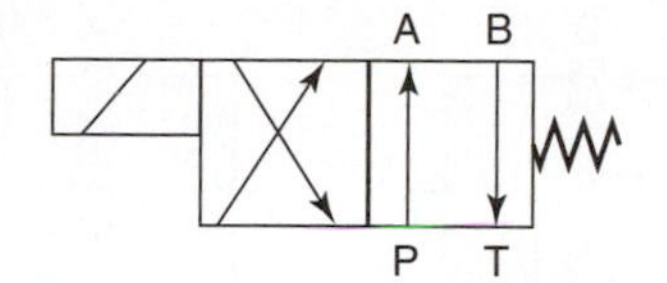

FIGURE 5-46 A two-position valve with four ports.

Figure 5-46 shows several important parts of a two-position directional control valve. This valve is actuated by an electric solenoid. The solenoid symbol is shown on the left side of the valve as a slash positioned diagonally across a rectangle. If the valve is operated manually, it has the symbol of the manual operator at the end of the valve. You will learn more about valve actuators later in this section.

The symbol diagram of the two-position valve shows two boxes. The number of positions the valve has is indicated by boxes that show the interconnections of the valve ports. This two-position valve shows two boxes, and a three-position valve will show three boxes. The symbol diagram also shows the number of ports for the valve. It is important to understand that you should count the number of ports shown in only one of the boxes to determine the number of ports the valve has. Each box that shows the different positions for the valve show the valve ports again, so the diagram for a two-position, four-port valve looks like it has four ports in the box showing the position on the left side of the valve and four ports in the box showing the position of the valve on the right side. In reality, this valve has just four ports, but we must show all four ports in the box on the left and all four ports in the box on the right to indicate the positions for the valves. The valve in the figure shows the valve ports identified as P, T, A, and B. P stands for the port connected to the pressure source, T stands for the port that is connected to the tank, and A and B indicate the valve ports that are the output for the valve. The spring return for the valve is shown on the right side of the valve. Only the ports on one box (position) are identified, and the position that has the ports identified is considered the initial valve position. The initial position is usually the position the valve is in when no power is applied to the valve. In Figure 5-46, the ports on the right side that are controlled by the spring are identified, so the spring side of the valve is considered the initial position for the valve. Even though the ports on the left side of the diagram are not identified, they carry the same letters as the ports on the right side.

Because this is a two-position valve, Figure 5-46 shows each position in its own box. The lines inside each box indicate how the **valve spool** interconnects the ports to ensure that fluid flows from one port to the other as indicated. When the valve is activated by the solenoid on the left, the spool moves to the position that shows the pressure port (P) connected to port B and the tank port (T) connected to port A. When the solenoid is not energized, the spring on the right side of the valve takes over, moves the valve spool

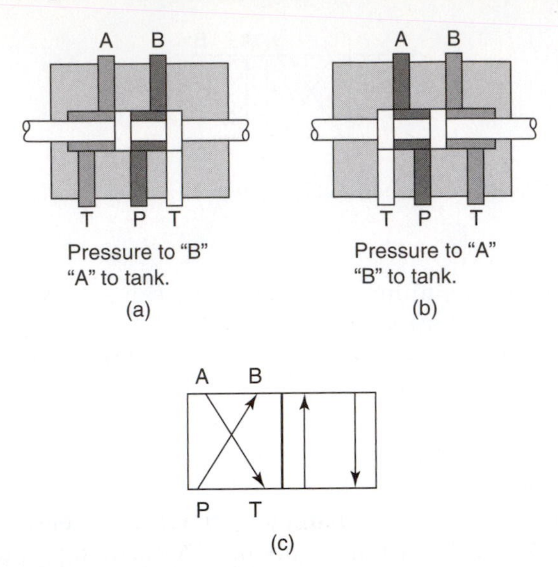

FIGURE 5-47 Internal flow pattern between pressure and port A and pressure port and port B. (a) Spool is set so pressure is sent to port A. (b) Spool is set so pressure is sent to port B. (c) Hydraulic diagram of directional valve with left position box showing pressure directed to port B, and right position box showing pressure directed to port A.

to the right, and the connections shown in the box on the right become active. The valve diagram shows that port P is connected to port A and port T is connected to port B when the spring has control and the spool is moved to the right.

Figure 5-47 shows the inner working of the spool and valve for the two-position valve. The spool is designed so that parts of it are cut away to allow fluid to flow around the segments of the spool, and other parts are full size to block fluid. When the spool shifts so that a cut-away part is aligned with two ports, fluid flows through those ports. When the full-size part of the spool is aligned with two ports, fluid flow is blocked between those ports. This diagram also shows how the spool for a two-position, four-way valve allows fluid to flow between port B and the pressure port (P) and between port A and the tank port (T) when it is shifted to the right. Figure 5-47b shows the spool shifted to the left so that the pressure port and port A have flow and the tank port (T) and port B have flow. Figure 5-48 shows a cutaway diagram of a directional control valve so you can see the design of the spool and the inside of the valve.

Figure 5-49 shows a three-position, four-way valve that is a operated manually. The three-position

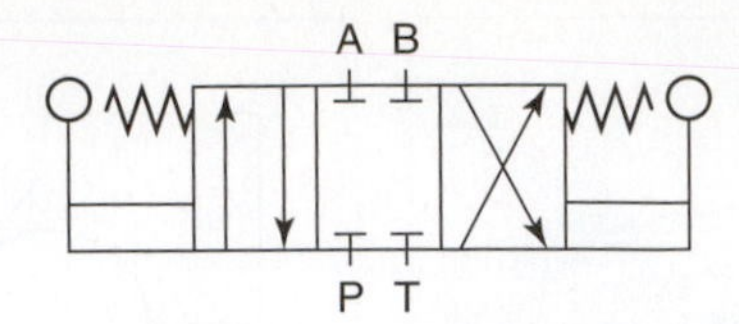

FIGURE 5-49 A three-position valve with four ports that is operated by a lever.

valve has three distinct positions to which its spool can be moved to cause the direction of fluid flow to change. The diagram for the three-position valve shows the connections of the ports in each of the three positions. The left side of the diagram shows what happens in the valve when the actuator on the left is activated. The arrows in this part of the diagram indicate that the pressure port (P) is connected to port A, whereas port B is connected to the tank port (T). When the actuator on the right side of the valve is activated, the flow through the valve is reversed so that the pressure port is connected to port B and port A is connected to the tank port. When the lever is not activated to the left or the right, the springs move the valve to the center position, which shows all four ports blocked. Figure 5-50 shows a the lever-operated valve and a cutaway diagram of this three-position valve. Notice the symbol in Figure 5-49 showed that the valve has two levers (one on each end), but the valve really has only one lever, as shown in Figure 5-50, which can move the valve spool in either direction.

Other Types of Valve Actuators and Other Types of Passageways Through Directional Valves

Hydraulic valves can have many types of actuators, including pushbuttons, rollers, and pushbuttons with a detent. Figure 5-51 shows the symbols for these types of actuators. Hydraulic valve actuators are sometimes called valve operators.

Directional valves can have two, three, or four ports. Directional valves can also have one, two, or three positions, and each position can have its ports configured

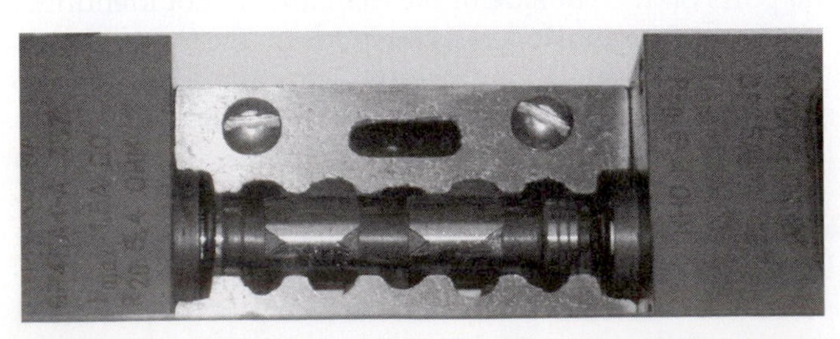

FIGURE 5-48 Cutaway view of the spool of a directional control valve.

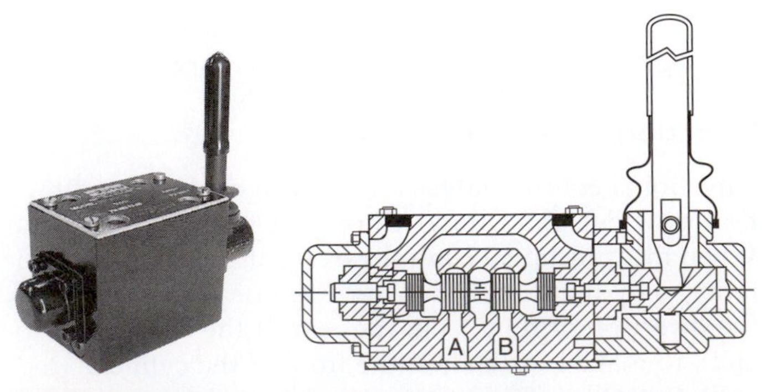

FIGURE 5-50 A lever-actuated valve and a diagram of the inside of a lever-actuated valve. (Courtesy of Parker-Hannifin Corporation.)

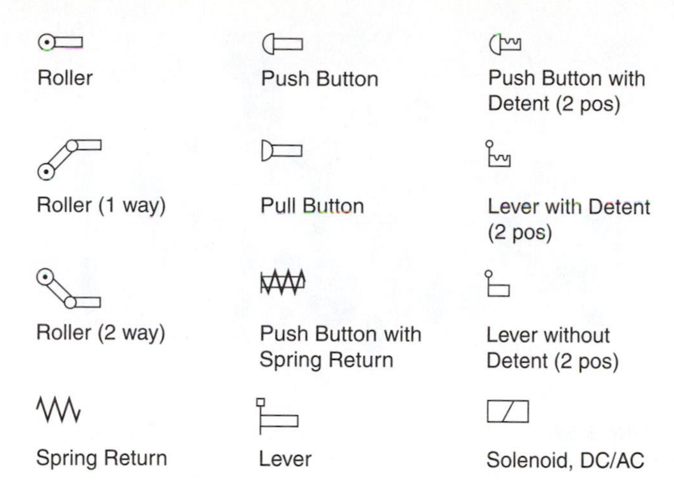

FIGURE 5-51 Types of actuators (operators) used with directional control valves.

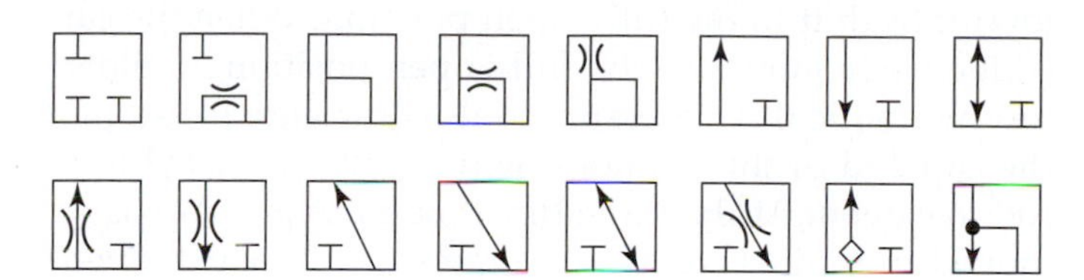

FIGURE 5-52 Common configurations for three-port directional valves.

differently. Figure 5-52 shows a group of common configurations for three-port directional valves, and Figure 5-53 shows a group of common configurations for four-port valves. You will not be required to select the types of valve ports, but you will need to be aware when you are replacing a hydraulic valve that there are many different types.

Pilot-Operated Hydraulic Valves

Some types of hydraulic directional control valves are so large that they cannot be operated with levers or solenoids. One way to operate these larger valves is to use hydraulic pressure called **pilot pressure**. In a hydraulic diagram, the pilot pressure lines are shown with a dashed line. The pilot pressure can be created from the main system pressure through a pressure-reducing valve, or with a separate smaller hydraulic pump designed to create pilot pressure. Figure 5-54 shows a pilot-operated directional control valve, and Figure 5-55 is a diagram of

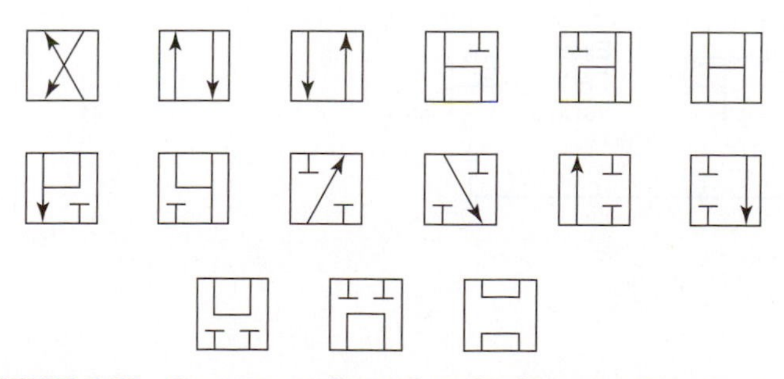

FIGURE 5-53 Common configurations for four-port directional valves.

FIGURE 5-54 A pilot-operated hydraulic valve. (Courtesy of Parker-Hannifin Corporation.)

a pilot-operated directional control valve. In Figure 5-54, the large valve is shown on the bottom of the valve stack, and its spool is activated by pilot pressure that is controlled and directed by the smaller directional valve mounted directly on top of it. Sometimes a pilot-operated directional control valve is called a master-and-slave valve. In the hydraulic diagram of Figure 5-55, you can see the smaller directional valve directly below the larger

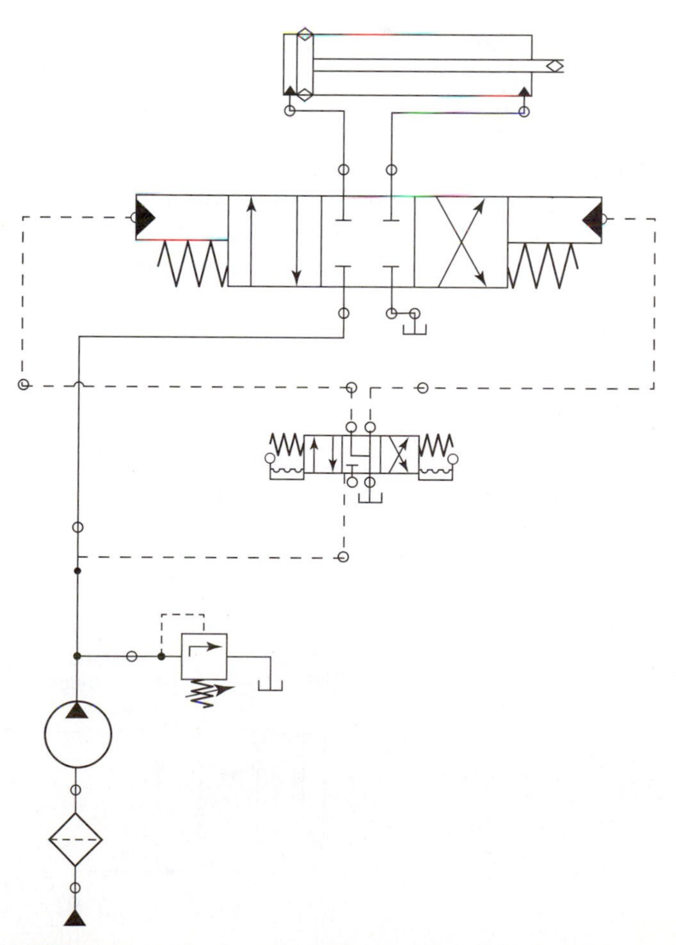

FIGURE 5-55 Hydraulic circuit with pilot-operated valve. The pilot pressure lines are shown as dashed lines.

directional valve that is connected to the larger cylinder. When the lever on the smaller directional valve is moved to the left, the pilot fluid moves through the smaller valve into the pilot control on the left side of the larger valve. When the pilot fluid reaches the pilot operator on the valve, it creates a pressure that is large enough to make the spool in the larger valve shift, causing the cylinder to extend its rod. If the lever on the smaller valve is moved to the right, the pilot pressure is directed to the right side of the larger valve, causing the spool to shift and make the cylinder rod retract.

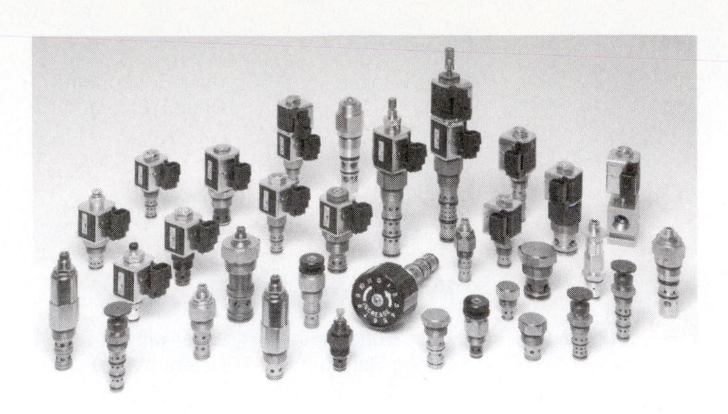

FIGURE 5-56 Cartridge valves screw into a manifold block. (Courtesy of Parker-Hannifin Corporation.)

Cartridge Valves

Another type of hydraulic valve used on wind turbines is the cartridge valve. A **cartridge valve** may be used for directional control, pressure control, or flow control. Unlike the types of valves we have discussed so far, cartridge valves are mounted in a manifold block instead of as stand-alone units. Being mounted in a manifold block limits the types and extent of leaks the valve can experience. It also makes the valve assembly more compact. Figure 5-56 shows a manifold block and screw-in cartridge valves. Figure 5-57 is a hydraulic diagram showing cartridge valves. Each cartridge valve has a main input port on its side, near the bottom of the valve. The valve has an output port at the bottom of the valve. The valve has a pilot port at the top of the valve. The internal part of the cartridge valve moves to the closed position when pressure is applied to it pilot port. When you want the valve to move to the open position, the pilot pressure must be removed from the pilot port so the valve will open when any fluid is present on the main input port on the side of the valve. Figure 5-57 shows four cartridge valves. When the manual valve lever is moved to extend the cylinder, the directional valve relieves pilot pressure from cartridge valve 1, which allows it to shift to the fully open position. When the cartridge valve moves to the fully open position, it allows full hydraulic fluid pressure to flow through it and into the cap end of the cylinder, which causes the cylinder rod to extend. At the same time, the pilot pressure is removed from the extend cartridge valve 2, which opens and allows fluid to flow from the rod end of the cylinder back to the tank. When both valves are open, the fluid makes the cylinder rod extend. The directional valve also provides hydraulic pressure to the pilot port of retract cartridge valve 2 on the left side of the diagram, which prevents oil from getting back to the tank through this valve. Pilot pressure is also provided to retract cartridge valve 1, which closes it and prevents main hydraulic oil from reaching the rod end of the cylinder. When pilot pressure is present on the pilot port of the cartridge valves, the pilot pressure and internal spring inside each valve send the valve back to the closed position so the rod stops moving.

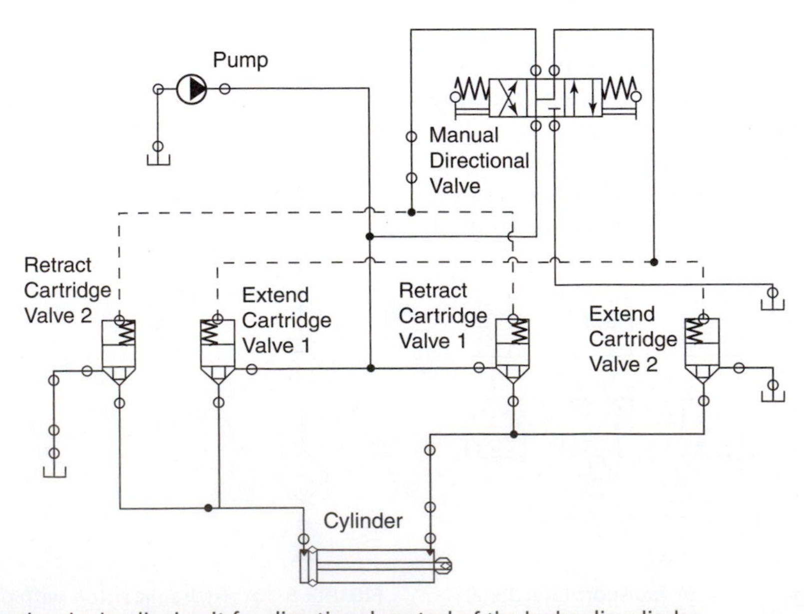

FIGURE 5-57 Cartridge valves in a hydraulic circuit for directional control of the hydraulic cylinder.

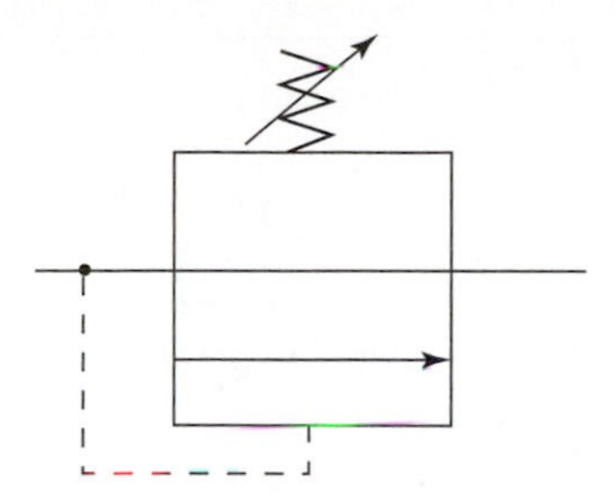

FIGURE 5-58 Hydraulic symbol for a pressure-relief valve.

When the manual directional valve is moved to the retract position, pilot pressure is sent to retract cartridge valve 1, which opens and sends full hydraulic oil pressure to the rod end of the cylinder. The same pilot pressure is also sent to retract solenoid valve 2, which opens and allows fluid to flow from the cap end of the cylinder back to the tank. When both valves are open, the cylinder rod retracts.

Pressure-Control Devices (Relief Valves)

Hydraulic systems need to have a means of controlling the pressure in the system. As you know, the hydraulic pump provides fluid flow to the system. When the flow of the fluid is restricted, pressure builds up in the system. When fluid is allowed to flow back to the tank, the pressure drops because the tank is not pressurized. **Pressure control** allows more or less fluid to bypass the system and return to the tank. Pressure-control, valves are also called relief valves because they relieve the pressure in the system.

Figure 5-58 shows the symbol for the pressure-relief valve. The symbol shows that the valve's relief pressure is controlled by spring tension and the direction of fluid flow through the valve when the pressure setting is exceeded. The arrow through the spring indicates that this valve is a variable-type relief valve. Figure 5-59 shows a pressure-relief valve, in which you can see that the valve has a handle that allows the spring tension on the valve

FIGURE 5-59 A pressure -relief valve. (Courtesy of Parker-Hannifin Corporation.)

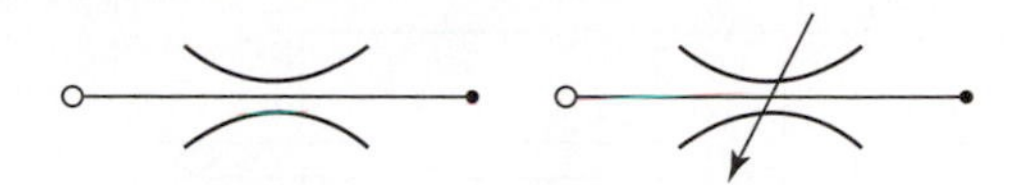

FIGURE 5-60 Symbols for hydraulic flow controls. The arrow through the second symbol indicates that the flow control is variable.

to be set manually. When the handle is turned clockwise, spring tension is increased, and the amount of hydraulic pressure must increase before the valve relieves pressure by allowing flow to bypass to the tank. When the valve handle is turned counterclockwise, the pressure setting is decreased. The spring tension opposes the fluid pressure in the pressure-relief valve, so where the spring tension is set determines the amount of pressure the system will need to exceed it and dump the remainder of the pressure back to the tank. The spring constantly adjusts the valve opening to allow more or less fluid to return to the tank to keep the fluid pressure constant in the system.

Flow-Control Valves

The fluid flow in a hydraulic system can be metered to control the speed at which a hydraulic cylinder extends or retracts. **Flow control** allows minor adjustments to be made in various parts of the system to balance the speed of its moving parts. The left side of Figure 5-60 shows the hydraulic symbol for a fixed flow-control valve, and the right side of the figure shows an arrow through the symbol, which indicates that the flow control valve is a variable flow-control valve. The symbol looks like the hydraulic line is being squeezed down to restrict flow. The flow-control device is connected in series with the component it controls. This means that whatever the flow is in the flow-control device, the same flow will move through the cylinder or hydraulic motor with which the device is in series, because the fluid has only one path available. Flow control is used in wind turbine systems to control the speed of actuators such as those that control the blade pitch, and also the system that controls the yaw movement of the nacelle.

Figure 5-61 shows the symbols for a flow-control device with a check valve added in parallel. The symbol on the left is for a fixed flow control, and the symbol on the right is for a variable flow control. The check valve blocks fluid flow in one direction and forces all fluid flow to move through the flow-control valve. This allows the fluid flow to be controlled by the valve setting. When the fluid flow is reversed, the check valve unseats and allows fluid flow at full flow to bypass the flow-control valve. This type of valve allows the flow to be controlled when a cylinder is extending so that its speed

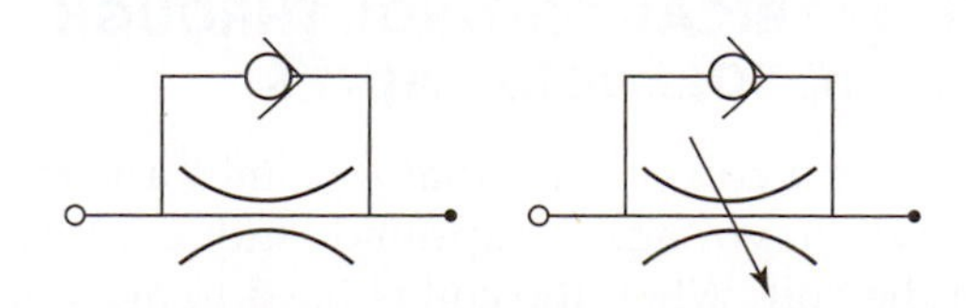

FIGURE 5-61 Symbols for flow controls with check valves.

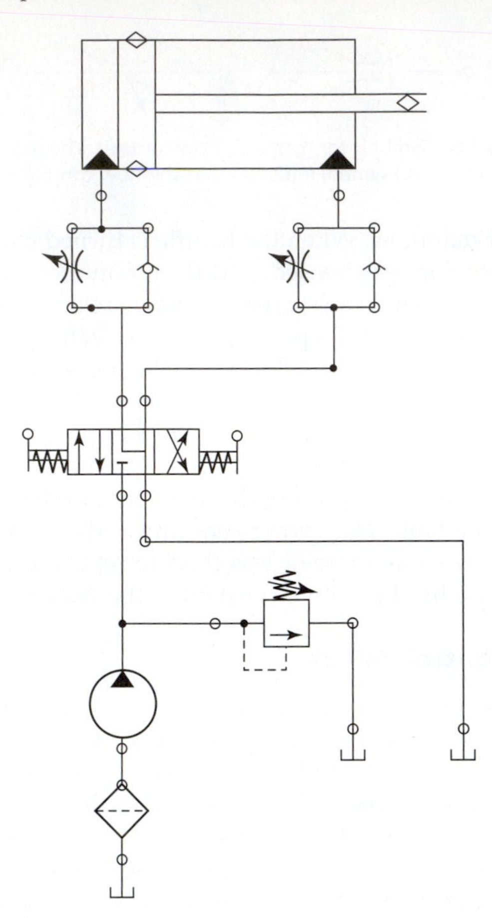

FIGURE 5-62 Diagram with flow control and check valves used to control the speed of the cylinder.

can be controlled. When the flow is in the opposite direction, it is allowed to bypass the flow-control valve by flowing through the check valve. When two of these flow controls with check valves are used together, they can allow the speed of extension of the cylinder to be different than the speed of retraction of the cylinder.

Figure 5-62 shows a hydraulic circuit diagram with flow control valves with check valves in a circuit that controls the speed of a hydraulic cylinder. The flow control on the left side of the cylinder controls fluid flow into the cap end of the cylinder, which controls the extend speed of the cylinder. The flow control and check valve on the right side of the cylinder control the speed at which the cylinder retracts. This type of configuration of the flow control and check valve is called *meter in control*. If the direction of the check valves were reversed in the circuit, the circuit would be called *meter out control*—the flow control on the left side would control the retract speed and the flow control on the right would control the extend speed.

5.30 ELECTRICAL CONTROL THROUGH HYDRAULIC SOLENOID VALVES

A solenoid is a coil of wire that becomes a very strong magnet when voltage is applied and current flows through the coil. When the coil is used to activate a hydraulic valve, the strong magnet pulls a metal armature

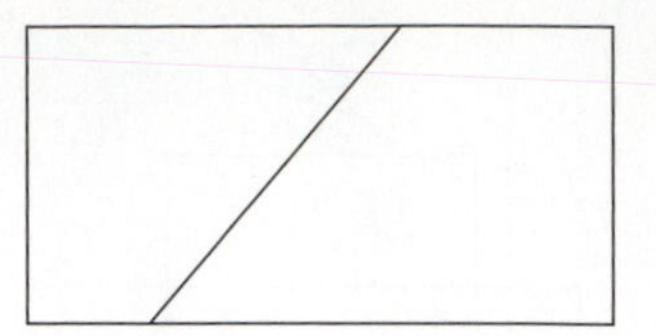

FIGURE 5-63 Symbol for solenoid-operated valve.

that makes a spool inside the valve shift. When the spool of the directional control valve shifts, it changes the way the ports are connected through the valve. The solenoid hydraulic valve has a spring loaded against the spool, so when the electric current stops flowing in the solenoid coil, the magnetic field stops, and the spring shifts the spool back to the original position. Solenoid valves are used to control directional, pressure-, and flow-control valves in hydraulic systems.

Figure 5-63 shows the symbol for a solenoid valve. You can see the symbol is a diagonal line through a rectangular-shaped box. Many types of hydraulic components can be controlled by a solenoid, so this symbol is added to the hydraulic symbol to indicate that the component is controlled by a solenoid. Because the solenoid valve is controlled electrically, the PLC controller can easily turn the valve on or off and provide fine control of the wind turbine.

5.31 HYDRAULIC CYLINDERS AND HYDRAULIC MOTORS

In every hydraulic system, an actuator is used to convert the energy in the hydraulic fluid into linear motion with a hydraulic cylinder, or to rotary motion with a hydraulic motor.

Hydraulic Cylinders

Hydraulic cylinders are used in wind turbines to control the pitch of the blades, and hydraulic motors are used to rotate the yaw positioning mechanism. Figure 5-64 shows four types of hydraulic cylinders. The first cylinder is a single-acting cylinder with a spring return. Fluid is sent to the cap end of the cylinder, and when fluid is removed, the spring makes the rod retract. The second cylinder is called a single-acting cylinder with spring extend. This cylinder takes fluid in on the rod end to make the rod retract; when fluid is not present, the spring makes the rod extend. The third type of cylinder is called a double-acting cylinder. In this type, fluid is put into the cap end to make the cylinder extend, and fluid is put into the rod end to make the cylinder retract. The fourth type of cylinder is called a double-ended cylinder, in which the rod extends out both ends of the cylinder. Fluid is put into the rod end or cap end to make the rod move left or right as with the double-acting cylinder.

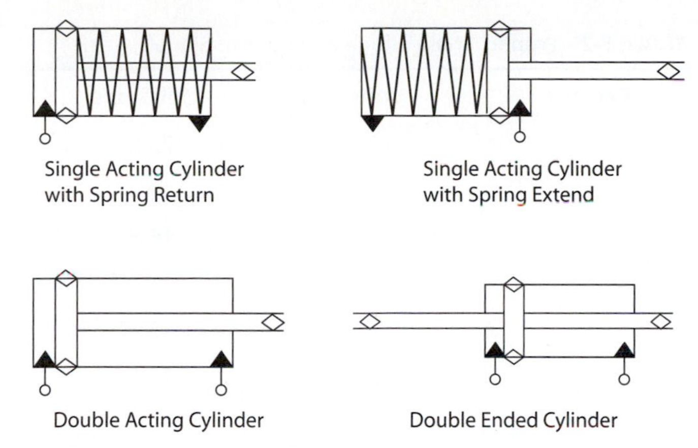

FIGURE 5-64 A single-acting cylinder with spring return, a single-acting cylinder with spring extend, a double-acting cylinder, and a double-ended cylinder.

Figure 5-65 shows a cutaway view of a hydraulic cylinder. The cylinder has seals to ensure that fluid does not leak around the cylinder rod. When the cylinder rod extends, the seal flexes to ensure that it does not leak. When the cylinder rod changes direction and retracts, the seal flexes in the other direction.

Hydraulic Motors

A hydraulic motor takes fluid so that the shaft in the motor rotates as the fluid moves through the motor and back to the tank. The hydraulic motor allows the hydraulic fluid to be converted into rotary motion. The flow-control valves in a hydraulic circuit can control the volume of fluid moving through the hydraulic motor, which controls the speed of the motor. Hydraulic motors are used in the yaw control system to rotate the direction in which the nacelle is pointing. Figure 5-66 shows a typical hydraulic motor. The motor has a port where fluid enters the motor and a port where fluid leaves the motor. If you change the direction in which fluid moves through the motor, the direction its shaft turns reverses. The hydraulic motor's shaft is splined so that it can be fitted into a gearbox to transfer the rotary energy of the hydraulic fluid into mechanical motion and move gears or the main gear ring of the yaw motion mechanism.

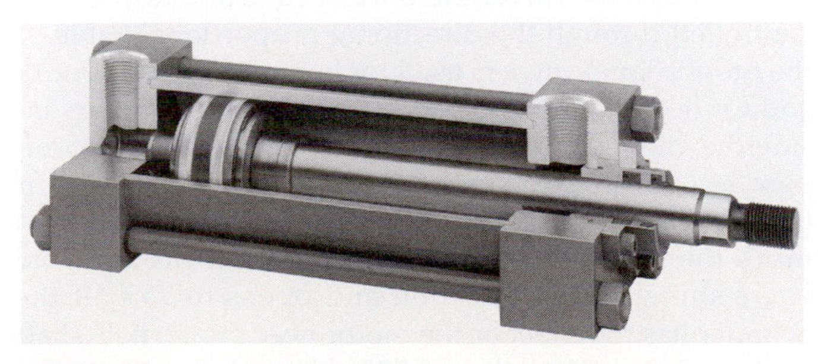

FIGURE 5-65 Typical hydraulic cylinder. (Courtesy of Parker-Hannifin Corporation.)

FIGURE 5-66 A hydraulic motor. (Courtesy of Parker-Hannifin Corporation.)

5.32 HYDRAULIC PROPORTIONAL CONTROL VALVES

The hydraulic cylinder explained in the previous section had two positions—fully extended or fully retracted. Some hydraulic applications need the cylinder to operate at some point between fully extended and fully retracted. Also, when flow-control valves are set manually, they remain in that position and the system operates at that speed. A simple example of this type of control is any light that works via a switch that is either on or off. Because the light can only be on or off, this type of control is called *on/off control*. Proportional control is analogous to a light fitted with a dimmer switch, which allows the light to be set at 10%, 20%, 50%, or 75% as well as at 0% and 100%. Another example of proportional control is the water valve on a kitchen sink: You can open the valve to provide any amount from a very small flow through 100% flow.

In a hydraulic system, proportional control allows an actuator or control valve to operate at any value between 0 and 100%. Proportional control is used on systems such as the blade pitch control system, so the blade pitch can be set at any position between 0 and 100% as needed. For example, if the blade pitch needs to be set at 20% to collect more wind, hydraulic proportional control can extend the hydraulic cylinder to 20%. If the wind changes and the blade pitch needs to be changed to 50%, the proportional controller can adjust it to 50%.

When a hydraulic motor is used to set the yaw movement of the nacelle, the hydraulic flow valves that control the motor's speed and direction are **proportional valves**. The PLC or computer sets the electrical signal to a value between 0 and 100%, and this signal controls the proportional valve to a position between 0 and 100%. The proportional valve is basically an electrical solenoid valve that has a more precise spool and more powerful

FIGURE 5-67 A hydraulic proportional valve. (Courtesy of Bosch Rexroth Group.)

coil. The coil receives variable voltage, 0–10 V, or variable current, 4–20 mA, to control the position of the spool of the valve. Because the voltage or current must be varied, an electronic amplifier is used between the hydraulic proportional valve and the PLC or computer that sends the electrical signal.

Figure 5-67 shows a proportional valve, and Figure 5-68 shows the electrical amplifier that controls it. Figure 5-69 shows the symbol for the hydraulic proportional valve. In this figure there is an arrow through the solenoid symbol, which indicates that the electrical solenoid valve receives variable voltage or variable current.

FIGURE 5-68 Amplifier for a hydraulic proportional valve. (Courtesy of Bosch Rexroth Group.)

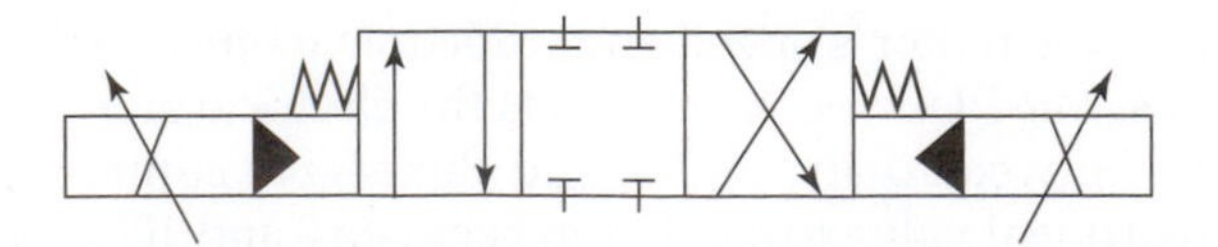

FIGURE 5-69 Hydraulic diagram for a sproportional valve. Notice the arrows through the solenoid symbols on each end of the valve.

TABLE 5-2 Proportional Voltage and Current Signals

Percent	Volts	Milliamps
100	10	20.0
90	9	18.4
80	8	16.8
70	7	15.2
60	6	13.6
50	5	12.0
40	4	10.4
30	3	8.8
20	2	7.2
10	1	5.6
0	0	4.0

Typical Voltage and Current Signals

Typical voltage and current signals for proportional hydraulic valves are listed in Table 5-2. In this table the voltage is 0–10 V and the current is 4–20 mA. Both signals are similar in that they can provide accuracy of 1/1000 if needed, and cost is not a problem. The higher the resolution and accuracy, the more expensive the system becomes. The table shows the value in 10% increments. It is important to understand that electrically, the voltage system and the milliamp system are very similar. The major difference is that the milliamp signal provides a 4-mA offset. This means that the zero signal is set at 4 mA. If a wire in the system becomes broken or if the amplifier becomes broken, the signal will go all the way to 0 mA, which the system can interpret as a broken signal. However, the voltage signal uses 0 V as the lowest voltage, so it is impossible to determine whether 0 V means a wire is broken.

Proportional Directional and Flow Control

The proportional directional valve can adjust its spool inside the valve from 0 to 100%, which also adjusts the amount the valve can open. This means that the directional valve can also be used to control flow of hydraulic fluid through the valve, and the direction and flow can be controlled through the directional proportional valve. If the proportional valve is used for both directional control and for flow control, the flow-control symbol appears inside the valve symbol. This makes the system less complex, as one valve can control both direction and flow. If the flow has to be set at 75% to extend the cylinder or to move the shaft of the hydraulic motor, the proportional valve shifts in that direction and opens to 75%. If the cylinder has to retract or the motor needs to turn its shaft in the opposite direction at 50%, the valve shifts to the other direction and moves to the 50% position.

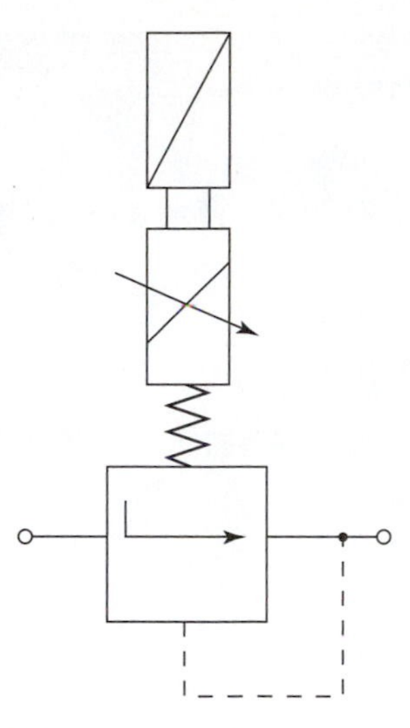

FIGURE 5-70 Symbol for a hydraulic proportional pressure valve.

Proportional Pressure Control

Another type of proportional hydraulic valve is a proportional pressure-control valve. A proportional pressure valve can be used as a pressure-relief valve, a pressure-sequencing valve, or a pressure-reducing valve in hydraulic circuits for wind turbines. Figure 5-70 shows the hydraulic symbol for a proportional pressure valve. The main part of the symbol is the pressure valve, and it has a spring control, a variable solenoid control, and a solenoid control. The arrow across the solenoid symbol indicates that the pressure-control valve is variable and adjusts when variable voltage or current is supplied to the valve. The PLC can be used to measure various sensors around the wind turbine and then send a variable voltage or current signal to the proportional pressure valve to set the hydraulic pressure for the system. The pressure can be adjusted upward when more torque or power is needed, and it can be reduced as needed by adjusting the amount of voltage or current that is sent to the valve.

Hydraulic Servo Control Valves

A **hydraulic servo control valve** is a proportional valve that has a feedback sensor built into the valve. When a proportional valve receives an electrical voltage or milliamp signal, it moves as an open-loop controller so that the signal is sent to the valve and the valve moves by some percentage. A feedback sensor is added to the proportional valve to make it a servo valve that operates in a closed loop. The feedback sensor measures the amount of movement the valve spool moves and sends this information back to the valve amplifier, which adjusts the amount of signal voltage to make the valve placement more accurate, which in turn makes the hydraulic control very accurate. The servo valve has the ability to make the hydraulic control of pressure or flow more accurate than a simple proportional valve, and the feedback feature also allows the servo valve to react much more quickly than a proportional valve. The feedback sensor in the servo valve allows the valve to continually check its actual position against the setpoint position, and makes adjustments quickly until the valve is positioned correctly. Hydraulic servo valves allow the blade pitch positioning system and the yaw positioning system to control the wind turbine to adjust continually to make the wind turbine as efficient as possible.

FIGURE 5-71 A hydraulic servo valve. (Courtesy of Bosch Rexroth Group.)

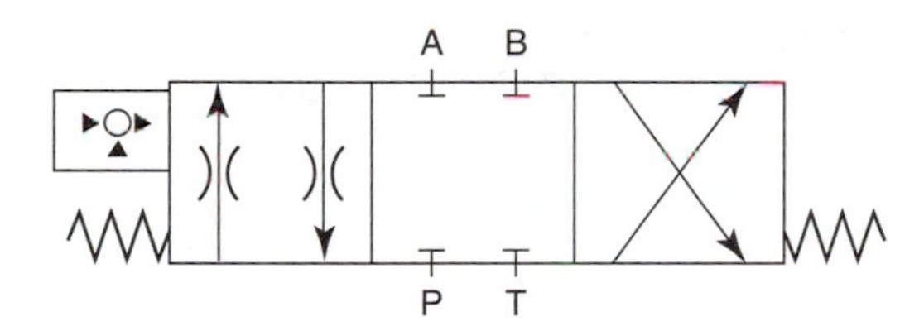

FIGURE 5-72 Symbol for a hydraulic servo valve.

Figure 5-71 shows a hydraulic servo valve, and Figure 5-72 shows the hydraulic symbol for the hydraulic servo valve. Notice the symbol on the operator box at the left side of the valve. The arrow pointing in from the left represents the setpoint, the arrow pointing up from the bottom represents the feedback signal, and the arrow pointing to the right represents the output signal for the servo. Notice the flow control symbol on the two arrows in the left-hand box (position). This indicates that this servo valve acts as a directional valve and a proportional flow control.

5.33 BRAKES ON WIND TURBINES

Wind turbines require control over the blade rotation speed for several reasons. First, if any maintenance is to be done on the blades, gearbox, or generator, the blades must be locked in place so they do not rotate. Brakes must also be available if the wind speed exceeds the safe limit or the electrical load suddenly decreases, which takes the load off the generator. The yaw motion control also requires. The brakes for the yaw control are set whenever the nacelle is pointed in the optimum position or when the yaw needs to be locked for maintenance. This section explains how mechanical and hydraulic brakes operate to control wind turbine. On some smaller wind turbines, brakes can be controlled by springs

FIGURE 5-73 Drive train and yaw drive brakes. (Courtesy of Rexroth Bosch Group.)

and mechanical systems. On larger wind turbines, these brakes are controlled by hydraulic systems.

On larger wind turbines there are three basic places that brakes are applied. Figure 5-73 shows the basic parts of the wind turbine and the places where brakes are used. The first set of brakes shown is the yaw motion brake system, which is located on the yaw ring gear. The yaw brakes set a series of interlocking pins into the yaw mechanism that secures it so it cannot move. When the yaw motion mechanism needs to move, the brake pins are pulled back to allow the nacelle to move. The yaw brakes on some smaller wind turbines are caliper brakes. These caliper brakes are always set by spring tension, and the hydraulic system overrides the spring pressure on the brakes to open the pads to open and allow the yaw mechanism to move. When the yaw mechanism has moved the nacelle to the desired position, the hydraulic pressure is released from the calipers and the spring pressure puts them back into the set position where they make contact to the yaw ring. When the brakes are set, they take some of the strain off the yaw motion system. These caliper brakes are similar to the calipers and pads on front wheel brakes for automobiles, and they are easily released when the yaw control needs to move the nacelle again. The yaw brakes are set and released many times when the wind direction is changing.

The next brakes shown in Figure 5-73 are the brakes on the rotor directly behind the rotor hub at the front of the wind turbine. The brakes on the rotor are also caliper brakes. Another set of brakes is on the high-speed shaft between the gear box and the generator. A large rotor plate is attached to the shaft, and the caliper brakes are mounted so that they can close and stop the rotor plate from rotating. When the brakes are set to stop the blade rotation, the brakes on the rotor plate directly behind the blades are set and the brakes on the high-speed shaft between the gearbox and generator are also set. The brakes on the high-speed shaft may also be set whenever the gearbox transmission is set to idle.

FIGURE 5-74 Hydraulic caliper brakes for a rotor.

Figure 5-74 shows the caliper brakes on the rotor plate of a smaller wind turbine. The calipers close and make friction contact to the rotor place, just as the disc brakes on the front wheels of a car do. The rotor disk provides a very large surface where the brake pads can provide pressure and create sufficient friction to make the shaft stop rotating. Figure 5-75 shows the large hydraulic cylinder that extends its rod upward to move the caliper mechanism to the open

FIGURE 5-75 A hydraulic cylinder that extends to make the brake pads open. Strong springs make the brakes pads press tightly against the rotor.

position. The brakes are kept in the set position by heavy springs, and when the calipers need to open to allow the rotor disc to rotate, the hydraulic cylinder extends and makes the caliper pads open.

Hydraulic Brakes

On some wind turbines that use hydraulic systems to control the brakes, the brakes are called *fail-safe brakes* because spring pressure keeps the brakes in the set position. The hydraulic system causes the brakes to open, so if there is a hydraulic failure and all hydraulic pressure is lost, the springs will return the brakes to the set position automatically, which will safely apply the brakes. When the brakes need to be released or opened, hydraulic fluid causes a hydraulic cylinder rod to extend and move the brake caliper mechanism a small distance to make the two pads of the disc brakes open. When the hydraulic fluid pressure is removed, strong spring tension closes the brakes against the rotor surface again, creating friction that makes the rotor stop rotating. When the pressure from the hydraulic fluid is turned off for any reason, the spring pressure causes the disk brakes to return the pads to a position where they are touching the rotor and creating enough friction to stop the rotor from turning. The distance between the brake pads and the rotor is kept to a minimum, usually less than a half an inch. Because the distance is very short, the brakes can be applied and the rotor can be stopped very quickly. The hydraulic system uses a directional valve, flow control, and pressure control to ensure that the right amount of hydraulic pressure is applied to keep the brakes open far enough to allow the rotor to turn. The springs keep pressure on the calipers to maintain friction on the rotor. The hydraulic pressure can be controlled by a proportional or servo hydraulic system, which makes its application very accurate.

Mechanical Brakes

Small wind turbines may not have a hydraulic system and may activate their brakes through mechanical means. These brakes are set for a fail-safe condition to be energized. This means that the springs in the brakes apply pressure all the time, unless the wind turbine blades need to be allowed to turn. When the wind begins to blow enough for the blades to turn, a cam activates the brakes and causes them to open and release their grip on the rotor and allow the blades to rotate. If the wind speed drops below the minimum needed to turn the blades, the cam moves back and allows the springs to take over and apply the brakes again. If any problems occur with the blades, the springs always cause the brakes to be applied and stop the blades from turning. This is called a fail-safe condition, because the blades are always protected against overspeed.

Mechanical Parking Brakes

Wind turbines need a parking brake to lock the blades in position when the blades must be secured for maintenance to be performed or if the blades need to be stopped to protect them from high winds. The parking brake may be a separate brake or it may be part of the braking system. OSHA requires that a parking brake be installed in the wind turbine that can be activated to stop the blade from rotating. The parking brake can also be set as fail-safe control to ensure that the blade can be quickly stopped or brought under control to ensure a safe working area.

Dynamic Braking

Another method of controlling the speed of a wind turbine is through dynamic braking. *Dynamic braking* is a process that sends energy produced by the generator to a resistive load that makes the generator increase its load and slow down. When the generator produces electricity, it can be used to charge batteries or power any electrical loads. The voltage and current flowing from the generator creates a back pressure or reverse toque on the generator shaft that is constantly trying to stop it from turning. This reverse torque also tries to stop the high-speed and low-speed shafts from turning, which in turn slows the blades' rotational speed and tries to stop the blades from turning. The balance of wind energy trying to make the blade turn and the load on the generator trying to make the blade stop turning become a balanced system in which the blades cause the generator to turn at sufficient speed to produce electrical energy.

This same theory can be used to control the speed of the wind turbine through a wide variety of wind loads. For example, if the wind is blowing very hard but the electrical load that is using electrical power is at its minimum, the back torque the generator produces is minimized and the speed of the blades will begin to increase to a faster and faster speed, which could allow the blades to spin faster than their safe speed. When this occurs, the mechanical brakes need to be applied, or a dynamic braking system can be employed. The dynamic braking system puts a small electrical resistance load called a *load bank* in the circuit with the generator output, which causes the generator to load up just as though a normal electrical load was applied. This resistance load causes the generator to put the back torque on its shaft and bring the wind turbine blades back to a safe speed. As the regular electrical load picks up again, the resistive load is removed, allowing all the electrical power to go directly to the regular electrical load.

When the resistive load is put in the generator circuit, it produces a large amount of heat as it consumes the electrical current. If dynamic braking is used, a method to cool the resistors in the load bank must be provided. If the wind turbine is used to supply voltage and current to the grid, this type of load problem very rarely occurs, as the load to the grid is fairly constant. If the wind turbine is used to charge batteries or supply voltage to a residential application or small business, this loading problem occurs more often; in this case, dynamic braking is used more frequently to control the smaller loads and keep the blades from overspeed conditions.

Questions

1. Explain the four things that occur when the PLC processor scans its program.
2. Explain how a technician can use a PLC to determine whether an input switch or output solenoid is working correctly.
3. The input module has a status indicator for each input circuit. Explain how the status indicator is used in troubleshooting.
4. Explain what an internal control relay in a PLC is, and why it is different from a hardware-type relay.
5. Explain the operation of a hydraulic flow-control valve, a directional control valve, and a pressure-control valve.
6. Describe the four basic parts of any programmable controller.
7. Explain the operation of a proportional hydraulic valve.
8. Identify the basic parts of a horizontal wind turbine.
9. Explain why the pitch of the blades of the wind turbine may need to be adjusted.
10. Identify the places on a horizontal wind turbine where brakes are used.

Multiple Choice

1. A latch coil
   a. Maintains its state when power is interrupted and turned back on
   b. Is only sealed in and returns to reset (off condition) when power is interrupted
   c. Requires an unlatch coil with the same address to get it to turn off
   d. Both a and c
2. The analog signal for a PLC
   a. Represents values between 0 and 100%
   b. Is only an on/off signal
   c. Is always 110 V
3. The hydraulic proportional valve
   a. Is an on/off valve
   b. Is a 0–100% valve
   c. Allows fluid to flow in only one direction
4. When the PLC processor is in the run mode,
   a. It executes its scan cycle.
   b. It does not execute its scan cycle.
   c. It is impossible to tell whether the processor is executing its scan cycle, because you are not online.
5. A hydraulic servo valve
   a. Is a proportional valve with a feedback sensor in it
   b. Is a valve that allows fluid to flow in one direction
   c. Stores fluid for the hydraulic system
6. The basic parts of a closed-loop diagram are
   a. The feedback, the process variable, the input, and the sensor
   b. The setpoint, the summing junction, the process variable, the controller, the amplifier, and the output
   c. The input, the output, and the math algorithm
   d. All of the above
7. The nacelle yaw position is moved in order to
   a. Move the blades into the wind to ensure that they harvest the most wind
   b. Move the blades out of the wind when the wind is blowing too strongly
   c. Move the blades to a position to stabilize them out of the wind during maintenance
   d. All of the above
8. Wind turbine blades are adjusted
   a. To their maximum position to ensure that the blades capture as much wind as possible when the wind is blowing at normal conditions
   b. To their minimum position to protect the blades when the wind is blowing at overspeed condition
   c. To a position to help the rotor to start turning when the wind speed is light
   d. All the above
9. The image register in the PLC
   a. Provides values from 0 to 100%
   b. Holds a copy of the status of input and outputs as 1 when it is on and 0 when it is off
   c. Holds a copy of the status of input and outputs as 0 when it is on and 1 when it is off
   d. All the above
10. The SCADA system consists of
    a. Serial and parallel control data acquisition
    b. Supervisory central area data acquisition
    c. supervisory control and data acquisition

CHAPTER 6

Generators

OBJECTIVES

After reading this chapter, you will be able to:

- Identify the basic parts of a DC generator and explain its operation.
- Explain how to change amount of voltage a DC generator produces.
- Explain how brushes and commutator segments create DC voltage in a generator.
- Explain the difference between series, shunt, and compound DC motors.
- Identify the basic parts of a three-phase AC motor and explain its operation.
- Explain the operation of a rotating armature and a stationary stator AC generator.
- Explain the operation of the rotating stator and the stationary armature AC generator.
- Explain why an AC alternator needs only slip rings and brushes, not commutator segments and brushes.
- Explain the terms *sine wave* and *frequency* with regard to alternating current.

KEY TERMS

AC voltage
Alternating current
Alternator
Amps
Armature
Asynchronous AC generator
Brushes
Commutator segments
Current
DC voltage
Direct current
Doubly fed induction generator
Electromagnet
Flux lines
Frequency
Generator
Ground
Induced current
Induction generator
Induction motor
Laminated steel rotor
Magnet
Magnetic coil
Permanent-magnet generator
Resistance
Rotating magnetic field
Rotor
Self-excited generator
Separately excited generator
Slip
Slip rings
Squirrel-cage rotor

Stator
Synchronous AC generator
Three-phase generator (alternator)
Three-phase voltage
Torque
Volts
Wound rotor

OVERVIEW

Wind turbines blades turn a shaft that is ultimately connected to either an alternating-current (AC) or a direct-current (DC) generator. By definition, a **generator** is a machine that converts mechanical energy to electrical energy by rotating its shaft; the electrical energy can be either **alternating current** or **direct current**. An **alternator** is an energy converter (generator) that converts mechanical (rotational) energy to AC current only. The AC or DC generator can be used as a stand-alone generator that provides electricity to a home or small business, or to a battery pack in which voltage can be stored for use at a later time; or the generator can be used to produce large amounts of electricity for the grid, just like any generator at a coal-fired or nuclear-powered electric utility plant. Voltage from the generator can also be generated at any frequency, converted using an inverter to 60 Hz AC, and fed directly into the grid.

This chapter will explain how the AC or DC electricity can be created by turning the shaft of a generator. It will begin with an overview of magnets and the theory of electromagnets. You will learn how a very simple DC generator operates and then how more complex AC alternators work. You will learn the basic parts of each of these machines and how they operate. The DC generator is an energy converter that creates DC electrical energy when its shaft is rotated by mechanical means, and the alternator produces AC electrical energy when its shaft is rotated.

In order to fully understand all the parts and theories of AC and DC generators, you will need a review of the basic operation of DC and AC motors as well as an understanding of the names and functions of all the motor components. The information in this section will explain the difference between series, shunt, compound and permanent-magnet DC motors as well as different types of AC induction and synchronous motors. You will need to understand motors because many of the same operational and control functions are used in generators and alternators. In fact, many AC and DC generators are simply called *machines*, because you can put electricity into the machine and it will run as a motor, or you can rotate its shaft and the same machine will produce electrical power.

6.1 OVERVIEW OF AC AND DC ELECTRICITY

As a beginning to learning about AC and DC electricity, it is important to understand the definitions of the basic electrical terms—voltage, current, and power—and their units. These terms are often used interchangeably when you are first learning about electricity. By definition, **current** is the flow of electrons, and **voltage** is the force that causes the electrons to flow. Current is measured in amperes, usually called simply **amps**, and voltage is measured in **volts**. More voltage can move more current (amps). Power is the rate at which electrical energy is transferred in an electric circuit, and it is calculated by multiplying the voltage by the current in the circuit. The unit for power is watts.

Before we start our discussion of magnetic theory, we need to review the terms AC and DC as they refer to voltage and current. Direct current (DC) is the flow of electrons in one direction. A battery is a good source of **DC voltage**. When a battery is used as a power source, the electrons, which are negative, move from the negative terminal of the battery through the circuit and back to the positive terminal. In some applications, such as automobile electricity, the flow of current has been traditionally or conventionally described as flowing from the positive terminal to the negative terminal. This description of electrical flow is called *conventional current flow,* and it became very common because the automotive electrical system has only one positive terminal and uses the metal frame of the vehicle as the negative side of the circuit, called the **ground**. Because there is only one positive connection and many negative (grounded) points, it became easier to explain the circuit as though voltage starts at the positive terminal and moves about the circuit to any one of the negative points. The flow of electrons is called electrical current, and the electrons are negative, so current actually flows from the negative terminal of the battery to the positive terminal. The main point to remember is that when DC is used as the power source, the electron flow is in only one direction.

6.2 WHAT IS ALTERNATING CURRENT?

In an alternating-current (AC) circuit, the electrons travel in one direction and then change direction and move in the other direction. This movement by the electrons can best be shown in its characteristic waveform. The AC waveform is shown in Figure 6-1 and is called a sine wave. The sine wave has a positive half-cycle during which the electrons flow in one direction and a negative half-cycle during which the electrons flow in the reverse direction.

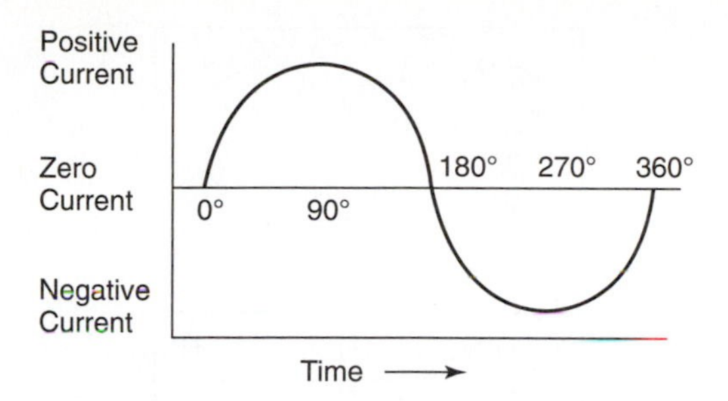

FIGURE 6-1 AC sine wave moving through 360 degrees.

6.3 FREQUENCY OF AC VOLTAGE

AC voltage generated at 60 hertz has a sinusoidal waveform that is positive for 1/60th of a second and then is negative for 1/60th of a second. This oscillation from positive to negative is called **frequency**, and it is the most important feature of AC voltage that makes it different from DC voltage. Frequency is the rate at which the electrical current changes from flowing in the positive direction to flowing in the negative direction. The electrical current flows in the positive direction and then in the negative direction in AC voltage because it is produced by moving a wire through a magnetic field first in one direction and then in the opposite direction. This is accomplished by forming the wire in the shape of a coil and pressing the coil onto a rotating shaft. When the shaft is rotated, the wire in the coil automatically moves in one direction, cutting the magnetic lines of force when the shaft rotates through 180 degrees, and the wire moves down through the magnetic field. As the shaft continues to rotate through the remainder of one revolution, from 180 to 360 degrees, the wire moves back up through the magnetic lines of force and electrons move in the opposite direction. This cycle continues as the shaft rotates, so the electron flow is created in the positive direction and then in the negative direction. The typical frequency for AC voltage in the United States and much of the rest of North America is 60 Hz. The frequency of 60 Hz is determined by the rotating speed of the alternator when the voltage is generated. Frequency is measured as the number of cycles (sine waves) that occur in 1 s. Figure 6-2 shows a number of sine waves occurring in 1 s.

Frequency = Number of cycles in 1 second

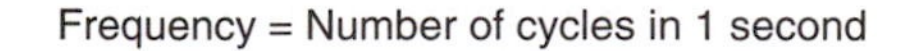
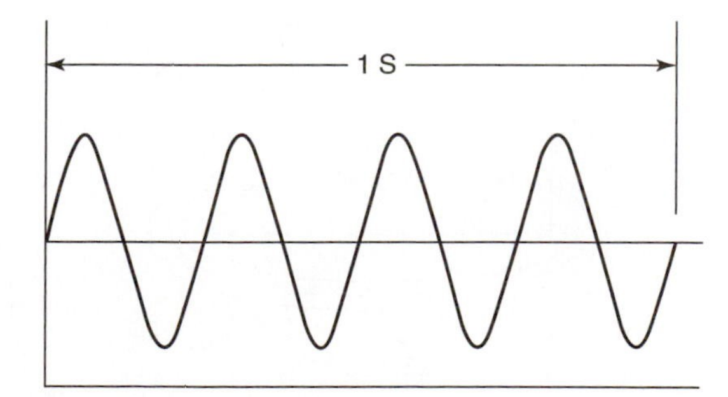

FIGURE 6-2 The frequency of AC voltage is calculated from the number f cycles that occur in 1 s.

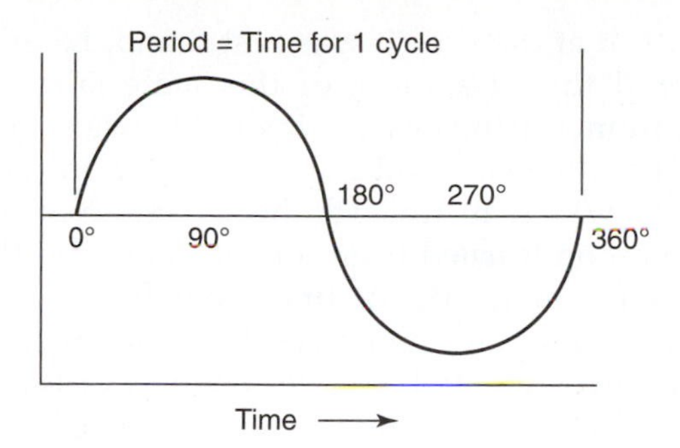

FIGURE 6-3 The period of AC voltage is calculated as the time it takes for one cycle to occur.

The *period* of a sine wave is the time it takes one sine wave to start from the zero point, pass through 360 degrees, and return to the zero point, as seen in Figure 6-3. A period represents one complete cycle, which is one complete revolution of the alternator. Because the shaft of the alternator rotates one full circle to produce the sine wave, we equate one complete cycle to 360 degrees. Thus the sine wave can be described in terms of 360 degrees. In Figure 6-3 you can see that the sine wave starts at 0 degrees, reaches a positive peak at 90 degrees, returns to zero at 180 degrees, reaches a negative peak at 270 degrees, and finally returns to 0 at 360 degrees. The number of degrees will be used to identify points of the sine wave in future discussions.

Mathematically, the period (P) of a sine wave can be described as

$$P = \frac{1}{\text{frequency}}$$

and the frequency (F) can be defined as

$$F = \frac{1}{\text{period}}$$

6.4 INTRODUCTION TO MAGNETIC THEORY

The operation of all types of DC and AC generators can be explained using several simple magnetic concepts. As a technician or maintenance person who works on wind turbine generators, you will need to fully understand magnetic theory in order to understand how these magnetic components operate. You must understand how a generator is supposed to operate before you can troubleshoot it or perform tests to determine why it has failed. Understanding magnetic theory will make this job easier. It is also important to understand that some of these concepts of magnetic theory rely on AC voltage. These concepts are introduced in this part of this chapter; more detail about AC voltage and magnetic concepts that use AC voltage follows in the last part of this chapter.

Magnet is the name given to material that has an attraction to iron or steel. This material was first found naturally about 4000 years ago as a rock in a city called Magnesia. The rock was called magnetite, but there was no use for it at the time it was first found. Later it was discovered that if a piece of this material was suspended from a string or wire, it would always orient itself so that the same end always pointed in the same direction, which is toward the earth's North Pole. Scientists soon learned from this phenomenon that the earth itself is magnetic. At first the only use for magnetic material was in compasses. It was many years before the forces created by two magnets attracting or repelling each other were utilized as part of a control device or motor.

As scientists gained more knowledge and as equipment became available to study magnets more closely, a set of principles and laws evolved. The first of these stated that every magnet has two poles, called the north pole and the south pole. When two magnets are placed end to end so that similar poles are near each other, the magnets repel each other. It does not matter if the poles are both north or both south, the result is the same. When two magnets are placed end to end so that the south pole of one magnet is near the north pole of the other magnet, the two magnets attract each other. These concepts are called the *first and second laws of magnets.*

When sophisticated laboratory equipment became available, it was found that this phenomenon is due to the basic atomic structure and electron alignments. By studying the atomic structure of magnets, scientists determined that the atoms in the magnet are grouped in regions called *domains,* or *dipoles.* In material that is not magnetic or that cannot be magnetized, the alignment of the electrons in the dipoles is random and usually follows the crystalline structure of the material. In material that is magnetic, the alignments in each dipole are along the lines of the magnetic field. Because each dipole is aligned exactly like the ones next to it, the magnetic forces are additive and are much stronger. In materials in which the magnetic forces are weak, it was found that the alignment of the dipole is random and not along the magnetic field lines. The more closely this alignment is to the magnetic field lines, the stronger the magnet is. Today we refer to a piece of soft iron in which all dipoles are aligned as a *permanent magnet.* The term *permanent magnet* is used because the dipoles remain aligned for very long periods of time, which means the magnet will retain its magnetic properties for long periods of time. Figure 6-4a shows a diagram of nonmagnetic metal, in which the dipoles are randomly placed, and Figure 6-4b shows a piece of metal that is magnetic, in which the dipoles align to make a strong magnet.

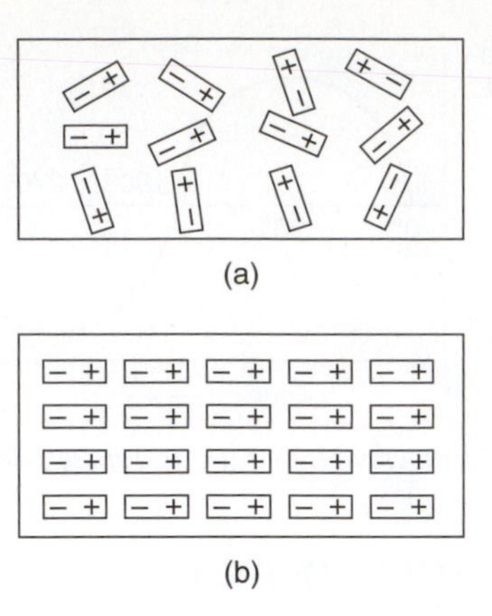

FIGURE 6-4 Example of metal in which dipoles are randomly placed, which makes a very weak magnet. (b) Example of metal in which the dipoles are aligned to make a very strong magnet.

A Typical Bar Magnet and Flux Lines

Figure 6-5 shows a bar magnet that is made of soft iron that has been magnetized. The magnet is in the shape of a bar, and its north and south poles are identified. Because the bar remains magnetized for a long period of time, it is a permanent magnet. The magnet produces a strong magnetic field because all its dipoles are aligned. The magnetic field produces invisible **flux lines** that move from the north pole to the south pole along the outside of the bar magnet. Figure 6-5 shows that these flux lines are lines of force that form a slight arc as they move from pole to pole.

You can perform a simple experiment that will allow you to see that the flux lines do exist and what they look like as they surround the bar magnet. For this experiment you will need a piece of clear plastic film, such as the plastic sheets used for overhead transparencies, and some iron filings. Place the plastic sheet over a bar magnet, making the plastic as flat as possible, and sprinkle iron filings over it. The filings will be attracted by the invisible flux lines as they extend in an arc from the north pole to the south pole along the outside of the magnet. Because the flux lines begin at one pole and stretch to the other, the highest concentration of flux lines will be near the poles. The iron filings will also concentrate around the poles, but you will be able to see a definite pattern of flux lines along each side of the bar magnet. If an overhead projector is available, the image of the flux lines can be projected onto a projector screen or blackboard so that they can be seen more easily. The pattern of these filings will look similar to the diagram in Figure 6-5. The number of flux lines around a magnet is related directly to the

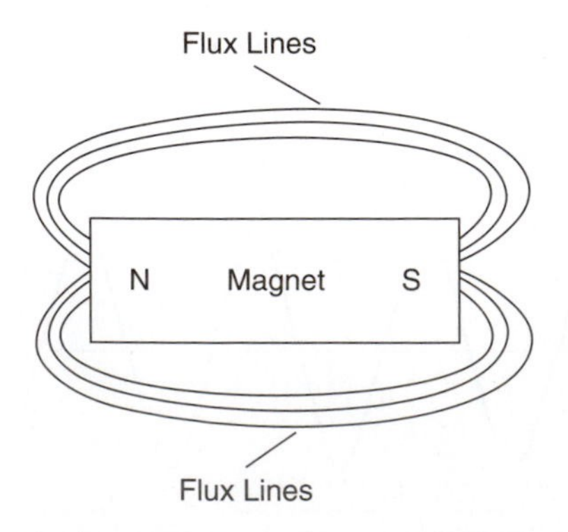

FIGURE 6-5 A bar magnet. The poles are identified as north (N) and south (S). Flux lines emanate from the north pole to the south pole.

strength of the magnet. A stronger magnet will have more flux lines than a weaker magnet. The strength of a magnet's field can be measured by the number of flux lines per unit area. Because the strength of a magnet's field is based on the alignment of the magnetic dipoles, the number of flux lines will increase as the alignment of the magnetic dipoles increases.

Some materials, such as Alnico and Permalloy, make better permanent magnets than iron, because the alignment of their magnetic domains (dipoles) remains consistent even after repeated use. You may find these materials used in some expensive controls and motors, but normally the permanent magnet will be made of soft iron. The reason permanent magnets are useful in many types of controls, especially in motors and generators, is because the soft iron produces residual magnetism for long periods of time over many years. Permanent magnets have several drawbacks, however. One of these is that the magnetic force of a permanent magnet is constant and cannot be turned off if it is not needed. This means that if something is attracted to a magnet, it will remain attracted until it is physically removed from the force of the flux lines. Another problem with a permanent magnet's flux field being constant is that it cannot easily be made stronger or weaker if circumstances so require.

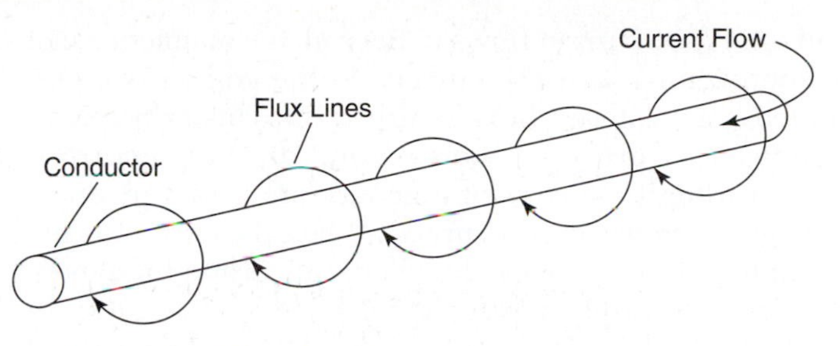

(a) Few flux lines around conductor that is not coiled

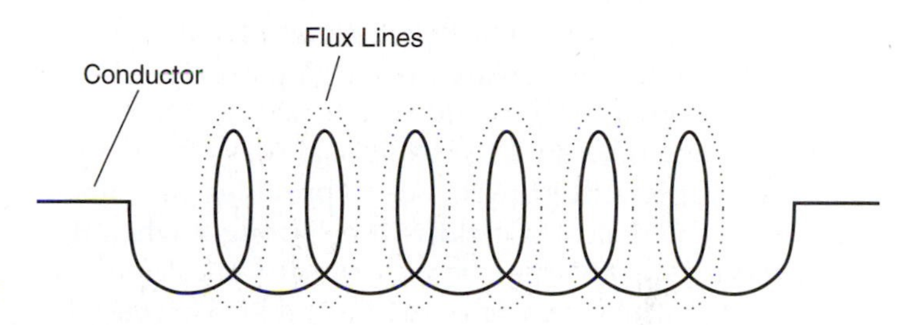

(b) Flux lines become more concentrated when wire is coiled

FIGURE 6-6 (a) Flux lines around a wire that is carrying current. (b) Flux lines around a coil of wire that is carrying current. Notice that the number of flux lines increases when the wire is coiled.

Electromagnets

An **electromagnet** is produced when current flows through a coil of wire. The electromagnet is also called a **magnetic coil**. One type of electromagnet is made by connecting a coil of wire to an electric cell (battery). The electromagnet has properties that are similar to those of a permanent magnet. When a wire conductor is connected to the terminals of the battery, current begins to flow, and magnetic flux lines form around the wire like concentric circles. If the wire is placed near a pile of iron filings while current is flowing through it, the filings will be attracted to the wire just as if the coil were a permanent magnet. Figure 6-6 indicates the location of magnetic flux lines around conductors. Figure 6-6a shows that flux lines occur around any wire when current is flowing through it. You can set up several simple experiments to demonstrate these principles. In one experiment you can insert a current-carrying conductor through a piece of cardboard and place iron filings around the conductor on the cardboard. When current is flowing in the wire, the filings will settle around the conductor in concentric circles, showing where the flux lines are located. As the amount of current is increased, the number of flux lines will also increase. The flux lines will also concentrate closer and closer to the wire until the current reaches *saturation*. When the flux lines reach the saturation point, additional increase of current in the wire will not produce any more flux lines.

When a straight wire is coiled up, the flux lines concentrate and become stronger. Figure 6-6b shows flux lines around a coil of wire that has current flowing through it. Because the flux lines are much stronger in a coil of wire, most of the electromagnets that you will encounter will be in the form of coils. For example, coils are used in transformers, relays, solenoids, and motors.

One advantage that an electromagnet has over a permanent magnet is that the magnetic field can be energized and de-energized by interrupting the current flow through the wire. The strength of the magnetic field can also be varied by varying the strength of the current flow through the conductor that is used for the electromagnet. This concept is perhaps the most important concept of magnetism, because it is used to change the strength of magnetic fields in motors, which causes the motor shaft to produce more torque so it can turn larger loads or for a generator to produce more voltage. When extra current flows through the coil, the magnetic field increases until the current causes the magnetic field in the wire to reach saturation. When the saturation point is reached, any additional current flowing in the wire will not produce additional flux lines.

Another important concept concerning electromagnetic coils is that the magnetic field can be turned off by interrupting the current flow through the coil. When a switch is added to the coil circuit, the magnetic field can be turned on and off by turning the switch on and off to interrupt the flow of current in the coil. When the switch is opened, current is interrupted and no flux lines are produced, so no magnetic field exists.

Components such as generators, relays, and solenoids use the principle of switching the magnetic field on and off. When current flows in the coil, the magnetic field is energized. When the current to the coil in is interrupted, the magnetic field is turned off. This principle is also used to turn generators on and off. When current flows through the coils of a generator, it will generate voltage when its shaft is turned. When the current is interrupted, the magnetic field will diminish and the generator will stop producing current.

Adding Coils of Wire to Increase the Strength of an Electromagnet

Another advantage of an electromagnet is that its magnetic strength can be increased by adding coils of wire to the original single coil of wire. The increase of the magnetic field occurs because the additional coils of wire provide a longer length of wire, which provides additional flux lines. The magnetic field will be stronger when the coil is more tightly wound, because the flux lines are more concentrated. Thus, very fine wire is used in some electromagnets to maximize the number of coils. As smaller wire is used, however, the amount of current flowing through it must be reduced so that the wire is not burned open.

You will learn that this principle is used in some motors to increase their horsepower and torque ratings. These motors have more than one coil that can be connected in various ways to affect the torque and speed of the motor's shaft. **Torque** is the amount of rotating force available at the shaft of a motor. You will also learn that coils can be connected in series or in parallel to affect the torque and speed of a motor.

Using a Core to Increase the Strength of the Magnetic Field of a Coil

The strength of a magnetic field can also be increased by placing material inside the helix of the coil to act as a core. The farther the core is inserted into the coil, the stronger the flux field becomes. When the core is removed completely from the coil, it is considered to be an *air coil magnet,* and the magnetic field is at its weakest point. If a soft iron is used as the core, it will strengthen the magnetic field, but it also creates a problem because it has excessive residual magnetism, which is unwanted. Residual magnetism means that the core will retain magnetic properties when current is interrupted in the coil, which will make it like a permanent magnet. This problem can be corrected by using laminated steel for the core. The laminated steel core is made by pressing sheets of steel together to form a solid core. Figure 6-7 shows an example of layers of laminated steel pressed together to form a core. When current flows through the conductors in the coil, the laminated steel core enhances the magnetic field in much the same way as the soft iron; and when current flow is interrupted, the magnetic field collapses rapidly because each piece of the laminated steel does not retain sufficient magnetic field.

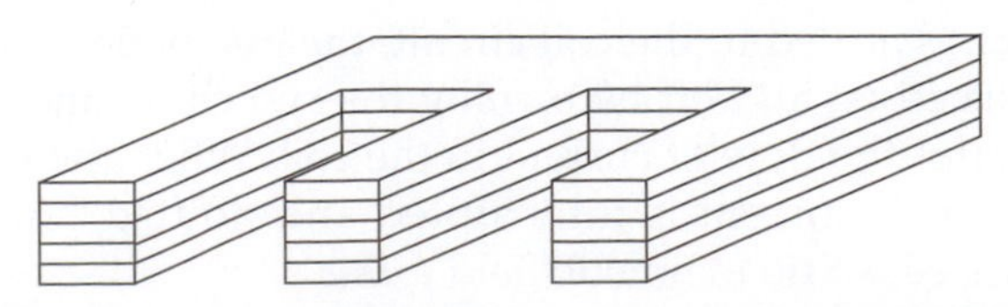

FIGURE 6-7 Thin strips of laminated steel pressed together for use in a coil.

Reversing the Polarity of a Magnetic Field in an Electromagnet

When current flows through a coil of wire, the direction of the current flow through the coil determines the polarity of the magnetic field around the wire. The polarity of the magnetic field determines the direction in which the motor shaft turns in an AC or DC motor. If the direction of current flow is reversed, the polarity of the magnetic field is reversed, and the direction in which the motor shaft turns is also reversed. In some motors, such as yaw drive motors and hydraulic pump motors, the direction of rotation is very important. In these applications you may be requested to change the connections for the windings in the motor or the supply voltage for a three-phase motor to make the motor rotate in the opposite direction. The changes you make take advantage of changing the direction of current flow through a coil or changing the polarity of the supply voltage with respect to the other phases so that the motor reverses its direction of rotation.

Another way to think of reversing the polarity is when a wire is moved through a magnetic field to generate a voltage as in a generator or alternator. When a wire is passing through a magnetic field in one direction, it makes current flow in that direction. If the wire moves back through the magnetic field in the opposite direction, the current flow in the wire is reversed. This is the operational theory that is used in generating current with a generator or alternator.

6.5 DC GENERATORS

A generator produces voltage when its shaft, also called an **armature**, is rotated. A DC generator has the same parts as a DC motor. The major difference is that a motor takes electrical energy in and produces rotating energy (torque) and a generator creates electrical current when its shaft receives rotating energy from a source such as wind turbine blades. The DC motors explained earlier can be used as DC generators if an energy source such as wind turbine blades is used to turn the shaft rather than voltage being applied to the terminals. When a DC machine is used as a generator, electrical current is available at the armature terminals when the armature, or **rotor**, is rotated rapidly.

The voltage produced by a DC generator is used to charge batteries, or the DC power can be sent through a power electronic frequency converter (PEFC) or electronic inverter, where it is turned back into 60-Hz AC voltage. Even though you may not work on many

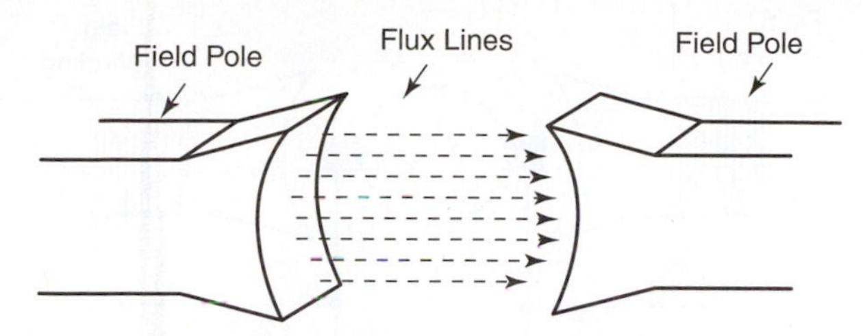

FIGURE 6-8 Field poles and flux lines of a simple DC generator.

DC generators, it is important to understand their theory of operation because the theory applies to a large variety of other electrical devices such as AC generators (alternators).

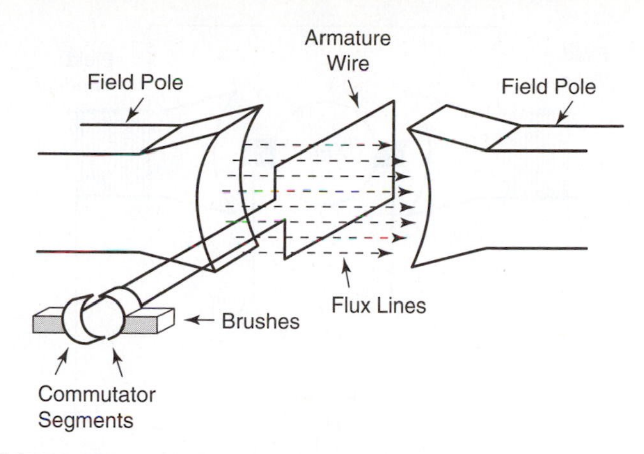

FIGURE 6-9 Armature wire cutting through the flux lines of a DC generator.

DC Shunt Generators

The basic parts of a DC shunt generator include the field winding, which is mounted in the stationary part of the generator, called the **stator**; the armature, which is the rotating part with its commutator segments; and the shaft. The **brushes** make contact between the stationary part of the generator and the **commutator segments**, which are electrically connected to the armature winding. The end plate also has bearings that allow the armature shaft to rotate more easily. The armature shaft extends out from one end of the generator to provide a place to mount the pulley sheave or coupling that is used to transfer mechanical rotating energy to the armature shaft. Figure 6-8 shows the two stator segments of a DC generator. The dashed lines between the two stator segments represent the flux lines of the magnetic field. In this figure, a small magnetic field is produced by residual magnetism from the iron in the stator. The magnetic field in the stator can be increased by the current flowing through the field winding.

At this point, we need to clarify some simple electrical theory. A generator produces voltage, which is a form of potential energy. Current flow depends on the amount of resistance (load) that is connected to the potential voltage source and the amount of voltage that is applied. **Resistance** is any opposition to current flow. In this section you will see that at times the generator will produce a voltage; and when a load is connected to the generator, it will cause current to flow. When the load is not connected, the generator is not loaded, and it will not provide a current flow.

When a wire passes through a magnetic field, current flows in the wire. In a generator, the wire is part of the armature, which is the rotating part of the generator. When the armature begins to rotate, the wire in its windings passes through the magnetic field, which creates current flow in the winding. The stronger the magnetic field or the faster the armature turns, the more current is produced in the armature.

Figure 6-9 shows a wire moving through a magnetic field. Because the armature rotates in a circular motion, you can assign values of 0 to 360 to represent the degrees of a circle. You can then plot the output waveform from the generator's armature as the armature turns through one rotation. As the wire in the armature moves from 0 degrees through 90 degrees, the sine wave displays 0 V to peak positive. As the wire in the armature moves from 90 to 180 degrees, the sine wave displays peak voltage back to zero. Each armature winding has one commutator segment connected to each end of the winding. The generator in Figure 6-9 shows one armature winding, two commutator segments connected to the ends of the armature winding, and two brushes making contact with the commutator segments. This means that the commutator segments rotate in the same 360-degree pattern as the armature and the brushes, which are permanently mounted so that they make contact with the commutator segments for 180 degrees, or one-half a revolution. The important point to remember about the relationship between the brushes and the commutator segments is that the brush identified with a positive sign (+) makes contact with the end of the armature coil and the commutator segment that is positive at any instant of time, and the brush that is identified with a negative sign (−) always touches the commutator segment that is connected to the end of the armature wire that is negative at that instant in time. At first this may be difficult to understand, but on closer inspection you will see that as the armature rotates and the wire in its coil moves down through the magnetic field, a positive half of a sine wave is produced. When the armature rotates through the remaining 180 degrees, the commutator segments have also moved 180 degrees, which results in another positive half of a sine wave being produced and sent through the positive brush. This causes the generator to produce a series of positive half-waves with respect to the positive brush. Because all the half sine waves have the same polarity, the output voltage of the generator is identified as DC voltage.

The brushes and commutator segments provide two important functions. First, the brushes and commutator segments provide an electrical contact between a stationary part (the brushes) and a rotating part (the

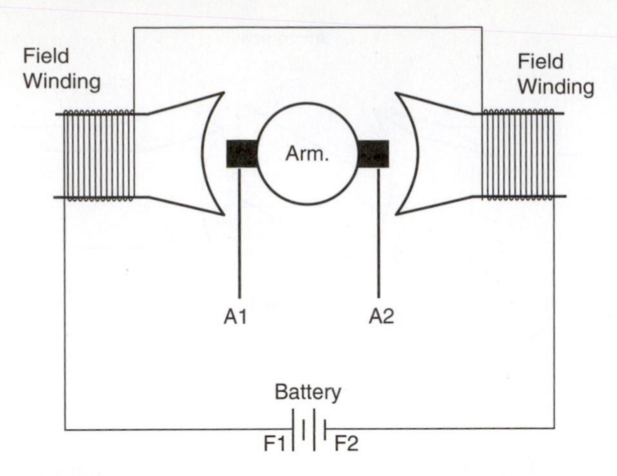

FIGURE 6-10 Electrical diagram of a separately excited generator. Notice that the exciter voltage comes from a battery.

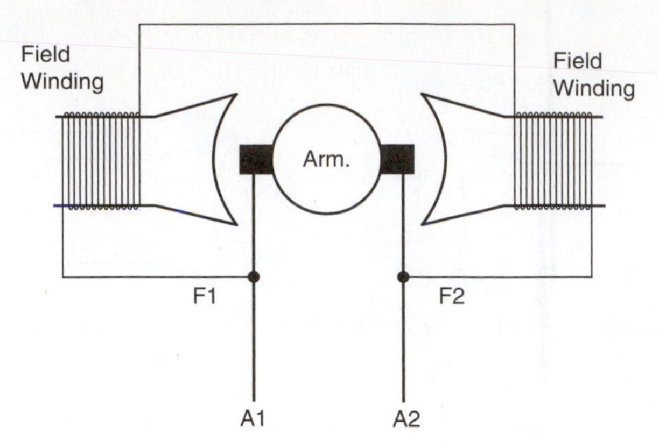

FIGURE 6-11 Self-excited generator. Notice that the exciter voltage comes from the armature of the generator.

commutator segments). The second function the brushes and commutator provide is timing, so that the voltage produced by the armature is always pulsing DC.

Separately Excited Shunt Generators

Figure 6-10 shows a **separately excited generator**. In this figure you can see that the armature is now represented as a circle with the notation *Arm.* inside it, and a brush is shown on each side of the armature. From this point on, the armature for each type of generator will be shown using this symbol. The two wires connected to the brushes will be identified as Al and A2 to indicate that they are armature wires.

In a separately excited generator, an outside voltage (called a battery voltage) is used to increase the current flow in the field, which in turn creates a stronger magnetic field. The separately excited generator exists only where a second power source is available. The point to remember about the separately excited generator is that the amount of voltage that is used to control the magnetic field is relatively small compared to the amount of voltage and current produced at the armature of the generator.

Self-Excited Shunt Generators

Figure 6-11 shows a **self-excited shunt generator**. In this figure you can see that the Fl terminal for the field winding is connected to the Al terminal of the armature, and the F2 terminal for the field winding is connected to the A2 terminal of the armature. This connection allows a small amount of voltage from the armature to be routed back through the field winding to create the magnetic field. As in the separately excited generator, the actual amount of voltage that the field needs is usually less than 2% of the output of the generator at the armature terminals.

Because the self-excited generator requires field current to create armature voltage, a small amount of residual magnetism is needed to get the generator to begin producing voltage when the armature first begins to turn. The small amount of residual magnetism will provide sufficient magnetic flux lines to produce a small amount of armature voltage. As the armature voltage increases, it produces more field current, which in turn allows the generator to produce more armature voltage. You should remember that the field winding is essentially a fixed resistance, so any increase in voltage applied to the field winding will result in an increase in field current. The increase in field current will strengthen the magnetic field up to the point at which the magnetic field becomes saturated. Saturation is the amount of field current that creates the strongest magnetic field; any further increase in field current will not create an increase in magnetic strength.

If a self-excited generator has gone for a long period of time without generating a voltage, it may lose its residual magnetism. If this occurs, an outside source of voltage, such as a battery or rectified voltage from the grid, must be used to provide the voltage for the field to begin to create the magnetic field again. When an outside voltage source is used to establish the field current, it is called *flashing the field*. It is important to observe the polarity of the external voltage and the field polarity so that you do not reverse the polarity of the generator field, which will cause the armature voltage to be reversed with respect to the markings on the generator terminals. If this occurs, you can re-establish the proper polarity by flashing the field again with a voltage source that has the reverse polarity of the previous voltage. It is important to control the polarity in the field winding with respect to the armature windings by ensuring that the positive field terminal is connected to the positive armature terminal. This means that Fl should be connected to Al and F2 should be connected to A2.

In the self excited DC generator, the field winding and the armature winding are connected in parallel. This type of configuration is called a shunt generator. The term *shunt* means that the windings are connected in parallel with each other. It also means that if this type of generator is turned at a constant speed, the armature voltage will remain fairly constant. Because the field is connected in parallel with the armature, the voltage applied to the field will also remain fairly constant. This equilibrium will continue as long as the load is constrained. If the load resistance goes down, such as when more electric lights are turned on in a residential application, the armature amperage will increase and cause the generator to produce more electrical power. When this occurs in a shunt generator, the output voltage will decrease slightly. This means that a shunt generator produces the most voltage when its electrical load is low (its electrical resistance is high), and it produces the least voltage when its electrical load is high. If the voltage is used for charging a bank of batteries, the small change in voltage is not a problem.

Compound Generators

A compound generator is shown in Figure 6-12. In this figure you can see that the generator has both a series field and a shunt field connected. The series field consists of a few turns of very large wire, and it is connected in series with the armature so that all the current the armature produces when a load is connected to the generator also flows through the series field winding, which creates a stronger magnetic field. The main feature of this type of generator is that it produces a fairly constant voltage regardless of its load.

Figure 6-12 shows several examples of how a compound generator can be connected. Figure 6-12a shows an example of a short-shunt generator, Figure 6-12b shows a long-shunt generator, and Figure 6-12c shows a short-shunt differential generator. When the shunt field winding (Fl–F2) is connected in parallel with only the armature in the compound generator, it is called a *short shunt*. If the field winding is connected in parallel with both the armature and the series field (S1–S2), it is called a *long shunt*. If the series field is connected so that its polarity is reversed with reference to the armature, the generator is called a differential generator. The short shunt is the most popular way to connect these windings, because it provides the most constant voltage under varying loads. A differential generator provides fairly constant voltage when its load is low (its electrical resistance is high), but its voltage drops significantly when a load is applied.

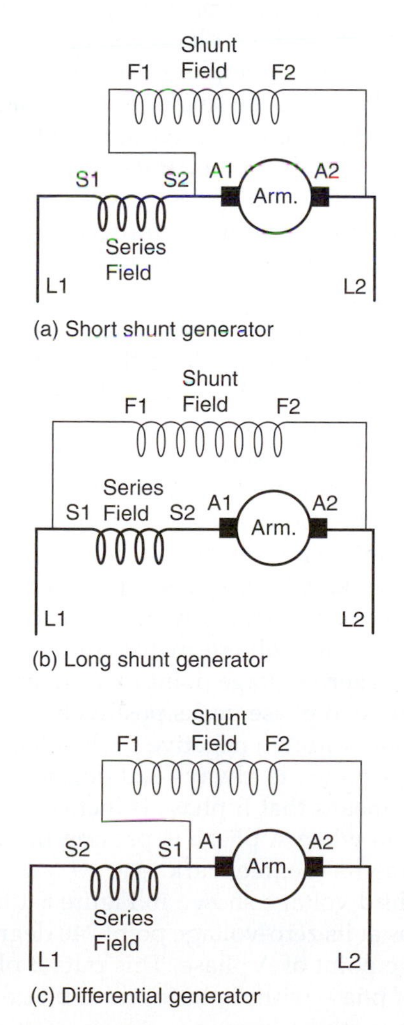

FIGURE 6-12 (a) Generator connected as a short-shunt generator. (b) Compound generator connected as a long-shunt generator. (c) Compound generator connected as a differential generator.

Series Generators

If a DC machine is connected so that only the series field winding is connected in series with the armature and the shunt field is not used, it is called a series generator. A series generator is very seldom used on its own, but you need to understand its basic operation so you will fully understand the series winding in a compound DC generator. In a series generator, the series field, armature, and electrical load connected to the generator are all connected in series. This means that as the electrical resistance in the load decreases, the electrical current increases, so the generator will need to produce more electrical current. When the electrical resistance in the load is high, the electrical current is low, so the load on the generator is relatively low. This also means that when the current flow from the generator is low, its voltage is also low; and as its current increases, its voltage increases. Because this produces a variable voltage that can range from near 0 V to the maximum of the generator, the series generator is not very useful in most applications.

Controlling the Amount of Voltage and Its Polarity in DC Generators

The output voltage of a DC generator can be controlled in either of two ways. First, the speed of rotation can be controlled at a constant rate. This is usually accomplished by controlling the speed of the diesel or gasoline motor that is turning the generator. If the DC generator is turned by an AC motor, its speed will also tend to be constant. The second way to control the voltage of a DC generator is to control the field current. This can be controlled automatically by an electrical circuit called a

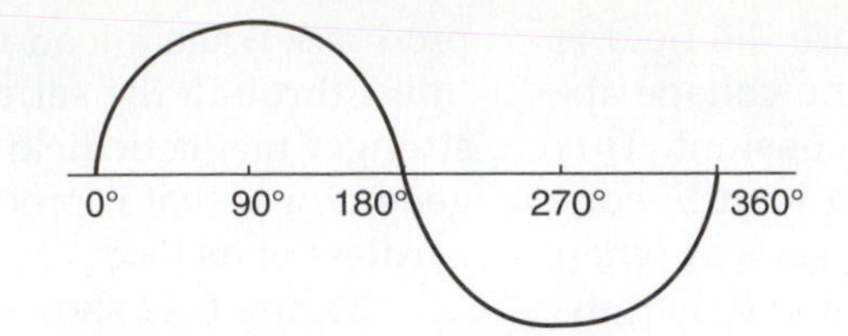

(a) Single-phase sine wave consisting of 360°

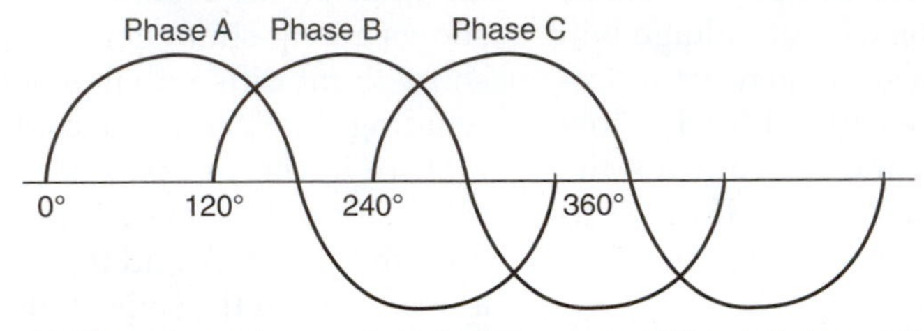

(b) Three-phase sine wave; each sine wave is 120° out of phase with the next

FIGURE 6-13 (a) Single-phase and (b) three-phase voltage.

voltage regulator. When the speed is kept constant and a voltage regulator is used, the output voltage of the DC generator can be accurately controlled. If the voltage regulator uses solid-state devices such as insulated gate bipolar transistors (IGBTs) or operational amplifiers (op-amps), the voltage can be controlled automatically, even as the load changes. If the voltage regulator is an older style, you may need to adjust the field current manually to set the voltage to the level that the application requires.

The polarity of a DC generator can also be controlled in either of two ways. First, the direction of rotation can be changed, so the polarity of the armature voltage is reversed. If the generator is driven by a diesel or gasoline engine, this may not be an option. If the generator is driven by a three-phase AC motor, the motor can be reversed fairly easily by swapping any two of the three incoming voltage lines to the motor.

If the direction of rotation cannot be changed, you can change the polarity of the field winding with respect to the armature winding. In some cases you will have trouble getting the polarity the way you want it, and you may need to flash the field with an external voltage to change the polarity.

6.6 AC MOTORS

To better understand AC generators, it is important to understand the theory of operation and the basic parts of AC induction motors. The reason for this is that parts of the AC generator actually operate as an AC motor, and without the theory of operation for the AC motor, it will be difficult to understand the operation of AC induction generators.

Characteristics of Three-Phase AC Voltage

To understand the basic operation of a three-phase AC induction motor, it is very important to understand the operation of AC voltage. The AC induction motor is designed specifically to take advantage of the characteristics of the three-phase voltage that it uses for power. Figure 6-13a is a diagram of single-phase voltage and Figure 6-13b is a diagram of three-phase voltage. In the diagram that shows three-phase voltage, you should notice that each of the three phases represents a separately generated voltage like the single-phase voltage. The three separate voltages are produced out of phase with each other by 120 degrees. The units of measure for this voltage are electrical degrees. The sine wave has 360 degrees. The sine wave may be produced once during each rotation of the generator's shaft rotating through 360 degrees. If the sine wave is produced by one rotation of the generator's shaft, 360 electrical degrees are equal to 360 mechanical degrees. All of the discussions in this section will be based on 360 electrical degrees being equal to 360 mechanical degrees.

The first voltage shown in Figure 6-13b is called A phase, and it is shown starting at 0 degrees and peaking positively at the 90-degree mark. It passes through 0 V again at the 180-degree mark and peaks negatively at the 270-degree mark. After it peaks negatively, it returns to 0 V at the 360-degree mark, which is also the 0-degree point. The second voltage in Figure 6-13b is called B phase, and its zero-voltage point is 120 degrees later than that of A phase. B phase peaks positively, passes through 0 V, and passes through negative peak voltage as A phase does, except that it is always 120 degrees later than A phase. This means that B phase is increasing in the positive direction when A phase is passing through its zero voltage at the 180-degree mark.

The third voltage shown in Figure 6-13b is called C phase. It has at its zero-voltage point 240 degrees after the zero-voltage point of A phase. This puts B phase 120 degrees out of phase with A phase and C phase 120 degrees out of phase with B phase.

An AC motor takes advantage of this characteristic to provide a **rotating magnetic field** in its stator and rotor that is very strong because three separate fields rotate 120 degrees out of phase with each other. Because the magnetic fields are induced from the applied voltage, they will always be 120 degrees out of phase with each other. The induced magnetic field is 180 degrees out of phase with the voltage that induced it, which occurs naturally with induced voltages. The 180-degree phase difference is not as important as the 120-degree phase difference between the rotating magnetic fields, which helps create torque for the motor shaft.

Because the magnetic fields are 120 degrees out of phase with each other and are rotating, one will always be increasing in strength when one of the other phases is losing strength as it passes through the zero-voltage point on its sine wave. This means that the magnetic field produced by all three phases never fully collapses, so its average is much stronger than that of a field produced by single-phase voltage. A detailed discussion of three-phase voltage and transformers is provided at the end of this chapter.

Parts of a Three-Phase Motor

AC motors are available that operate on single-phase or three-phase supply voltage systems. Most single-phase motors are less than 3 hp; although some larger ones are available, they are not as common. Three-phase motors are available up to several thousand horsepower, although motors that are less than 50 hp are most common.

The AC induction motor has three basic parts: the stator, which is the stationary part of the motor; the rotor, which is the rotating part of the motor; and the end plates, which house the bearings that allow the rotor to rotate freely. This section provides information about each of these parts of an AC motor. Figure 6-14 shows an exploded view of a three-phase motor. This figure provides information about the location of the basic parts of a motor and how they work together.

Stator

The stator is the stationary part of the motor and is made of several parts. Figure 6-15 shows the stator of a typical induction motor. The stator is the frame for the motor housing the stationary winding with mounting holes for installation. The mounting holes for the motor are sized according to National Electrical Manufacturers Association (NEMA) standards for the motor's frame type. Some motors also have a lifting ring in the stator to provide a means for handling larger motors. The lifting ring and mounting holes are actually built into the frame or housing part of the stator.

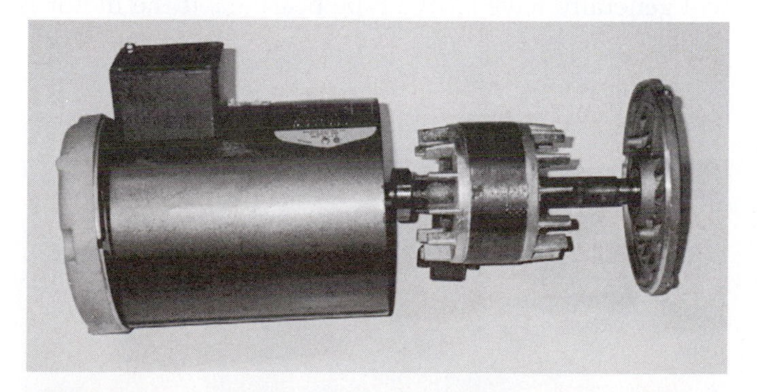

FIGURE 6-14 Exploded view of a three-phase induction motor. Notice the squirrel-cage rotor with its shaft resting in the bearing in the end plate.

FIGURE 6-15 The stator of an AC motor.

An insert in the stator provides slots for the stator coils to be inserted. This insert is made of laminated steel to prevent eddy current and flux losses in the coils. The stator windings are made by wrapping a predetermined length of wire on preformed brackets in the shape of the coil. These windings are then wrapped with insulation and installed in the stator slots. A typical four-pole, three-phase motor has three coils mounted consecutively in the slots to form a group. The three coils are wired so that they each receive power from a separate phase of three-phase power supply. Three groups are connected together to form one of the four poles of the motor. This grouping is repeated for each of the other three poles so that the motor has a total of 36 coils to form the complete four-pole stator. It is not essential that you understand how to wind the coils or put them into the stator slots; rather, you should understand that these coils are connected inside the stator, and 3, 6, 9, 12, or 15 wires from the coil connections are brought out of the frame as external connections. The external connection wires can be connected to allow the motor to be powered by 208/230 or 480 V, or they allow the motor to be connected to provide the correct torque response for the load. Other changes can also be made to these connections to allow the motor to start so that it uses less locked-rotor current, which means it will not draw as much current when it is starting.

After the coils are placed in the stator, their ends (leads) are identified by a number that is used to make connections during installation. The coils are locked into

FIGURE 6-16 A wound-rotor armature with two slip rings.

the stator with wedges that keep the coils securely mounted in the slots but that allow them to be removed and replaced easily if the coils are damaged or become defective as a result of overheating.

Wound Rotor

The rotor in an AC motor can be constructed from coils of wire wound on laminated steel, or it can be made entirely from laminated steel without any wire coils. A rotor with wire coils is called a **wound rotor**, and it is used in a wound-rotor motor. The wound-rotor motor can produce more torque than a similar-size induction motor because it uses brushes and slip rings to transfer current to the rotor. Because the wound-rotor motor has brushes and slip rings, however, it requires more maintenance than an induction motor. Figure 6-16 shows the two **slip rings** on the armature of a wound-rotor motor.

Laminated Steel Rotor

The AC induction motor that use a **laminated steel rotor** is called an **induction motor** or a *squirrel-cage induction motor.* The core of the rotor is made of die-cast aluminum or copper in the shape of a cage. A diagram of a **squirrel-cage rotor** is shown in Figure 6-17. You can see why the rotor is called a cage rotor, because the aluminum or copper bars are held in place by an end ring. The cage rotor is sometimes called a *squirrel-cage rotor* because the overall

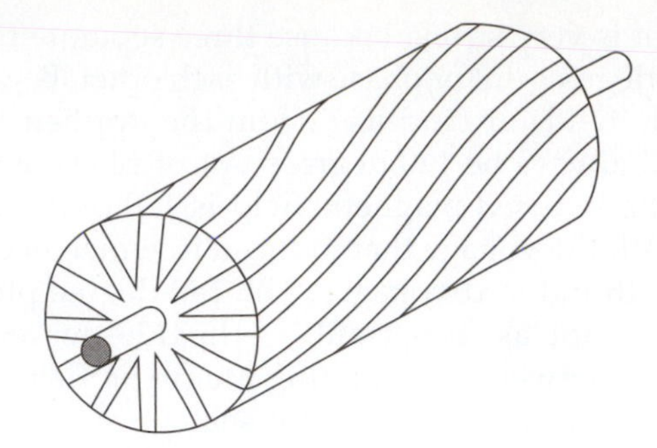

FIGURE 6-17 Squirrel cage rotor.

shape of the rotor is the same as that of an exercise wheel used for a hamster or squirrel.

Laminated sections are pressed onto this core, or the core is molded into laminated sections when the squirrel-cage rotor is manufactured. Figure 6-18 shows a squirrel-cage rotor, in which you can see the laminated steel sections pressed onto the skeleton of the cage core. This laminated steel is a thin piece of steel that has insulation coating it to prevent it from making electrical connection with the piece of steel next to it. The thin pieces of steel are pressed onto the rotor bars, rather than the rotor bar being a solid piece of iron, because the individual pieces of steel are easily magnetized to one polarity, then demagnetized and remagnetized again to the opposite polarity. The changing magnetic field is very important when AC current is used to create a magnetic field in the rotor. If the rotor was made of solid iron or steel, it would magnetize easily, but it would not give up the magnetic field easily, which would be a problem when AC current is applied to the rotor.

Fins or blades are built into the rotor to provide a means of cooling the motor. It is important that these fan blades are not damaged or broken, because they provide all of the cooling air for the motor and they are balanced so that the rotor spins evenly, without vibration.

Motor End Plates

The bearings for the motor shaft are located in the end plates. If the motor is a fractional-horsepower motor, it will generally have sleeve-type bearings; if the motor is

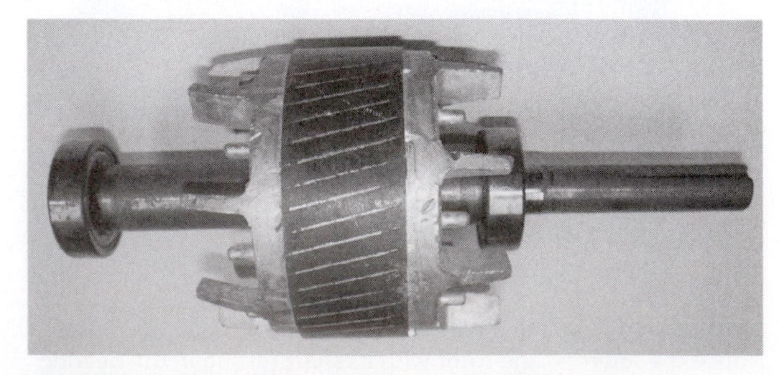

FIGURE 6-18 Squirrel-cage rotor for an induction motor.

one of the larger types, it will have ball bearings. Some ball bearings on smaller motors are permanently lubricated, whereas the bearings on larger motor require periodic lubrication. All sleeve bearings require a few drops of lubricating oil periodically.

The end plates are mounted on the ends of the motor and held in place by long bolts inserted through the stator frame. When nuts are placed on the bolts and tightened, the end plates are secured in place. If the motor is an open type, the end plates have louvers to allow cooling air to circulate through the motor. An access plate may also be provided in the rear end plate to allow field wiring if one is not provided in the stator frame.

If the motor is not permanently lubricated, the end plate has an oiler tube or grease fitting for lubrication. It is important that the end plates are mounted on the motor so that the oiler tube or grease fitting is above the shaft, so that gravity will allow lubrication to reach the shaft. If the end plate is rotated so that the lubrication point is mounted below the shaft, gravity will pull all of the lubrication away from the shaft and the bearings will wear out prematurely. If you need to remove the end plates for any reason, they should be marked so that they can be replaced in the exact position from which where they were removed. This also helps to align the holes in the end plate with the holes in the stator so that the end plates can be reassembled easily.

Operation of a Wound-Rotor AC Motor

The wound-rotor motor has a set of coils that are mounted in the stationary part of the motor called the stator. The rotor for the wound rotor motor is made with coils of wire pressed onto the rotor section. The terminal ends of each of these coils are connected to a slip ring, and a carbon brush rides on the slip ring. Brushes and slip rings are used to get electrical current from the stationary part of the motor through the brushes to the slip ring, which is connected to the coil on the rotating part. If the wound-rotor motor is single-phase, it will have two slip rings; and if it is a three phase motor, it will have three slip rings.

Because electrical current is provided to the rotor directly from the brushes to the slip rings, the magnetic field in the rotor is very strong. When power is applied to the motor, electrical current flows through the stationary windings and creates a strong magnetic field, and electrical current flows from the brushes to the slip rings through the rotor coil, which also becomes an electromagnet. Because the current supplied to the stationary windings is AC, it makes the magnetic field move around the sets of coils in a rotating fashion. The polarity of the coils in the rotor is opposite those in the stator, so the rotor's magnetic field is attracted to the magnetic field that is rotating through the stationary coils. This makes the rotor begin to spin, or chase the rotating magnetic field in the stator. The speed of the rotor is set by the frequency of the AC voltage and the number of field coils in the stator.

TABLE 6-1 Motor Speeds at 60 Hz

Number of Poles	2 Poles	4 Poles	6 Poles	8 Poles	10 Poles	12 Poles
rpm	3600	1800	1200	900	720	600

Table 6-1 shows the speed of a 2-pole through a 12-pole motor when 60-Hz AC voltage is applied.

The main drawback of a wound-rotor motor is that the brushes wear down and slip rings become rough as a result of electrical contact and electrical arcing. This means that the brushes have to be replaced and the slip rings resurfaced periodically.

Operation of an AC Induction Motor

The principle of operation of an induction motor is based on the fact that the rotor receives its current by induction rather than with brushes and slip rings as in the wound-rotor motor. The rotor in the induction motor is made of laminated steel and rotates in close proximity to the stator, so it can easily be magnetized. When current is flowing through the stator coils, a strong magnetic field is created, and when the laminated steel sections of the rotor pass through this magnetic field, current is induced into the laminated steel rotor, causing it to become magnetized. Just as in the wound-rotor motor, the magnetic field in the laminated steel rotor is the opposite polarity of the magnetic field in the stator. Because the magnetic field in the stator is created by AC current flowing through it, it causes the magnetic field to rotate from coil to coil around the stator. The magnetic current that creates the magnetic field in the laminated steel rotor will be 180 degrees out of phase with the current that produced it. This will cause the magnetic field in the rotor to have polarity that is the opposite of the polarity of the magnetic field rotating through the stator. Because the magnetic field in the rotor has the opposite polarity to the magnetic field rotating through the stator, it will cause the rotor magnetic field to be attracted to the stator magnetic field, which in turn causes the rotor to begin to spin.

Because the induction motor uses **induced current** to create the magnetic field in the rotor, there are some extra losses that are not found in the wound-rotor motor. Therefore, the induction motor is slightly less efficient than the wound-rotor motor. However, because the induction motor does not need brushes or slip rings, it does not need as much maintenance.

Using Slip to Create Torque in the Induction Motor

When the rotor in the induction motor begins to turn, it creates a voltage that is out of phase with the voltage that is applied to the stator. This voltage is called *electromotive force* (EMF); and because it is out of phase with the applied voltage, it is called *counter-EMF* or CEMF. The amount of electrical potential difference between the applied voltage and the CEMF that the rotor creates determines the amount of current that the motor draws. The amount of current that the motor draws in turn determines the amount of torque the motor shaft (rotor) provides. The amount of CEMF is determined by the amount of slip the rotor has. **Slip** is the difference between the rated speed and the actual speed of a motor; and the more slip the motor has, the more torque it can create at its shaft, so larger load it can move.

Theoretically, the rotor and the induction motor should spin at the same rate as in a wound-rotor motor, and the speed should be determined by the number of poles and the frequency of the applied voltage. In reality, if the motor rotor turned at its rated speed, the rotor would produce a CEMF equal to the applied voltage, so the amount of current the motor would draw would be zero. If the amount of current the motor is drawing is zero, the motor shaft will not produce any torque and will not be useful. In order for the motor to have torque at its shaft, there must be a difference between the applied voltage and the CEMF created by the rotor. This difference can occur only when the speed of the rotor is slower than its rated speed. It is this difference of speed in the rotor that is called slip. The larger the difference in speed, the more slip occurs, and the more torque the AC induction motor will produce. Because the speed of the AC motor will be slightly less than its rated speed, it will be less efficient than a wound-rotor motor.

Synchronous AC Motors

Another type of AC motor is called a synchronous motor, and it combines the strengths of the wound-rotor AC motor and the induction-type AC motor. The synchronous motor has a wound rotor, which means it has coils of wire in the rotor and it has slip rings and brushes. It also has a small squirrel-cage winding embedded in the wound rotor. When AC voltage is first applied to the synchronous motor, it starts as an induction-type motor because the small squirrel-cage winding receives an induced current that causes the rotor to begin to turn and come up to speed. When the speed of the rotor is at about 95%, a small amount of DC current is applied to coils in the wound rotor. This small DC current flows through the brushes and slip rings, and they cause a very strong magnetic field to be created in the rotor coils. Because the magnetic field in the rotor coils is very strong, the rotor speed synchronizes with the rotating field that the AC voltage creates in the stator winding. This synchronization causes the rotor to spin at its rated speed based on the number of poles and the frequency of the AC voltage without any slip. The synchronous motor shaft thus provides maximum torque and is more efficient. Because, like the wound rotor motor, the synchronous motor has brushes and slip rings, it will require some periodic maintenance.

6.7 BASIC AC ALTERNATORS (GENERATORS)

An electrical generator converts mechanical energy of a turning shaft into electrical energy and can produce AC or DC electricity. An AC alternator is actually an AC generator, because it produces AC voltage when its shaft is turned, but it is also called an AC alternator because it produces AC (alternating current) when its rotor is rotated through a magnetic field as its shaft is turned.

There are several variations of AC generators that you will encounter when working with wind turbines. The AC generator can have a wound rotor, with brushes and slip rings, or it can have a squirrel-cage laminated steel rotor. The AC generator can be designed so that the large current generated is produced in the rotor, or it can be designed so that the large current generated is produced in the stationary winding. If the large current generated is produced in the rotor, slip rings and brushes must be used to get the large current off the rotor and out to the frame of the generator. The part of the generator that produces the large current is the armature. If the large current is produced on the rotor, the generator is called a rotating armature machine; and if the large current is produced in the stationary part of the machine, the generator is called a stationary armature machine. It is important to remember that for this reason the stationary part of the machine is always called the stator, and the rotating part of the machine is always called the rotor. Figure 6-19 shows a generator mounted to the gearbox shaft of a wind turbine. You can see the electrical box for the generator connection on the front of the generator.

Electromagnetic Theory Used to Operate a Basic Generator to Create AC Voltage

You learned about electromagnetic theory in earlier sections of this chapter, so you should be able to understand how electromagnets are used in a basic generator to create AC voltage. This section provides an overview of

FIGURE 6-19 A wind turbine generator.

how AC voltage is generated in a basic generator. As you know, AC stands for alternating current. This term is derived from the fact that the voltage alternates, being positive for half of a cycle and then negative for the other half of the cycle. You should remember the characteristic AC waveform as shown in Figure 6-1 is a sine wave, and both the voltage and the current have sine wave characteristics. It is also important to understand that the generator produces voltage, and when the voltage is applied to the grid or to a load, current begins to flow. The combination of voltage and current is called electrical power. Electrical power is calculated by multiplying voltage times current.

Because the AC generator creates an AC sine wave that alternates between positive and negative, we will refer to the AC generator as an *alternator* for the remainder of this section. We use the term *generator* to mean a machine that creates voltage when its shaft is rotated, and *alternator* to mean a machine that creates AC voltage.

The simple alternator has two basic parts: the rotor, which is a rotating coil of wire mounted on a shaft, and the stator, which is a stationary coil that has current flowing through it so that it creates a magnetic field. Sometimes the stationary coil is just called the *field*. A voltage is applied to the stationary coil (field) to ensure that a current is flowing through it to create a very strong electromagnetic field.

The coils of wire in the rotor are mounted to the shaft of the alternator. This shaft is rotated by an energy source such as the blades of a wind turbine. When the rotating coil of wire passes through the magnetic field that is created by current flowing through the stator, an electron flow is created in the coil of wire. Because the coil of wire rotates, it pasess the positive magnetic field and then the negative magnetic field during each complete rotation of 360 degrees. This action causes the voltage sine wave to be produced, and a complete sine wave occurs in 360 degrees. In the United States, the speed of rotation of the alternator is maintained at a constant rate so that the sine wave will have a frequency of 60 cycles per second, called 60 Hz.

Because the rotor shaft is constantly turning, a set of carbon brushes that conduct electricity is used to make contact between the stationary part of the alternator and the slip rings on the rotor. The brushes are held in a mechanism that uses springs. The springs ensure that sufficient tension is placed on the brushes to make contact with the rotor shaft so that the generated voltage can be moved from the rotating part of the alternator to the stationary frame, where it can be insulated and connected for use by the loads that are connected to the alternator. It is also important to understand that the amount of voltage that is applied to the stator can be adjusted, to adjust the amount of voltage that is produced by the alternator.

It may seem strange to use a voltage force that is applied to the stator coils just to get voltage out of the armature. Why, you might wonder, would you not just use the small amount of voltage applied to the field coil as the output voltage instead of putting this voltage into the alternator? The reason is that the alternator is a very good energy converter, and so only a very small amount of voltage is needed to cause the stator to create the current to make the magnetic field, whereas the actual amount of voltage produced by the alternator may be several hundred volts. The actual power (volts amps) that the alternator produces will be nearly equal to the amount of power that is put into its shaft. For example, if a large amount of wind energy is harvested by the wind turbine blades and is used to turn the shaft of the alternator, it can produce several hundred times the amount of electrical voltage that is used to excite the alternator to create the magnetic field.

Theory of Operation of the Stationary-Armature Alternator

The stationary-armature alternator is very widely used, because it produces its large current in its stationary coils that are mounted in the stator. This means that the electrical connections to the stator to get the large current off the machine can be made directly to the ends of the coil as a permanent connection, because the stator does not rotate. On the stationary-armature machine, the connection is permanent, and on the rotating-armature type machine, this connection is made through slip rings and brushes.

The stationary-armature alternator has a small amount of electrical current called *exciter current* that flows through its rotor coil, where it produces a magnetic field in the rotor. The small amount of voltage needed to create the exciter current can be transferred into the rotor through induction or through a small set of slip rings and brushes. If the exciter current is transferred into the rotor through slip rings and brushes, it must come from an external source such as rectified (DC) voltage from the grid or from a battery. Because the rotor is turning, the exciter current for the rotating coils comes through brushes and slip rings that are connected to the end of each rotor coil. The amount of current needed to produce the magnetic field in the rotor is relatively small, so it does not wear down the brushes or pit the slip rings as quickly as if the large current were produced on the rotor.

Once the exciter current is flowing through the rotor, the rotor becomes a very strong magnet, which produces a strong flux field. As this flux field moves past the coils of wire in the stator, the alternator produces AC voltage. The amount of exciter current can be increased or decreased to increase or decrease the amount of voltage the alternator produces. Wind turbine blades turn at low speeds, so a gearbox is used to increase the speed of the shaft connecting the gearbox and generator at very high speeds, usually above 1800 rpm. In most wind turbines, this speed is maintained in a fairly constant range because the blades are pitched or the yaw is rotated. Because the speed is fairly constant, the exciter current is adjusted as needed to keep the voltage output at 690 V.

When the AC generator uses voltage from an outside source, it is called a *separately excited machine*. The exciter voltage can also come from its own stator, in which case it is called a *self-excited machine*. The self-excited machine must have a permanent magnet or some other means to start the original magnetic field in the rotor so the generator can begin producing electricity. Once the stator is producing electricity, it can use its own voltage to excite its rotor.

On alternators in which the exciter current comes from induction, a small amount of residual magnetism in the rotor provides sufficient magnetic flux to get the stator to begin producing a small amount of AC voltage. As the voltage begins to build up, the rotor begins to induce a small amount of AC voltage in it as it rotates through the magnetic field produced by the stator as it produces voltage. The rotor can induce an AC voltage because the clearance between the rotor and stator is very small, and the magnetic field created when the stator is producing voltage is very strong. Remember that whenever a wire (coil) moves through a magnetic field, an induced voltage occurs in the wire. The rotor is designed as a coil, so it will induce a small AC voltage as it passes through the magnetic field.

In a stationary-armature alternator, in which the magnetic field is produced in the rotor, DC or AC current is supplied from an external source to the brushes and on to the slip rings. If the alternator is small and produces only a single phase of AC voltage, it needs only two slip rings, because the rotor has only one coil of wire with two ends. One slip ring is connected to each end of the coil. If the alternator is larger and produces three-phase current, it will have three separate slip rings to provide voltage to create exciter current in each of the three rotor coils. The magnetic field in the rotor increases its magnetic field as the small amount of current is supplied to it until the field reaches saturation. The amount of voltage the AC alternator produces increases when the speed of the rotor increases or when the amount of exciter current flowing to the rotor increases and causes the magnetic field in the rotor to become stronger. This means that a field regulator, a computer, or a programmable logic controller (PLC) can be used to control the field current and keep the output voltage of the alternator constant when the speed at which the rotor turns varies as the wind speed changes.

In summary, when exciter current flows through the rotor winding, it produces a magnetic field. Because the rotor is turned by an outside energy source (the wind turbine blades) and it rotates, it produces a rotating magnetic field. When the rotating magnetic field passes the stationary coils of the stator, a large voltage is produced in the stator. When the alternator stator terminals are connected to the grid, the alternator begins to produce current. The amount of voltage the alternator produces is controlled by varying the amount of exciter current in the rotor that causes the rotating field and by the speed at which the rotor is turning. The voltage produced by the alternator is alternating current (AC), so it will have a frequency, and that frequency is a result of the speed of the rotor's rotation. This means that the rotor speed must be kept fairly constant to keep the frequency constant at 60 Hz, so in order to adjust the voltage output of the alternator, the voltage supplied to the alternator field must be adjusted or regulated, which in turn adjusts the field current. Later in this chapter we will explain other ways to control the frequency of the voltage produced by the alternator. Here we provide more detail about how the alternator produces power.

AC Voltage in the Stator

When the rotating magnetic field in the rotor passes the stator winding, the polarity of the magnetic field determines the polarity of the voltage that is produced in the stator. Figure 6-20 shows the rotating field as it passes the stator winding and the resulting AC voltage that is produced. You can see in Figure 6-21a that when the flux lines of the positive end of the magnetic field pass across the wires in the stator winding, a positive voltage begins to be produced. Only a few flux lines are passing through the winding at first, so the amount of voltage is small. As the rotor continues to move the magnetic field more and more, flux lines continue to cut across the stator coil until the maximum number of flux lines are moving across the coil and the amount of voltage produced is at its maximum. When the rotor completes its movement from 90 to 180 degrees, fewer and fewer positive flux lines cut across the stator winding. This causes the amount of voltage being produced to decrease until the amount of voltage reaches zero again. You should recognize the waveform that is produced as the positive part of the AC sine wave.

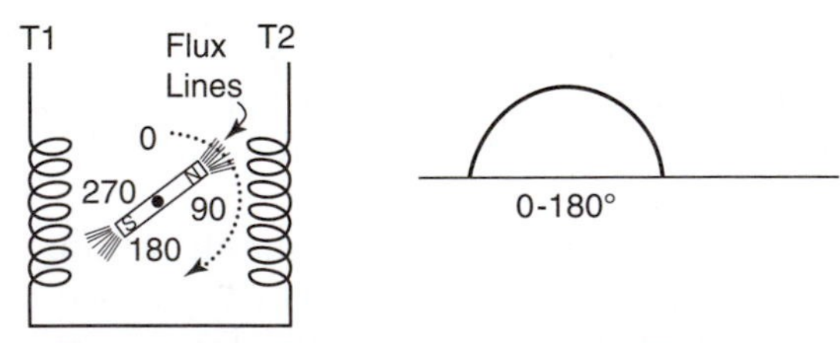

(a) The positive half of an AC sine wave 0-180

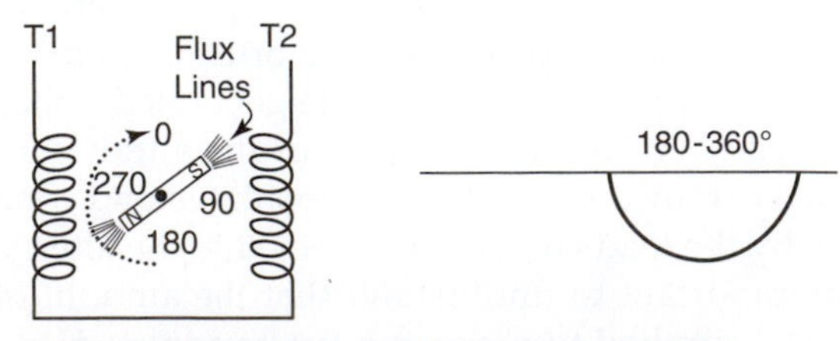

(b) The negative half of an AC sine wave 180-360

FIGURE 6-20 (a) The positive half of the AC sine wave is produced as the magnetic flux lines of the north pole of the rotor move down past the stator coil on the right. (b) The negative half of the AC sine wave is produced as the magnetic flux lines of the north pole of the rotor move up past the stator coil on the left.

When the rotor moves from 180- to 270 degrees, the flux lines from the opposite end of the rotating magnetic field begin to cut across the stator winding. These flux lines are created from the north magnetic pole, so the voltage that is produced in the stator is negative. This voltage starts at zero and builds continually, until the maximum number of flux lines are cutting across the stator winding when the rotor reaches 270 degrees. At this point the voltage is at its negative peak. As the rotor continues to rotate from 270 to 360 degrees, the voltage is continually reduced, until it reaches zero again. In Figure 6-20b you can see that this voltage waveform is the negative half of the sine wave. This means that when the rotor moves through one complete rotation (360 degrees), the alternator will produce a waveform that is a sine wave with one positive and one negative peak. The time it takes the rotor to make one revolution, which produces one sine wave, is called the *period*. If the alternator is to produce voltage at 60 Hz, it must make one rotation in 0.016 s. This can also be described in terms of the alternator rpm. If the alternator produces 60 sine waves in 1 s, it must be turning at 3600 rpm (60 Hz/s × 60 s/min). The frequency of an AC alternator is important, and one way to control the frequency is to control the speed at which the alternator shaft rotates. On larger wind turbines, in which the alternator has multiple coils, the shaft speed is usually regulated at approximately 1800 rpm to produce voltage with a frequency of 60 Hz.

Producing Three-Phase Voltage

The alternator produces some amount of AC voltage, which causes current to flow in a load. The amount of AC power that is created by the voltage and current can be calculated. One way to design an alternator so that it produces the most energy in the smallest physical size is to create three individual stator windings, so that each produces its own set of sine waves. This type of alternator is called a **three-phase alternator**, and it produces three distinct voltages waveforms. The three-phase alternator is sometimes called a **three-phase generator**. The stator windings are placed in the alternator so that the three waveforms are 120 degrees apart. This type of voltage is called **three-phase voltage**. Figure 6-21 shows a three-phase voltage waveform. Each winding may be made of multiple coils of wire to provide more power.

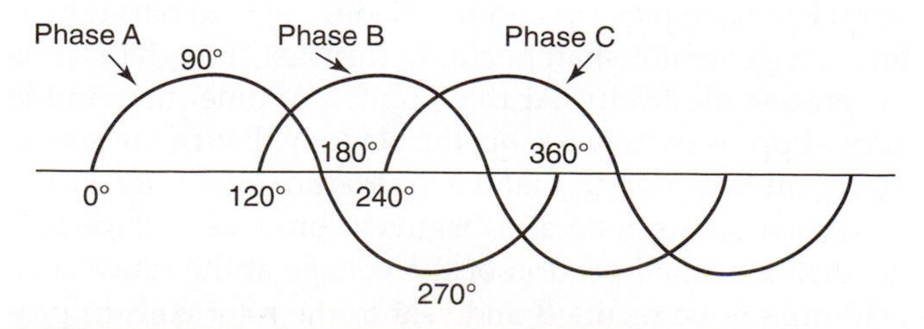

FIGURE 6-21 Three-phase voltage waveform.

Theory of Operation of a Rotary-Armature AC Alternator

The second type of AC alternator is the rotary-armature alternator. This machine is more like a DC generator in that the stationary winding is used to create the magnetic field, and when the wind turbine blades rotate and cause the alternator rotor to spin, voltage is produced in the rotor. Because the large voltage is produced on the rotor, slip rings and brushes are needed to get the voltage off the rotating member. The brushes and slip rings must be large enough to carry the large voltage and current.

The coils of wire in the rotary-armature AC alternator cut through the magnetic flux lines produced by the stationary coils as current flows through them. The current that flows through the stationary coils is the exciter current, and it is created when exciter voltage is applied to the field coils. The more voltage is applied to the exciter coils, the larger the exciter current becomes, which causes more flux lines to be produced. When the rotor coils are rotating, they create more voltage as more flux lines are produced and as the rotor spins faster. At some point the speed of the rotor may need to be controlled to ensure that the alternator produces exactly 60 Hz.

The maximum amount of voltage and current the rotary-armature alternator can produce is limited by the size of the slip rings and brushes. Because all of the voltage and current produced by the alternator must pass through the slip rings and brushes, their size limits how much voltage and current can be produced. The brushes and slip rings also wear, so they must be periodically checked and replaced as they wear down. For this reason the rotary-armature alternator is not as widely used as the stationary-armature alternator.

6.8 ASYNCHRONOUS AC GENERATORS

The three-phase **asynchronous AC generator** is an example of a rotating-field type of generator, and it has a cage-type laminated steel rotor. This type of AC generator is also called an **induction generator** and generates alternating current. The asynchronous AC generator is actually an induction motor whose shaft is turned by the wind turbine blades, and it produces a large amount of voltage and current. Asynchronous AC generators for larger wind turbines produce three-phase AC voltage at 690 V. This voltage is then sent through a transformer, where it is increased to 12,470 V and connected to the grid.

The asynchronous AC generator is an induction generator and has a squirrel-cage rotor made of laminated steel. The large windings for the stator coils are mounted in the stationary part of the generator, and the terminal ends of the coils are connected to the generator case. Because these connections are made through the stationary part of the generator through its case, electrical connections can be made permanently, without any losses. There are no slip rings or brushes, so the rotor gets voltage to create its magnetic field through induction as it turns past the stationary coils. Slip rings are pieces of conducting metal that are mounted to the rotating member. Carbon brushes ride on the slip rings, which creates an electrical connection between the rotating member and the stationary part of the generator. The asynchronous AC generator needs to be connected directly to the grid or other power source to get current through the stator coils so that voltage can be induced into the rotor as it turns. The wind turbine blades turn the low-speed shaft of the gearbox, and the high-speed shaft of the gearbox turns the generator rotor at very high speeds, above 1800 rpm for a 60-Hz machine and above 1500 rpm for a 50-Hz machine. As the rotor is turned by the turbine blades, it moves past the magnetic field created when power is applied to the stator to get the machine started. Once the rotor begins to receive induced voltage, it creates its own magnetic field, which begins to cut across the coils of the stator and starts to generate voltage on a large scale. After the wind turbine blades begin to turn from sustained winds (usually above 6 mph), the generator is on its own, generating voltage at 690 V. The PLC or dedicated controller can adjust the amount of field current to ensure that the generator produces voltage as close to 690 V as possible.

Generator Slip in the Asynchronous Generator

The asynchronous generator has been selected for use in wind turbines because it runs at a fairly constant speed across its complete range of loads when it is connected to the grid at 60 Hz. When the turbine blades are under a minimal load, the speed will be about 1800 rpm. When the wind picks up and the blades begin to produce a maximum load, the speed will be about 1820 rpm. The difference between the actual speed and the rated synchronous speed is called slip. As the wind turbine blades harvest more or less wind, the gearbox and the design of the generator help the generator shaft stay within this rpm range. Because the speed at which the shaft turns stays within a minimal range, the forces on the blades, shafts, gearbox, and generator stay fairly constant, which limits wear and tear on the system over time. The other advantage of staying within this narrow range of speed is that the generator frequency will always be within tolerance for 60 Hz.

Controlling the Frequency of the Voltage from the Generator

Controlling the frequency of voltage from the generator is very important if the generator output is connected directly to the grid, or if it is used as a source for AC voltage to a home or small business. This can be done in several ways. If the generator output power is fairly small, such as for an application in which it is the source of AC power for a small business or home, all the power can be run through a power electronic frequency converter (PEFC), which is also called an inverter. An inverter is an electronic controller that uses diodes to change the AC voltage to DC voltage, and then uses insulated gate bipolar transistors (IGBTs) or other power electronic devices to convert the pure DC voltage back to single-phase or three-phase AC voltage as a quasi–sine wave. You will learn more about power electronic frequency converter (inverters) in Chapter 8, where you will learn how electricity is connected to the grid. In this type of application, in which all the power is run through an inverter, the original frequency of the generated voltage is not a concern, because the AC voltage at any frequency is converted to DC voltage in the inverter, prior to being converted back to AC voltage at exactly 60 Hz. For this type of system, the inverter must be large enough to carry all the power that is generated by the generator.

Another way to ensure that the generator is producing voltage at exactly 60 Hz is to ensure that its speed is controlled as close to 1800 rpm as possible. If the speed varies slightly, the frequency will vary slightly, but when the generator is connected to the grid, the larger grid voltage will force the generator frequency to match 60 Hz. This type of speed control can be accomplished by adjusting (trimming) the wind turbine blades, and moving the nacelle incrementally to ensure that the generator rotational speed is as close to 1800 rpm as possible. You will learn about other ways to control the frequency of the generator in the next sections, where synchronous generators and double-fed induction generators are explained.

6.9 SYNCHRONOUS GENERATORS

The **synchronous AC generator** is another example of a rotating-field type of generator. The rotor for this type of generator is normally a wound rotor that has coils connected to slip rings. Brushes ride on the slip rings to provide a path for exciter voltage to reach the coil as it is turning. This type of generator can also be considered a synchronous motor. Most commonly, the synchronous generator is used in wind turbine applications that are connected directly to the grid. One of the reasons it is connected directly to the grid is that when it first begins to turn, the generator actually receives voltage from the grid, because it is not generating voltage. When this occurs, the synchronous generator actually operates as a synchronous motor, and the voltage from the grid helps the motor come up to near-synchronous speed. When AC voltage from the grid is applied to the synchronous generator, a small portion of the AC voltage is rectified through diodes and sent to the rotor as DC voltage to provide the rotor current. When the DC current flows through the rotor coil, it produces a very strong magnetic field and causes the rotor to begin to turn as a motor. As the wind begins to pick up and the blades begin to harvest energy, the generator shaft begins to turn fast enough that it is generating electricity. At this point, any time the wind is above approximately 6 mph, the blades will turn the generator shaft fast enough that it can produce electricity. When the synchronous generator begins to produce voltage as a generator, a small portion of the voltage at the stator coils continues to be rectified and sent to the rotor coils to produce the strong magnetic rotating field that the generator needs to continue to generate voltage. The level of field voltage can be controlled to ensure that the generator produces the rated voltage at any speed at which the shaft is turning. When the field voltage is adjusted, it automatically changes the field current, which adjust the amount of output voltage the generator produces.

One of the features of the synchronous generator is that having DC current provided to its rotating field coil produces such a strong magnetic field that the synchronous generator has virtually no slip. This means that if the generator is connected correctly to the grid, it's shaft will run at near its design speed at all times, which will ensure that it produces voltage with a frequency near its rated 60 Hz.

If the synchronous generator is used in an application in which it is not tied directly to the grid, the speed of the rotor will vary slightly, which will cause the frequency to vary slightly from its rated 60 Hz. In these applications, the wind turbine can be allowed to accept larger gusts of wind, which will turn the blades faster and allow the generator to produce more electrical energy. Because this application is not connected directly to the grid, an inverter must be used to provide an output voltage at a constant 60 Hz. In this type of application, the inverter needs to be large enough to convert all the AC electricity generated to DC, and then back to 60-Hz AC. The advantage of this type of design is that the wind turbine is allowed to harvest as much energy as possible from the wind, regardless of its speed and gusts. This makes this type of wind turbine installation slightly more efficient than one where the generator shaft speed is controlled by trimming the blades and moving the nacelle.

6.10 DOUBLY FED (DOUBLE-EXCITED) AC INDUCTION GENERATORS

Another type of induction generator that is used with wind turbines is called a **doubly fed induction generator**. This type of rotating-field AC generator has DC voltage supplied to the rotor and AC voltage supplied to the stator. For this reason, this type of machine is sometimes called a *double-excited induction generator.* The majority of doubly fed generators have multiphase wound rotors that have brushes that ride on slip rings that are connected to the rotor coils. The slip rings and brushes allow the exciter current to be provided directly to the rotor, without the losses experienced when the rotor receives its voltage through induction. A small number of doubly fed generators have a laminated steel squirrel-cage rotor that receives its exciter voltage through induction. The doubly fed induction generator is usually a three-phase generator.

When the wind turbine blades turn the generator rotor at speeds below its rated synchronous speed, the stator is generating power, but part of it has to be fed back to the rotor. When the wind blows more strongly, the turbine blades rotate the generator rotor above synchronous speed, and both the rotor and stator produce power to the grid. This means that the doubly fed generator is more efficient than any of the singly fed generators discussed previously.

One of the important control factors that makes the doubly fed induction generator widely used in wind turbines is that an inverter can be used to control the frequency of the voltage fed to the rotor at exactly 60 Hz. When the stator is tied to the grid at 60 Hz, the generator will constantly produce 60 Hz regardless of the speed at which the generator shaft is turning. Because the inverter is needed only to convert the exciter current, it does not have to be very large, compared to it being sized to convert all of the output power. Remember that the amount of current the rotor uses as exciter current is very small compared to the total output power the generator produces. For this reason, the inverter for the exciter current will be much smaller than if it were sized to convert the entire output voltage.

One advantage of using the doubly fed induction generator is that it can produce power with a constant 60-Hz frequency in wind speeds from 6 mph to 50 mph. The doubly fed induction generator can control the output frequency by applying 60-Hz exciter current through an inverter, which make the output generated power run at 60 Hz as long as the gearbox keeps the high-speed shaft that turns the generator at an rpm above the synchronous speed of 1800 rpm. This allows the wind turbine to accept gusting winds and allows the blades to harvest the extra energy, which in turn improves the wind turbine's efficiency. If the wind turbine is very large (2 MW or larger), the control system can incorporate individual wind turbine blade adjustments and nacelle directional yaw adjustments to harvest the maximum amount of wind available.

6.11 PERMANENT-MAGNET SYNCHRONOUS GENERATORS

The **permanent-magnet generator** relies on a very strong permanent magnet (PM) to provide the original rotating magnetic field for the generator. The rotor in this generator is a very strong permanent magnet that continually puts out a very strong magnetic field. The stator is made of coils of wire that are mounted in the stationary part of the generator. When the wind turbine blades begin to turn the rotor, the magnetic field cuts across the stator coils, which causes a voltage to be generated in the stator coil. Because the stator coils are in the stationary part of the generator, there are permanent electrical connections at the terminal end of each coil that allow voltage to be taken off the stator coils. Because the rotor is a permanent magnet, and the voltage is generated in the stator coils, the PM generator does not need brushes and slip rings or brushes and commutator segments.

Early permanent-magnet generators were small generators suitable for use only to charge batteries or, used with a small inverter, to provide a very small output AC voltage at 60 Hz. Today, many larger wind turbines use PM generators because of their simple design, which requires vary little maintenance. Most of the wind turbines that use PM generators are direct-drive generators, and because the PM generator is a low-speed generator, it does not produce voltage at 60 Hz. Instead, it

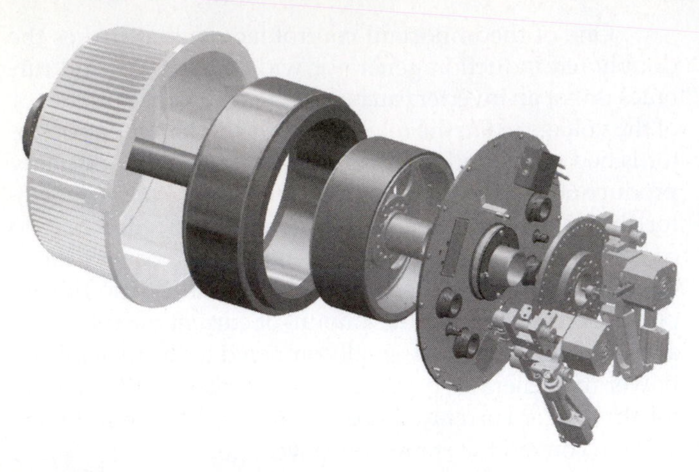

FIGURE 6-22 Permanent-magnet (PM) generator. (Courtesy of Northern Power Systems Inc.)

provides low-voltage AC that can be rectified and used for battery charging or voltage that can be sent through an inverter and transformers to provide 60 Hz for immediate use or for grid-tied systems. Figure 6-22 shows the internal parts of a permanent-magnet generator used in a wind turbine.

6.12 USING AN ALTERNATOR TO PRODUCE DC VOLTAGE

DC generators have many design issues that require a large amount of preventative maintenance, so AC alternators are now commonly used to provide DC voltage and current. This is easily accomplished by placing two diodes in each winding of the alternator. A diode is an electronic component that allows current to flow in only one direction, and if an AC sine wave flows through a diode, only the top half or the bottom half will be allowed to flow through it. The direction in which the diode is connected in the system determines the polarity of the voltage flowing through it, because it will block either the top half or the bottom half of the sine wave. A three-phase alternator uses six diodes (three sets) to rectify the AC voltage to DC. The diodes must be sized to carry the maximum working voltage and current of the alternator. If one diode in a set fails, the amount of voltage produced by that winding will be approximately half of the rating of that coil. If two diodes in any set burn out, all the voltage produced by that coil will be lost. Bad diodes are the primary cause of these types of alternator failures.

Questions

1. Identify the basic parts of an AC alternator.
2. Explain the operation of the rotary-armature type of AC alternator.
3. Explain how voltage is taken off the rotating part of a DC generator.
4. Explain two ways you could get more voltage from the alternator.
5. Draw a sine wave and identify 0, 90, 180, 270, and 360 degrees.
6. Identify two ways in which exciter voltage can be supplied to the rotor of an alternator.
7. Explain how a doubly fed alternator operates.
8. Explain the difference between commutator segments and slip rings.
9. Explain the difference between a wound-rotor armature and a squirrel-cage rotor.
10. Explain why laminated steel pieces are used in a squirrel-cage rotor.

Multiple Choice

1. The moving part of the alternator is called the
   a. Stationary winding
   b. Rotor
   c. Field
   d. Slip rings
2. The DC generator uses ________ to get voltage from its rotating part.
   a. Brushes and commutator segments
   b. Brushes and stator
   c. Brushes and slip rings
   d. All of the above
3. A three-phase alternator
   a. Has three individual stator windings that each produces a sine wave that is 120 degrees out of phase with the others
   b. Has three separate rotors that spin simultaneously and produce sine waves that are 120 degrees out of phase with each other
   c. Has three separate shafts for the wind turbine gearbox to turn
   d. All of the above
4. The voltage that the alternator produces can be increased by
   a. Increasing the speed of the rotor rpm
   b. Increasing the amount of exciter current
   c. Adding more commutating segments
   d. All of the above
   e. Only a and b
5. The strength of an electromagnet coil can be increased by
   a. Increasing the current that flows through the coil
   b. Increasing the number of coils or the amount of wire in the coils
   c. Providing an iron core for the coil
   d. All the above
   e. Only a and b

CHAPTER 7

Gearboxes and Direct-Drive Systems

OBJECTIVES

After reading this chapter, you will be able to:

- Explain the operation of a gearbox for a wind turbine.
- Explain why wind turbines need gearboxes.
- Identify five types of gears and explain their functions.
- Explain how a direct-drive wind turbine system is able to produce voltage with 60-Hz frequency.
- Explain the function and operation of a bearing.
- Identify five types of bearings.

KEY TERMS

Axial force
Ball bearing
Bearing race
Bevel gear (miter gear)
Bushing
Double helical gear
Drive train
Driven gear
Driver gear
Gear backlash
Gearbox
Gear ratio
Helical gear
Herringbone gear
High-speed shaft
Journal
Low-speed shaft
Needle bearing
Pillow block
Planetary gear
Rack
Radial force
Roller bearing
Speed reducer
Spur gear
Thrust load
Thrust bearing
Worm gear

OVERVIEW

This chapter will cover gearboxes and direct-drives systems for wind turbines. In a wind turbine, the turbine blades are connected to a hub that allows them to rotate when the wind blows through the blades. The hub is connected to a shaft that transmits the power to the generator. In this chapter you will learn that in some wind turbines the turbine blade is connected directly to the generator, and in other, larger systems the shaft from the turbine blades and hub is connected to a gearbox containing a number of gears that adjust the speed of the turbine blade to the speed required at the generator. In Chapter 6 you learned about generators and that, in the United States, the generator needs to turn at a specific speed to produce voltage at exactly 60 Hz to connect to the grid. In this chapter you will learn how the gears in the gearbox provide the means to adjust the speed of the high-speed output shaft to provide a constant speed to the generator regardless of the speed at which the turbine blade and rotor are turning.

In this chapter you will also learn that some generators are designed to be connected directly to the turbine blade through a drive train and are allowed to run at any speed. The frequency of the voltage produced by this type of generator is not controlled; instead, in this type of design, the voltage produced by the generator is sent through an electronic inverter, where it is converted to an output AC voltage of exactly 60 Hz.

This chapter you will show how gears, gear ratios, gearbox, and drive trains are used to transmit power from the turbine blade to the generator. You will learn how to install these components and how to troubleshoot and maintain them. You will also learn about the need for periodic maintenance on gearboxes and the shafts that are connected to them.

7.1 WHY GEARBOXES ARE NEEDED

Grid-tied wind turbines must produce AC voltage at a very constant frequency and at a constant voltage level. Because the wind blows at varying speeds throughout a day, the turbine blades will turn at varying speeds. For years this has created problems for wind turbine designers. The designers have been challenged to design turbine blades that harvest as much of the wind as possible even at low wind speeds. When wind speeds are low, the turbine blade turn at much slower speeds than when wind speeds are higher. If the turbine blades are connected directly to the generator, the variation in turbine rotational speed means that the generator also turns at various speeds. As you learned in Chapter 6, the speed at which the turbine blades turn the generator directly affects the voltage level and the frequency of the voltage the generator produces.

If the output voltage of the wind turbine is used in an application in which the AC voltage produced by the generator is converted to DC voltage, the frequency of the generated voltage becomes irrelevant, because an electronic converter can be used to change the AC voltage to DC voltage. In some smaller wind turbines that are used as stand-alone power supplies for homes or buildings in remote locations, the DC voltage is stored in batteries and used as needed. The DC voltage can only be used in lighting and heating applications or with appliances that have DC motors. In this type of application the generator can be connected directly to the wind turbine blades and allowed to run at any speed at which the turbine blades turn.

In grid-tied applications, the power that is produced by the wind turbine must match the electrical voltage and frequency of the grid to which it is connected. In this type of application the percentage deviation of frequency and voltage is tightly controlled to ±0.5%. In grid-tied applications, wind turbines use electronic inverters, which change the AC voltage without regard to voltage level or frequency to DC voltage and then back to AC voltage at a very tightly controlled voltage and frequency. The inverter must be designed to handle the largest amount of voltage and current that the wind turbine is allowed to produce when its blades are turning at maximum speed. The maximum rpm at which the turbine blades turn usually occurs when wind speed is at its highest. In these types of wind turbines the maximum speed at which the blade is allowed to turn must be controlled so that overspeed does not cause the generator or the inverter to overload. If the maximum speed of the wind turbine is not controlled, it is possible for the blades to turn at such a high speed that the generator produces too much voltage and current, which may cause the generator and or the inverter to overheat and become severely damaged.

One way to control the speed at which the generator turns is to use a **gearbox** that acts as a transmission between the turbine blades and the generator. The gearbox has internal gears designed in a specific ratio so that the slower speed of the input shaft from the turbine blade is increased to the much higher speed needed for the generator. Internal gears are gears that are on the inside of the gear train. The generator needs to turn at a relatively high speed to produce voltage at 60 Hz. For example, a two-pole generator must turn at 3600 rpm to produce an output frequency of 60 Hz. A four-pole generator must turn at 1800 rpm to produce an output frequency of 60 Hz. The turbine blade rpm for larger commercial wind turbines may be very slow, between 10 and 20 rpm at optimum load. Some smaller wind turbines designed for homes may turn at speeds as high as 300–600 rpm in high winds. In all wind turbines, these speeds are well below the 1800 rpm needed to turn the generator at a sufficient speed to produce the specified frequency.

Given the problem of variable wind speeds and the low rpm of the turbine blades, a gearbox that can adjust the gear set, much like an automatic transmission in an automobile, is used to keep the engine rpm at a fairly constant speed. By constantly changing gear ratios, the gearbox in a wind turbine keeps the speed of the output shaft to the generator at a nearly constant rpm regardless of the changing speed of the input shaft from the turbine blades. For this reason the gearbox is a necessary component of a wind turbine. Figure 7-1 shows the location of the gearbox in relation to the turbine blade and the generator and identifies the basic parts. The blades and rotor can be seen on the front of the wind turbine, and the gearbox transmission and the generator are near the rear of the nacelle. The anemometer wind vanes are on the very

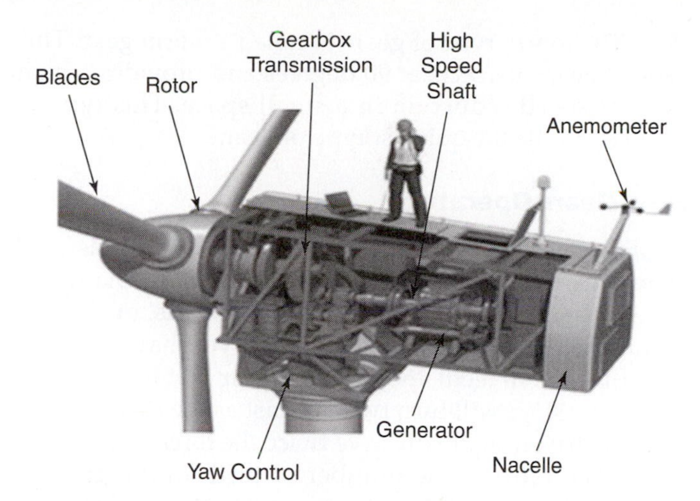

FIGURE 7-1 An elementary diagram showing the relative locations of the turbine blade, gearbox, and generator. It also shows the location of the nacelle where it is mounted on the tower. (Courtesy of Fotolia © Stephen Sweet.)

top, at the rear of the nacelle. The anemometer and other wind instruments determine the direction and speed of the wind and report this information to the main computer or controller. The yaw controls consist of a ring gear and yaw drive motor that are used to rotate the nacelle into or out of the wind direction. The entire nacelle sits on top of the tower

Now that you know why a gearbox is installed between the turbine blade and the generator, it is important to understand the theory of operation and the function of each individual type of gear that is used in the gearbox. Gearboxes and transmissions are not new, as they have long been used in industrial applications such as large motor-driven or engine-driven pumps and generators. Gearboxes have also been used for many years in large earth-moving and mining equipment that uses a large diesel engine to power one or more sets of wheels, as well as in transmissions in large trucks and other types of vehicles. The technology used in the design of these gearboxes has steadily improved over the years, so that they are now very reliable and can run for long periods of time without significant maintenance or repair. The major reasons for this are that the manufacturing technologies used in the production of the gears and the gearboxes have steadily improved, and the tolerances of the gears havesimilarly improved markedly. Improving the tolerance of the gears improves the life of the gears, because there is less wear and tear on the gears. Lubrication and cooling processes for gearboxes have also improved tremendously over the years, helping to extend the life of the components inside.

Gearboxes are also lasting longer and have become are more reliable because newer equipment ensures more accurate alignment between the turbine blade rotor and the generator, which results in less vibration due to misalignment. As you learn more about gears and gearboxes, you will learn that gears provide several functions, such as change of direction, change of speed, change of torque, and transfer of energy.

7.2 ADVANTAGES OF GEAR RATIOS

When a number of gears are put together in a transmission or gearbox, the number of teeth in each gear creates a ratio called the **gear ratio**. The gear that provides the input energy is called the **driver gear**, and the gear that it moves and that receives the energy is called the **driven gear**. In some gear trains the driven gear is also called the *follower*. Gears are used to provide a difference of speed, a difference of force, and a change of direction between the shaft of the driver gear and the shaft of the driven gear.

If the driver gear has 10 teeth and the driven gear has 30 teeth, as shown in Figure 7-2, the gear ratio is 3 to 1; for every complete revolution the driver gear makes, the driven gear will make three complete revolutions. Another way to determine the gear ratio between two gears is to divide the number of teeth in the driven gear by the number of teeth in the driver gear. If the driver gear has 10 teeth and the driven gear has 30 teeth, you get 3 when you divide 30 by 10, so the gear ratio is again 3 to 1, which is expressed as 3:1.

The gear ratio is important because tells you know how much torque is being transmitted from the driver gear to the driven gear, and if you know the speed of the driver gear, it tells you the speed of the driven gear. For example, if the gear ratio is 3:1 and the driver gear is turning at 300 rpm, the driven gear will move at 100 rpm. If the torque of the driver gear is 200 newton-meters (N-m), then the output torque of the output driven gear will be 600 N-m.

Sometimes the number of teeth in the driver gear and the number of teeth in the driven gear will not calculate to an even number such as 2:1 or 6:1. If the driver

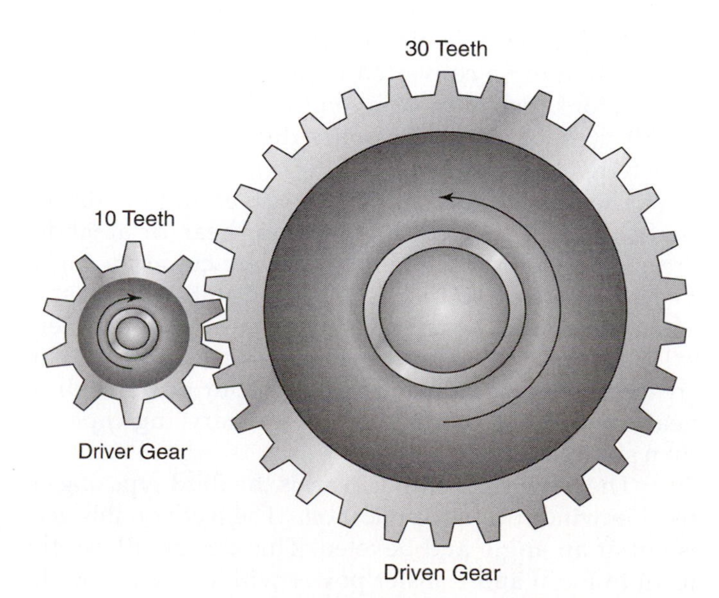

FIGURE 7-2 This driver gear has 10 teeth, and the driven gear has 30 teeth, so the gear ratio is 3:1.

gear has 8 teeth and the driven gear has 30 teeth, the gear ratio is 15:4, or it could be expressed as 30:8. This gear ratio can be shown with reference to 1, so it will include the fractional part of the number. For example, if the driver gear has 30 teeth and the driven gear has 8 teeth and you divide 30 by 8, the gear ratio can be expressed as 3.75:1.

If the driver gear has 200 teeth, and the driven gear has 10 teeth, then the gear ratio is 1 to 20 or 1:20. In general, if the gear ratio is a ratio such as 2:1 or 5:1, the gear train will produce less speed at the output driven gear and more torque (force). If the gear ratio is 1:10 or 1:20, the output driven gear will have more speed and less torque at the output driver gear.

If the drive train has three gears, the driver gear meshes with the center gear, called the idler gear, which turns in the opposite direction. The third gear becomes the output gear: It receives its energy from the idler gear and its rotation will be in the opposite direction from the idler gear but in the same direction as the driver gear. If you want the driver gear and the driven gear to turn in the same direction, the transmission needs to have an odd number of gears; and if the driver gear and the driven gear are to turn in opposite directions, the transmission needs to have an even number of gears.

7.3 TYPES OF GEARS

Four common types of gears are generally used in gearboxes and in wind turbine applications and machines. They include spur gears, helical gears, bevel and miter gears, and worm gears. The teeth on the **spur gear** are cut straight and are parallel with the shaft. The number of teeth in the drive gear and the number of teeth in the driven gear are selected to create a gear ratio that will cause a specific speed or force. The drive gear is the gear where energy is put into the system, and the driven gear is the gear that the drive gear moves. Another type of spur gear is used as a pinion with a rack. This allows rotary motion to be converted to linear motion for use in applications such as motion control. The number of teeth in this type of gear determines the speed or the force transferred to the rack.

The **helical gear** allows the driver and the driven shafts to meet parallel, like the spur gear, or at 90 degrees. The teeth of the helical gear are cut at an angle. The numbers of teeth in the driver and driven gears again determine the relationship of speed or force. Helical gears provide improved tooth strength and increased contact ratio compared to spur gears. Helical gears also can provide greater load-carrying capacity than spur gears.

The **bevel gear (miter gear)** is the third type of gear used in wind turbine applications. The teeth on this gear is cut at an angle and beveled. This design allows the teeth to mesh and transfer power with less wear on the teeth, and this type of gear is quieter than other types. The bevel gear can run at much higher speeds and with less wear than other types of gears.

The fourth type of gears is called a **worm gear**. This type of gear intersects at 90 degrees and provides a high ratio of speed reduction in a small space. This type of gear should be the quietest type of gear.

How Gears Operate

Gears are used in machines to cause a positive transfer of power from one shaft to another, provide a change of speed, change direction of rotation of shafts, or gain an advantage of torque. For example, if you have a driver gear that has 10 teeth and a driven gear that has 20 teeth, the driver shaft will turn twice as fast as the driven shaft, and the driven shaft will have twice the force as the drive shaft. The ratio of the number of teeth in the gears is equal to the ratio of the speed and the ratio of force between the gears. Another important point to remember is that if two gears (the driver and driven) are used, the rotation of the drive gear and the driven will be in opposite directions. Each time another gear is added, the direction of rotation of that gear is changed. The driver shaft and the driven shaft can be mounted parallel to each other or at some angle to each other, depending on the application. A more detailed discussion and examples of all of these gears are given in the next section. You will learn that all gears are designed for maximum power transfer with a minimum of friction.

Categories of Gears

Gears used in industry today can be generally classified into three categories, each of which includes a number of different types of gears. Table 7-1 lists the three categories of gears: *parallel-axis gears*, in which the drive shaft and driven shaft are mounted parallel to each other; *intersecting-axis gears*, in which the drive shaft and driven shaft intersect at some angle (such as 90 degrees); and *nonparallel and nonintersecting gears*.

TABLE 7-1 Categories of Gears and Their Efficiencies

Category	Types of Gears	Efficiency (%)
Parallel-axis gears	Spur gear Rack Internal gear Helical gear Helical rack Double helical gear	98–99.5
Intersecting-axis gears	Bevel gear Spiral bevel Spur bevel	98–99
Nonparallel, nonintersecting gears	Worm gear	30–90
	Screw gear	70–95

FIGURE 7-3 A set of spur gears in which the driver gear turns in one direction and causes the driven gear to rotate in the opposite direct. Notice that the larger driver gear on the left will cause the smaller driven gear on the right to rotate faster. The speed ratio is equal to the ratio of the teeth in the gears. (Courtesy of Power Transmission Solutions, Emerson Industrial Automation.)

Spur Gears

Spur gears are gears that transmit power between two shafts that are parallel with each other. Figure 7-3 shows two spur gears. You can see that the teeth for these gears are cut at 90 degrees across the gear face. The gears must match with each other exactly so the teeth of one will mesh correctly with the teeth of the other. The spur gear is a low-cost gear that is used for general-purpose loads at medium speeds (up to 4000 rpm). If the speed becomes too high, this type of gear will become very noisy. Because it is low-cost and requires little maintenance, the spur gear is the most widely used type of gear.

Figure 7-4 shows a third gear, called an *idler gear,* added to the transmission so that the shaft rotation of the driven gear will be the same as that of the driver gear. If two gears are used, the rotation of the shafts is reversed. If a third gear is used, the first and last gear turn the same direction, and the idler gear between them rotates in the opposite direction. If the set of gears is designed to increase force, torque, or speed, the number of teeth in each gear can be varied, too.

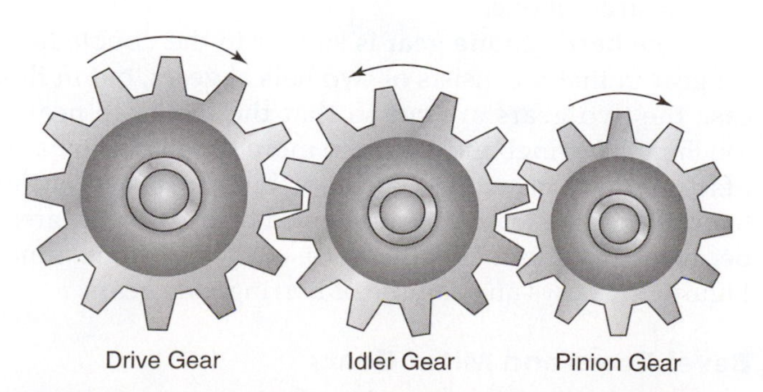

FIGURE 7-4 Three spur gears used to create a gear ratio that allows the final shaft rotation of the pinion gear to be in the same direction as that of the input drive gear. The drive gear turns clockwise and causes the idler gear to rotate counterclockwise. The idler gear causes the pinion gear to rotate back in the same clockwise direction as the input drive gear.

FIGURE 7-5 Spur gear with rack. As the spur gear rotates, the rack moves in a linear direction. This gear set converts rotational motion to linear, or vice versa. (Courtesy of Power Transmission Solutions, Emerson Industrial Automation.)

Spur Gears with a Rack

Another type of spur gear that you will encounter in some applications is called a *gear and rack.* The rack is used to convert rotary motion into linear motion. When the shaft with the driver gear rotates, the teeth of the spur gear mesh with the teeth of the rack, which causes the rack to move in a linear motion. When the rotation of the drive shaft is reversed, the motion of the rack is also reduced. Figure 7-5 shows an example of a spur gear and rack. This type of gear configuration can also be used in smaller instruments to convert linear motion into rotary motion. For example, the shaft of a potentiometer or the shaft of an encoder can be connected to the round spur gear. When the linear rack moves with a linear movement, the rotary spur gear rotates and in turn rotates the shaft of the encoder or potentiometer, which will produce a signal that indicates the amount of travel. This arrangement is used in wind turbine instruments to indicate wind direction.

In some wind turbines, this type of gear application is used to rotate the yaw mechanism. The nacelle is mounted on a large circular gear that acts like the rack, and the spur gear is connected to the shaft of the yaw control motor. When the yaw motor rotates the spur gear, the large circular gear moves and causes the nacelle to rotate the wind turbine into the wind.

Helical Gears

Helical gears are used in applications where the drive shaft and the driven shaft are mounted parallel to each other. Figure 7-6 shows two helical gears. Helical gears are cut on an angle across the gear face. The angle at which the teeth are cut on the gear is called the *gear pitch.* Helical gears are available with pitch (angle) from a few degrees to 45 degrees, with the average being about 20 degrees. Because the teeth are cut on an angle, helical gears are a little longer than spur gears and can

FIGURE 7-6 Two helical gears. Notice that the teeth are cut at an angle across the face of the gear. (Courtesy of Power Transmission Solutions, Emerson Industrial Automation.)

transfer more load than spur gears. Helical gears also run much more quietly than spur gears. When two helical gears are used together, they can each have a different number of teeth to provide a gear ratio that allows for a change of speed or torque. The other important feature of helical gears is that they create a **thrust load**, which is a force along the length of the shaft. Special bearings are used with helical gears to counteract the thrust load.

Crossed Helical Gears

Crossed helical gears are similar to helical gears in that the teeth are cut diagonally across the face. The In crossed helical gears, however, the two gears are perpendicular to each other, which allows for the drive shaft and the driven shaft to intersect at 90 degrees. The ratio of the number of teeth in the drive gear to the number of teeth in the driven gear determines the speed advantage and torque advantage for these gears. Figure 7-7 shows an example of crossed helical gears.

FIGURE 7-7 Crossed helical gears allows two shafts to intersect at 90 degrees. (Courtesy of Power Transmission Solutions, Emerson Industrial Automation.)

FIGURE 7-8 Double helical gears. Notice the space between the two sets of teeth.

Double Helical Gears and Herringbone Gears

Another way to counteract the thrust load is to use a **double helical gear**. Figure 7-8 shows an example of double helical gears. In this figure you can see that one set of helical gears is cut at one angle across the gear face, and the other set is cut at an angle in the opposite direction. The result is that the thrust created by one set of teeth is counteracted by the thrust in the opposite direction created by the other set of teeth. The helical gear can also be used with a rack to provide a means of transferring rotary motion to linear motion. Helical gears can provide more power transfer than typical spur gears with a rack.

Double helical gears consist of two helical gears of opposite hands on a common gear blank with a space between the meeting of the two gears. The advantage of a double helical gear is that it eliminates axial thrust (thrust that is parallel to the shaft). The double helical gear is better at transferring force because it has twice the number of teeth and the face surface is nearly double that of traditional gears with a single set of teeth. This type of gear can be made out of one piece of metal if the clearance in the space between where the two gears meet is wide enough. If the space is not wide enough, the double helical gears may be made out of two pieces of metal. The double helical gear spreads the load across two sets of gears instead of one.

The **herringbone gear** is similar to the double helical gear in that it consists of two helical gears, but in this case the two gears are cut so that the teeth are nearly touching. Herringbone gears are more expensive to manufacture than other types of gears, but the design of the teeth allows the load to be dispersed over twice the area, because two sets of teeth are used at the same time. Figure 7-9 shows an example of herringbone gears.

Bevel Gears and Miter Gears

Figure 7-10 shows two examples of a **bevel gear**. These gears are used when the drive shaft and the driven shaft intersect at approximately 90 degrees. The bevel gear provides the most efficient means of transmitting power of all the gear designs, but it may also be necessary to

FIGURE 7-9 Herringbone gears. Herringbone gears are similar to double helical gears, except that there is no space between the two sets of teeth.

(a)

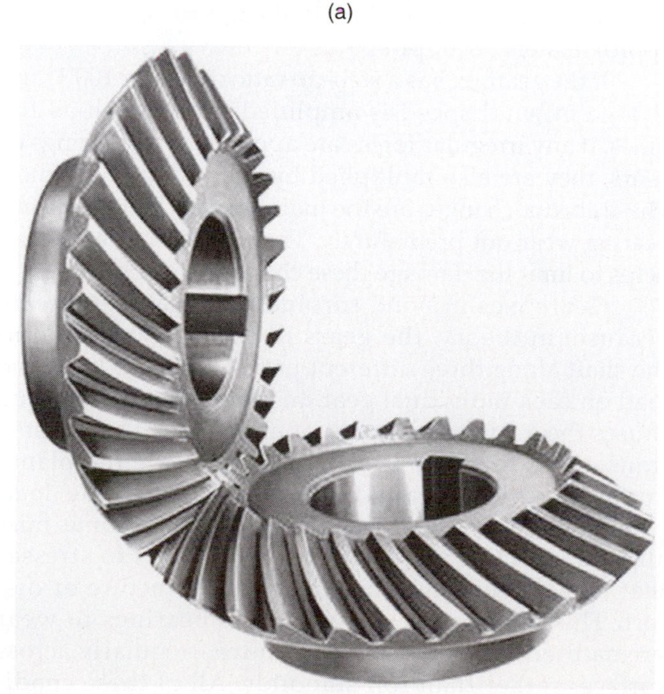

(b)

FIGURE 7-10 (a) Bevel gear. (b) Spiral bevel gear. (Courtesy of Power Transmission Solutions, Emerson Industrial Automation.)

specify the bevel gear where space is limited and the shafts must be mounted at 90 degrees. The intersection of the shafts in a bevel gear does not have to be at exactly 90 degrees. In some applications the shafts intersect at angles less than 90 degrees, and in others the shafts intersect at angles greater than 90 degrees. If the two gears in the bevel gear are of equal size and are mounted at exactly 90 degrees to each other, the gear can be referred to as a *miter gear.* Thus you may encounter a transmission that some people call a miter gear and others call a bevel gear. If the gears are of equal size and are mounted at exactly 90 degrees to each other, they can be called either a bevel or miter gear.

Worm Gears

Figure 7-11 shows an example of a worm gear. The worm gear consists of a cylindrical worm that meshes with a wheel gear. In this type of application the worm is generally the driver and the wheel gear is the driven gear. One application of the worm gear is called a **speed reducer.** The major features of this type of gear are the large amount of speed reduction (100:1) and the small package. The screw part of the transmission provides a larger mechanical advantage than other types of gears, but this also creates a loss of efficiency. The other reason a worm gear is used is because it has more surface area that stays in contact between the worm and rotating gear, which makes it run quietly and allows it to absorb higher shock loads. Worm gears are used in the yaw directional control system. The wheel gear is mounted to the nacelle base and the cylindrical worm gear is mounted to the drive motor. When the direction of the nacelle and blades is required to be changed to keep the blades into the wind, a signal is sent to the drive motor, which causes it to rotate and causes the wheel gear to move the nacelle several degrees. The computer or programmable logic controller (PLC) compares the change of direction in which the nacelle moves with the signal from the anemometer, which indicates the true wind direction. The yaw control can be

FIGURE 7-11 A worm gear allows rotary motion from the driver gear to be converted into linear motion in the worm gear.

used to move the nacelle into the direction of the wind, or out of the wind if overloading or overspeed occurs.

The efficiency of the worm gear is typically very low, which means you will lose some of the horsepower of the driver. For example, if you need 1 hp to move a load and you need to use a worm gear, the driver motor may need to be larger than 2 hp. Other problems with the worm gear are that it produces some axial loading forces, which require more expensive bearings to offset them, and that it has a tendency to produce large amounts of heat that must be carried away through additional lubrication.

7.4 HELICAL PLANETARY GEARS

Figure 7-12 shows a large **planetary gear** that might be used in a wind turbine. Figure 7-13 is a diagram of the planetary gear with all the parts identified. The inside gear, called the *sun gear,* is engaged with the outside gear, called the *ring gear,* through a number of smaller planetary gears. Notice that all the gears have teeth that are cut diagonally across the face of the gear, which makes the gears helical gears. Planetary gears get their name because the ring gear causes the planetary gears to rotate around the sun gears, much as the planets rotate around the sun in our solar system.

Planetary gears are used in wind turbine gearboxes because they provide several advantages. Planetary gears allow the input shaft to the gearbox and the output shaft from the gearbox to be parallel and run in the same direction, which means that the gearbox and generator can be mounted in a straight line. Another reason planetary gears are used in gearboxes is that they are specifically designed to handle the extreme loads that are constantly changing in wind turbines. At one instant the wind may be blowing at 10 mph, but a few minutes later, gusts of wind in excess of 30 mph can subject the turbine blades to very severe and changing loads. This abrupt change causes extreme loading forces to be placed on the blades, shafts, and gearbox. In the largest wind turbines, the large blades and rotors can be over 300 ft in diameter and provide 1–2 million ft-lb of torque. The planetary gears in these gearboxes provide 75:1 or 100:1 step-up ratios. The wind turns the rotor blades at a slow speed of 10–20 rpm. The rotor is connected to the low-speed shaft, which is connected to the input side of the gearbox. The step-up ratio of the gearbox provides an output shaft connected to the generator shaft that can provide speeds of approximately 1800 rpm.

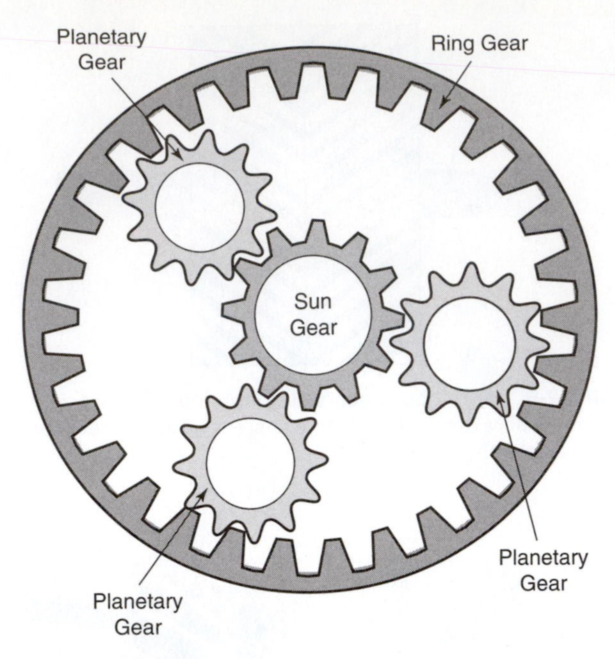

FIGURE 7-13 Planetary gears with the sun gear, planetary gears, and ring gear identified.

FIGURE 7-12 Planetary gears. The inside gear, called the sun gear, is engaged with the outside gear, called the ring gear, through a number of smaller planetary gears. (Courtesy of NKE Bearings.)

If the gearbox has a step-up ratio of 75:1 or 100:1, any change in wind speed is amplified by as much as 100 times. If any irregular forces are applied to the bearings or gears, they are also multiplied by as much as 100 times. These abrupt changes are the main reasons that gears and bearing wear out prematurely. The planetary gear design helps to limit the damage these changing forces do.

Gearboxes in wind turbines use planetary gears because, in theory, the gears divide the torque from the shaft along three different paths, which reduces the load on each individual gear. In reality, the high torque causes the gears to twist out of alignment. This twist is transmitted to shafts bearings gears inside the planetary set, which means they do not share the load equally. The result is that when gears become misaligned, shock loads and other forces lead to stresses that may eventually cause the gear to fracture or distort. This misalignment also causes bearings to wear prematurely, because they may move irregularly across surfaces rather than roll smoothly. All of these conditions cause the gearbox to wear prematurely, and complete failure can result if periodic maintenance does not

TABLE 7-2 Advantages and Disadvantages of Planetary Gearboxes

Advantages	Disadvantages
Provide high gear ratios needed to connect the slow-speed shaft from the blade to the high-speed shaft going to the generator.	Possibility of premature wear on bearings and gears if misalignment occurs.
Spreads torque across multiple gears instead of just one gear.	Requires additional periodic maintenance.
Gear set is more compact than other types of gears of similar size.	Possibility of additional noise if wear exists.
Increased efficiency when turbine blades turn at slow speeds.	
More reliable than other types of gears.	

identify and fix the problems. The gearbox is mounted in the nacelle at the top of the tower, so periodic maintenance is expensive and time-consuming because technicians must climb the tower and secure themselves before maintenance can begin.

Advantages and Disadvantages of Planetary Gearboxes

Planetary gearboxes solve many problems for wind turbine gearboxes, but they also present several problems. Table 7-2 shows the advantages and disadvantages of planetary gearboxes.

Other Types of Gears Made Specifically for Wind Turbines

Newer designs of wind turbine gearboxes have incorporated new technology to combat some of the problems of misalignment and premature wear of planetary gear sets. Several companies manufacture specialized gears for wind turbine shafts. These bearing try to share the loads and eliminate misalignment, which dramatically improves wind turbine reliability.

Planetary gears equalize the loads on planets by anchoring them onto a planetary carrier to limit the side-to-side load. The planetary gear ultimately equalizes forces on the planets, even while transmitting varying levels of torque. Figure 7-14 shows a bearing used in the gearbox to keep the gears precisely aligned. This type of bearing is designed specifically for the rough environment of changing loads and torque that occurs in wind turbines. These types of bearings hav vastly extended the life of wind turbine gearboxes. More details about bearings are given in the next section.

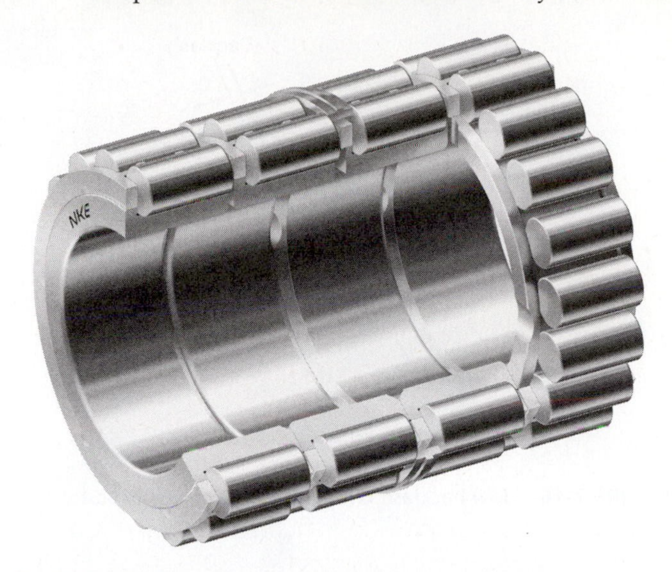

FIGURE 7-14 Bearing used with wind turbine gearbox shaft. (Courtesy of NKE Bearings.)

Gear Meshing and Backlash

One of the major problems you will encounter as a maintenance technician is working on a set of gears that will not mesh correctly or that have excessive backlash. **Gear backlash** is the small amount of space that remains after two teeth mesh, which is normal and necessary. Figure 7-15 shows a typical set of gears as they mesh. In this figure you can see how the gear teeth fit together. The depth of the gear must fit the mating gears exactly, with a small amount of clearance between the tip of the gear and the other gear. If there is not sufficient clearance, the gear teeth will wear down prematurely.

Figure 7-16 shows how backlash is measured. In this figure you can see that a small amount of backlash is permissible, but as gears wear, the amount of backlash will increase until it reaches a point at which it is unacceptable. As the amount of backlash increases, the gears will also begin to vibrate excessively when the gears

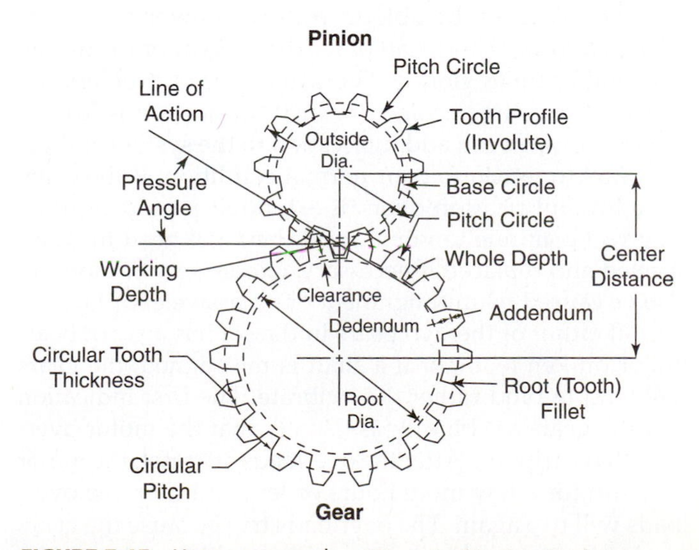

FIGURE 7-15 How gears mesh. (Courtesy of Boston Gear.)

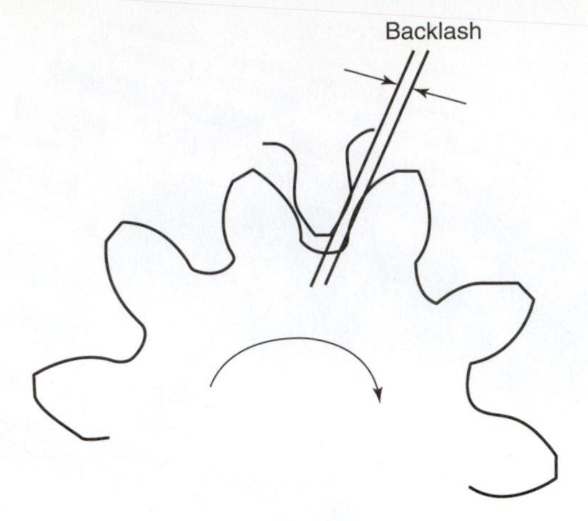

FIGURE 7-16 How backlash is measured. (Courtesy of Boston Gear.)

change speed or change direction of rotation. If the gears are connected to a motor that is controlled by a servo system, the computer on the servo system will try to compensate for the excessive backlash and will actually make the gears vibrate even more excessively. The controller must have a way to work with the backlash and make corrections to the torque transfer that will allow the gear set to continue to be used even though the gears are wearing out.

The inspection procedure for the gears will include a process to measure the backlash to ensure that it is within specifications. If the backlash is beginning to increase, you may be able to add shims or thrust washers to help reduce the amount of backlash. These devices help reduce the extent of the backlash, but generally they are only temporary, and the gears will need to be replaced as soon as new gears can be obtained.

Troubleshooting Gears

In working with gears and transmissions, you will sometimes find problems with worn and noisy gears. When gears begin to wear, their teeth will not mesh correctly and they will not be able to transfer power smoothly. When this occurs, the shaft of the driver gear or the driven gear will begin to vibrate. This same type of problem can occur if the gears become dry or if lubrication is low. In some cases, you can add lubrication to the system and get the gears to return to their normal condition. If the gears have low lubrication over an extended period of time, they will continue to vibrate and they will need to be inspected and replaced if necessary. Excessive vibration can also be caused by misalignment or excessive endplay.

If either of the two gears in the set has a worn bearing or broken teeth, or if a shaft is misaligned, the gears will tend to bind rather than vibrate. The first indication that the gears are binding is usually that the motor overloads have tripped. After the overloads are reset, the motor may run for a few more hours or less and then the overloads will trip again. The overloads trip because the gears are beginning to bind. You will be able to detect the binding by observing the drive motor and the machine that is receiving the power: You will notice that the transfer of power is not smooth. You may also be able to measure the excess current draw of the motor as the motor works harder to move the gears. If you try to turn the shaft by hand, you will find some places where it turns smoothly and other places where it does not turn smoothly. If you turn the shaft to the point where the gears begin to bind, you should be able to see the teeth that are causing the binding. You will not be able to repair the teeth that are binding; you will have to remove and replace the gears.

7.5 BEARINGS

Bearings are necessary to ensure that gears are kept in the correct position and to minimize friction between moving parts. Bearings also help minimize vibration when they are used with shafts. The motion between moving parts may be linear or rotary. Anyone who works with mechanical applications, such as wind turbines, will encounter bearings. You must be able to identify each type of bearing so that you can understand why it is being used, as well as how to remove and replace bearings. This section introduces the various types of bearings used in wind turbine applications. It also explains methods of evaluating bearings to ensure that they are operating correctly, and it shows the proper way to remove and replace a bearing.

Important Terms Used with Bearings

Several important terms that apply to bearings will be introduced here so that you will be able to understand some of the features of bearings that are designed to counteract the forces they encounter. Figure 7-17 is a diagram of a bearing and shaft that identifies several of these important terms. The **journal**, indicated by "a," is the part of the bearing or shaft where the bearing rides. In terms of the bearing, the journal is the inside part of the bearing that mates with the shaft. In terms of the shaft, the journal is the part of the shaft where the bearing

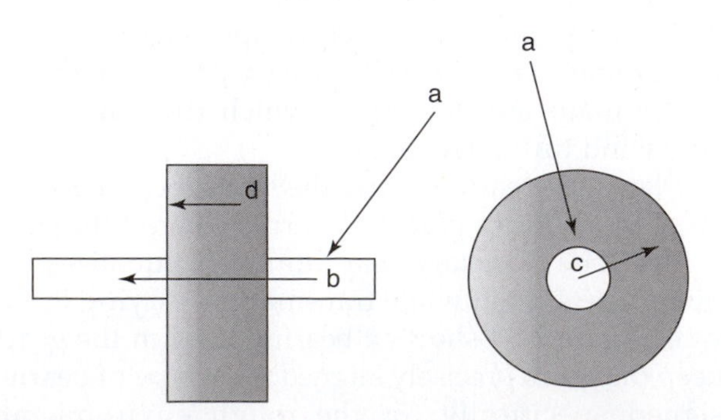

FIGURE 7-17 The journal (a) is the part of the bearing or shaft that supports the bearing. The axis (b) is an imaginary line that runs through the shaft. Radial force (c) is a force that emanates from the center of the bearing outward, like a spoke in a wheel. Thrust (d) is a force that acts against the side of the bearing.

mates with it. In some cases the journal on the shaft is polished to ensure that the bearing fits tightly with the shaft. The *axis,* which is an imaginary line through the shaft, is shown at "b." **Axial force** refers to the direction from which forces are applied to a shaft. An axial force is a force that moves along the length of the shaft. **Radial force** is indicated at "c." A radial force is a force that originates at the center of a bearing or shaft and moves outward, like the spoke of a wheel, to the outside of the bearing. *Thrust* is a force that is applied against the side of the bearing, is shown at "d." The arrow is pointing away from the bearing in this example, but the force could also be an external force that is applied inward against the arrow to the side of the bearing.

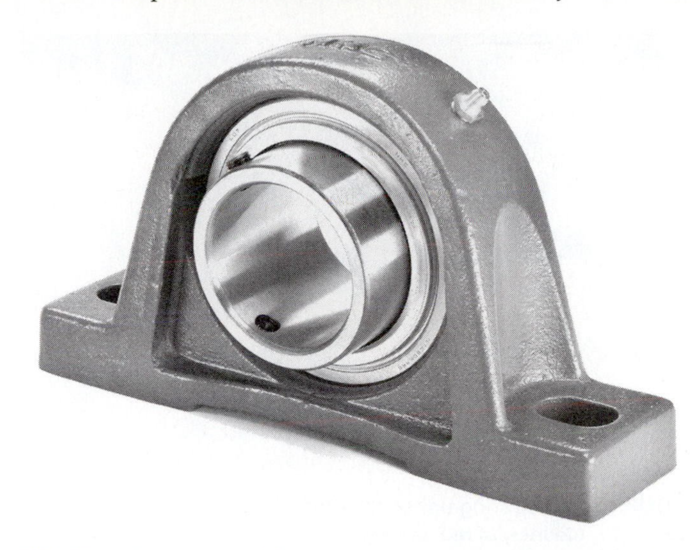

FIGURE 7-19 A pillow block that supports a bushing. Notice the lubrication port on the pillow block. (Courtesy of Boston Gear.)

Plain-Type Bearings (Bushings)

As you work with gears and bearing in wind turbines, you will find bearings that are designed for sliding applications and bearings that are designed for rolling applications. Bearings designed for sliding loads can be classified by the load for which they are designed, such as *axial (side) loads* and loads that emanate out from the center of the bearing, called *radial loads*. Bearings that are designed for rolling loads can be classified by the type of bearing device that is used in the bearing, such as ball bearings, roller bearings, and needle bearings.

The first type of bearing designed for sliding loads that we will discussed is is called a *plain bearing*. This type of bearing is also called a **bushing**. Typically, these types of bearings are used for lighter loads. Figure 7-18 shows an example of this type of bearing. In the figure you can see that the bearing is a solid piece of metal that supports the shaft. The plain bearing is a rather simple design in that it has no moving parts. It uses a thin film of lubrication that exists between the bearing surface and the shaft surface. Because the lubricant is used to limit the amount of friction between the two surfaces, it is of the utmost importance that the proper amount of lubrication is always present. All plain bearings have some means of ensuring that lubrication is properly distributed across all bearing surfaces on a continual basis. Bushings have grooves to allow the lubrication to reach all areas of the bearing and shaft surface. The only way these types of bearings fail is when the lubrication is not present.

FIGURE 7-18 Examples of bushings. (Courtesy of Boston Gear.)

The plain bearing (bushing) is generally mounted in a **pillow block**. A typical pillow block is shown in Figure 7-19. The pillow block is designed to hold the bushing in place against all types of axial and radial loads. The pillow block has two mounting holes (or slots) that allow it to be mounted or secured. The mounting holes may be slotted to allow for proper tensioning of belts or chains that may be mounted to the shaft on which the bearing is mounted. The mounting slots also allow for proper positioning of the pillow block, which helps to minimize vibration of the shaft and ensure proper alignment of the shaft. Figure 7-20 shows a *bearing unit,* which is basically a pillow block housing that will accept a bearing insert to fit the application shaft. The bearing unit makes it easy to change out worn bearings during periodic maintenance or whenever a bearing fails.

Another type of pillow block is designed as a split unit. A split pillow block is made of a top section and a bottom section, which are connected together by bolts. This type of pillow block is used in applications in which the pillow block is taken apart frequently to inspect the bearing or to remove and replace the bearing. When you need to remove and replace the bearing, you can remove the mounting bolts from the pillow block and remove the top of the pillow block. Once the top half of the pillow block

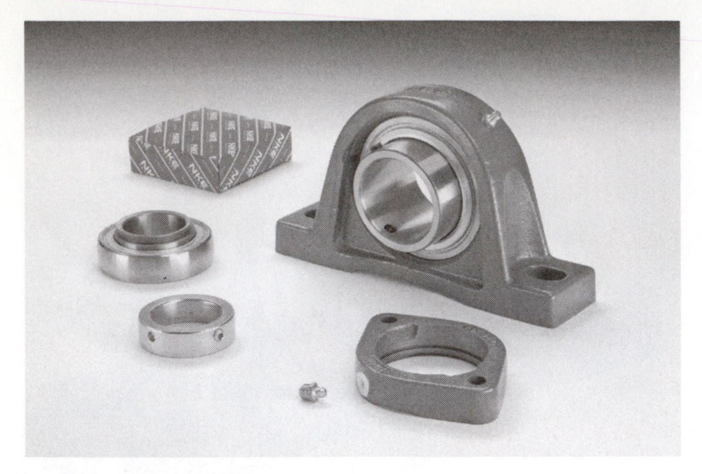

FIGURE 7-20 Bearing units consist of a housing and an insert bearing. (Courtesy of NKE Bearings.)

has been removed, the shaft with the bearing mounted on it is lifted out and a bearing puller is used to remove the bearing. When you have the new bearing installed on the shaft, you can replace the shaft and bearing into the bottom half of the pillow block. After the bearing is properly located in the pillow block, the top half of the pillow block can be replaced and the mounting bolts reinstalled. If a sheave or sprocket is also mounted on the shaft, the proper tension is set and the mounting bolts are tightened.

Rolling-Element Bearings

There are three basic types of *rolling-element-type bearing:* the ball bearing, the roller bearing, and the needle bearing. Each of these types of bearings has an inner ring and an outer ring, called **bearing races**. The inner race is designed to support the shaft on its inside and the bearing components, such as balls or rollers, on its outer edge. The outside race constrains the balls or rollers against the inner race. A bearing race is generally made from hardened steel.

Ball Bearings

Figure 7-21 shows typical **ball bearings**, and Figure 7-22 is a diagram of a typical ball bearing. From these figures

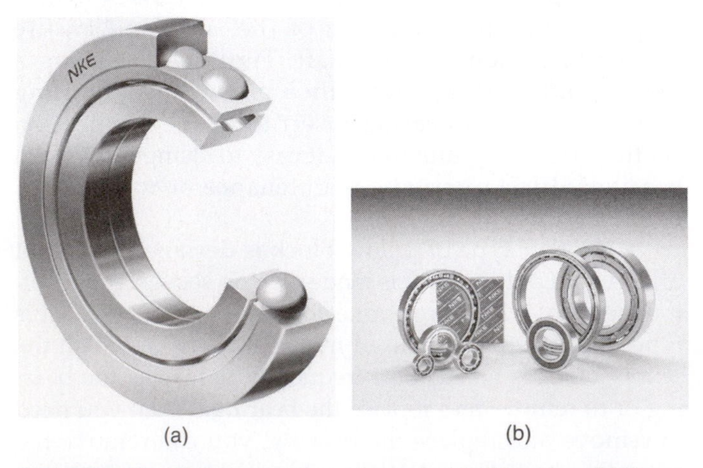

FIGURE 7-21 (a) Cutaway view of a typical ball bearing. (b) Examples of ball bearings. (Courtesy of NKE Bearings.)

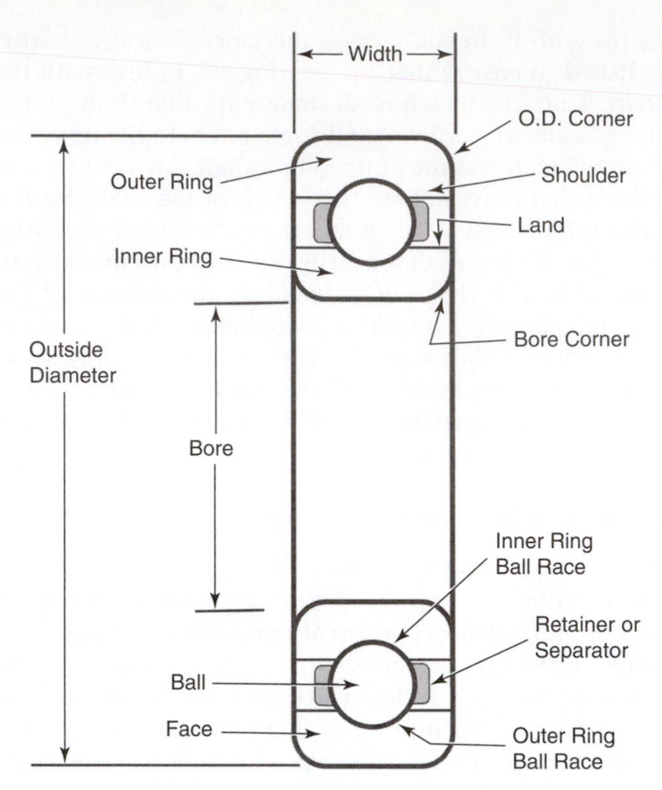

FIGURE 7-22 Diagram of a ball bearing showing the inner race, outer race, ball cage, and balls. (Courtesy of Boston Gear.)

you can see that the ball bearing consists of an inner race, an outer race, a ball cage, and balls. The inside dimension of the race is called the *bore*. The bore of the bearing must match the outside diameter of the shaft. The balls are kept in place in the bearing by a *bearing cage*. The cage is designed to allow each ball to make contact with the inner race and the outer race. It also keeps the balls from coming loose and falling out of the bearing. If the bearing cage is damaged, it may allow one or more of the balls to fall out of the bearing. The outside race keeps the balls in contact with the inner race; its size determines the outside diameter of the bearing.

The side of the bearing is called the *face*. At times you will use a tool to apply force from a mallet to the face of the bearing to properly seat the bearing. You must be sure not to damage the bearing face or allow it to be mashed into the balls in the bearing cage when the bearing is installed on the shaft.

A heavy-duty-type ball bearing is called a *double-row ball bearing*. This type of bearing uses a double row of balls that roll in two deep grooves. Figure 7-23 shows examples of double-row ball bearings. The double-row ball bearing can provide extra friction relief because it has twice the number of balls.

Roller Bearings

The **roller bearing** is different from the ball-type bearing in that the roller is made of solid steel rollers that look like small steel tubes. The main feature of roller bearings is that they provide a larger rolling surface, which can

FIGURE 7-23 Double-row ball bearings. (Courtesy of NKE Bearings.)

support stronger forces. The roller bearing can also bear higher-impact loads or shock loads in wind turbines, which occur when shafts are started and stopped frequently or change speed quickly. The roller bearing is made of an inner race, an outer race, and the rollers. A retainer ring ensures that the rollers stay on track inside the races. One problem with roller bearings is inconsistent loading and tracking of the rollers. Ball bearings, on the other hand, tend to track consistently, because their size and design are uniform. If the rollers are mounted in the race at a slight angle, they can support both axial and radial forces. When the rollers are mounted in the bearing at a slight angle, they are called *tapered roller bearings*. Figure 7-24 shows an example of a tapered roller bearing and its race. The race is the outer housing in which the bearing sits, and it provides the mating surface for the tapered bearing.

Another type of roller bearing is the *double-row roller bearing*. This type of roller bearing allows the two separate rows of rollers to align constantly under load. This causes less wear on the bearing and allows it to support larger loads. Figure 7-25 shows an example of a double-row, tapered roller bearing.

FIGURE 7-24 Tapered roller bearings. (Courtesy of NKE Bearings.)

Needle Bearings

The third type of rolling-element bearings is called the **needle bearing**. The needle bearing has an outside race, but it does not have an inside race. This arrangement allows the needles to be in contact with the shaft being supported. The rollers in this type of bearing are called *needles* because they have a much smaller diameter than a typical roller. The needle bearing has the ability to carry much larger radial loads at much higher speeds than other types of rolling-element bearings. It is important

FIGURE 7-25 A double-row tapered bearing. (Courtesy of NKE Bearings.)

FIGURE 7-26 Thrust bearing. (Courtesy of NKE Bearings.)

that the needle bearing have adequate lubrication, which must be checked frequently.

Thrust Bearings (Washers)

Another type of bearing is called a **thrust bearing**. An example of this type of bearing is shown in Figure 7-26, and examples of thrust washers are shown in Figure 7-27. From this figures you can see that the bearing in Figure 7-26 looks similar to other types of bearings, but it is specifically designed to absorb thrust loads. Thrust washers are special washers that can bear tremendous loads in the axial direction; they also limit side-to-side movement in bearings. Thrust bearings are generally used in conjunction with plain bearings, or bushings. The thrust bearing may be mounted on a shoulder that is cut into a shaft. In some applications, two or more thrust bearings are used to take up the required space between the shoulder of the shaft and the bushing-type bearing. If one of the thrust bearings (washers) begins to wear, additional thrust washers can be added. If you disassemble a shaft that has thrust washers, it is important to ensure that the thrust bearings are cleaned and well lubricated when they are replaced. You also need to put the thrust bearing back in the exact location on the shaft so it is installed correctly and can do its job properly.

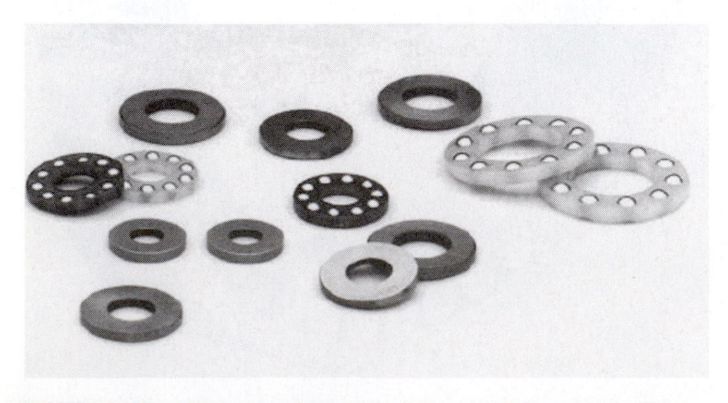

FIGURE 7-27 Examples of thrust bearings, which are sometimes called thrust washers. (Courtesy of Boston Gear.)

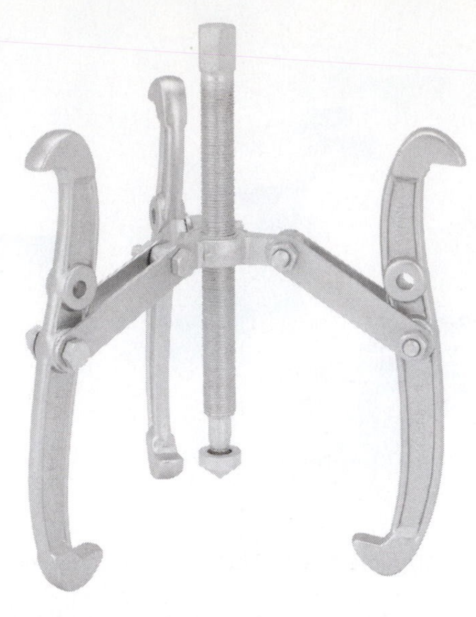

FIGURE 7-28 Bearing and gear puller. (Courtesy of Harbor Freight.)

Maintaining Bearings and Gears

When you are working with wind turbines, you will be called to check bearings that are noisy or worn. If you need to remove and replace a bearing, you will need to use a bearing or gear puller. Figure 7-28 shows a typical puller. Notice that the gear puller has a number of legs that are connected behind a bearing or gear, and a screw shaft in the middle. Figure 7-29 shows a hydraulic

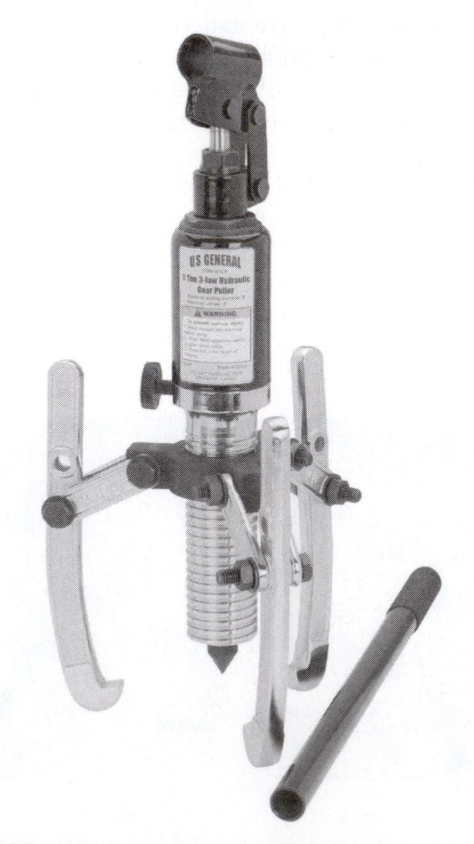

FIGURE 7-29 A hydraulic bearing and gear puller used to remove a gear from a shaft. (Courtesy of Harbor Freight.)

FIGURE 7-30 Hydraulic bearing press. (Courtesy of Harbor Freight.)

puller. Once the legs are in place, the hydraulic handle is pumped and the gear begins to move off the shaft in the direction of the puller. Different size legs are available to allow the puller to be used on larger and shorter shafts. In some applications, where a gear is on the shaft with the bearing, the gear must be pulled first to get at the bearing.

Bearing Presses

In some applications you will need a press to remove a bearing from a race. The bearing press may need hydraulic pressure to provide enough force to remove the bearing. A typical hydraulic bearing press is shown in Figure 7-30. If the bearing is smaller, an arbor press may be used to remove it. Figure 7-31 shows an arbor press. Each of these presses can also be used to reinstall a bearing onto a shaft. It is important to secure a press for removing and replacing a bearing from the shaft so that you do not score the shaft or damages surfaces of the bearing, as this will cause the new bearing to wear out quickly.

7.6 GEARBOX DIFFERENTIAL AND SPUR GEARS

Differential gears are used where two shafts intersect at 90 degrees rather than running parallel to each other. Figure 7-32 shows a typical differential gear set.

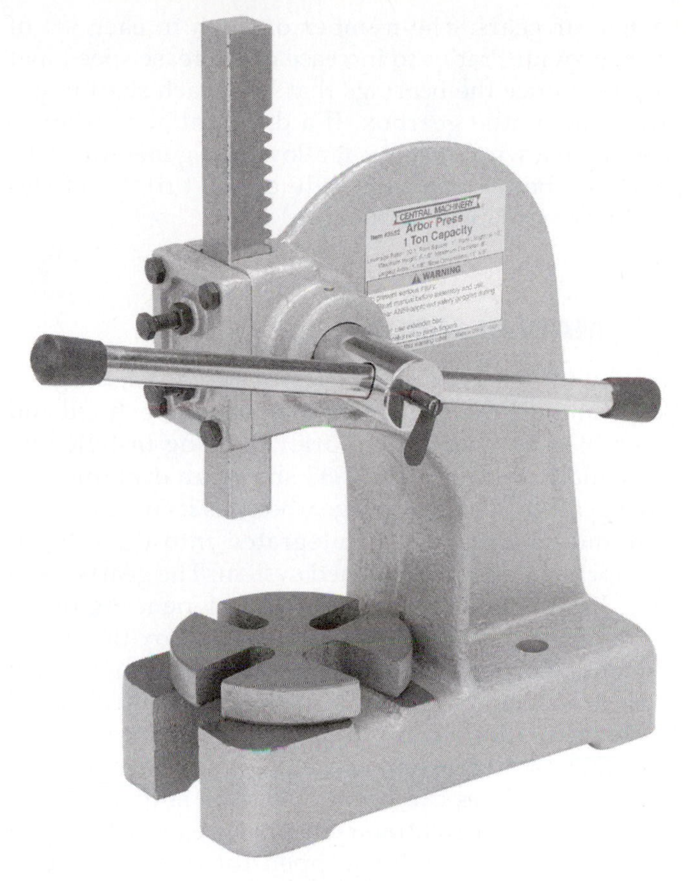

FIGURE 7-31 Arbor press used to remove and reinstall a bearing. (Courtesy of Harbor Freight.)

The gearbox cover has been removed so that you can see the input shaft at the bottom of the gearbox and the output shaft protruding out at the top left of the gearbox. Notice that the input shaft and the output shaft are at 90 degrees (right angles) to each other. Spur gears and helical gears are used to accomplish this setup. In some differentials, bevel gears are used rather

FIGURE 7-32 Gear ring on rotor. (Courtesy of EWT International.)

than spur gears. The number of teeth in each set of gears provides ratios to increase or decrease speed and torque. Notice the bearings that hold each shaft in position inside the gearbox. If a differential gearbox is used with a wind turbine, it allows the generator to be mounted below or to the side of the turbine blades rather than in line with them.

7.7 MAIN GEARBOX

In wind turbines in which gearboxes are used, the gearbox is an integral component that is manufactured and assembled as a subsystem prior to being installed on the wind turbine. Figure 7-33 shows an example of a large gearbox. This type of gearbox has its own lubrication and cooling system integrated into it so that it can operate as a self-contained system. The gearbox can have liquid cooling or air cooling, depending on its needs. Figure 7-34 shows another gearbox that has a large rotor plate mounted on the front that can be used with a mechanical braking system. The mechanical brakes provide friction pads that come into contact with this rotor plate and cause the gear system to slow down. The brakes can be applied with sufficient force to stop the shaft rotation completely, especially when the turbine needs to be stopped for maintenance or when it is in danger of becoming damaged from overspeeding.

Figure 7-34 shows a gearbox for a wind turbine. The gears are mounted inside the gearbox and transmit power to the high-speed shaft, where the generator connects. The planetary bearing and the high-speed shaft bearing are important parts of the main gearbox to ensure that it transmits power with minimal vibration.

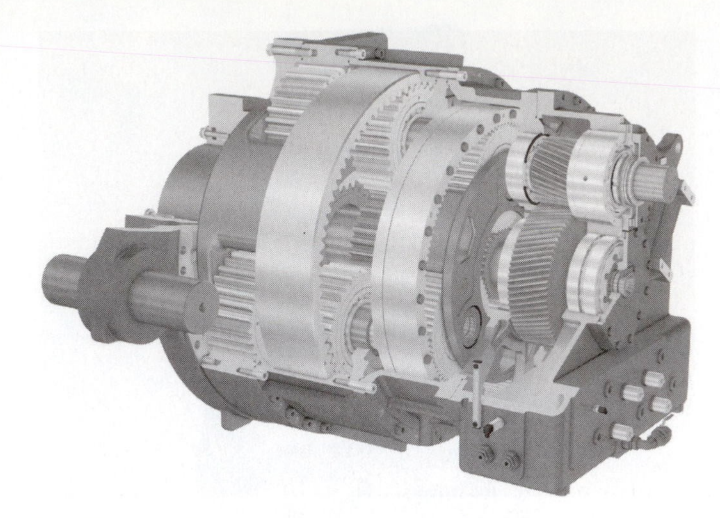

FIGURE 7-34 Cutaway view of a gearbox for a wind turbine. (Courtesy of Bosch Rexroth Group.)

Figure 7-35 shows the gears from a main gearbox so that you can see them as though you were inspecting them through the inspection cover during periodic maintenance. The helical gears in this gearbox allow the slow-speed shaft and the high-speed shaft to be mounted parallel to each other, which allows the turbine blades, low-speed shaft, main gearbox, high-speed shaft, and generator to all be mounted in a line. The low-speed shaft is the shaft that connects the rotor to the gearbox, and the high-speed shaft is the shaft that connects the gearbox to the generator. Figure 7-36 shows a diagram of this type of gear set. The input shaft enters the gear set on the right side, and the output shaft is connected on the left side. You can see the larger planetary gear on the far left, the helical gears in the middle and to the right side of the gearbox.

FIGURE 7-33 A complete wind turbine gearbox. (Courtesy of Bosch Rexroth Group.)

FIGURE 7-35 Helical gears from a wind turbine gearbox. (Courtesy of Kidwind Project®.)

FIGURE 7-36 Gears from a wind turbine gearbox. (Courtesy of Bosch Rexroth Group.)

7.8 DRIVE TRAINS

The **drive train** in a wind turbine consists of the turbine blades, rotor, low-speed shaft, gearbox, high-speed shaft, and generator. This set of components allows the wind turbine blades to harvest wind energy and transfer it through the shafts and gearbox to the generator, where it is converted to electrical energy. Figure 7-37 shows all of these components. A braking system may also be part of the drive train, to stop the generator and turbine blades when necessary.

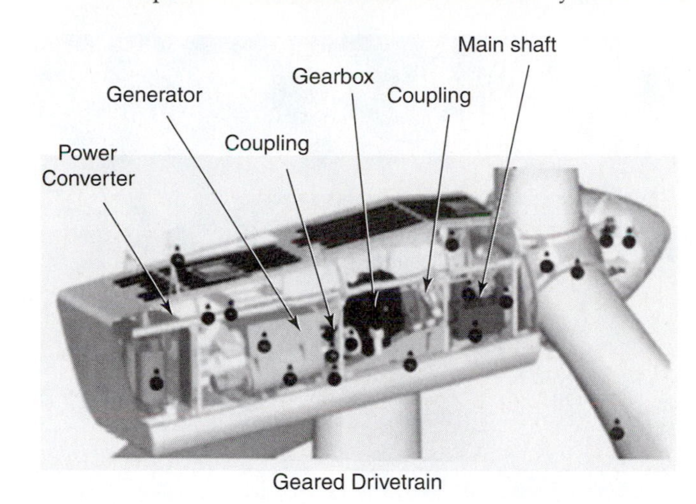

FIGURE 7-37 Drive train of a horizontal wind turbine.

Drive Train Mounting Arrangement Options

The drive train components are mounted inside the nacelle on the nacelle bedplate. The nacelle bedplate consists of a very large metal plate that has the strength to support all of these components and keep them in alignment. In many systems, the turbine blade rotor, gearbox, and generator are mounted to each other in a shop. After they are started up and tested, they are then lifted into the wind turbine nacelle as an assembled unit. This process ensures that the components are initially aligned and tested as an assembled unit and as much of the vibration is removed and other adjustments made while the system is in the shop. Some turbine manufactures use a test stand to test these components under operating conditions and fully loaded while they are in the shop, so they can make any adjustments while the unit is on the ground. Figure 7-38 shows a typical gearbox and generator mounted as a unit ready to be tested.

Drive Train Compliance

Drive train compliance is the correct alignment of all the components in the drive train. If any of the components do not align exactly perfectly, they will cause the shafts to rotate in irregular patterns, which will cause major vibration and therefore premature wear and tear on the components. When a shaft rotates in conjunction with transferring power from the turbine blades and rotor to the gearbox, it must rotate in a perfect circle. If either of the shafts becomes misaligned between the rotor and the gearbox, or between the gearbox and generator, it will

FIGURE 7-38 Gearbox and generator assembled together as complete unit. (Courtesy of NREL, U.S. Department of Energy.)

FIGURE 7-39 The top of the nacelle is open so the gearbox and generator can be worked on. (Courtesy of Siemens Wind Energy.)

rotate in an elliptical pattern, which will cause the shaft to whip and vibrate. This misalignment will be amplified at every point where energy is transferred from one component to the next. The shaft connections and shaft rotation must be aligned perfectly when the system is installed before the turbine blades are allowed to rotate at high speed.

Specialty equipment is used to ensure the perfect alignment and also to measure the amount of vibration due to any imbalance of weight on the shaft. Vibration analysis is used to determine the source of the vibration and its amount. Adjustments can be made to shaft alignments and weights added to specific points on the shafts to eliminate vibrations. Vibration monitors are used in some systems to detect any unwanted vibration as soon as it occurs. In other systems, the system must be checked periodically for any vibration, on a schedule, after a certain number of hours of operation. Drive train compliance must be checked periodically to ensure that vibrations in the low-speed shaft, gearbox, and high-speed shaft are kept to a minimum. Figure 7-39 shows a wind turbine with its nacelle opened so that the shafts, gearbox, and generator can be inspected. Figure 7-40 shows technicians working on a turbine drive train with shafts, gearbox, and generator being lowered into a nacelle. When you work on a wind turbine, you may need remove and replace the gearbox or the generator, and this work must be done through an open nacelle.

FIGURE 7-40 Turbine drive train with shafts and gearbox being lowered into the nacelle. (Courtesy of Nordex.)

7.9 DIRECT-DRIVE SYSTEMS

Some wind turbines do not try to control the speed of the shaft that supplies power to the generator. In these applications the generator is connected directly to the wind turbine blade rotor. If the direct-drive wind turbine does not use any transmission gears, the speed at which the wind turbine blade is turning will be the speed at which the generator is turning. Some direct-drive wind turbines use one set of gears called a transmission to increase the speed at which the generator turns. This means that the output frequency of the voltage from the generator for a direct-drive wind turbine is constantly changing as the wind changes the speed of rotation of the turbine blades. Because the output frequency of the voltage the generator produces is constantly changing, an electronic inverter must be used. The inverter changes the AC voltage from the generator into a constant frequency of 60 Hz at the output. This type of application is actually a little more efficient than those that use a gearbox. The important part of this application is that the inverter must be sized to handle the electrical load that the generator produces. The advantage of the system is that it has one less major mechanical component, the gearbox, which can break down or need periodic maintenance. Figure 7-41 shows an example of a direct-drive wind turbine.

Because direct-drive units do not need a gearbox, the blades are mounted directly to a rotor and the generator occupies the remainder of the space. The wires

FIGURE 7-41 A direct-drive wind turbine. (Courtesy of EWT International.)

from the generator are routed directly to the ground, and the inverter or storage device is located on the ground. This arrangement allows fewer parts to be mounted in the air on top of the pole, where maintenance is more difficult.

Figure 7-42 shows another type of direct-drive wind turbine. In this figure you can see that the rotor is connected directly to the generator shaft, which makes the nacelle very compact and much lighter. The main problem in the direct-drive system is that the generator rotates at speeds that vary over a large range, because the wind speed may vary over a range of from 10 up to 50 mph.

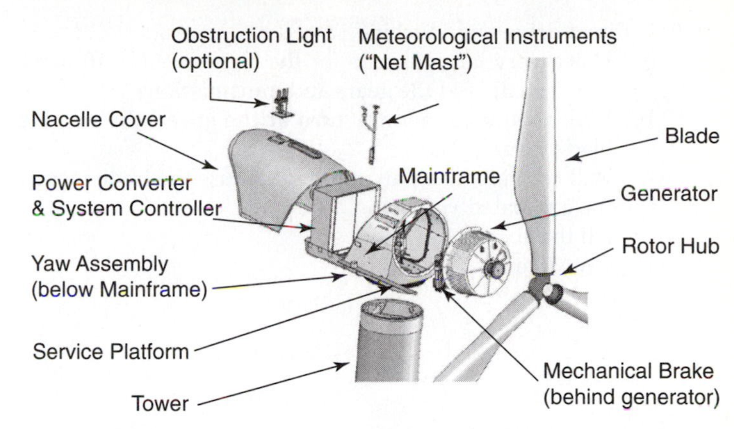

FIGURE 7-42 Diagram of a direct-drive wind turbine. (Courtesy of Northern Power Systems Inc.)

Questions

1. Explain the function and operation of a bearing.
2. Determine the gear ratio of a set of gears in which the driver gear has 20 teeth and the driven gear has 80 teeth.
3. Explain the operation of a gearbox for a wind turbine.
4. Explain why a wind turbine needs a gearbox.
5. Explain how a direct-drive wind turbine system is able to produce voltage at 60-Hz frequency.
6. Identify five types of gears and explain their functions.
7. Explain why a planetary gear is used in a gearbox.
8. Identify what type of gear is used in the yaw control.
9. Explain the function of a pillow block.
10. Explain why generator must turn at a fairly constant speed.

Multiple Choice

1. A gearbox is required on a wind turbine because
   a. The turbine blades turn at the same speed as the generator and the gearbox is used to store excess energy.
   b. The turbine blades turn at a relatively high speed and the generator needs to turn at a relatively low speed.
   c. The turbine blades turn at a relatively low speed and the generator needs to turn at a relatively high speed.
   d. The gearbox allows the generator to turn at variable speeds from high speed to low speed.
2. If the driver gear in a set of gears has 10 teeth and and the driven gear has 50 teeth, the gear ratio is
   a. 2:1
   b. 3:1
   c. 5:1
   d. 1:5
3. If the driver gear in a set of gears has 5 teeth and and the driven gear has 40 teeth, the driven gear will turn at _____ rpm when the driver gear turns at 800 rpm.
   a. 10 rpm
   b. 20 rpm
   c. 100 rpm
   d. 1600 rpm
4. If a set of gears uses the planetary gear,
   a. The inside gear is called the ring gear and the outside gear is called the sun gear.
   b. The inside gear is called the sun gear and outside gear is called the ring gear.
   c. The shaft to the input of the planetary gear is set at 90 degrees to the output shaft.
   d. All the above.
5. The drive train in a wind turbine
   a. Consists of the turbine blades, rotor, low-speed shaft, gearbox, high-speed shaft, and generator
   b. Has its components mounted inside the nacelle on the nacelle bedplate
   c. Transfers energy from the turbine blade all the way to the generator
   d. All the above
   e. Only a and c
6. The high-speed shaft
   a. Connects the turbine blades to the gearbox
   b. Connects the gearbox to the generator
   c. Is connected directly to the low-speed shaft
   d. All the above
7. Bearings
   a. Work with gears and shafts to ensure that they are in the correct position
   b. Minimize friction between moving parts
   c. Help to limit vibration when shafts rotate
   d. All the above
   e. Only a and b
8. A pillow block
   a. Is designed to support or hold a bearing securely in place
   b. Helps provide proper alignment for shafts to eliminate vibration
   c. May be designed as a solid unit or a split unit
   d. All the above
   e. Only a and c

9. A direct-drive wind turbine
   a. Has a very large nacelle for the high-speed shaft, low-speed shaft, and the gears and transmissions
   b. Will have its generator turn at the speed the turbine blade turns
   c. Will typically need an inverter because the frequency of the generated voltage will vary
   d. All the above
   e. Only b and c

10. Drive train compliance
   a. Is the correct alignment of all the components in the drive train
   b. Is important to ensure that vibration is limited as much as possible in the shafts that transfer energy from the turbine blades to the generator
   c. Must be checked periodically to ensure that all the components remain aligned
   d. All the above
   e. Only a and c

CHAPTER 8

The Grid and Integration of Wind-Generated Electricity

OBJECTIVES

After reading this chapter, you will be able to:

- Identify the three main sections of the electrical grid.
- Explain what the grid is and how it works to distribute electrical power to customers.
- Explain what the smart grid is and how it is different from the grid.
- Explain how a transformer works.
- Identify the two windings of a transformer.
- Identify the federal law that requires utility companies to purchase power from wind turbine generators and state in what year the law was passed.
- Explain what a substation is, and why it is important to the grid.
- Identify three things that the IEEE 1547-2003 standards address concerning technical requirements for power that is connected to the grid.
- Explain what a net meter is.
- Explain what power quality issues are.

KEY TERMS

Apparent power
Brownout
Capacitive load
Flicker power
Grid
Grid companies (Gridco's)
Grounding
Induced current
Induction
Inductive load
Institute of Electrical and Electronics Engineers (IEEE)
Kilovolt-ampere (kVA)
Low-voltage ride-through (LVRT)
National Electrical Code (NEC)
Net meter
Power factor
Power quality
Primary winding
Public Utility Regulatory Policies Act (PURPA)
Reactive power
Resistive load
Secondary winding
Service drop
Smart grid
Step-down transformer
Step-up transformer
Substation
Transformer
Transmission companies (Transco's)
True power
Turns ratios
Volt-ampere (VA)

OVERVIEW

This chapter will cover information about how electrical output from a wind generator is connected to the electrical grid. It is important to understand that if you are hired to work on the electrical systems of wind turbines, or on the electrical portions of the grid and any power lines, you will need to have a complete working knowledge of electricity and receive specialty training to be able to work around voltages over 200 V. You will also be required to wear special personal protective equipment (PPE) whenever you are in close proximity to high voltages. Special training in arc flash hazards is also required, as well as special clothing to protect you against arc flash. Arc flash is an electrical arc that may be emitted from switchgear or other electrical terminals, and this arc has the potential to cause severe burns or electrocution if you are not wearing protective gear. Personnel who work for utility companies work on these high-voltage lines every day, and they have a solid safety work record using this equipment over the years . The information in this chapter will give you a solid understanding of what the grid is and how wind turbines are connected to it.

The chapter will begin by explaining the grid at present in terms of size and potential for growth. Information about the new *smart grid* will also be provided, and the features of the smart grid will be explained. Information about the National Electrical Code and other rules and regulations pertaining to the grid will be identified and explained.

Another part of this chapter will include information about how electricity from a wind turbine is used to supply power directly to a commercial building or residence. The basic utility meter and the net meter will be explained. This section will also identify the switches and connections for power distribution for a grid-tied system. Additional information will be provided to explain how the electrical power from wind turbines may need to be cleaned up to match the voltage and frequency of the power on the grid. This can be accomplished by controlling the generator in the wind turbine and by controlling the wind turbine speed or with an electronic inverter to ensure that the frequency is the same as that of the grid. Other information about voltage control and reactive power control will also be given. Problems in connecting generated voltage to the grid will be discussed together with power quality issues, and safety circuits in the wind turbine generator, such as low-voltage ride-through protection, and flicker power quality will be explained.

This chapter will also explain how a wind turbine electrical system is grounded, and how the power lines from the generator are connected to the substation or grid power source on site. This information will include why the cable installation goes underground and overhead, and the use of transformers. If the wind turbine is part of a larger wind farm, it will be tied in to the wind farm substation, a topic that will be explained at the end of this chapter.

8.1 UNDERSTANDING THE GRID

This first section of the chapter will review the information in Chapter 1 concerning the grid. The **grid** is the name that is applied to the entire electrical power distribution system in the United States and the rest of North America. If a wind turbine is designed to produce power for the grid, it must be connected to an existing section of the grid and its power must meet the specifications for that section of the grid. The grid is used to distribute electricity across the United States and is actually made up of miles of interconnected wires that move electricity from where it is generated to where it is consumed. When electricity is generated, it may be at values of less than 5000 V. This voltage must be stepped up with transformers to higher voltages of 545–765 kV for long-distance transmission, and it may be stepped up to 345–545 kV to be transmitted short distances. When the electricity reaches the area where it will be used, it is transformed back down at a substation near the city or the industrial site to a level of about 12,470 V and then routed around town or the site to where it is finally consumed. Service transformers on the power poles outside residences, or at the smaller substations for an industrial application, step the voltage down again, from 12,470 to 120 V per leg for residential customers and to 240 V or 480 V for industrial users. Figure 8-1 shows the distribution system from the generating site through transmission lines to a residential customer.

Subsections of the grid include generation sites, transmission systems, subtransmission systems, substations, feeders, service systems, and customers. Generation sites may include coal-fired steam plants, nuclear

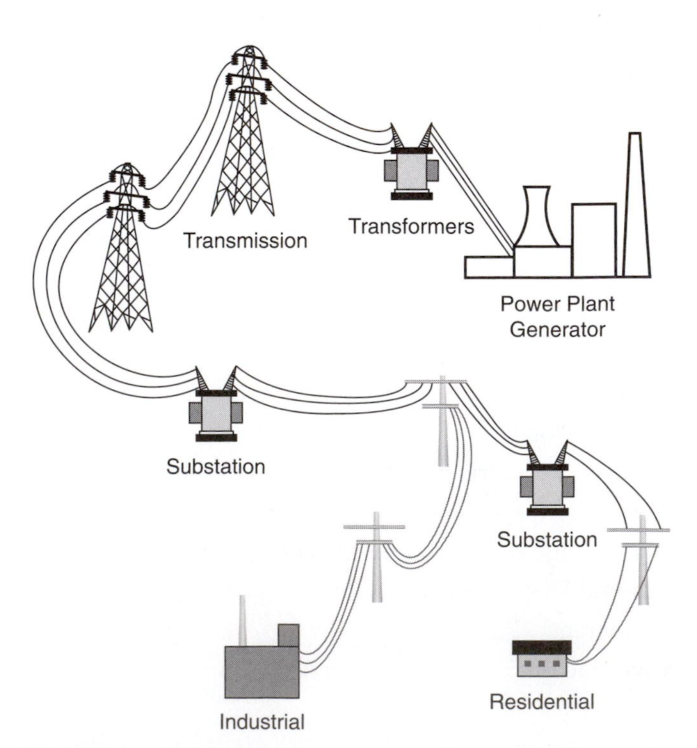

FIGURE 8-1 Electrical power distribution from generation station to residential customer.

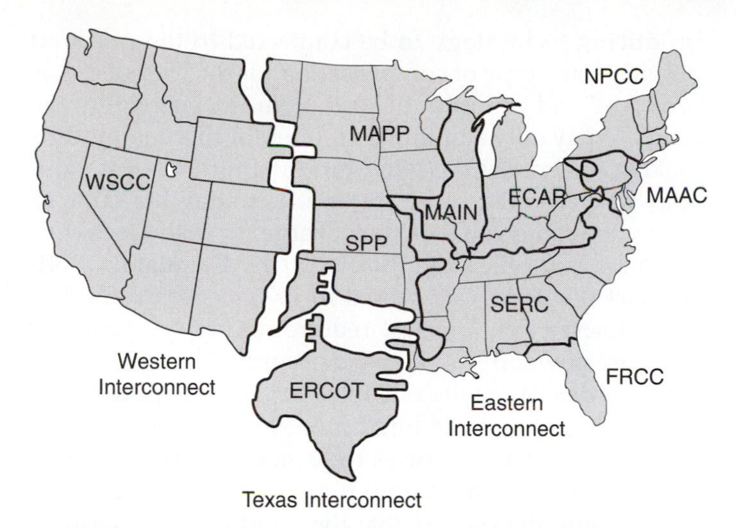

FIGURE 8-2 Map of the United States showing the Eastern Interconnect, the Western Interconnect, and the Texas Interconnect of the grid. (Courtesy of U.S. Energy Information Administration.)

power plants, hydroelectric plants, wind turbines, and photovoltaic (solar) sites. The transmission and subtransmission lines include the high-voltage and extra-high-voltage transmission lines that carry power across the country over longer distances. A **substation** is a large set of transformers located at the edge of a city or near an industrial site, which steps down the electrical power transmitted from the generation site at very high voltage. Feeder lines provide voltage throughout cities and residential areas. Service systems include the transformers and wiring to provide the service drop to industrial, commercial, and residential systems. The customer portion of the grid includes any wiring around the customer site, such as a large industrial application.

In Chapter 1 you also learned that the grid can be illustrated by the map shown in Figure 8-2, where you can see how it is connected in each state. The grid is divided into three major grid sections: the Eastern Interconnect, the Western Interconnect, and the Texas Interconnect. The Eastern Interconnect is further divided into subsections: the Northeast Power Coordinating Council (NPCC), which includes Maine, Vermont, New Hampshire, New York, Massachusetts, and Rhode Island; the Mid-Atlantic Area Council (MAAC), which includes Maryland, Delaware, New Jersey, and parts of Pennsylvania; the East Central Area Reliability Coordination Agreement (ECAR), which includes Michigan, Indiana, Ohio, West Virginia, and Kentucky; the Southeastern Electric Reliability Council (SERC), which includes the southern states except Florida; the Florida Reliability Coordinating Council (FRCC), which includes all of Florida; the Mid-America Interconnected Network (MAIN), which includes Wisconsin and Illinois; the Mid-Continent Area Power Pool (MAPP), which includes the Dakotas, Nebraska, Minnesota, and Iowa; and the Southwest Power Pool (SPP), which includes Kansas and Oklahoma. The Texas Interconnect includes the Electric Reliability Council of Texas (ERCOT), which includes all of Texas; and the Western Interconnect includes the Western Systems Coordinating Council (WSCC), which includes all of the western states from New Mexico, Colorado, Wyoming, and Montana westward.

Table 8-1 shows the demand, supply, and capacity margin for the United States as of 2009. The capacity margin is the amount of electrical power that is produced compared to the amount of electrical power that is consumed. It is important that the amount available be more than the amount consumed, to meet potential industrial and residential needs. In 2009 the capacity margin of the Eastern Interconnect Grid was 19.3%, for the Texas Interconnect Grid it was 13.6%, and for the Western Interconnect Grid it was 22.0%. The capacity margin for the entire United States was 19.4%. It is very important to understand how much power is needed in each section of the grid, and how much power is available to supply the need in those sections, because this is the amount of electrical power that must be produced every hour of every day, and wind turbines and other alternative energy must produce this amount if they are going to replace traditional energy sources. You will learn that energy produced by wind turbines and other alternative energy sources can provide only a small percentage of this larger demand.

The grid provides electricity for each of the areas shown in the map, and if you are providing electricity to the grid from a wind turbine, you must know which part of the grid you will be connecting to and be familiar the specifications for that part of the grid. You will also need to file for permits to connect to the grid.

In general, the term *grid* refers to the voltage transmission portion of the system. This portion extends from the point at which voltage is generated, such as at a wind turbine or on a wind farm, to the point at which voltage will be used in an industrial, commercial, or residential

TABLE 8-1 U.S. Capacity Margins, 2009

	Eastern Grid	Texas Grid	Western Grid	U.S. Total
Demand (MW)	554,412	63,376	136,441	754,229
Supply (MW)	688,783	72,204	174,978	935,965
Capacity margin (%)	19.3	13.6	22	19.4

Source: U.S. Energy Information Administration.

area. Sometimes the term *grid* is used to refer to a section of the grid that serves an area of the United States or North America. The actual transmission lines, towers, transformers, and switchgear are owned by individual companies called **transmission companies (Transco's)** that manage this hardware. Companies that manage the grid function that interconnects and routes electricity through the hardware (cables) so that no area has a brownout or a blackout due to insufficient power are called **grid companies (Gridco's)**. Some parts of the grid hardware and grid management may be owned by local or regional utility companies.

The electricity that flows through this hardware is treated as a commodity and is bought and sold on short-term and long-term contracts. Some of the ownership of the equipment and the sale of the electricity that flows through the lines is regulated by the states in which the equipment exists and where the electricity is being consumed. Other parts of the system, because it moves electricity between two or more states, is covered by interstate regulations. This is why it is important to understand what part of the grid you will be connecting your wind turbine to, and what laws and regulations apply. For example, some states require utility companies that produce power and companies that transmit power to purchase power generated from wind turbines. Other states have not reached this level of cooperation between wind turbine owners and the utility companies. In most areas you will be required to fill out an application and request a permit to connect your system to the grid, and to get reimbursed for the power the wind turbine produces.

The main voltages on the grid are separated into four different voltage levels for long-distance transmission. One of the highest voltages used to transmit voltage within the United States grid is 765,000 V. Some higher-transmission lines transmit voltage from larger hydroelectric generation stations. Other lines transmit voltage at 500,000, 345,000, or 230,000 V. Typically, the higher the voltage, the longer the distance the voltage is being transmitted.

The grid is divided into three distinct sections: the Western grid, the Eastern grid, and the Texas grid. The three major sectors are isolated to the extent that a problem in one grid sector can be contained so it cannot affect the other two. There are large control centers in each of the subsectors that monitor the grid continually to ensure a sufficient amount of electrical energy is available, and that blackouts are limited to very small areas if they do occur. In the remaining sections of this chapter, you will learn about the pros and cons of the grid system as it exists today. You will also see how the grid is being updated to something called the "smart grid."

Now that you know a little about what the grid is and how it works, the next important item is to understand how to connect a wind turbine to the grid. Approximately 42 states currently have laws that allow or require utilities to allow wind turbines and other electrical producing technology to be connected to the grid and provide some type of compensation for the energy that is produced and put into the grid. Some states require the utility to pay only a minimal amount for this energy, thus making the energy the wind turbine produces less valuable, whereas others require utilities to pay the same or nearly the same value as they charge for their electricity. A federal law called the **Public Utility Regulatory Policies Act (PURPA)** was passed in 1978 as part of the National Energy Act. This law requires utility companies to purchase power from independent providers, but the law does not determine the rate to be paid for the power.

One way of providing connection to the grid uses a utility meter that measures the amount of power going into the grid or coming from the grid. Because this meter can measure the voltage that the wind turbine produces or the voltage that is used, the meter is called a **net meter**. Some versions of the net meter allow the excess generated electricity to be banked, or credited to the customer's account, which basically means the electricity that is generated is sold at the same rate as electricity that is used. The net meter allows the electric meter to spin both forward and backward, which simplifies the process of measuring how much electricity was produced or how much electricity was consumed.

8.2 THE SMART GRID

The **smart grid** is basically an update of the present-day grid. Currently, information and energy on the grid go in one direction, from the point where it is generated to the point where it is consumed. The power company that generates electricity, or the wind turbine that produces electrical power, provides power to the grid, over which it flows together with other electrical current to the point where it is used by consumers. Any data or information that might assist energy producers currently flows in only one direction—from the energy producer to the end user, the consumer.

The smart grid is based on the concept that information on the grid should flow in both directions. For example, it would be useful if the electric meter in the residence or commercial location had the ability to report back to the generating company or the transmission company the details of how much energy was being consumed at that point and at all different times from that location. This data could be recorded and analyzed so that predictions could be made about the times at which energy would be consumed in the future. Additional information could be provided through this meter to let consumers know the cost of the energy they are consuming at any given time, and whether the energy would be more expensive or less expensive at a different time during the day. For example, excess electrical power is available on the grid late at night, at times called low-peak or off-peak times. Electricity-generating companies are able to provide electricity at a cheaper rate during these times, if consumers can use the power at those times. For example, it may be possible to heat water or bricks in a storage unit for a furnace during these off-peak times, thereby saving the consumer a considerable amount of money.

The smart grid will identify each step in the grid transmission process and each end user with a computer address that allows all the equipment at that location to send and receive data about the amount of energy being used, the time of day, and other information such as temperature or other environmental conditions. Because each point in the grid will have its own specific address, it will be essentially a large computer network in which information can flow to and from each node in the network. The information can then be displayed on a small display monitor or on a computer in a home or commercial establishment. This display of information will help consumers understand how much energy they are using at any given time, and whether there are alternatives available at a later date or at a lower price. For example, if you have an electric vehicle that needs recharging from time to time, the smart grid will be able to indicate when the electricity for recharging will cost the least.

Since the electrical meter, and any switchgear to each end user, can be monitored, it can also be controlled. For example, the individual loads in a home could be connected to a controller through which they could be individually monitored as well as turned on or off for a short period of time to shed power during a peak power period. The controller would send and receive control information to the air conditioning system, the electric hot water heater, the electric clothes dryer, and the electric range. The smart grid could monitor these components to determine how often they were used and at what time they were used. The utility company could then offer the homeowner several pricing packages that would fit with the peak times the power company has. The power company would make the electrical power most expensive during peak times and least expensive during off-peak times. Power is less expensive during off-peak times because there is more generating capacity available than power is being consumed at that time. This is usual during late-night hours because customers are asleep, and the number of factories and commercial establishments that use large amounts of energy at night are few. The homeowner could then program some of the electrical loads, such as heating hot water, recharging an electric car, or heating a thermal electric furnace at that time to ensure they were using power when the lowest-priced electricity was available. This single function would allow electrical energy consumers to better match the production of the electrical energy producers.

Today, power-producing utilities try to predict the amount of load for any given period, and then burn the energy source such as coal or nuclear power to produce the electricity that will be consumed during the next several hours. Because this process of consuming energy to create electricity is time-delayed, some of the energy may be wasted if it is not used several hours later. You can see that because the current system does not monitor in real time, estimating usage patterns and changes in the system because of weather and other external conditions is very inefficient.

The smart grid, with its residential and commercial controllers, will also allow end users to enter into contracts with energy producers to agree to have some of their largest loads disconnected or turned off for short periods of time when the grid is becoming overloaded, or if a brownout is occurring. A **brownout** is a period of time when more grid power is being used than is being produced, so that a low-voltage condition occurs in which the voltage drops 10–20%. If the brownout continues for a longer period of time, large electrical loads that have motors, such as air conditioning compressors and pumps, may be severely damaged or overheated. One method of fixing a brownout condition is to allow the distribution control for the grid to begin to shed larger loads and heavy users until the voltage level comes back up. The smart grid will allow large industrial users to make agreements with their power company to have large sections of their electrical equipment disconnected for short periods of time, in return for a better rate for the electricity they purchase. This system has been around for over 10 years for large industrial users, and with the smart grid this pricing arrangement will be available to residential users as well. For example, a residential user might agree to have his air conditioning shut off during high-peak periods or in the event of a brownout, and the power company would agree to offer a reduced rate for electricity to that consumer.

The smart grid will also allow the controller to lock out certain loads and only allow them to be energized during the lowest-peak periods of the day,or week, depending on the electrical load.

Because the smart grid will need to send and receive information over a computer network, it will need to connect to homes through existing computer networks, or the signals can be sent in packets over the high-voltage lines. Computer communications and networking equipment will be able to filter this information off the single-phase and three-phase high-voltage lines so that the computer data can be sent and received through the high-voltage lines.

On the transmission side, the smart grid will be able to determine which areas of the country and which areas of the grid sector are using the most power. This will help in understanding more quickly how to redistribute electrical power to the sectors that need it most at that time. It will also allow the interconnection system to identify problems and to isolate and disconnect small sections faster, to ensure that they do not affect the entire grid. You might remember a problem in August 2003, when lightning struck a power line in the Cleveland, Ohio, area, and the series of events that followed caused a large section of Ohio, Pennsylvania, New York, Michigan, and Ontario, Canada, to be blacked out. The blackout occurred because the grid was not able to disconnect the intermediate sectors fast enough and prevent the blackout from spreading across several states. The new technical equipment that will be used in the Smart grid will be able to limit the blackout area to a much smaller region, possibly within several blocks of where the lightning strike occurred, instead of allowing it to spread across several states.

Another problem with the present-day grid is that it is very inefficient in its transmission capability. Some experts believe that as much as 30% of the power in the grid today may be lost to inefficiency, either through heat loss from older wiring, or as a result of poor equipment and connections. The smart grid will be able to identify the most efficient means of transmitting power from where it is produced to where it is consumed.

Because modern wind turbines have programmable logic controllers (PLCs) or dedicated computer controls, they will be able to integrate with the smart grid very easily. The amount of voltage and current the wind turbine is able to produce can be closely monitored at all times, and the amount available to send into the grid will be able to be measured at all times.

Some of the largest energy companies in the United States, such as General Electric, IBM, and Siemens Electric, are working to improve the smart grid and get it integrated as quickly as possible. One of the problems in the integration of the smart grid is the high cost of converting residential and commercial users to new equipment as well as changing out transmission equipment and switchgear. The smart grid is currently being created and integrated, and this integration process will last for many years. Many experts compare the creation and integration of the Smart grid to the period when electricity was first being introduced to all parts of the United States. At that time it took many years to get electricity to all the places where it was needed.

8.3 TRANSFORMERS, TRANSMISSION, AND DISTRIBUTION INFRASTRUCTURES

To understand the transmission and distribution infrastructures for electricity, you need to understand the theory and concepts of a transformer, and of groups of transformers that make substations. A **transformer** is an electrical component that takes voltage from one level, such as 900 V AC, and changes that voltage to another level, such as 12,470 V, for transmission. The transformer consists of two coils of wire that are positioned close to each other inside a metal case, but the two windings are not connected to each other in any way. The two coils are called the **primary winding** and the **secondary winding**. AC voltage is applied to the primary winding, and it is induced into the secondary winding. The voltage applied to the primary winding is AC, so it has a characteristic sine wave that starts at zero, increases to a positive peak, returns to zero, then goes to a negative peak, and finally returns to zero again. Each time the voltage increases from zero to its peak, it will saturate current into the primary winding, which creates a very strong magnetic field around the wires in the winding. When the voltage decreases from its peak back to zero, the magnetic field and its flux lines collapse and the flux lines in the magnetic field cut across the wires in the secondary winding. Any time flux lines cut across the wires of the any coil, an electrical current is induced into that coil. This means that each time the primary winding of the transformer receives the AC voltage, increases to the positive peak and returns to zero, then decreases to the negative peak and returns to zero, a corresponding AC waveform is created in the secondary winding of the transformer. The waveform in the secondary winding is out of phase with the voltage in the primary winding by 180 degrees. Because the primary winding and the secondary winding are not connected, the voltage in the secondary winding is induced because the collapsing flux lines of the magnetic field from the voltage in the primary winding cut across the wires in the winding of the secondary winding.

The amount of voltage in the secondary winding is determined by the ratio of the number of coils in the primary winding to the number of coils in the secondary winding. If the secondary winding has twice as many coils as the primary winding, the secondary voltage will be twice as large as the primary voltage, and the secondary current will be half as large as the primary current. The amount of power that goes through the transformer is measured in units of **volt-amperes (VA)**, and the amount is usually in thousands of volt-amperes or **kilovolt-amperes (kVA)**. This means that if you know the voltage for a 10-kVA transformer, you can divide the kVA rating by the voltage and determine the amount of current.

If the secondary voltage for transformer is larger than the primary voltage, the transformer is called a **step-up transformer**; if the secondary voltage is smaller than the primary voltage, the transformer is called a **step-down transformer**. When a number of transformers are used together at the edge of a community, it is called a substation.

Overview of Transformers

Transformers are used to step up or step down voltage levels. When voltage is generated, it is stepped up to several thousand volts and in some cases several hundred thousand volts so that its current will be lower when it is transmitted. If the current is lower, the size of wire used to transmit the power can be smaller. Smaller wire weighs less, so smaller transmission towers can be used.

After voltage is transmitted on the grid over hundreds of miles at thousands of volts, it may be sent to a city where it will be consumed. The high voltage needs to be stepped down to approximately 12,470 V as it flows around the city so that it is less dangerous. This level of voltage is sufficient to transmit power throughout a city. When voltage arrives at a factory, it must be further stepped down to 480, 240, or 208 V and 120 V for all the power circuits in the factory, including the office. When voltage is transmitted to residential areas, the voltage will be 240 V line to line and 120 V line to neutral. Transformers provide a means of stepping up or stepping down this voltage.

Operation of a Transformer and Basic Magnetic Theory

The transformer consists of two windings (coils of wire) wrapped around a laminated steel core. The winding where voltage is supplied to a winding transformer is the primary winding. The winding where voltage comes out of the transformer is the secondary winding. Figure 8-3 is a diagram of a typical transformer with the primary and secondary windings identified.

A transformer works on a principle called **induction**. Induction occurs when current flows through the primary winding and creates a magnetic field. The magnetic field produces flux lines that emanate from the wire in the coil as current flows through it. When this current is interrupted or stopped, the flux lines collapse, and the action of the collapsing flux lines causes them to pass through the winding of the secondary coil that is adjacent to the primary coil. You should remember that the AC sine wave starts at 0 V, increases to a peak value, returns to 0 V, and then repeats the wave form in the negative direction. The process of increasing to peak and returning to zero provides a means of creating flux lines in the wire as current passes through it and then allowing the flux lines to collapse when the voltage returns to zero. Because the AC voltage follows this pattern naturally 60 times a second, it makes the perfect type of voltage to operate a transformer.

When the AC voltage returns to zero during each half-cycle, the flux lines that were created in the primary winding begin to collapse and start to cross the wire that forms the secondary coil of the transformer. When the flux lines from the primary winding collapse and cross the wire in the coil of the secondary winding, the electrons in the secondary winding begin to move, which creates a current. Because the current in the secondary winding begins to flow without any physical connection to the primary winding, the current in the secondary winding is called an **induced current**.

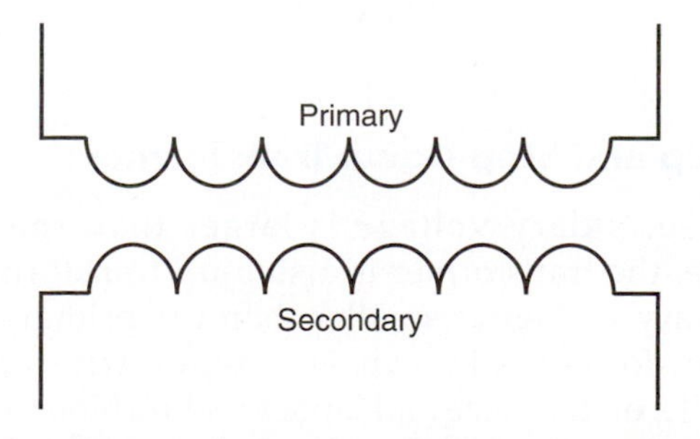

FIGURE 8-3 Primary and secondary windings of a transformer.

The following provides a more technical explanation of the relationship between the AC voltage and the windings of the transformer. The magnetic field in the primary winding of the transformer builds up when AC voltage is applied during the first half-cycle of the sine wave (0 to 180 degrees). When the sine voltage reaches its peak at 90 degrees, the voltage has peaked in the positive direction and begins to return to 0 V by moving from the 90-degree point to the 180-degree point. When the voltage reaches the 180-degree point, the voltage is at 0 V, and it creates the interruption of current flow. When the sine-wave voltage is 0 V, the flux lines that have been built up in the primary winding collapse and cross the secondary coils, which creates a current flow in the secondary winding.

The sine wave continues from 180 to 360 degrees, and the transformer winding is energized with the negative half-cycle of the sine wave. When the sine wave is between 180 and 270 degrees, the flux lines are building again, and when the sine wave reaches 360 degrees, the sine wave returns to 0 V, which again interrupts current and causes flux lines to cross the secondary winding. This means that the transformer primary energizes a magnetic field and collapses it once in the positive direction and once again in the negative direction during each sine wave. Because the sine wave in the secondary is not created until the voltage in the primary moves from 0 to 180 degrees, the sine wave in the secondary is *out of phase* by 180 degrees to the sine wave in the primary that created it. The voltage in the secondary winding is called *induced voltage* because it is created by induction.

The induced voltage is developed even though there is total isolation between the primary and secondary coils of the transformer. Figure 8-4 shows the four stages of voltage building up in the transformer and the flux lines building and collapsing at each point as the sine wave flows through the primary coil.

Transformer Voltage, Current, and Turns Ratios

The amount of voltage a transformer will produce at its secondary winding for a given amount of voltage supplied to its primary is determined by the ratio of the number of turns in the primary winding compared to the number of turns in the secondary. This ratio is called the **turns ratio**. The amount of primary current and secondary current in a transformer also depends on the turns ratio.

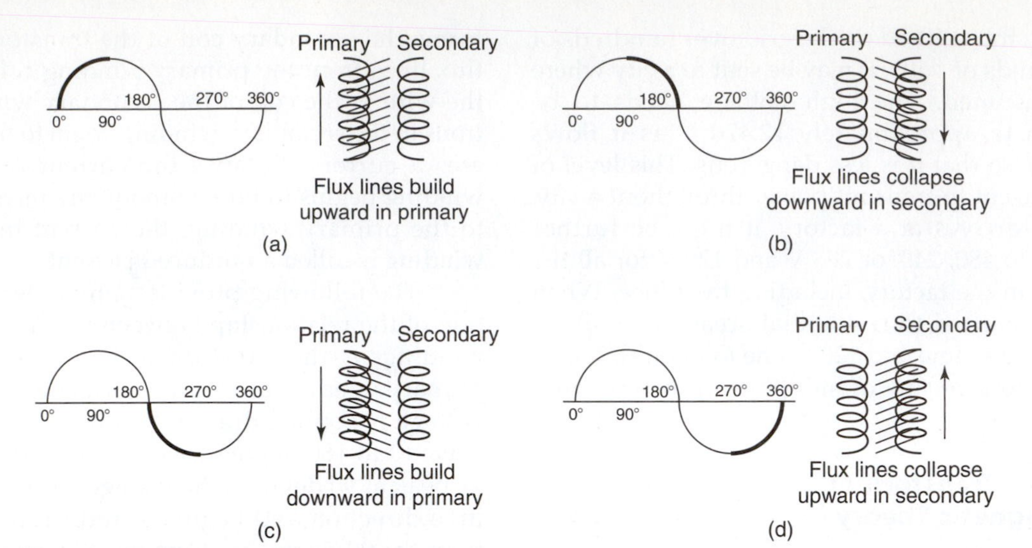

FIGURE 8-4 (a) Flux lines build upward in primary winding. (b) Flux lines collapse downward in secondary. (c) Flux lines build downward in primary. (d) Flux lines collapse upward in secondary.

Figure 8-5 indicates the primary voltage (E_p), the primary current (I_p), the number of turns in the primary winding (T_p), the secondary voltage (E_s), the secondary current (I_s), and the number of turns in the secondary winding (T_s). The primary and secondary voltages, the primary and secondary currents, and the turns ratio can all be calculated from formulas. The turns ratio for a transformer is calculated from the following formulas:

$$\text{Turns ratio} = \frac{T_s}{T_p} \quad \text{or} \quad \frac{V_s}{T_p} \quad \text{or} \quad \frac{I_p}{I_s}$$

The ratio of primary voltage, secondary voltage, primary turns, and secondary turns is

$$\frac{E_p}{E_s} = \frac{T_p}{T_s}$$

From this ratio you can calculate the secondary voltage using the formula

$$E_s = \frac{E_p \times T_s}{T_p}$$

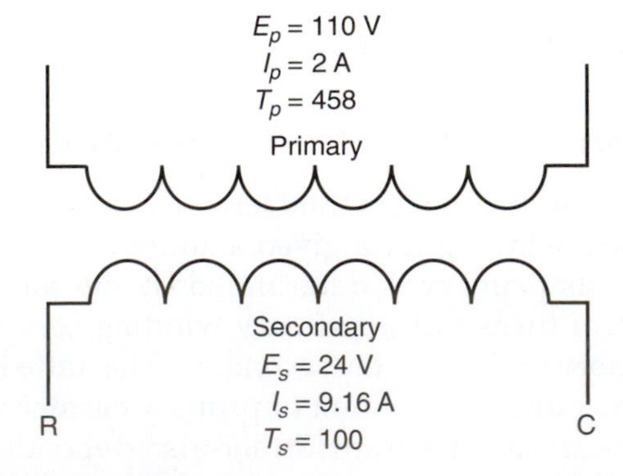

FIGURE 8-5 Transformer with primary voltage of 110 V and secondary voltage of 24 V.

The ratio of primary voltage, secondary voltage, primary current, and secondary current is

$$\frac{E_p}{E_s} = \frac{I_s}{I_p}$$

(Notice that the ratio of voltage to current is an inverse ratio.) Using this ratio, you can calculate the secondary current:

$$I_s = \frac{E_p \times I_p}{E_s}$$

If you know that the primary voltage is 240 V, the primary current is 2 A, the number of primary turns is 458, and the number of secondary turns is 229, you can easily calculate the secondary voltage and the secondary current using these formulas:

$$E_s = \frac{E_p \times T_s}{T_p}$$

$$\frac{240\text{ V} \times 229\text{ turns}}{458\text{ turns}} = 120\text{ V}$$

$$I_s = \frac{E_p \times I_p}{E_s}$$

$$\frac{240\text{ V} \times 2\text{ A}}{120\text{ V}} = 4\text{ A}$$

Step-Up and Step-Down Transformers

If the secondary voltage is larger than the primary voltage, the transformer is a step-up transformer. If the secondary voltage is smaller than the primary voltage, the transformer is known as a step-down transformer. Typically, on mid-size and large wind turbines, the generator creates voltage at approximately 900 V, and a step-up transformer is used to get the voltage to12,470 V so it is compatible with the grid. A step-down transformer is used when the voltage arrives at the substation in a community at 12,470 V and must be stepped back down to 480 V for use in factories and 240 V so it can be used in residential areas.

VA Ratings for Transformers

The VA (volt-ampere) rating for a transformer is calculated for either the primary side or the secondary side of the transformer. The VA rating calculation can be accomplished by multiplying the primary voltage and the primary current, or by multiplying the secondary voltage and secondary current. If the primary voltage is 240 V and the primary current is 4 A, the VA rating for the transformer is 960 VA. The VA rating indicates how much power the transformer can provide or handle. As a technician, you can say that the primary VA is equal to the secondary VA, and not worry about the losses. If you were a design engineer, you would need to be more precise, and you would find that the secondary VA is slightly less than the primary VA because of transformer losses. It is important that the VA rating of the transformer be large enough for the application you are using it for. If the VA rating is too low, the transformer will be damaged and fail prematurely because the extra current will cause it to overheat. If you need to replace a transformer, you must be sure that the voltage ratings match and that the VA size of the replacement transformer is equal to or greater than that of the original transformer.

On large transformers that are used to transmit large amounts of power from a wind turbine that produces 1.8 MW for the grid, the secondary voltage from the step up transformer will be12,470 V, and the current could be as high as 140 A. The transformer for this application would have to be approximately 1800 kVA.

Troubleshooting a Transformer

It is important to understand that you will need special training and special personal protective equipment to work on transformers with voltages greater than 200 V. Larger voltages create an electrical hazard called *arc flash,* which has the potential to cause injuries as well as death if an arc occurs when power is turned on or off or at other times. The information in this section about troubleshooting applies to small transformers that are disconnected from an electrical circuit. The information is provided for background so that you will better understand the workings of a typical transformer. If the transformer you are testing is not connected to power, you can test each of the windings with an ohmmeter for continuity. Each of the coils should have some amount of resistance that indicates the amount of resistance in the wire in each coil. If you measure infinite resistance (∞), it indicates the winding has an open, and if you measure 0 ohms, it indicates that one of the windings is shorted.

Another way that you can test a small transformer is by applying power to its primary winding. It is important that you complete the continuity test first to detect any shorts before you apply power. If the transformer has a short, do not apply power. It is also important to understand the safety rules when working around live voltage and to follow the rules during this test. If the transformer windings are not shorted, you can apply any amount of AC voltage that is equal or less than the primary rating of the transformer. If the transformer is working correctly and it is operational, a voltage will be present at the secondary terminals of the transformer. If no voltage is present at the secondary, be sure to check all the connections to ensure that they are correct. If voltage is still not present at the secondary, the transformer is defective and should be replaced. It is important to remember that the transformer has a primary winding and a secondary winding that are placed in close proximity to each other; when AC voltage is applied to one winding, induction will cause voltage to be available at the other winding.

Common Voltages Used in Power Distribution Systems in Cities

Some city and suburban distribution systems use transformers to create a range of voltages, such as 7,200/12,470, 7,620/13,200, 14,400/24,940, and 19,920/34,500 V. The higher voltages are needed to keep the wire size as small as possible as power is distributed around and through a city. The transmission lines within a city will be a combination of overhead lines and construction utilizing traditional utility poles and wires. In some newer installations there may be some locations that include underground construction with cables and indoor or cabinet substations. However, underground distribution is significantly more expensive than overhead construction. Typically, overhead power lines and power poles are used to distribute power in rural areas.

Substation and Power Distribution Around the City

A substation is a group of large transformers located in one area where power is brought into the city. The high voltage from the grid is transmitted across country through overhead wires that are supported on large steel towers. Figure 8-6 shows a series of high-voltage towers on which power is transferred across country and eventually to a substation. Notice the size of the towers, and how they are arranged in straight lines, to provide the shortest path between where power is generated and where it reaches a substation.

The power from the high-voltage lines feeds the input of a large group of transformers at the first substation outside the city. Figure 8-7 shows the large number of transformers mounted on the ground, and the structural steel that provides support for all of the cables that

FIGURE 8-6 High-voltage power lines that feed voltage from where it is generated to a substation just outside of a city.

FIGURE 8-7 A large substation at the edge of a city. Notice that the transformers are located on the ground, and a large steel structure overhead is used to ensure that the wires feeding the transformers do not touch each other.

interconnect the transformers. You will also notice lightning protection at the very tops of each tower. The function of the substation is to break the large volume of high voltage and current down into subsections to feed different parts of industrial, commercial, and residential areas of the city.

After the large voltage and current is brought into the large substation, it is broken down and sent to areas of the city that have large industrial applications, commercial applications such as malls and large stores, and then to residential areas. Figure 8-8 and Figure 8-9 shows examples of smaller substations that are found around a city, closer to industrial applications, commercial applications, and residential applications. You can see a trailer that has several portable transformers mounted on it inside the substation connected to the grid. The portable transformers can be quickly installed in case of transformer failure, or for additional load capability. Figure 8-10 shows three transformers that provide three-phase voltage to a small commercial application, such as several stores located on the same property. In residential areas, one power pole may have a transformer that services 8–10 homes. In some residential areas, the power lines are buried and the transformer sits in the backyard of one home but services several homes.

FIGURE 8-8 Substation for a small commercial or small industrial application.

FIGURE 8-9 A small substation for residential applications. Notice the portable transformers on a trailer on the far right side of the picture.

FIGURE 8-10 Three transformers provide three-phase voltage to a small commercial application.

8.4 GRID CODE RULES AND REGULATIONS

Rules and regulations apply to any connection that is made between a producer of electricity such as a wind turbine generator and the grid. These rules and regulations include nontechnical issues such as agreements for interconnection between the wind turbine generator and the grid that address the exchange price of power and purchase agreements between the owner of the wind turbine and the operator of the electrical grid in that area. Other rules and regulations cover technical issues that concern safety and power quality, and other conditions affecting electrical connections and isolation between the wind turbine generator and the grid.

These rules and regulations address electrical safety and electrical power quality issues. The rules and regulations may fall under any number of major codes from safety organizations that publish interconnection codes and standards. These may include the National Fire Protection Association (NFPA), which publishes the National Electrical Code (NEC), Underwriters Laboratory (UL), and the Institute of Electrical and Electronics Engineers (IEEE) (pronounced I triple E). Two federal laboratories, the National Renewable Energy Laboratory (NREL), which is a national laboratory of the U.S. Department of Energy, Office of Energy Efficiency and Renewable Energy that is operated by the Alliance for Sustainable Energy, and Sandia National Laboratories also work closely with the NFPA, UL, IEEE, and the distributed generation (DG) community on code issues and equipment testing. The labs are not responsible for issuing or enforcing codes, but they do serve as valuable sources of information on interconnection issues. The technical and safety issues are part of codes and standards for the interconnection process, and they provide some consistency and standards that ensure safety in any grid connection.

8.5 THE NATIONAL ELECTRICAL CODE AND OTHER REQUIREMENTS FOR THE GRID

National Fire Protection Association (NFPA) writes the **National Electrical Code (NEC)**. The latest codebook was published in 2008, and it is revised every three years. The National Electrical Code specifies code requirements for electric wiring to ensure that it does not cause a fire hazard or an electrocution hazard. The National Electrical Code establishes sizes for current-carrying conductors and identifies requirements for grounding and other safety issues. Because the generator for a wind turbine produces electrical power, all connections, interconnections, installation, and sizing of the wiring must be completed according to the National Electrical Code. Other important issues such as grounding, and installation practices for the wiring in the tower and in the nacelle, must also meet code requirements. You can purchase a copy of the National Electrical Code and have it with you to ensure that you are following the code.

The Institute of Electrical and Electronics Engineers (IEEE) has written a standard that addresses all grid-connected distributed generation, including renewable (wind) energy systems. IEEE Standard 1547-2003 provides technical requirements and tests for grid-connected operation, regardless of where the electrical power originates. These standards were written in cooperation with the NREL. The standards include interconnection systems and interconnection test requirements for interconnecting distributed resources (DR) with electric power systems (EPS). This ensures that when a wind turbine generator is connected to the grid, it follows all standards and regulations that apply. Writing and implementing the standards and regulations helps move all sectors that connect electrical power into the grid closer to the new standards for a smart grid. The standards are written for all distributed energy resources (DER), which is defined as small-scale electrical generation that is connected to distributed generator (DG) systems, which include wind turbine generators. This includes both wind turbine generators that connect directly to the grid and those that do not connect to the grid but that connect directly to a residence or small commercial establishment.

The interconnection technical specifications and requirements include

- Voltage regulation
- Integration with area EPS (electrical power system) grounding
- Synchronization (to ensure that phase is close)
- DR (distributed resources) on secondary grid spot networks
- Inadvertent energizing of the area EPS
- Monitoring provisions
- Isolation device (interconnecting switchgear)
- Interconnect integrity

Interconnection technical specifications and requirements also cover

- Area EPS faults
- Area EPS reclosing coordination
- Voltage levels
- Frequency
- Loss of synchronism
- Reconnection to area EPS

8.6 SUPPLYING POWER FOR A BUILDING OR RESIDENCE

When a wind turbine is designed to provide power directly to a building or residence, it may need to follow some of the same standards and regulations that apply to grid connection. However, because the power being fed to the building does not need to comply with other existing voltage, it may be able to withstand some slight variations in voltage and frequency without causing damage to the equipment in the building. If the power from a wind generator is used mainly for resistive loads such as heating and lighting, the issues of voltage levels and frequency control are not as important. If the wind turbine generator provides part of the power for the building and the remainder of the power comes from the grid, then all connections, equipment, and the power itself must meet the same standards as if it was connected directly to the grid.

In some small applications, these issues can be taken care of by a single piece of equipment that includes an inverter and a safety switch. The inverter will take care of voltage and frequency issues, and the safety switch will take care of the interconnect and disconnect issues.

8.7 SWITCHES AND CONNECTIONS FOR POWER DISTRIBUTION

Voltage from wind turbines needs a point where it is connected to the grid. Figure 8-11 shows a connection where the power from two wind turbines connects to the grid. You can see the interconnection and the switch at the top of the power pole. Power is fed underground from the wind turbine, and conduit protects the wire as it is routed to the top of the pole.

When a wind turbine generator is connected to the grid, it must use safety equipment that protect the equipment from being damaged and ensures that the electricity being produced cannot harm people. One of the major components is the safety disconnect switch. The safety disconnect automatically or manually disconnects and protect the wiring and components of the wind turbine generation system from power surges and other equipment malfunctions by disconnecting and isolating the equipment from the grid. The disconnect safety switch also ensures that the wind turbine generator can be shut down safely and isolated from the grid for maintenance and repair. The safety disconnects also ensure that wind turbine generating equipment can be isolated from the grid so that it does not send any power back into the grid. This is a very important to ensure that power from the generator does not enter the grid when the grid needs to be isolated and powered down for maintenance, such as after severe high winds, ice storms, or hurricanes, when some grid transmission lines may have become damaged and must be worked on.

FIGURE 8-11 Electrical connection between wind turbines and the grid. The switchgear for the connection is near the top of the pole.

The safety equipment must include grounding equipment. **Grounding** includes using grounding rods that are inserted directly into the earth to a depth of about 8 ft. The grounding rods provide a low-resistance path from the generation system to the earth ground to protect the system against current surges from lightning strikes or equipment malfunctions. The grounding system also protects the metal services around the wind turbine from causing electrical shock hazards to any humans working on or touching metal surfaces. If a live wire touches any metal surface and the surface is not grounded, any person who touches the metal surface could receive a severe electrical shock. If the metal surface is connected to the ground system, as soon as the live wire touches the metal, the ground circuit will cause the current in the live wire to increase quickly to a point at which it causes a fuse or circuit breaker in the circuit to open and turn off voltage to the circuit. In this way, the ground circuit ensures that any unwanted or unsafe voltage creates a short circuit that causes the fuse or circuit breaker to blow.

Other types of safety equipment include surge-protection equipment. This type of equipment prevents electrical power surges from reaching the generating equipment. A power surge can occur when a wind turbine generator becomes out of phase with the power in the grid, or it may be caused by a lightning strike. The surge protection can open and clear the surge, or it can divert the excess energy directly to ground.

The wind turbine generator must have a switch that connects and disconnects its circuits to the grid. When the wind turbine is stopped for maintenance or needs to be disconnected from the grid for any other reason, the switch is moved to the open position, which disconnects all grid voltage from the wind turbine generator. When the wind turbine generator is producing voltage at the correct voltage level and frequency, the switch will close and the wind turbine output will be fed to the grid. If the amount of wind is too low to produce voltage that is the right amount at the right frequency, the wind turbine generator is electrically disconnected from the grid, but the generator is allowed to turn at idle speed.

If the wind turbine generator was allowed to remain connected to the grid when its blades are not turning fast enough to produce the correct amount of output voltage, the generator would actually consume electrical power and run as a motor. Once the wind is strong enough to turn the generator at a speed high enough to produce power that has the correct voltage and frequency, the output power is ready to be connected to the grid.

One last important factor called voltage phase must be monitored, so that the connection to the grid takes place when the phase of the voltage in the wind turbine generator is nearly identical to the phase of the voltage on the grid. If the electrical circuit from the generator is connected to the grid and the phase of the generated voltage is not close to the phase of the voltage on the grid, a large surge current will be sent through the generator, which can cause severe damage to it.

The switches that are used to connect the wind turbine generator circuit to the grid are similar to the switches used by utility companies to connect and disconnect the power their generators produce, and these switches have been utilized for many years. This means that most switch problems have been corrected and the switches have been modified over the years so that they perform the switching process seamlessly. The only problem with voltage that is produced by a wind turbine generator is that the voltage tends to have more irregular levels of voltage and more changes of frequency than voltage and frequency that is produced by larger energy producers such as nuclear power and coal power energy producers.

Disconnect switches are also used when sections of the grid need to be powered down and isolated so that line personnel can work on them. For example, after a severe ice storm or after a hurricane, large portions of the electrical distribution wiring for the grid can become damaged in a given geographic area. When this occurs, that section of the grid is disconnected from the rest of the grid, and all voltage is removed so that technicians may work safely on any downed wires. This type of disconnection is called *power isolation,* and any electrical energy-producing equipment such as wind turbine generators must be disconnected and isolated from the grid to ensure that voltage is not fed into a section of the grid that might endanger a technician.

Other safety equipment includes a protection system that lets the wind turbine ride through grid faults such as low-voltage problems without tripping or adding to the problem. Safety equipment may also cover frequency that drops below or raises above the standard 60 Hz.

8.8 UTILITY GRID–TIED NET METERING

Net metering is a method of metering the energy consumed or produced when a wind turbine is connected to a home or business. When a net meter is used to measure power, it records the amount of energy the home or business uses when the wind turbine is not operating and all the electrical power is coming from the grid. This is like a typical household or business that does not have a wind turbine and uses only electrical power from the grid. The electric meter measures the amount of power used, and the customer is billed for that amount. When the wind turbine begins to produce electrical power, the residence or business will begin to use only the power from the wind turbine generator, so the electric utility meter will not record this power and the customer will not be charged for this power usage. If the home or business uses less power than the wind turbine is producing, the

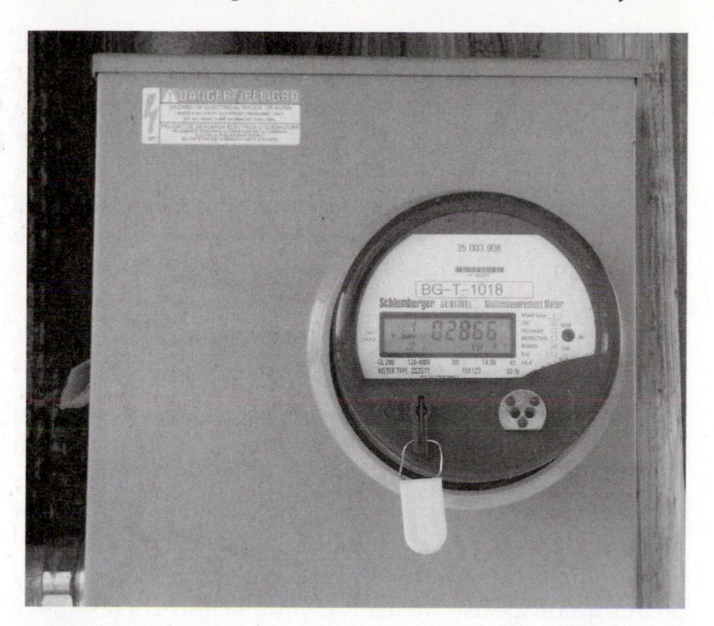

FIGURE 8-12 A net meter that measures the power delivered from the electric utility or the power fed back to the utility.

excess power the wind turbine is producing is fed back into the grid. The net meter will spin in the opposite direction and record the amount of power the wind turbine generator is putting into the grid. Figure 8-12 shows a residential net meter.

Some utility companies have a different, lower rate that they pay customers for the power their wind turbine generator puts back into the grid. Other utilities basically allow the net meter to spin backwards when the wind generator is producing power, which provides the customer the full retail value of all the electricity the wind generator produces.

If the state or location where the customer is does not allow net metering, under existing federal law (PURPA, Section 210), the utility customers can use the electricity they generate with a wind turbine to supply their own needs, such as for lights and appliances. This basically offsets any electricity they would otherwise purchase from the utility at the utility's retail price. If the wind turbine produces any excess electricity, that is, more than what the customer needs for his or her own use, the net meter does not run backwards; rather, the electrical utility purchases that excess electricity at the wholesale or *avoided cost* price, which is generally much lower than the retail price. The excess energy is metered on a separate meter rather than on the net meter. Today, nearly 30 states require the utility companies under their jurisdiction to offer some form of net metering for small wind systems. It is important to understand that these requirements vary widely from state to state.

In most states, net metering rules were enacted by state utility regulators, and these rules apply only to utilities whose rates and services are regulated at the state level. In most of the states with net metering statutes, all utilities are required to offer net metering for some wind systems, although many states limit eligibility to small systems. If a net meter or an auxiliary meter is used, it must be installed at the customer's expense. Net metering simplifies billing by allowing the customer to use any excess electricity to offset electricity typically bought from the utility company. In other words, the customer is billed only for the net energy consumed during the billing period.

8.9 OVERVIEW OF POWER QUALITY ISSUES

The term **power quality** refers to issues dealing with the voltage and current frequency that is produced by a wind turbine generator or other power-generation source. Power quality issues also referred to things that filtering normally keeps out of the system, such as electrical noise, DC injection, and harmonics. The ideal power quality exists when the voltage and current have a sinusoidal waveform with a frequency of exactly 60 Hz. The voltage of ideal power stays relatively constant regardless of changing conditions in the generator and on the grid. The ideal power can be obtained from a wind turbine by using electronic controls and other types of controls to ensure the power has the highest quality. If power from a wind turbine has problems with voltage or frequency, it will be detected and must be corrected before the generator is allowed to produce power and connect to the grid.

8.10 FREQUENCY AND VOLTAGE CONTROL

Frequency and voltage on the grid is controlled by the individual power producers who put large amounts of electrical power into the grid. The grid receives voltage from large energy producers such as nuclear-powered generators, coal-fired generation systems, and hydroelectric generators, as well as from wind generators and other forms of alternative energy. The large energy producers have an easier time controlling voltage and frequency than operators of wind turbines do, because the energy source they use to turn the generators is more consistent, whereas the wind energy that moves wind turbine blades tends to be more variable.

Because the majority of the electrical energy on the grid has constant voltage and constant frequency from the larger energy producers, the grid can withstand some small variation of voltage and frequency in the energy produced by wind turbines. Wind turbines have a variety of methods to control the frequency of the voltage they produce. As you know, the frequency changes as the speed at which the generator shaft turns changes. Larger wind turbines use blade pitch control and yaw directional control to help keep the rotor shaft that has the blades attached to it turning at a fairly constant speed. These larger wind turbines also have variable-speed gearboxes that can compensate somewhat for different wind speeds and maintain the speed of the high-speed output shaft fairly close to the required speed to maintain a frequency of 60 Hz.

Other types of wind turbines have double-fed generators, and they use an inverter to control the frequency of the exciter voltage to exactly 60 Hz, which causes the generator to produce voltage at near 60 Hz under varying speed conditions, as long as the speed is within 5–10% of the rated speed needed to produce 60 Hz. These generators also produce voltage whose frequency is nearly 60 Hz, and when this voltage is fed into the grid, which has large volumes of voltage with exactly 60-Hz frequency, this larger volume will cause any voltage with a frequency other than 60 Hz to comply and operate at exactly 60 Hz. This effect will work as long as the amount of voltage that is not at the correct frequency is not too large with respect to the amount of voltage on the grid that is at exactly 60 Hz.

On small and mid-size wind turbines, an electronic inverter can be used to create an output voltage at exactly 60 Hz and the rated voltage required for the grid connection. In these applications, the inverter must be sized large enough to handle all the voltage and current the generator produces. The inverter functions by changing all the AC voltage, regardless of voltage level and frequency, to DC voltage. Capacitors and inductors are used in the inverter to remove any remaining ripple from the original AC voltage. The pure DC voltage is then sent back through a set of transistors, such as insulated gate bipolar transistors (IGBTs), where the transistors are turned on and off so that their output looks similar to an AC sine wave. The frequency of the voltage output and the voltage is controlled by the electronic control circuit in the inverter that controls the IGBTs. The electronic inverter does an excellent job of producing AC voltage at exactly 60 Hz and the voltage required for the grid, but it must be large enough to handle all of the voltage in all of the current the wind turbine generator produces. This limits the size of the wind turbine generator that can use an inverter of this type.

Adjusting the voltage in a wind turbine depends on several things. First, the speed at which the generator turns not only controls the frequency of the voltage that is produced, it also controls the amount of that voltage. The faster the wind turbine generator turns, the higher the voltage will be. The amount of voltage the generator produces is also affected by the amount of exciter voltage that is fed into its fields. The larger the exciter voltage, the more voltage generator will produce. This means that if the generator is producing a voltage that is above its rated value, a voltage regulator can decrease the amount of voltage provided to the exciter field, and the voltage level at the generator will decrease. If the amount of voltage the generator is producing is too small, the exciter voltage can be increased to make the generated voltage increase. A voltage regulator senses the voltage output from the wind turbine generator and adjusts the exciter voltage to make the output voltage higher or lower. If an inverter is being used with the generator, the inverter can control the output voltage as well as the output frequency.

8.11 VOLTAGE, TRUE POWER, AND REACTIVE POWER

As you know, voltage is the force that causes electrons (electrical current) to flow. The stronger the voltage, the more force moves the current to flow through the system. When you multiply the amount of voltage in a system by the amount of current, the result is electrical power, which is measured in watts. The current in the circuit can be caused by several different types of loads. A pure **resistive load** is a load such as an electrical hot-water heater element, an electrical resistance heating element in a furnace, or an incandescent light bulb. Another type of load is called an **inductive load**. Inductive loads include current consumed by motors, magnetic coils, and other inductors (coils). The third type of load is called a **capacitive load**. Capacitive loads include current that flows through capacitors used to correct the power factor, and they also occur in some computer power supplies.

True power, also called *real power*, is the voltage and current that flows only through resistive loads. In these loads the voltage and current waveforms are in phase with each other and there is no wasted energy. When voltage and current go through capacitive or inductive loads, the voltage and current waveforms are out of phase. In an inductive circuit, the current waveform lags behind the voltage waveform. In a capacitive circuit, the voltage waveform lags behind the current waveform. The power that flows through the inductive loads and the capacitive loads is called **apparent power**. The ratio of true power (TP) to apparent power (AP) is called the **power factor**. The formula for power factor (PF) is PF = TP/AP. Because true power is always less than apparent power, the power factor is always less than 1.00. The closer the PF is to 1.00, the more efficient the circuit is and the less the losses are.

Any time the voltage and current are out of phase, power losses begin to occur. If a grid circuit is connected to a factory that has a large number of large motors, the circuit will have losses due to the inductive loads. The circuit losses can be corrected by adding an equal amount of capacitance. When capacitance is added to a circuit to correct the inductive losses, it is called power factor correction. The capacitance for the power factor can be added at the end users that are creating the large inductive loads (the factories), which will give them a better rate, or the power factor can be corrected where the power is produced. This means that if the inductive loads are scattered throughout the grid circuit, such as a large number of air-conditioning compressor motors in residential areas, capacitance cannot be added at each home, so the capacitance is added by the electric utility where the electrical power is produced. This means that any electrical power produced by wind turbines must be able to add capacitance to correct the power factor of the area of the grid to which the turbines are connected. Every wind turbine must meet the standards required by the grid to correct the power factor for the power it is producing.

Some wind turbines that have squirrel-cage induction generators will have an additional problem in that the squirrel-cage induction generator will cause the power generated to be reactive and have inductive reactance. This power is used to excite the generator, and the amount of **reactive power** depends on how much current the exciter generator uses and how fast the generator shaft is turning. This is not a large problem if a single wind turbine is used in a stand-alone application, but when multiple wind turbines on a wind farm are all using squirrel-cage induction generators, the amount of reactive power is sufficient to cause a problem. In these wind farm applications, banks of capacitors must be available to be added to the generated load to correct the power factor where these generators tie into the grid.

8.12 LOW-VOLTAGE RIDE-THROUGH

Low-voltage problems in the grid may occur for several AC voltage cycles or may last for several hours. Such problems may arise in one phase of a three-phase system or in all three phases. Low voltage is a condition that exists whenever the voltage on the grid drops 10% or more below the rated voltage level.

When low voltage occurs in the grid system where a wind turbine is connected, the wind turbine must have the capability to *ride through* this condition. It does not matter whether the wind turbine is causing the low voltage or whether some other factor is causing the low voltage in the grid where the wind turbine is connected. The safety system for this problem is called **low-voltage ride-through (LVRT)**. This means that when a low-voltage fault occurs in the grid, electrical instrumentation records the severity and the duration of the problem. The wind turbine generating system must have a system to protect it through the low-voltage condition and not trip off. If the wind turbine generator is causing the low-voltage condition, it must be able to correct the condition quickly or disconnect the generator from the grid. The choices may include completely disconnecting from the grid and automatically reconnecting after the low-voltage dip has cleared, or remaining connected, staying in operation, and riding through the low-voltage condition. The third choice is to stay connected to the grid and help correct the problem by adding reactive power from the generator's capacitive banks, which will help alleviate the low-voltage condition. The method the wind turbine is programmed to use will depend on the types of problems the grid to which it is connected is subject.

8.13 FLICKER AND POWER QUALITY

Flicker power is a short-lived voltage variation in the electrical grid, which might cause a load such as an incandescent light to flicker. This problem is more noticeable when a wind turbine is connected directly to a residential or commercial application as the sole source of power. When the wind blows more strongly for a short time, the generator will produce slightly more voltage, which will cause a light bulb to glow more brightly during that period. Eventually the controls on the wind turbine will sense that the voltage has increased as a result of the increase in the generator speed, and it will lower the exciter current, which will bring the voltage back to the normal level. The sequence to identify the overvoltage and bring it under control may require several seconds, during which the increased brightness of the light bulb will be noticeable. This problem can be controlled in residential applications by using an inverter to control the voltage and frequency directly with the wind turbine, or the wind turbine generator output can be sent to a bank of batteries, and the inverter can be used to smooth out the voltage to a constant level coming from the batteries as it is turned back into AC voltage.

If the wind turbine is connected directly to the grid, tighter controls have to be added to the generator to ensure that problems with the voltage are kept to a minimum in both duration and amount. Some wind turbines have mechanical means to try and control excess speed from the wind turbine blades caused by gusty wind speeds to maintain tighter voltage and frequency. These wind turbines may use any combination of blade pitch controls and yaw directional controls to keep the blades turning at their specified speed.

Power Quality

Several factors in the production of electrical power affect power quality. These include voltage level, frequency, phase shift, and power factor correction. The wind turbine electrical system must have controls to correct deficiencies in each of these factors. For example, the electronic voltage regulator may be used to control the amount of exciter current to ensure that the voltage level is fairly constant regardless of the speed at which the shaft is turning. Double-fed generators have several methods of controlling the frequency of the exciter voltage to ensure that the frequency of the output voltage from the generator is held within a few percent of 60 Hz. Phase-shift problems occur when the power from the generator is connected to the grid and there is a difference of phase angles between the two voltages. This means that the sine wave for one of the voltages has started a few microseconds before the other, so the sine waves from the two voltages do not start at 0 degrees at the same exact instant. Complex electronic circuits monitor the phase of the active voltage on the grid continually and ensure that the phase of the voltage being generated by the wind turbine is similar to that of the grid before the switch connecting the wind turbine generator to the grid is closed. If the phase difference between the voltage that the wind turbine produces and that of the grid is too large, adjustments will be made to the wind generator exciter voltage to bring its phase relationship closer to the voltage at the grid, so that the sine wave of the generated voltage and the grid voltage both start at their 0-degree point at exactly the same time. When the phase relationship of the wind turbine generator is identical to the voltage on the grid, the connection switch is closed, allowing the wind turbine to be connected to the grid so voltage from the wind turbine can begin to enter the grid.

Electronic circuits are also used in wind turbines to measure the power factor for the voltage from the wind turbine generator and the voltage on the grid where the connection is made. If the power factor meters indicate that the circuit has too much inductive reactant, capacitance from the banks of capacitors is connected to the circuit, which will correct the power factor. The banks of capacitors are continually connected and disconnected from the grid as required to correct the power factor. It is important to understand that the inductive motors at the factories that are causing the inductive reactance to occur in the circuit may be switched on and off a number of times during the day, which will cause the amount of reactance in the circuit to vary continually. The electronic circuit that constantly monitors the power factor will monitor the changes in the inductive reactance in the circuit and immediately adjust the amount of capacitance from the capacitor banks to ensure that the power factor on that sector the grid is continually corrected.

Islanding

If a small section of the electrical grid becomes disconnected from the main grid, this condition is referred to as *islanding*. Islanding may occur when lightning strikes, or a short circuit occurs in one of the areas of the grid, and the circuit breakers trip in one or more sections of the grid and cause the isolation. Islanding may also occur if an automobile strikes a power pole, or another problem occurs that causes the circuit breakers to trip. In some cases it is possible for a wind turbine that is connected to the isolated area to continue to produce enough voltage to support consumers and loads in the isolated section of the grid. The problem that arises during this time is that the phase of the voltage from the wind turbine may drift out of phase from the voltage in the remainder of the grid. If the condition that caused the disconnection is cleared, and the small section of grid is connected back to the main grid while the voltage is out of phase, a phase differential problem will arise, which can cause a strong current surge in the wind turbine generator and severely damage it. For this reason, it is important that the electronic detection system be used to continually monitor the main grid voltage and ensure that the phase differential between it and the wind turbine voltage is minimal when the reconnection is made. The phase detection circuits continually monitor the phase differential between the wind turbine and the main grid, and if it becomes too large, the wind turbine is disconnected from the grid through its disconnect switch. When the phase in the wind turbine is adjusted and the differential between the wind turbine voltage and the grid voltage becomes small again, the wind turbine is reconnected to the grid.

8.14 SYSTEM GROUNDING

The grounding system for a wind turbine consists of one or more copper grounding rods that have been inserted into the earth to a depth of 8 ft. Heavy copper conductor is used to connect the metal frame of the tower to the rod. When the metal frame is connected to the rod that is in the ground, the metal of the wind turbine tower is not at the same electrical potential as the earth ground. Any metal in the system will be bonded to the metal frame, which means it may be mechanically connected with nuts and bolts, or it may have a wire conductor connecting to the frame of the tower. Bonding ensures that all of the metal parts of the tower and system have a low-resistance path to the earth ground. If more than one rod is used, copper conductors connect the copper ground rods together.

The ground system is used for two separate safety systems in wind turbines. First, the ground provides a low-resistance electrical path to the earth between any metal in the wind turbine and the ground rods. If any electrical wiring that has voltage in it becomes chafed and allows electrical voltage to come into contact with the metal frame of the tower or other parts of the wind turbine system, this voltage will immediately be taken to the earth ground, which will cause an extremely high current to occur. The high current will cause a circuit breakers or fuses protecting the high-voltage lines to open and interrupt the circuit. If the system was not grounded, and the circuit breaker did not trip, anyone who came into contact or touched any metal parts could receive severe electrical shock or even be electrocuted. If you find a circuit breaker that is open, you must test the system to see if a short circuit has occurred. When you close the circuit breaker, if it immediately opens again, you should suspect a short circuit somewhere in the system and you will need to begin isolating wires until you find the wire that is the problem.

The second reason the system is grounded is to help protect against lightning strikes. If the wind turbine blades or the wind turbine tower is struck by lightning, the large amount of electrical energy in the lightning bolt must be quickly dissipated into the earth where it will not cause damage. If the wind turbine and tower are not grounded properly and the lightning strike enters the equipment, it may be severely damaged. The ground system must be checked periodically to ensure that it is connected correctly and that all connections provide a low-resistance path to the earth.

8.15 UNDERGROUND FEEDER CIRCUITS

Most voltage lines from wind turbines are run underground through conduits to the point of connection to the grid. These wires are run underground to help protect them from the elements and also for esthetic reasons. Also, if the wires were run above ground, power poles and other hardware would have to be used that might become a problem during wind turbine maintenance, when large cranes may be brought in to work on the system. If the power lines are run overhead on power poles, crane operators have to be continually aware of their locations and maintain a secure distance from them.

The underground conduits are generally poured into any concrete that the tower uses as its pad, and additional conduit is run between the wind turbine and the point where it is connected on the power pole. In Figure 8-13 you can see where three separate conduits are used to bring the power out of the ground and up to the top of the pole where the connection to the grid is

FIGURE 8-13 Electrical connection between wind turbines and the grid. The switchgear for the connection is near the top of the pole. The interconnection wires between the wind turbine and the power pole are buried underground. Notice the lighter-colored conduit that protects these wires as they come above ground at the base of the pole and rise to the top of the pole.

made. Today it is possible to run underground conduits for thousands of feet with a piece of equipment that tunnels through the ground under the surface at approximately 3 ft or deeper in order to go below any surface obstructions. This piece of equipment is called a horizontal directional drill, and it uses a hydraulic system to power a drill bit on a horizontal path approximately 3 ft or more below the ground. The horizontal directional drill uses 10-ft sections of pipe to push the drill farther and farther toward its target. The technician checks the location, direction, and depth of the drill head every 10 ft as another section of pipe is added. The operator uses a small joystick to adjust the head of the drill bit to move left, right, up, or down to ensure that the drill and pipe continue on in the exact direction and depth. A small hole approximately 18 in. in diameter and 3 ft deep or deeper is dug at the target location, and the pipe and drill are pushed into the ground until it emerges in this hole. Once the drill head reaches the target hole, a large spool of plastic conduit is connected to the drill head and is pulled back through the tunnel the drill created. The pipe for the horizontal directional drill is pulled back, and each time a 10-ft section becomes free, it is unthreaded from the long section and placed back on the magazine for the directional drill so it will be ready to be used for the next session. When the last section of pipe is pulled from the tunnel, the plastic conduit is inserted in place from point to point, waiting for final connection. The plastic conduit has a pull cord inserted through its complete length when it is manufactured, and this cord is used to pull the wires through the conduit to complete the installation of the underground service. The directional drill can tunnel under driveways, trees, or other things to bury conduit and wire underground. Underground installation of wiring is more expensive than aboveground installation, but it is able to withstand problems such as ice storms, tornadoes, and other events that may bring down power poles.

8.16 CABLE INSTALLATION

The cables that connect a wind turbine generator to the grid conductors must be run either underground or overhead. If these cables are run underground, connections are made at service points along the extent of the conduit, where they can be opened and inspected later if there is a problem. If the wire is run overhead, the connections are made at power poles and transformers. Utility companies and companies that provide installation and service of high-voltage lines have been in business for many years and are very good at this job. If they are making service connections overhead, they will use trucks with hydraulic lifts and buckets to lift the service technicians to the level where the connections are made. They also have service equipment that is specialized to help make underground installation simpler. Many times this type of cable installation work is contracted out to companies that have specialized equipment and technicians with specialized training.

FIGURE 8-14 Wooden power poles and metal electrical transmission towers bring power into and out of electrical substations.

8.17 OVERHEAD FEEDER CIRCUITS

Wooden power poles and steel electrical transmission towers are used to bring high-voltage power lines into and out of substations. Figure 8-14 shows an example of this mixture of power poles and transmission towers. The overhead feeder circuits must be high enough above the ground to allow service vehicles to move freely underneath without touching them. These overhead power lines must also be supported between the poles and towers so that the lines do not group or touch each other in strong winds. Another reason the lines are run overhead in free air is that the air surrounding the wires will help keep them cool, which allows them to be loaded to the maximum extent.

Overhead feeder circuits consist of electrical wire that is supported by power poles. The power poles typically run close to streets and highways, where the utility company owns a small strip of land called a right-of-way. Any trees that are located in this right-of-way under the power poles must be trimmed and maintained to ensure that they do not bring down power lines in a storm or in heavy ice and snow. Figure 8-15 shows power lines connected to poles for support. In this figure you can also see how close the power poles are to the roadway. Each residence or commercial application that needs electrical power will have a connecting line from the main wires to the building. The wires that connect from the building to the main power lines are called a **service drop**. You can see a service drop connection right below the three transformers on the power pole. Originally these poles were called telephone poles because telephone lines were connected to these poles as well. Today, electrical power lines, service drops, telephone lines, and cable television lines all may be mounted on the same pole at different heights. The feeder circuits run electrical wire from the substation throughout the area so that each residence, commercial establishment, and industrial site can be connected to the electrical grid.

FIGURE 8-15 Overhead power lines are run along streets and highways where utility companies own right-of-way strips of land.

8.18 WIND FARM SUBSTATIONS

When a large number of wind turbines are located on a wind farm, an electrical substation must be provided for the step-up transformers needed to boost the voltage for long-distance transmission. The substation will be near where the wind turbines create electrical power. Figure 8-16 shows a substation on a wind farm. All the electrical wiring connections between the wind turbines and the substation are made through underground conduits. The high-voltage lines leaving the substation connect to the grid above ground, on overhead wires. The high-voltage lines then use existing high-voltage transmission towers and lines to distribute the electrical power to the cities and industries that need the power. The substation at the wind farm may also house all of the electrical disconnect switches as well as safety fuses and circuitry.

FIGURE 8-16 Substation for wind farm. (Courtesy of John Fellhauer, jfellhauer@surenergy.us.)

8.19 CONNECTING TO RESIDENTIAL OR COMMERCIAL SINGLE-SOURCE POWER SYSTEMS

When a wind turbine is connected directly to a residence or commercial location, it is called a *single-source power system.* Figure 8-17 shows the electrical hardware for a single-source power system. The power lines from the wind turbine are installed in underground conduits and come up into the switch and metering boxes at the bottom. If backup power from the grid is provided, it comes in overhead, as a normal residential or commercial location would. You can see the switchgear, metering boxes, and overhead service from the grid in the picture in the figure. The switchgear and net meter must be provided when the wind turbine is installed, and need to be included in the cost of installation for the wind turbine. The net meter will run in one direction when the wind turbine is providing power to the grid, and it will run in the opposite direction when the application uses electrical power from the grid rather than from the wind turbine.

FIGURE 8-17 Electrical disconnect and metering components for a commercial application. The power from the wind turbine comes in underground, and electrical service from the grid comes in overhead.

Questions

1. Explain what the grid is and how it works to distribute electrical power to customers.
2. Explain what the smart grid is and how it is different from today's grid.
3. Identify the three main sections of the electrical grid.
4. Identify the federal law that requires utility companies to purchase power from wind turbine generators, and state what year the law was passed.
5. Explain how a transformer works.
6. Identify the two windings of a transformer.
7. Explain what a substation is and why it is important to the grid.
8. Identify three things that the IEEE 1547-2003 standards address for technical requirements for power that is connected to the grid.
9. Explain what a net meter is.
10. Identify two things that are addressed in power quality issues.

Multiple Choice

1. The grid is
   a. The entire electrical power distribution system in the United States and the rest of North America
   b. The electrical connection between a wind turbine generator and the wind turbine tower
   c. An electrical connection that exists only between wind turbines in a wind farm
2. The smart grid
   a. Is exactly like the regular grid and can only deliver electrical power
   b. Has the ability to send and receive information to and from each customer along with electrical power
   c. Filters and purifies the voltage as it is delivered
3. Transformer windings include
   a. The first and second windings
   b. The input and output windings
   c. The primary and secondary windings
   d. All the above
4. The turns ratio for the transformer affects
   a. The number of turns in the primary and secondary windings
   b. The ratio of voltage in the primary and secondary windings
   c. The ratio of the current primary and secondary windings
   d. All the above
5. Concerning step-up and step-down transformers, __________
   a. The voltage in the step-up transformer is higher in the primary winding than in the secondary winding.
   b. The voltage in the step-down transformer is lower in the secondary winding than in the primary winding.
   c. The voltage in the step-down transformer is higher in the secondary winding than in the primary winding.
   d. All the above.
6. How do the terms *power factor, true power,* and *apparent power* relate to each other?
   a. The power factor is the true power divided by the apparent power.
   b. The power factor is the apparent power provided by the true power.
   c. The power factor is the true power multiplied by the apparent power.
   d. All the above.
7. Low-voltage ride-through (LVRT) is
   a. The ability of an electrical generation system such as a wind turbine not to fault out when the voltage on the grid gets too low
   b. A special circuit in a wind turbine generator that ensures the proper voltage is supplied when the wind turbine is not connected to the grid
   c. A special circuit in a wind turbine generator at causes it to trip off when voltage gets too high
   d. All the above
8. Flicker occurs
   a. When a wind turbine generator speeds up or slows down with wind gusts and increases or decreases voltage
   b. When the voltage in the grid causes the wind turbine to speed up or slow down
   c. When the power switch that connects the wind turbine generator to the grid turns on and off quickly
9. Grounding is
   a. A copper rod inserted into the earth up to 8 ft down.
   b. An electrical connection of all metal parts of a wind turbine to a system of copper conductors that terminates at copper rods inserted into the earth
   c. A safety circuit that helps to protect the wind turbine from lightning strikes and short circuits
   d. All the above
10. The disconnect switch for the wind turbine generator
    a. Connects or disconnects the electrical circuits of the wind turbine generator to the grid
    b. Adjusts the amount of mechanical power that is transferred from the gearbox to the generator
    c. Connects or disconnects the mechanical power to the generator in a wind turbine

CHAPTER 9

Types of Towers, Tower Designs, and Safety

OBJECTIVES

After reading this chapter, you will be able to:

- Explain how a large monopole steel or cast concrete tower is assembled.
- Identify the safety gear that people are required wear if they work above the ground.
- Explain what a monopole wind turbine tower looks like.
- Explain what a lattice wind turbine tower looks like.
- Explain the areas of safety that OSHA covers for wind turbines.
- Explain what guy wires are and how they are installed.
- Explain how the gin pole and tilt-up tower work to erect a wind turbine pole.
- Identify the steps required to erect a lattice wind turbine tower.
- Explain why a crane must be taller than the wind turbine tower it is assembling.
- What is the role of the FAA in deciding where a wind turbine should be located.
- List several ways you could get to the top of a monopole wind turbine tower.

KEY TERMS

Automatic descent
Carabineer
Climbing safety
Cube-type wind turbine
Electromagnetic interference
Evacuation
FAA
Fail-safe system
Fall protection
Gin pole
Ground-fault protection
Grounding
Guy wire
Lanyard
Lattice tower
Lightning arrestor
Monopole tower
OSHA
Rescue equipment
Safety brakes
Safety harness
Service platform
Shut-down brakes
Surge protection
Tilting-type tower

OVERVIEW

This chapter will provide information about towers used for wind turbines. Since the wind typically blows harder at greater heights, the taller the tower, the more energy the turbine can produce. For example, suppose that a wind turbine at the top of a 60-ft tower produces 50 kWh. If the same wind turbine is moved to the top of a taller, 100-ft tower, it might produce 65 kWh, which is 30% more power. A wind turbine on a taller tower can capture more wind, but it also puts more stress on the tower and the foundation that supports the tower. Also, it is more difficult to perform maintenance on wind turbines set on taller towers. This chapter will describe the different types of towers that are used in the industry today and will explain how they are mounted and maintained. Some of the pro's and con's for each type of tower will also be presented. This chapter will provide information about towers for small, medium, and large wind turbines, as well as for wind turbines located at sea or above other water bodies.

You will also see how technicians learn to work in and around these towers. Information will be provided about the cranes and other equipment used to work on wind turbines mounted at varying heights. You will see the safety gear that is used by technicians when they are climbing and working above the ground, and you will learn about the rules and regulations that pertain to working around towers and above the ground and the regulatory organizations that maintain control of these issues. You will also learn about rules and regulations that address where wind turbines can be located.

In this chapter you will also learn about the types of maintenance that require technicians to climb a tower to get to the nacelle or to service parts of the system that are above ground. You will learn about guy wires that are used to make towers more sturdy, how to check the guy wires periodically, and which towers do not need guy wires. Towers and tower maintenance is not new or unique to wind turbines. Very high towers have been used for years and are used today to support equipment for electrical power transmission towers, radio and TV transmission, and cellular telephone service. Some of the same safety issues carry over from these types of towers, and similar technical equipment is used for maintenance on these towers and on wind turbine towers.

FIGURE 9-1 A lattice wind turbine tower.

9.1 TYPES OF WIND TURBINE TOWERS

Types of towers for wind turbines range from simple tubular steel pipe for smaller wind turbines to larger pipe for midsize towers. Some towers for small residential wind turbines are small enough that they can be erected with the wind turbine already installed on them using a set of pulleys and ropes. Towers for larger residential or small commercial wind turbines are larger, and a small tower truck or small crane is required to erect them. Because some smaller wind turbines must be lowered to the ground to perform maintenance, a tilting type of tower may be used. The **tilting-type tower** is designed with a tilting mechanism that secures the pole at its base and allows the tower to be lifted into place until it is upright. The tower is then secured with **guy wires**. When the wind turbine needs inspection or maintenance, the tower is lowered until the unit is on the ground, where it can be worked on safely.

Another type of tower, called a **lattice tower**, and is usually a three-legged tower that is put together in sections, with the wind turbine mounted on its top section. Figure 9-1 shows an example of a lattice tower. The lattice tower is an open tower that provides a steplike structure that is easy to climb when maintenance or inspection is required. The lattice tower does not usually need guy wires to support it, because its wide base keeps it from moving even in strong winds. The largest wind turbines usually do not use lattice towers, however. Instead, they use **monopole towers** (monopole = one pole), which are much taller and may range from 60 to 100 m in height.

Tubular Steel Towers for Small Wind Turbines

The tower for a small wind turbine is generally constructed from tubular steel that is between 1.5 and 3 in. in diameter. The larger the wind turbine blades and generator, the larger the pipe diameter must be. Figure 9-2 shows an example of this type of tower. For small wind turbines, the height of the tower is usually between 30 and 90 ft. The height depends on the location and any

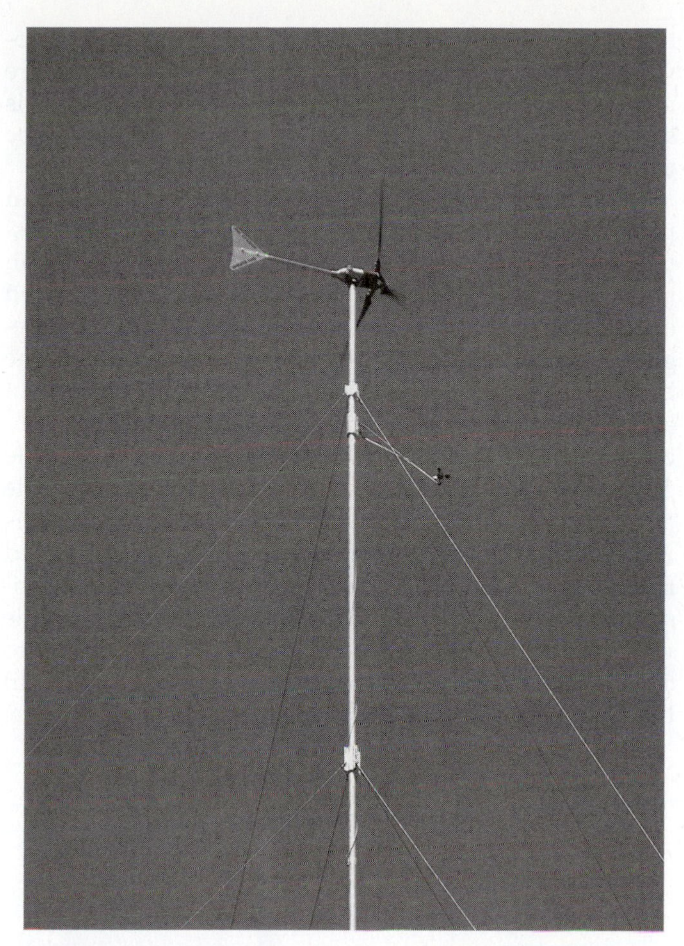

FIGURE 9-2 A tubular steel tower for a small wind turbine. Notice the number of guy wires attached along the length of this tower to keep it secured in the strongest winds. (Courtesy of Southwest Windpower.)

restrictions imposed by zoning. The wind turbine is mounted as high as possible, or as allowed, to harvest the maximum amount of wind.

In Figure 9-2 you will notice that guy wires are connected to the tower at various levels on the tower. The guy wires are pulled outward and down until they are connected to the ground to provide stabilization for the tower and wind turbine. The guy wires are connected at the top and middle of the tower to provide optimum support to the tower. There may be t four or more guy wires attached to each section of the tower to provide support, so that the tower does not bend or move during the highest winds. In the figure you can see three sets of guy wires connected along the length of the pole. The guy wires must be inspected periodically to ensure that they have not come loose or begun to corrode. The guy wires and tower must also be inspected periodically to ensure that the tower is positioned as close to the vertical position as possible. If one set of guy wires becomes loose, it will allow the tower to move out of it plumb vertical position. It is also important that the guy wires are protected on the ground so that people will not trip over them when they are working in the area, and so that lawn mowing machinery does not run over them. In some cases this means that a fence must be placed around the perimeter of where the guy wires are mounted into the ground.

Since the height of some of these towers is close to 100 ft, the pipe for the tower must be put together in sections before it is erected. Generally, for smaller wind turbine installations, the wind turbine is mounted securely to the tower while it is on the ground, and then a crane or block and tackle is used to pull the tower into the upright position. When the tower is upright, it is positioned into its base. The tower base and concrete pad must be laid out and constructed several days in advance of mounting the tower in place, to allow the concrete slab and base to set up so they do not crack when the tower is put in position. It is important to understand the amount of stress that is put on the base once the tower is put in position and the wind begins to blow against the wind turbine. One way of thinking about this is that if the wind turbine is harvesting wind energy at 10 kWh, this means that that amount of mechanical energy is pushing against the blades and tower and trying to push it over. The only thing that is holding the wind turbine into position is the tower, the base and slab, and the guy wires. If these are not of sufficient size and stronger than the strongest winds, the wind turbine will sustain damage when the wind is too strong. It is also important to understand that the tower is designed for the strongest winds that the wind turbine will sustain.

Cylindrical Pipe for Wind Turbine Applications

For the majority of wind turbine installations, the pipe that the wind turbine pole is made of is designed specifically for the wind turbine application. Some small residential systems have used any pipe that was available to keep the cost down, and this created problems when the pipe was not strong enough to hold the wind turbine in position during the strongest winds, even with guy wires attached. Today, with the increase of smaller residential wind turbines, pipe designed specifically for wind turbine applications is available.

The pipe comes in sections of 8–10 ft and are threaded together so that you can make the tower any size from 20 to 100 ft. The wind turbine generator, with the blades connected to the hub, is mounted on the end of the last section of pipe. When the wind generator is being connected to the last section of pipe, the pipe needs to be positioned about 5–8 ft off the ground, far enough that the blades do not touch the ground or become damaged when they are being attached to the top of the pole. Once the generator and blades are secured to the end of the pipe, the end of the pipe that goes into the ground is moved into or near the base so that the crane or lifting device can erect the tower. After the tower is pulled into a vertical position, the guy wires can be connected and tensioned to ensure that the tower is secure.

Tilting-Type Towers

Tilting-type towers are designed specifically for applications where a smaller tower and wind turbine are going to be erected and later the system will need to be lowered for periodic maintenance. The tilting mechanism is mounted into a slab that has been constructed specifically for the size of the tower it is to support. When the slab is poured with concrete reinforcement bar, the bolts that the tilting mechanism will use to secure it to the slab are positioned accurately. This ensures that when the concrete slab is poured, and cures up, the bolts will be in the correct position to secure the tilting mechanism. After the concrete has cured for several days, the tilting mechanism can be mounted to the bolts that have been poured into the slab, and nuts can be screwed down on the bolts and torque to ensure that the mechanism is held tightly in position. Later the tower base is bolted or securely attached to the tilting mechanism.

Some larger tilting-type towers may need the aid of a small crane to lift the tower to the vertical position. Other, smaller tilting-type towers are designed to be raised to the vertical position with a winch, block and tackle, or other mechanical means after the tilting mechanism is secured to the bolts in the concrete slab or base. The next step is to locate the tower into position in the tilting mechanism and bolt it securely to the tilting mechanism. Once the bottom of the tower is secured into the tilting mechanism, the tower can be raised slightly to provide clearance when the wind turbine and its blades are mounted to the top of the tower. After the wind turbine is secured to the top of the tower, a winch or crane can be used to pull the tower into the vertical position.

Another type of tilting mechanism is called a **gin pole**. Figure 9-3 and Figure 9-4 show two examples of raising a small wind turbine tower with a gin pole. The gin pole is an arrangement where an additional pole is placed on the base of the main wind turbine tower pole at 90 degrees. When the main wind turbine pole is moved into the base, it will be located parallel to the ground and the right angle gin pole will be pointing upward. A winch or vehicle can be used to pull the right-angle arm of the gin pole downward, which will cause the main wind turbine pole to move into the vertical erect position. When the tower is in the vertical position, the final hardware can be inserted in the tower in the tilting mechanism and tightened to ensure that the tower stays in a vertical position. After the wind turbine pole is in the vertical position and bolted into the concrete base, the gin pole is removed and stored until the wind generator needs to be lowered for maintenance.

FIGURE 9-3 A gin pole is used to move a small wind turbine on its pole into the vertical position. (Courtesy of NREL, U.S. Department of Energy.)

FIGURE 9-4 A gin pole is used to pull a wind turbine on its pole into its vertical position. (Courtesy of NREL, U.S. Department of Energy.)

If the tower needs guy wires, the guy wires are attached to the wind turbine pole while it is still on the ground. After the tower is erected, the guy wires are pulled into their final positions and secured to their ground hardware. Once the wind turbine pole is nearly vertical, the final adjustments to the guy wires can be made to position the tower in an exact vertical position. Also, the guy wires can be tensioned properly to ensure that they hold the wind turbine in the proper position during the strongest winds.

Guyed and Nonguyed Lattice Towers

A lattice tower is an open-type tower and is generally a three-legged tower. Figure 9-5 shows a lattice-type tower. The term *lattice* comes from the appearance of the metal structure of the tower and cross members that make it structurally sound. You can see that this tower looks similar to a large radio or TV transmission tower, and it provides a framework that technicians can use to climb the tower to work on the wind turbine at the top. The lattice-type tower has three rigid polls that are interconnected with metal braces, which give it its strength. The braces tie together to form triangular shapes that go from top to bottom, which provide the most rigid construction between the three poles.

Figure 9-6 shows another example of a lattice tower, which you can see is very wide at the base and becomes narrower toward the top. Since the tower is mainly framework, it is one of the lighter towers used to support small and mid-sized wind turbines. It also provides a very sturdy support platform for the wind turbine, while its framework with its open sections ensures that it does not become a obstruction to the wind the way a solid tower might. This means that the wind basically goes straight through the tower and does not cause meet much resistance, which could cause maintenance and wear problems on the tower over its lifespan.

One drawback of the lattice-type tower is that it is easy to climb and therefore may pose a risk of liability if it is not secured and unauthorized persons try to climb it. Its open structure also provides locations for birds to nest, or it may become a platform for predatory birds

FIGURE 9-5 A lattice-type wind turbine tower.

FIGURE 9-6 Example of a lattic-type tower where you can see that the base of the tower is the widest part and the tower becomes narrower at the top.

FIGURE 9-7 A lattice type tower that has been assembled at the site where it will be erected. Notice the feet that provide the mounting connection for the tower. (Courtesy of John Fellhauer, jfellhauer@skurenergy.us.)

FIGURE 9-9 Electrical cable is installed inside the lattice tower while it is still on the ground. (Courtesy of John Fellhauer, jfellhauer@surenergy.us.)

such as hawks and other hunter birds, which like to wait in open areas so they can easily see their prey. Even though bird kills from wind turbines are not as big a problem as bird kills from cats that roam loose in the same areas, it is still a concern to environmental groups. Some insurance companies may require large fenced areas around the perimeter of these towers to ensure that unauthorized persons do not try to climb the tower.

The lattice-type tower is generally assembled completely at the site where it will be erected, or it may be assembled in sections and brought to the site, where final assembly of the larger sections is completed. The sections are normally bolted together to provide simpler installation. Figure 9-7 shows a lattice-type tower that has just been assembled and is still lying on its side. This picture shows the feet of the tower that have mounting hardware welded to them.

Figure 9-8 shows technicians mounting a wind turbine on the top of the tower just prior to erecting the tower.

FIGURE 9-8 Technicians are mounting a wind turbine without its blades to the top of the lattice tower. The blades will be installed while the tower is still on the ground. (Courtesy of John Fellhauer, jfellhauer@surenergy.us.)

The small wind turbine on top of the tower is lifted several feet in the air to provide space to mount the wind turbine and its blades. A brace structure is used to hold the top section of the tower about 5–8 ft above the ground while the wind turbine is mounted at the top of the tower. The blades can be connected to the turbine before it is connected to the tower or after it is connected to the tower. You can see in this figure that the wind turbine is mounted to the tower and secured without the blades installed. The blades will be installed after the wind turbine is secured to the top of the tower, while it is still on the ground. On some models, the blades are attached to the wind turbine before the wind turbine is attached to the tower.

While the lattice-type tower is still on the ground, other installation tasks can be completed, such as feeding the power cable from the wind turbine through the tower to the base. Figure 9-9 shows the electrical power cable after it has been placed inside the tower while it is on the ground. Since the wind turbine and the lattice tower are lying on the ground, this is a simple task. If the lattice tower was already erected, this task would be much more difficult because the weight of the power cable would have to be supported as someone climbed the tower and carried it to the top. It is also easier to make the electrical connections and other hardware connections such as tightening nuts and bolts while the tower is on the ground. Once the tower has been erected, these simple tasks become much more complex and time-consuming. The power cable can also be secured to the sides of the tower while it is on the ground prior to being erected. At this time a restraining cable is placed around one or more of the wind turbine blades and secured to the top of the tower to prevent the blades from rotating and becoming damaged while the crane is lifting the tower into the vertical position.

After technicians have installed the wind turbine to the tower and have attached the blades to the rotor hub, they will place a lifting harness or sling around the

FIGURE 9-10 The technician secures the lifting harness around the wind turbine and the top of the lattice tower to prepare it for lifting by the crane. (Courtesy of John Fellhauer, jfellhauer@surenergy.us.)

FIGURE 9-12 Technicians quickly install nuts and washers on the bolts in order to secure the feet of the lattice-type tower once it is in the vertical position. (Courtesy of John Fellhauer, jfellhauer@surenergy.us.)

wind turbine and the top of the tower and make it ready to be lifted to its vertical location. Once the lifting harness is secured, it will be connected to the cable on the crane and then will be ready to be lifted into position. Figure 9-10 shows a technician securing the lifting harness to the wind turbine and the top of the tower to make it ready to be lifted into place by the crane. Figure 9-11 shows the crane beginning to lift the top of the tower

FIGURE 9-11 The crane begins to lift the lattice-type tower into the vertical position. Notice that the wind turbine is mounted on the top of the tower. (Courtesy of John Fellhauer, jfellhauer@surenergy.us.)

and the wind turbine to get the lattice type tower into a vertical position. Once the crane has the tower completely vertical, technicians will use cables or ropes to guide the tower over its mounting bolts. Once the tower is in close proximity to its mounting bolts, the technicians will signal the crane operator to begin to lower the tower. Technicians will make minor adjustments to the position to ensure that the legs of the tower line up over the bolts, and that the bolt holes in the mounting brackets on the feet of the tower are correctly positioned over the bolts that have been secured in the concrete pad. As the tower is lowered into position, technicians install nuts and washers over the bolts and begin the process of tightening them down and lowering the tower into its final resting position. Figure 9-12 shows technicians installing the nuts and washers on the bolts. The nuts are tightened and torque to ensure the feet are secured to the concrete pad.

After the tower has been securely bolted to the pad, a technician will climb the tower and remove the lifting harness from the wind turbine at the top of the tower. At this time he will also remove the restraints from the blades that were used to keep the blades from rotating while the tower was being erected to the vertical position.

Some lattice-type towers are not wide enough at the base to provide a rigid platform for the wind turbine. This means that in strong winds, the top of the tower may move slightly when the blades are harvesting the maximum amount of wind energy. Guy wires are used in these applications to make the tower more rigid and to allow the wind turbine to operate in stronger winds. Figure 9-13 shows an example of a lattice-type tower with guy wires. This type of tower is called a guyed lattice tower. The guy wires are connected to the lattice tower while it is still on the ground, and they are pulled out into position as the tower is erected into the vertical location. If the tower has 8 or 12 guy wires, several technicians will be

FIGURE 9-13 A guy lattice tower. The guy wires make the lattice tower become more rigid. (Courtesy of NREL, U.S. Department of Energy.)

FIGURE 9-14 Large horizontal-axis wind turbines mounted on monopole towers.

needed to help keep the guy wires from becoming tangled while the tower is being raised to the vertical position. The ends of the guy wires that are connected to the ground must be pulled outward near to their final connection position to keep them from tangling with the wind turbine blades or the lattice tower. Once the lattice tower is near its vertical position, the guy wires can be connected to the hardware that is embedded in the ground. At this time the guy wires must have enough slack to allow the tower to be moved slightly to its perfect upright position. Tensioning the guy wires may take several hours to ensure that the tower is perfectly positioned upright, and that the cables are providing sufficient tension to ensure that the lattice tower provides a stiff base for the wind turbine. These guy wires will need to be periodically inspected for corrosion and to ensure that they are at the right tension. Over time, the wind will cause the tower to move slightly, which will cause the guy wires to stretch slightly. When this occurs, the guy wires will have to be retensioned to ensure that they are providing structural integrity to the tower.

Monopole Towers

Larger wind turbines must be mounted at heights of approximately 50–80 m (164–262 ft) in order to harvest the maximum amount of wind energy. These types of wind turbines use a tower called a *monopole tower* ("mono" means singular or one). Figure 9-14 shows several wind turbines mounted on monopole towers. The size of these wind turbines is approximately 1.8 MW, and they are mounted at over 250 ft in the air, which is high enough above the tree line to ensure that they harvest the maximum amount of wind. Since these towers must support the weight of the large nacelle and blades of the wind turbine as well as keep the wind turbine in position during strong winds as it harvests large amounts of wind energy, the tower must be very strong and sturdy from its base to its top. Some monopole towers are made of steel and are hollow inside so that an access ladder or stairwell can be installed to allow technicians to climb to the top of the tower to enter the nacelle for maintenance. Other monopole towers are made of cast concrete.

Figure 9-15 shows another a monopole tower. In this figure you can see that the base of the tower is larger than the top of the tower. The base is wider to provide more rigid support for the tower to keep it in place in strong winds. The top of the tower is narrower to make the tower weigh less overall yet still provide strength and stability at the very top of the tower.

Larger Monopole Towers

The largest wind turbines in the world are mounted on monopole towers. One of the larger wind turbines in Europe is the Enercon E-126. This turbine has a rotor blade width of 126 m (413 ft) and is installed in Germany. The tower height is over 131 m (429 ft), and it is made of 36 segments of precast concrete. The overall height of the wind turbine is 198 m (649 ft). Larger monopole towers can be made of precast concrete or structural steel. LINK Figure 9-16 shows the tower sections being lifted into place by a very large crane as a monopole tower is being built on site. Figure 9-17 shows the relative size of the tower's base compared to the height of a man

FIGURE 9-15 The base of the monopole tower is much wider than the top of the tower. The wide base provides a sturdy structure to ensure that the wind turbine is able to handle the strongest winds.

FIGURE 9-17 The base of the tower for a large wind turbine. Notice the size of the base compared to the technicians working near it. (Courtesy of NREL, U.S. Department of Energy.)

walking nearby. Figure 9-18 shows the large crane used to place the rotor on the nacelle, which is already mounted on the top of the tower, and Figure 9-19 shows the installation of a wind turbine that is finished, with the rotor and blades mounted on the nacelle, all at the top of the tower.

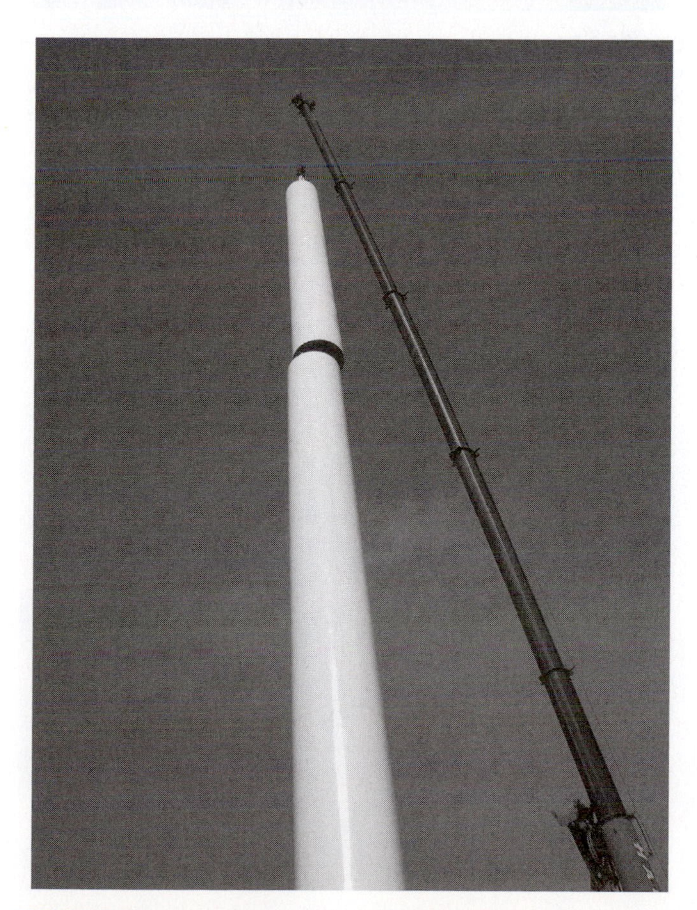

FIGURE 9-16 Large crane used to assemble large tower. (Courtesy of Northern Power Systems Inc.)

FIGURE 9-18 Blade being installed on a very large wind turbine. (Courtesy of Siemens Wind Energy.)

FIGURE 9-19 Large horizontal wind turbine and tall tower. (Courtesy of Northern Power Systems Inc.)

FIGURE 9-20 A large crane is used to place sections of steel tower on each other. Each section is mated to the one below so that the holes in the flanges line up for bolts and nuts to secure the two sections. (Courtesy of NREL, U.S. Department of Energy.)

Larger Steel Tower Sections Need to Be Bolted Together

If the tower is made from steel, it must be placed section upon section, and each section needs to be bolted together to secure it. Figure 9-20 shows a large crane placing a section of a tower on top of another section. When the two sections are fitted together, they each have a flange that lines up with a large number of bolts holes. Figure 9-21 shows the inside of a steel tower. The arrow points to the bolts that hold the flange of each tower section together. The bolts are placed through the holes and have nuts threaded on them.

If the sections of the tower are not connected together correctly, the tower may collapse or not provide the proper amount of support for the wind turbine. If the tower is not designed correctly or if it is not installed correctly, it may fail and damage the blades and nacelle. The bolts and nuts that are used to connect some steel towers must be inspected periodically for corrosion, lack of tightness, or any other problem which might allow them to fail. After the tower is installed and the sections are connected together, the nacelle is mounted to the tower, the rotor is lifted into place at the front of the nacelle, and it is connected to the low-speed shaft. When the rotor has been secured, the blades can be lifted into place one at a time and connected to the rotor. Figure 9-22 shows a crane lifting the blades into place during an installation, which will then be bolted to the rotor. You will learn more about the equipment inside the tower, such as the stairs or ladder and the equipment that secures the

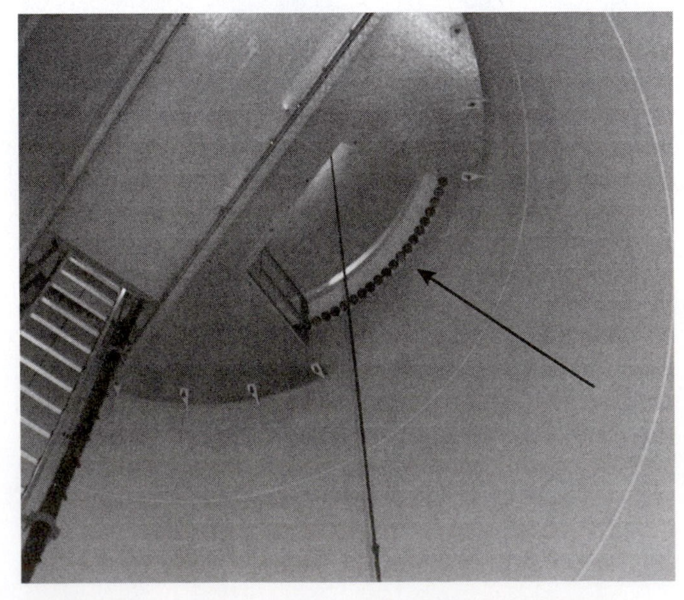

FIGURE 9-21 The arrow points to the bolts that connect the upper and lower sections of this steel tower.

FIGURE 9-22 Large crane used to lift blades to rotor. A technician bolts the blades to the rotor when it is in position. (Courtesy of Nordex.)

power cables as they connect from the generator down to the base of the tower, later in this chapter.

Lifting Equipment and Cranes for Installing and Erecting a Tower

When tall towers are used to support the largest wind turbines, they require tall cranes that can reach the top of the wind turbine are strong enough to lift the nacelle and blades into place. Figure 9-23 shows a tall crane that has a middle section that is operated by a hydraulic boom. The top of the crane is connected to the boom section, and when the crane is in position, the boom is extended by a large hydraulic cylinder until the top of the crane is above the wind turbine. The crane has a set of cables that do the actually lifting of the blades or maintenance parts to the top of the wind turbine tower. The crane must be so large because it needs to be taller than the wind turbine it is working on and it must be capable of lifting the heavy tower sections, nacelle, and turbine blades into position at great heights.

FIGURE 9-23 A tall crane with a lower-section boom that is extended with hydraulics to reach the top of a tall wind turbine.

When the crane has finished its work, it must be lowered and taken apart so that it can be transported by highway to the next site. Some previous figures, such as Figure 9-18 Figure 9-20, show large cranes, and you can see how much larger than the wind turbine the cranes must be. It is important to understand that these larger cranes are very specialized and very expensive to move from site to site. The cost of leasing a crane for installation or maintenance may be set by the job, or its cost may be determined by the day or hour. On larger wind turbine farms the crane may be designed to remain on site continually, as it is used on many different wind turbines on the same site.

Figure 9-24 shows a mid-size hydraulic crane that has the height and the power to lift small and medium-size monopole or lattice-type wind turbine towers into place. The monopole-type tower is a little heavier than the lattice-type tower, so a crane that has a bigger lifting capacity is required to lift the monopole-type towers. Figure 9-25 shows a smaller crane that is used to lift a small lattice-type tower. Since the lattice tower is much lighter, the crane's lifting height will be more important than its ability to lift weight. When you are ready to lift a tower into place, you will need to hire a crane that has the ability and size to complete the lift safely. You will need to show the crane operator the data for the weight of the

FIGURE 9-24 A mid sized hydraulic crane, which has the height and power to lift small and medium-size monopole wind turbine towers. (Courtesy of John Fellhauer, jfellhauer@surenergy.us.)

FIGURE 9-25 Smaller crane used to lift a lighter lattice-type wind turbine tower. (Courtesy of John Fellhauer, jfellhauer@surenergy.us.)

equipment you intend to lift into place, or you will need to gather the data from the technical specifications for the tower and take them to the crane operator when you put in the request for the crane. It is very important that the crane has excess capacity when it comes to lifting a tower into place. When the crane's boom is fully extended and is tilted at a 45-degree angle, the weight of the load will put the maximum amount of torque on the boom. You will have to trust the judgment of the crane operator on these matters so that you do not put the wind turbine at risk while you are erecting it. If your calculations are incorrect, the crane can easily tip over, or lose the tower during the lift, which will severely damage the wind turbine equipment.

FAA Tower Height Restrictions

As wind turbine towers become taller, they pose several hazards to aircraft. The Federal Aviation Administration (**FAA**) has oversight of any object that is tall enough that it could have an effect on the navigable airspace or communications/navigation technology of aviation. This means that if the wind turbine tower is too tall, it may pose a threat in that an airplane may inadvertently fly into it, or it may create a problem with radar or radio communications. If an area is used for multiple wind turbines, it can pose a problem for radar, since the turbines might create a shadow that radar cannot see behind.

For this reason the FAA requires that a Notice of Proposed Construction (Form 7460-1) be filed for any object that will extend more than 200 ft above ground level (or less in certain circumstances, for example, if the object is closer than 20,000 ft to a public-use airport with a runway more than 3200 ft long). This means that any wind turbine project that has a height that is above 200 ft must file Form 7460-1. As wind turbine heights have increased during the past couple of decades, this filing requirement has applied to increasing numbers of projects. The FAA considers it important to study several types of impacts on airspace around any wind turbine: (1) imaginary surface penetration, (2) operational impacts, and (3) electromagnetic interference.

After the Form 7460-1 is filed for a given project, the FAA undertakes an initial study within the relevant FAA region and issues a report that states its findings. If the wind turbine does not pose a problem, the finding will be a Determination of No Hazard to Air Navigation (DNH), and the project will receive a "green light" and can proceed immediately. If there are potential problems with the height or the placement of the wind turbine, the finding will be issued as a Notice of Presumed Hazard (NPH). If an NPH is issued, the FAA will then initiate a more in-depth technical analysis (commonly called an extended study). The technical analysis will explain any possible problems that the FAA has become aware of with regard to the impact on either air operations or on navigation. This process may include a public comment period during which the company that is requesting to install the wind turbine can present data about the situation. The FAA will explain its concerns and try to negotiate an acceptable height and location for the project. If there is still a disagreement between the FAA and the company wanting to place the wind turbine, the FAA will issue a Determination of Hazard (DOH), which can be appealed to the FAA in Washington. If the information and data in the appeal do not change the findings and secure a Determination of No Hazard (DNH), the company that intends to install the wind turbine can bring the issue before a federal court.

The three things the FAA worries about are issues concerning *imaginary surfaces,* which is the path an aircraft uses for landing or takeoff, or when circling an airport; *operational impacts,* which include issues that will impact an aircraft using visual flight rules or instrument flight rules; and **electromagnetic interference**, which will impact the operation of radar or other communications. The issues concerning imaginary surfaces deal mainly with areas where the aircraft typically fly during takeoff and landing, which are called corridors, and areas around airports where aircraft must fly circular paths at various heights as they wait their turn to land. If the proposed wind turbine location affects any of these areas, the height of the wind turbine is very important. Operational impacts are very similar to the imaginary surfaces because the rules that pilots follow during flight, called visual flight rules (VFR) or instrument flight rules (IFR), tell pilots where they are supposed to fly any time they come close to an airport or other aircraft. The concern here is that even though a proposed wind turbine site is not directly in the flight path leading to an airport, pilots may need to divert from the airport pathways slightly if they encounter other aircraft. During these diversions, the FAA does not want pilots to have to worry about wind turbines that may be too tall and pose a hazard to the aircraft. The problem that wind turbines pose to radar and communications equipment when they are located in certain areas is that they may cause the radar signal to degrade, or they may cause electromagnetic interference to show up on the radar screen. The problems with radar tend to be limited to wind turbines that are within 5 miles of the line of sight of the radar system. Typically, if the wind turbines are smaller, or located away from aviation activity, they will not be considered a problem by the FAA.

In some cases, when a wind turbine is proposed for a site near aircraft activity, the proposers may refer to existing towers such as cell phone towers that may be in the area. If the FAA has given permission for a cell phone tower that is taller than the proposed wind turbine tower, it may speed up the application, since data for the area has been gathered for the cell phone tower. The FAA will still have to identify the impact of the wind turbine even though it may be close to these larger towers.

FIGURE 9-27 The cube-type wind turbine is mounted on a pole that is lower and closer to the roofline so there is easy access for maintenance. Notice the reinforced structure needed to support the five-bladed wind turbine.

Self-Supporting Towers (Cubes)

Another type of wind turbine that has just recently become popular is called a self-supporting tower wind turbine and is referred to as "the Cube." Figure 9-26 shows an example of a **cube-type wind turbine** mounted on a tower that is built into the superstructure of the building. The cube-type wind turbine is designed so that it has a cowling around the five wind turbine blades it uses to produce energy from the wind. Since this type of wind turbine has five blades, it can harvest more energy from existing wind than a three-bladed or four-bladed wind turbine. However, since five blades are used, the structure for the turbine needs to be reinforced more than the structure for a three-bladed wind turbine. Figure 9-27 shows a close-up of the reinforced section for this wind turbine. In the figure you can see several ladders that are used for maintenance, which are removed when the wind turbine is in service. The housing for the wind turbine is mounted on the pole, and it can be designed either to point to a fixed direction or so that it can rotate the entire cube so that the turbine blades are positioned into the wind as the wind direction changes and thereby harness the maximum amount of energy. It is also possible at some locations to stack a second cube on top of the first to harvest additional wind energy.

FIGURE 9-26 A cube-type wind turbine with five blades is mounted on a large pool.

Notice that the diameter of the pole for this type of wind turbine is almost double the size of the pole for a traditional wind turbine, since this one has five blades. Also, notice that the height of this type of wind turbine is relatively low, since it sits just above the roof line of the building. The metal cowling around the turbine blades directs the wind through the blades and allows them to harvest slightly more wind than if the blades did not have a cowling. The tower for this type of wind turbine can be built into the structural metal of the building, or the tower can go directly into a concrete foundation in the ground, and the building and roof can be built around it. In some applications the tower is mounted directly into a concrete foundation that sits in an open space inside the building, and the roof of the building is not connected to the tower but instead provides access to a catwalk and ladders. One of the positive features of this type of wind turbine is that since it is mounted lower down, near the roof line, it is more easily accessible for maintenance.

Towers Located over Water

Research has indicated that the wind over many ocean or bay locations tends to be stronger and the duration of the winds more consistent. The problem in the past has been determining a method to locate wind turbines in these

FIGURE 9-28 Wind turbines located at sea. (Courtesy of Siemens Wind Energy.)

locations over water, where the depth of the water may be anywhere from 20 ft to several hundred feet. Some countries, such as England and Norway, have had many years of experience placing wind turbines offshore in such locations. Figure 9-28 shows one such wind farm, where multiple wind turbines are mounted in a line near each other. The towers for these wind turbines must be anchored on the seabed or use some type of floating application that allows them to be in the correct position to harvest the maximum amount of wind.

Figure 9-29 shows examples of how wind turbines may be placed near or over water. The first example in the left side of this drawing shows the current technology, where wind turbines are placed onshore but near open bodies of water, where they can receive the stronger winds. The next part of the drawing shows shallow-water installations, where the wind turbine is mounted in water depths up to 200 ft with a light anchor. You can see in this drawing that one way of mounting the tower is to place it into a light foundation on the seabed.

The next type of installation is called mid-depth, and this type of installation is in water depths between 200 and 1000 ft. This type of installation requires a very large foundation that is embedded into the ocean bottom. Since this type of installation is deeper, guy wires may be used to help keep the wind turbine tower in position. The base of the pole for the wind turbine is in a concrete foundation that is embedded into the seabed. If guy wires are used, the piers for them are attached to concrete that is also embedded in the seafloor.

The final type of installation shown in Figure 9-29 is for deep-water installations over 1000 ft in depth. Since this depth is significantly deeper, it is not financially sound to create a tower that goes all the way to the bottom of the ocean. In this application a floating platform is used, and wires are used to anchor the platform into place. The floating platform can be on the surface or it can be submerged. The anchors are placed deep into the ocean floor. On some systems, the cables for the anchors are constantly adjusted to make sure the floating wind turbine platform is stabilized so the wind turbine can harvest energy from the wind.

Figure 9-30 shows wind turbines installed in deep water over the ocean. Figure 9-31 shows a wind turbine

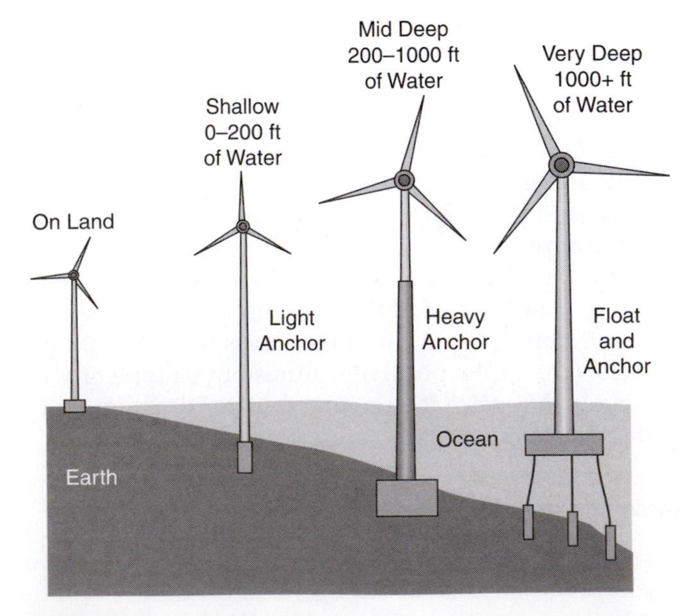

FIGURE 9-29 Placement of wind turbines near water and over water.

FIGURE 9-30 Wind turbine towers installed in 30–50 m of water. (Courtesy of Vestas Wind Systems A/S.)

FIGURE 9-31 Offshore wind turbine. (Courtesy of Siemens Wind Energy.)

mounted over water in a bay. This type of wind turbine is located in shallow water. Wind turbines can be located in all depths of water from very shallow to very deep. The reason water locations are viable is that the winds over water tend to be more continuous and dependable. One drawback of placing wind turbines over water, however, is that towers and anchors tend to oxidize more quickly, especially over salt water. Other problems include that some state codes and local codes require that the wind turbines be located *over the horizon,* so they are not visible from shore. Lake-shore and ocean-front property is very expensive, and most owners of these properties do not want to have wind turbines in their panoramic views.

Some of the problems that affect wind turbines that are located offshore include corrosion prevention and maintenance of the cables between the wind turbine and shore. Since these installations are over water or salt water, there is great concern about corrosion. Another concern for water installations is routing electrical cables between wind turbines that are located in or over water and the termination point on shore. Utility companies have a large amount of experience with underwater cables, since many power companies currently provide electric utility power through underwater cables to small islands, so some of this technology can be shared. The problems with underwater cables at sea include limiting the effects of corrosion and providing proper anchoring to ensure that cables do not move or stretch during storms.

Another problem for offshore installations is high winds due to strong storms or hurricanes. There are basically two ways of dealing with this problem: Build a wind turbine that is able to withstand very strong winds; build a wind turbine that can be quickly moved out of the path of a hurricane. The cost of either of these solutions affects the efficiency gained by placing the wind turbines over open water.

Commercial Applications Mounting Generators on Existing Lighting Poles

Another application for smaller wind turbines is to offset the cost of lighting in urban areas or rural areas. For example, automobile sales lots may need a large number of lights to illuminate the sales area in the evening. It is possible to connect a wind turbine to an existing light pole and use a battery pack and electrical inverter, or just the electrical inverter with a backup from the grid voltage. The electrical inverter converts the DC electricity from the wind turbine to AC. Figure 9-32 shows an example of a wind turbine mounted on a light pole. When the wind is blowing, the wind turbine will charge the batteries or provide voltage directly for the lights. Since the light pole is already erected and existing, the wind turbine may be able to be installed directly on the lighting pole if the pole is strong enough and has the capacity to support the wind turbine. A solar package may be added to this application to allow the batteries for the lights to be charged during the daylight hours. The lights can also be energy-efficient lights powered by DC voltage, and then an inverter is not necessary to convert the power stored in the battery back to AC voltage.

FIGURE 9-32 A wind turbine mounted directly on a lighting pole. (Courtesy of Southwest Windpower.)

If the light pole is not strong enough to support a wind turbine, or if the lighting hardware is not efficient, it may be more practical to install a new power pole with high-efficiency lighting and a battery pack for these types of applications. Some smaller lighting applications are available for residential and agricultural use where security lights are used. In these applications a small wind turbine can be used in addition to a solar pack to power the lights. Another application for this type of lighting is at toll plazas or at entrance and exit ramps from major highways as well as residential street lighting. It is possible to shift 70–80% of the energy required for street lights to wind turbine and solar packages. It may also be more practical to have one slightly larger wind turbine with a larger solar package to power four or five lights rather than have one wind turbine on every light pole.

Towers for Vertical-Axis Wind Turbines

Another type of wind turbine is the vertical-axis wind turbine. Figure 9-33 shows a wind farm that consists of only Darrieus-type vertical-axis wind turbines. The Darrieus-type wind turbines on this wind farm are installed as large wind turbines on large tower bases. The Darrieus-type wind turbine did not work out well as a large commercial electrical energy producer, so most of them have been taken down or are no longer used to produce commercial electrical power. Multiple guy wires hold the top of the tower in position. Figure 9-34 shows quietrevolution vertical-axis wind turbines. The wind turbines in this figure are mounted on monopoles that are approximately 25 ft tall, and these turbines do not need guy wires. These types of vertical-axis wind turbines do not need to be oriented into the wind to work efficiently, as their blades harvest the wind from any direction. The quietrevolution vertical-axis wind turbine also tends to operate more quietly than horizontal-axis wind turbines, so it can be used in residential or commercial applications.

FIGURE 9-33 Field of large Darrieus-type wind turbines. Notice the large tower base for each turbine. (Courtesy of U.S. Department of Energy.)

FIGURE 9-34 Quietrevolution vertical-axis wind turbines. (Courtesy of quietrevolution.)

9.2 FOUNDATIONS AND CONCRETE SUPPORT FOR TOWERS

The next area that you need to understand when working with towers is the part of the installation that supports the tower on the ground. In many small and medium-sized wind turbines that utilize a monopole, a concrete pad or pier can be used to support the tower. Figure 9-35 shows the metal forms that holds the bolts for the tower in the proper position and the wood forms that will provide the frame for the poured concrete. Notice the collar that is placed around the bolts to ensure that they will be located in the proper positions once the concrete is poured and cures. Engineering calculations determine the size and the depth that the concrete pad needs to be in order to support the size of the tower and wind turbine that will be bolted onto it. In some cases a soil sample may need to be taken to ensure that the soil will support the concrete pad and the tower. If the soil is sandy or soft, the concrete pad may need to be larger and deeper to provide the proper support. In some cases a deep concrete pier may be required to support the tower.

FIGURE 9-35 Metal form and bolts for a concrete pad for a wind turbine. Notice the metal ring that holds the bolts in the proper location while the concrete is poured. (Courtesy of Northern Power Systems Inc.)

FIGURE 9-37 Metal reinforcement rods are positioned and connected together prior to pouring the concrete for the pad. (Courtesy of NREL, U.S. Department of Energy.)

Figure 9-36 shows the same concrete pad as in Figure 9-35, but this photo shows the concrete pad after the concrete has been poured and cured. Notice the locations of the bolts and that they are exactly positioned to accept the tower. The bolts are held in place by a metal plate called a template, which is used to secure the bolts when the concrete is poured. The size of the pad may be deceptive in that only a small portion appears above the ground surface, but the entire depth will be sufficient to support the wind turbine tower.

Figure 9-37 shows the metal reinforcement material in place in the hole that has been excavated for the pad that will support a large monopole tower. The wire reinforcement bars are wired together or welded together to ensure that the foundation of this pad will be strong enough to support the tower in the strongest winds it will endure. Engineers have determined the proper depth for the soil where the tower will be located. In this large installation, the outside of the excavated area may be used to contain the concrete, or wood forms may be used. When the concrete is poured for this type of pad, it will be worked into the iron reinforcement bars thoroughly so that they will be able to provide a structure that is strong enough to support the tower during the strongest winds. Also notice in this figure that a metal ring is used to hold the bolts in the proper location so that they will match up exactly with the tower base that will be lowered onto them. It is vitally important that all of the bolts are aligned properly while the concrete is being poured, or the tower base will not fit properly onto the bolts. After the concrete is poured and has set up, the dirt around the pad will be filled in and tamped to ensure that the foundation will support the wind turbine tower. Figure 9-38 shows the finished poured pad for the monopole tower.

FIGURE 9-36 The finished pad after the concrete has been poured in the form that shown in Figure 9-35. Notice that the bolts are in the proper position to accept the tower base. (Courtesy of Northern Power Systems Inc.)

FIGURE 9-38 The finished pad, after the concrete has been poured. Only a small portion of the pad will be visible above the ground once dirt is filled in on the pad. (Courtesy of NREL, U.S. Department of Energy.)

FIGURE 9-39 The majority of the concrete pad for this monopole installation is below the surface. The small amount showing ensures that less of the concrete is exposed to the elements. (Courtesy of John Fellhauer, jfellhauer@surenergy.us.)

FIGURE 9-41 Footpad of the lattice-type tower aligned correctly over the bolts thatwere poured into the concrete pad. A metal template is used to hold the bolts in the proper location while the concrete is poured.

Figure 9-39 shows the monopole installation on a small concrete pad. It looks like the pad is very small for this large monopole, but in reality there is a large volume of concrete under the surface that holds the pole in place during the strongest winds. One reason for burying the majority of the concrete is to ensure that only a small amount is exposed to the elements. Having a smaller amount of concrete showing above the surface also makes the wind turbine installation more esthetically pleasing. Figure 9-40 shows a larger pad that is designed to support a lattice-type tower. In this figure you will notice that the electrical panel and the conduit for the electrical wiring have been buried and poured into the pad. The bolts that will line up with the feet at the bottom of the tower are also aligned and poured into the pad, so that when the tower is set on the pad, it will fit exactly over the bolts. Figure 9-41 shows the feet of the tower sitting over the bolts, and nuts are connected to the bolts to secure the lattice tower base into place. You can see why it is so important to have a template to secure the bolts in their exact position when the concrete is poured, because they cannot be moved after the concrete cures.

FIGURE 9-40 Concrete pad to support a lattice-type wind turbine tower. Notice that three sets of bolts have been poured into the concrete slab. The electrical connections for the wind turbine are also poured in place when the pad is poured. (Courtesy of John Fellhauer, jfellhauer@surenergy.us.)

9.3 CLIMBING TOWERS

At times it will be necessary to climb to the top of a wind turbine tower to work on the blades, generator, gearbox, or other equipment in the nacelle or on the top of the wind turbine. If the tower is a lattice-type tower, you will climb the lattice framework of the tower and tie yourself to the tower with a safety harness. If the tower is a midsize monopole, it may be large enough to have an interior ladder or stairway that allows you to climb to the top of the tower. On the very largest wind turbines, stairwells or ladders are provided for you to climb to the top to work in the nacelle. This section will provide information about climbing towers to the top and explain the types of safety harnesses and safety equipment that you will be required to use.

Inside Stairs and Ladders

Figure 9-42 shows the inside of a medium-sized monopole tower. This picture was taken at the base of the tower looking up toward the top of the tower. You can see a permanent ladder attached to the outside wall, and you will attach your safety harness to the outside railing of the ladder as you climb to the top. Figure 9-43 was taken from the top of the tower looking down the ladder. Most climbing systems such as ladders or stairs are located

FIGURE 9-42 View of an interior ladder, looking up from the base.

inside the towers. These climbing systems have resting platforms at various stages between the top and bottom. These platforms provide a place for personnel to stop their climb and rest if needed. They also become platforms where tools and other equipment can be stored for short periods of time as they are taken from the bottom of the tower to the top during the maintenance process.

If you plan to work on wind turbines as an installation or maintenance technician, you will need to spend some time inside and outside some wind turbine towers to make sure you are not afraid of heights or have problems working far off the ground in a safety harness. You will also need to practice climbing several types of towers to see if you have the physical strength to climb the inside ladders and stairs as well as ladders on the outside of a tower. Some people can climb the inside ladders but become fearful while climbing the outside, since they have a different reference of the height. Other people do not have any problems climbing to the top of a wind turbine as long as they are outside, but they have a problem climbing inside the tower since they cannot use other visual indicators as a reference. Some persons become fearful when looking up from the base of the tower, and others become fearful when they are at the top of the wind turbine and are looking down. If you find that you have problems with heights, do not be concerned, as there are many jobs available in the wind industry where you do not have to leave the ground. It is important to become aware of any limitations you might have as you decide the type of job you would like in the industry.

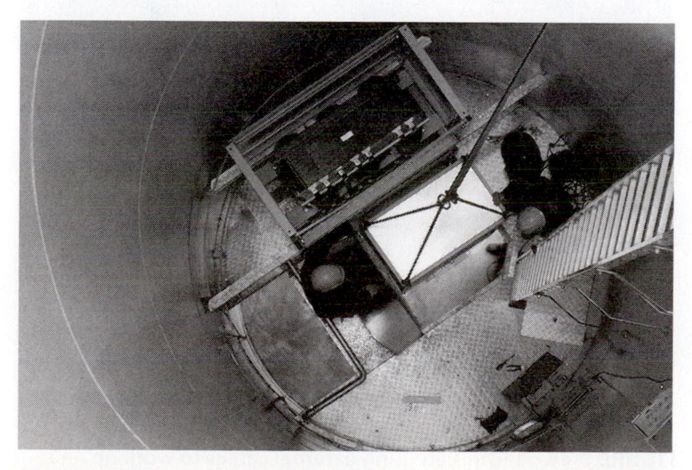

FIGURE 9-43 View of an interior ladder, looking down from the nacelle to the base. (Courtesy of EWT International.)

Electric Service Lift Inside the Tower

Some taller wind turbine monopole towers have an electrical service lift inside the tower that can lift personnel from the base of the tower to the top, where they have access to the nacelle. A service lift is designed for one person, and it is guided through the middle of the tower with guy wires that run from the base to the top of the tower to keep the lift basket correctly aligned. The service lift is pulled to the top with an electric lift motor that has an automatic safety brake that will stop the lift basket in case of emergency and hold it in place. The service lift basket is added to the tower as an additional method to ladders or stairs to allow technicians to get to the top of the tower.

The primary advantage of a lift is that it allows persons who may not have the fitness to climb the ladder to work on the generator, gearbox, and other parts of the nacelle. If these persons were required to climb the ladder or stairs, they might not be able to climb the ladder and still have the energy to work once they were at the top. Many times, older technicians are restricted from work at the top of the wind turbine in the nacelle because they do not have the fitness to climb the ladders. The service lift allows older workers, who might have more experience and skills than a younger worker, to safely reach the top of the wind turbine and work inside the nacelle without having to physically climb to the top. The service lift also allows technicians to work longer in the nacelle, because they do not have to conserve energy so they can climb down the ladder safely. In the event of a power failure, the service lift can be manually lowered to the base of the tower. The service lift is also an excellent method to move someone who is injured or otherwise unable to climb down from the top of the wind turbine.

FIGURE 9-44 Technicians use external ladder on the rotor to inspect the blades. (Courtesy of Nordex.)

FIGURE 9-45 Technicians use the access doors on the both sides of the nacelle to reach the small work platform on both sides of the nacelle. (Courtesy of Northern Power Systems Inc.)

Climbing Outside Ladders and Lattice-Type Towers

Most new wind turbine towers no longer have external ladders on the side of the tower, because it has been found that birds may use the ladder structure to make nests. A few older towers have external ladders that are used to get to the top of the wind turbine. Nearly all horizontal wind turbines do have external ladders on the top of the nacelle and near the rotor hub, or technicians use temporary climbing equipment so they can climb out of the nacelle and down over the rotor hub. Figure 9-44 shows technicians using this type of ladder to inspect the blades and rotor. Other types of horizontal-axis wind turbines have narrow work platforms on each side of the nacelle where technicians can stand while working on the nacelle. Technicians can reach these work platforms by climbing the ladder inside the tower and exiting an access door provided on each side of the nacelle. Figure 9-45 shows two technicians who have climbed the ladder inside the tower and exited through the access doors on either side of the nacelle to reach the small work platforms on both sides of the nacelle.

Some towers today have cable lines that technicians can use as a secondary escape method to get down from the top of the wind turbine from the outside of the tower. Figure 9-46 shows a technician using the cables on the outside of the wind turbine. Another method to get technicians safely to or from the top of the wind turbine tower, to work on the nacelle or the blades, is to use an external service lift. If technicians need to complete a large amount of work on the blades, they may use a special work platform called a blade maintenance platform. These systems use an electric lift motor that pulls the platform up the outside of the tower. The blade maintenance platform is large enough to carry all the equipment and one or more technicians to the top of the wind turbine tower. This type of platform is specifically designed with an opening to allow the blade to move through the middle of the platform as it rises to the top. This allows the work platform to fit all the way around the blade so that technicians can have access to all the surfaces of the blade. Figure 9-47 shows a technician checking the bolts on the blades where they connect to the rotor. Figure 9-48 shows technicians using a blade **service platform**, which allows them to work for extended periods of time high above the ground outside the wind turbine. If any maintenance is required

FIGURE 9-46 Technician using external cables to work up and down the outside of a wind turbine tower. (Courtesy of NREL, U.S. Department of Energy.)

FIGURE 9-47 Technician working on the outside of a wind turbine nacelle, inspecting the bolts where the blades bolt onto the rotor. (Courtesy of Nordex.)

on the blades, the platform can be stopped at the correct height and the work can be completed. These types of external platforms make it safer and more comfortable for technicians to work on wind turbines at extreme heights.

If the wind turbine tower is a lattice-type tower, technicians will need to climb the outside of the tower structure to get to the top to work on the wind turbine generator or blades. Figure 9-49 shows an example of a technician climbing this type of lattice tower. In this figure you can also see that the technician is wearing a safety harness that is tied off to the tower. The angular support pieces between the sections of the lattice-type tower provide a climbing surface that is fairly easy to use. It is important to understand that the lattice tower may become unsafe to climb if the wind is too strong, or if the tower is wet or has ice and snow on its surface. Technicians will report the conditions before they are ready to climb and verify with everyone that it is safe to climb the tower.

FIGURE 9-48 Inspecting turbine blades. (Courtesy of Siemens Wind Energy.)

FIGURE 9-49 Technician climbing a lattice-type tower. Notice that the technician is wearing a safety harness that is tied off to the tower. (Courtesy of John Fellhauer, jfellhauer@surenergy.us.)

9.4 SAFETY ISSUES WHEN WORKING WITH TOWERS AND CLIMBING SAFETY

The majority of the work around the wind turbine is similar to work at a construction site or a work site that has construction equipment, but it is different in that much of the work is performed above ground level. This means that technicians and engineers need to climb and work anywhere from several feet to several hundred feet above the ground. They also need to climb ladders inside or outside the wind tower, or use lifting equipment to reach these heights. **Climbing safety** is paramount on the work site, because this is where personnel are exposed to the most danger on the job. This section will discuss safety around wind turbines, climbing safety, and safety in general as it applies to wind turbines. You will learn about the safety issues and about the equipment that is required to be worn and used when you are on the job. The material in this section is not intended to be detailed enough to allow you to go straight to work; rather, it is an overview to give you sufficient information to become aware of all the detailed safety training you will need before you are ready to work on a site. Much of your safety training will be made available by your employer, and it will include detailed information as well as hands-on practice.

The first step in the safety process is to ensure that you have the best safety equipment and protection to help you prevent a fall, or stop you from falling in the event you slip. This equipment comes under the heading of *fall prevention and fall arrest equipment.* It includes a full body harness that allows you the freedom to move about and work, but that safely secures your body to minimize problems if you begin to slip or fall. This type of equipment will not necessarily stop you from falling, but it will protect you from being severely injured during a fall.

It is important also to have a backup system such as ropes and lanyards that tie you off to the workplace. This system will help give you the ability to limit the distance you fall if you slip. This type of equipment is called *anchorage equipment*, and it helps you to anchor yourself to a fixed part of the wind turbine tower.

Safety During Construction and Installation of Wind Turbines

There are several distinct processes while working with wind turbines when you will be exposed to safety issues. For example, while a wind turbine is being installed and the site is basically a construction site, you will have issues you must be aware of. Later, after the wind turbine is in use, you will have safety issues when dealing with wind turbine maintenance. And later, if you have equipment failure on the wind turbine or if you have equipment that must be changed out during an overhaul, you will have additional safety issues to be aware of.

Some of the safety issues that you will need to be aware of when you are working around a construction site include working with cranes, working with heavy machinery, working in inclement weather conditions, and working at heights. After the wind turbine is up and running, you will also need to be aware of how to work safely around rotating equipment and high-voltage electricity.

This means that you will need to adopt some basic safety practices, such as always having two or more persons on a crew when working on a tower or doing maintenance work. Never work alone, as you may get into a situation where you need help, or someone to call for additional help. You will need to become aware of current weather conditions and weather forecasts, such as for high winds, thunderstorms, rain, and snow, that may affect the safety conditions. You will also learn to have a safety plan developed for each of the procedures you are working on, and will go over this plan prior to starting the maintenance procedure.

U.S. and Canadian Standards for Safety

One of the things you will become aware of is that there are safety standards in place that cover working around construction equipment, cranes, and wind turbines. There are also organizations such as Underwriter Laboratories (UL) in the United States and the Canadian Standards Association (CSA) that test equipment in North America and provide approval if the test results meet their standards. Safety organizations in Europe, Asia, and Australia also cover these issues. You can check their websites to get a better understanding of what types of safety tests they perform and what their stamp of approval indicates. You should check all of your equipment and safety gear to ensure it has labels that indicate it meets or exceeds the standards of these organizations.

Safety Regulations for Wind Turbines

As you become more aware of safety issues that pertain to wind turbine construction, installation, and maintenance, you will find that federal, state, and local rules and regulations apply. In the United States, the rules for safety come mostly under the jurisdiction of the federal Occupational Safety and Health Administration (**OSHA**). Additional rules and regulations may be written by state occupational safety and health administrations as well. It is also possible in some large urban areas that cities and counties have additional safety regulations. You will need to become aware of which regulations pertain to you at the site you are working on. If you are working on wind turbines located in different states, you may have different safety regulations that you will need to be aware of and comply with, depending on which state you are working in at the time.

OSHA Regulations for Working Safely Around Wind Generator Towers

A number of OSHA standards apply to working around wind turbines during construction, installation, troubleshooting, and maintenance. These include Section 1910.12 for construction work, Section 1910.24 for fixed industrial stairs, Section 1910.29 for manually propelled mobile ladder stands in scaffold towers, and Section 1910, Subpart E, for means of egress, which includes escape routes and escape plans and fire prevention. This section includes compliance with NFPA 101-2000, the National Electric Code (NEC).

Additional areas that OSHA covers are detailed in Section 1910.95 on occupational noise exposure; Section 1910, Subpart I, which covers personal protective equipment (PPE) and head protection, foot protection, and electrical protective devices. Section 1910.145 covers specifications for accident prevention signs and tags, Section 1910.146 covers confined space, Section 1910.147 covers the control of hazardous energy lockout and tag out, Section 1910, Subpart L, covers fire protection, Section 1910.180 covers truck cranes, Section 1910.184 covers slings, Section 1910.219 covers mechanical power transmission apparatus, Section 1910, Subpart P, covers hand and portable power tools, and Section 1910, Subpart Q, covers welding, cutting, and brazing. The electrical portion of the wind turbine is covered under Section 1910.269 on electrical power generation, transmission, and distribution. You can refer to the OSHA website at www.OSHA.gov for more complete information about any of these areas.

It is also important to understand that everyone on an industrial worksite must take a safety course called the OSHA 10 Hour course. The information in the OSHA 10 Hour course includes an overview of the Occupational Safety and Health Administration, general safety and health provisions, as well as information concerning some of the following areas listed in the guidelines, such as electrical issues; fall protection; hand and power tools;

scaffolds, cranes, lifting devices, hoists, and elevators; personal protective equipment (PPE) and life-saving equipment; and stairways and ladders. This course can be taken online and includes a written test and certification that the student has passed the test. You may also take this course with an instructor, and a test will be provided to demonstrate that you understand the information in this course. You will learn more about OSHA standards and regulations when you reach the job and become more familiar with the work around a wind turbine.

ANSI/ASSE Z359 Fall Protection (Arrest) Code

The American National Standards Institute (ANSI) and the American Society of Safety Engineers (ASSE) are responsible for the Z359 Fall Protection Code, which has recently been updated. This code covers fall-arrest equipment and **fall protection**. Fall arrest and fall protection are considered two separate and distinct issues, and the equipment for each is very different. Fall-arrest equipment includes equipment that protects persons from falling and arresting equipment which stops or slows someone once they have begun to fall. The arresting system is designed to anchor you when you climb a wind turbine tower or a lattice tower framework, and if it is connected correctly, it will arrest (stop) your fall within 6 ft.

ANSI/ASSE Z359.1 covers safety requirements for personal fall-arrest systems, subsystems, and components. This part of the code establishes requirements for the design, performance, qualification, marking, instruction, training, inspection, use, maintenance, and removal from service of all of the following: connectors, lanyards, anchorage connectors, energy absorbers, full body harnesses, fall arresters, vertical lifelines, and self-retracting lanyards that include personal fall-arrest systems.

ANSI/ASSE Z359.2 covers the minimum requirements for a comprehensive managed fall protection program. This part of the code includes requirements and guidelines for a fall protection program that is managed by the employer, including duties and training, policies, fall protection procedures, rescue procedures, eliminating and controlling fall hazards, and incident investigations.

ANSI/ASSE Z359.3 covers safety requirements for positioning and travel restraint systems. This part of the code covers the requirements for the design, marking, qualification, performance, test methods, and instructions for use of lanyards and harnesses used for personal positioning and travel restraint systems.

Climbing Safety Equipment: Carabineers, Lanyards, and Safety Harnesses

Climbing and safety equipment is important to keep you safe on the job. This section will introduce and explain this equipment. A **carabineer** is a simple locking mechanism that is generally made of metal and allows quick connections between ropes and harnesses and other essential equipment. When technicians are climbing wind turbine towers, they need a locking mechanism that allows them to quickly connect to tie points on the tower, or to connect to tool buckets and other tools. There are three types of carabineers: nonlocking, self-locking, and manual locking. Figure 9-50 shows a locking carabineer.

FIGURE 9-50 Locking carabineer. Notice the clip that slides over the locking mechanism. (Courtesy of North Safety Products by Honeywell.)

A *nonlocking carabineer* is a general-purpose connector that is usually used to connect tools or other non–safety-related items to a tool belt or other component that needs to be secured while climbing and then disconnected quickly when they are needed for use. For example, some technicians connect wrenches and other tools to a tool belt with a nonlocking carabineer so they can quickly open the carabineer and use the tool, and then reconnect the tool back to the tool belt when they are finished. Nonlocking carabineers should not be used for safety or climbing.

Locking carabineers operate like the nonlocking variety but have the addition of a sleeve around the gate, preventing them from being released accidentally. Locking carabineers are able to handle larger weight loads. Generally, they are used connect the climber to safety tie points, and they are also used with the most important pieces of equipment on a harness.

Self-locking carabineers have a clasp that automatically closes and locks the carabineer. These carabineers are useful for nonessential purposes, but they should not be used for securing the climber or other essential safety functions. Carabineers are used to establish secure anchorage points and connections very quickly while technicians are climbing. They also allow technicians to tie themselves off safely in a working position when necessary.

There are several pieces of equipment that are designed to secure someone who is climbing on a wind

FIGURE 9-51 Safety harness used by technicians when climbing towers. (Courtesy North Safety Products by Honeywell.)

FIGURE 9-53 Shock-absorbing lanyard. (Courtesy North Safety Products by Honeywell.)

turbine tower. The most basic piece of equipment is called a **safety harness**, and an example is shown in Figure 9-51. You can see from the figure that this harness completely encases the climber's body and legs. There are several points for adjustment to ensure that the harness fits properly, and the harnesses come in different sizes. Once the harness is correctly fitted to the climber, safety anchor belts and other safety equipment such as lanyards are connected to the safety harness. The safety belt shown in Figure 9-52 is used to anchor technicians to the tower or other secure point when they are working. This belt must be unclipped while technicians are moving from point to point during the climb, but it is quickly connected when climbers reach a resting point or the point where they will be working. Figure 9-53 shows a shock-absorbing **lanyard** that climbers connect from their safety harness to a secure point on the tower. This type of equipment is used to protect climbers in the event that they slip or fall. Shock-absorbing lanyards will stop climbers from falling far, and the shock-absorbing function will ensure that they are not injured when they stop. The lanyard is always connected before the climber releases the safety belt to begin moving from point to point. The lanyard has enough slack to allow the climber to move. Once the climber is in position to begin work, the lanyard is connected to a safety tie point.

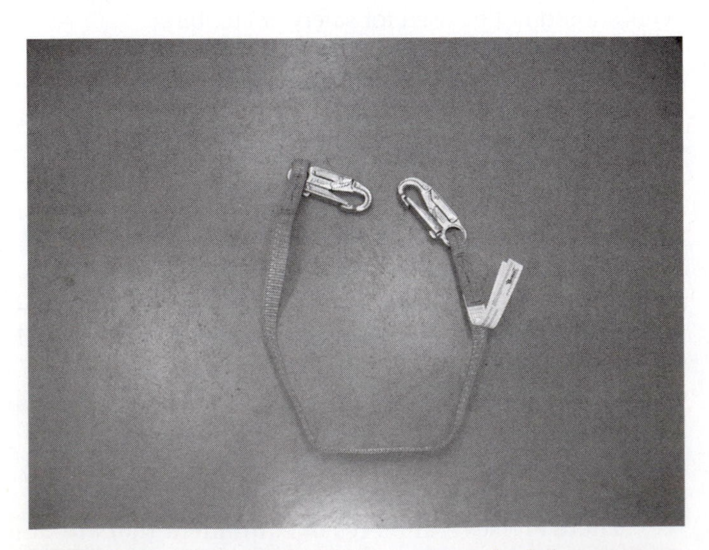

FIGURE 9-52 A belt that is used as an anchor for fall protection.

The next step in the overall safety of climbers is to inspect safety equipment frequently and to provide routine maintenance and care in keeping it clean and sound. It is important that all climbing safety equipment be thoroughly checked immediately prior to climbing, to ensure that it is not damaged in any way. The climbing equipment must also be maintained and cleaned to ensure that it does not wear out. If the climbing equipment is damaged in any way, it should be immediately replaced with new equipment. The final responsibility for safety and for determining the integrity and quality of the safety gear always rest with the climber. Figure 9-54 shows a technician wearing safety harness and other safety gear while working on a nacelle. Figure 9-55 shows a technician connecting his safety line to the frame of a lattice tower as he gets ready to climb the tower.

FIGURE 9-54 Technician wearing safety harness and other gear while working on a nacelle. (Courtesy of Northern Power Systems Inc.)

Rescue and Evacuation Strategies

Before technicians are ready to climb tower, they need to take a series of training classes on rescue and evacuation strategies. This information will provide them with the needed instruction to help other climbers who may be injured or in danger and may not be able to help themselves get down from the tower safely. At times the climbing gear or the equipment may have a problem that causes a climber to be stuck on the tower above the ground. If this should occur, the climber will indicate that he need help either through hand signals or radio, and personnel on the ground will put the rescue plan into action.

Rescue may include taking someone off the tower who is suspended somewhere between the top and the bottom of the tower. Another type of rescue may be required when some mechanical equipment has failed and someone has to get personnel safely to the ground. Each of these types of rescues is very complex, and you will need hands-on training to practice the techniques for correctly "packaging" victims and getting them to the ground safely.

FIGURE 9-55 A technician connects his safety line to the frame of a lattice tower. Notice that the technician is wearing a safety harness. (Courtesy of John Fellhauer, jfellhauer@surenergy.us.)

Another technique that you will learn is how to safely rescue yourself in the event that your equipment jams or becomes disabled during a climb. These techniques will also be presented so that you may practice them during a training session, under close supervision at heights where you will not become injured during practice. You will need to be evaluated to ensure that you can accomplish these techniques quickly and efficiently in an emergency.

Safe Evacuation from a Nacelle Using Ladders

Another important safety procedure that you will need to learn is safe **evacuation** from a nacelle in case of an emergency. If you are working at the top of a tower in the nacelle and an emergency occurs so that you or a co-worker needs to get to the ground as quickly as possible, you will need to implement evacuation and rescue techniques immediately. In the case of this type this type of emergency, speed and safety are paramount. For example, if there is a fire in a nacelle, you will need to get out of the area immediately and descend to the ground as quickly as possible. If a co-worker is injured and needs help, this will complicate the evacuation. You will need to be familiar with the evacuation procedures so that you can implement them immediately. You will be sent to training periodically to ensure that you fully understand these procedures and can implement them without error as quickly as possible. If an emergency is in progress and you need to implement the procedures, you will need to perform these steps automatically and safely. The equipment and devices for quick dissent will be discussed in the next section.

Automatic Descent Devices and Block-and-Tackle Rescues

There are several pieces of equipment that are designed specifically for quick **automatic descent** from the top of a wind turbine tower. You will receive training and become familiar with this equipment when you are on the job. This equipment may be located inside the tower in some systems, or it may be on the outside of other towers. During preparation for climbing, you will spend time identifying and reviewing procedures for automatic and quick descent. You will also review procedures for safe evacuation and rescue of yourself and co-workers if necessary.

Some equipment for automatic and quick descent may be as simple as a block and tackle or other type of rapelling system. The quick-descent system will always have a **fail-safe system** that will allow it to be implemented in case of a power failure, or if there is a fire or other type of emergency. Some larger towers provide a

FIGURE 9-56 A technician might evacuate the tower with help from the crane or, if the crane was not available, the technician could use cables to rappel down from the nacelle in an emergency.

quick-descent feature on the lift basket, which will allow you to move an injured co-worker to the ground safely and quickly. Figure 9-56 shows a crane near a wind turbine. In some cases a technician can use this type of equipment to lower himself from the nacelle.

Inspection, Care, and Maintenance of Rescue Equipment

It is very important that you have a schedule for inspection, care, and maintenance of your **rescue equipment**. Many federal and state regulations require this to be scheduled on regular basis and documented when it has been completed according to the schedule. It is important to understand that an emergency that requires the use of this equipment will come on suddenly and will not give you any warning or time to make adjustments to equipment that is not fit for use. Therefore it is important to review the equipment on a scheduled basis and identify the status of all equipment. If you find equipment that does not meet standards, it should be identified and removed from service immediately and replaced with operational equipment. It is also important to understand that you should never begin a climb or work above the ground if this equipment is not readily available and someone has not verified its condition.

Conducting Safety Exercises

Another part of the regulations that cover safety procedures and equipment requires conducting practice exercises on a scheduled basis. Understanding and correctly implementing safety procedures is vital to ensure that all personnel are capable of helping each other during a safety emergency. One way to ensure that everyone feels comfortable with the procedures and is able to carry them out quickly and efficiently is to conduct realistic safety exercises. These exercises must be planned and evaluated to ensure that all personnel understand their roles and are able to execute them without any problems. If deviations to the procedures or problems occur during the training exercise, follow-up training can be provided to ensure that the problems are corrected.

9.5 LIGHTNING SAFETY FOR WIND TURBINES

Since wind turbines are often the tallest structure in the area where they are located, they are often struck by lightning. Lightning is attracted to wind turbine towers, the nacelle where the generator is located, the wiring from the generator to the ground, and the turbine blades. Each of these parts of the wind turbine present a different type of problem in protecting the wind turbine from lightning. One way of ensuring that a wind turbine survives lightning strikes is to create design standards requiring blades to be capable of withstanding 98% of lightning strikes.

Lightning protection consists of proper grounding, **lightning arrestors**, and surge protectors. One way to protect wind turbines and wind turbine sites is to utilize a large number of lightning rods and lightning arrestors. The lightning protection system absorbs the lightning strike and transfers the large amounts of high voltage and high current to the ground, where it is safely dissipated.

Large wind turbine blades can be protected by embedding a lightning receptor (lighting rod) at the tip of the blade when the blade is built. This type of lightning control can direct the lightning into the proper wires and down to the ground system. On some larger wind turbines, more than one lightning receptor may be needed. Since the blades are rotating, a set of slip rings is required to transfer any energy in a lightning strike from the lightning rods in the rotating blade to the stationary part of the tower and down to the grounding system.

The nacelle can be protected from lightning strikes by steel mesh. The steel mesh is connected to the grounding system, so that any lightning strike directly to the nacelle will be dissipated and the energy sent to ground. Another part of the wind turbine that as susceptible to lightning strikes is the control signal wiring that runs between the computer or PLC and the various parts of the wind turbine. Since this wiring uses very low voltage, the high voltage from lightning strikes can be damaging if they enter the system. This wiring for the control signals can be protected with wire mesh so that any stray voltage during a lightning strike will be directed to ground.

If lightning does strike the blades, the surface of the blades may be damaged to the point where technicians will need to come in and make repairs in the field. Since the blades are made of fiberglass composites or other similar materials, technicians will need to work on the blades while they are in place. If the blades are damaged beyond repair, they will need to be removed and replaced with new ones.

Electrical Safety, Surge Protection, and Lighting Protection

The switchgear and electrical controls between the generator and the point where voltage ties into the grid are also susceptible to damage from lightning strikes as well as overvoltage conditions from grid voltage. If the wind turbine is tied directly to the grid, any variations of high voltage can be transferred from the grid back into the wind generator. This may occur because of equipment failure such as transformers failing or because of stray voltage.

One way of protecting the equipment is to use **surge protection**, which is commonly used in other electrical generating sites as well. The surge protection is designed to take any voltage that is above a safe value and route it directly to ground, where can be dissipated harmlessly. Surge protectors can be used in conjunction with fast-acting fuses that will open the conductors so that equipment is not damaged. Surge protection and other types of equipment can react within one-half of one sine wave, so that voltage does not exceed safe levels and damage the electrical equipment that is connected to the grid, such as the generator and inverters.

Grounding

Another way to protect wind turbine systems is to provide adequate **grounding**. A grounding system consists of copper ground rods that are driven into the earth at various locations to depths of 8 ft or more. If multiple ground rods are used, they are connected by heavy-duty conductors to ensure that the system has low resistance and can carry any lightning or ground currents into the earth where they can be safely dissipated.

It is very important to inspect the ground system on a wind turbine periodically to ensure that it can safely carry heavy currents into the earth. This means that the grounding system and its conductors must be checked with a megger or other type of instrument that can measure the resistance to ensure that it remains minimal throughout the ground system and that good connection to the earth is maintained. A megger is an ohmmeter that can measure resistance in millions of ohms (mega-ohms) and can detect very small failures in insulation so they can be repaired before they become a problem.

Ground-Fault Protection

Electrical circuits in the wind turbine can also be protected by the use of **ground-fault protection**. Ground-fault systems continually check the current that the generator produces and ensures that it is all going into the grid system. If any of the current strays from the path to the grid, indicating a problem, the ground-fault protection system creates an open in the main conductors, much as a fuse or circuit breaker does. When the ground-fault system detects a problem and opens, it interrupts all current from the generator to the point where it is tied to the grid. This type of system will also detect irregular voltage or currents that may be fed backwards into the system. The ground-fault system accurately measures all of the current leaving the generator and identifies any stray or extra current that enters the system, or whether any current has been diverted to ground. If either of these conditions has occurred, the ground-fault circuit determines the condition as a problem and opens the conductors in the main circuit. Typically, ground-fault circuits can be reset from a control system, or they may need to be reset manually by a technician who must inspect the system for problems.

Other Protection for Electrical Components

Since the electrical system of the wind turbine is connected to the grid, it must be completely protected against a variety of problems that commonly affect electrical generating equipment. For example, it must be protected against synchronization, which means that any time the frequency of the voltage and the phase of the voltage are not in sync with the voltage on the grid, the generator may need to be tripped and taken out of the system temporarily. If the difference between the frequency that is generated and the grid frequency is minimal, the grid frequency will override the generator frequency and force it to the frequency of the grid. If the problem is with the phase differential, the grid cannot correct that deficiency, and the generator will have to be disconnected from the grid and reset when the phase is correct.

Other protection circuits that are needed on the generator include undervoltage detection, overvoltage detection, overfrequency protection, underfrequency protection, and low-voltage ride-through protection. You may need to review the material in Chapter 8 about electrical connections to the grid if you need more information about these issues.

Safety Connection Switches

The electrical system on the generator and the wind turbine must have a specialized electrical switch that controls when the voltage from the generator is connected to the grid. As you learned in Chapter 8, the connection to the grid is very important, and the connection must be made at just the right time for the system. The safety connection switches are also there to make sure that the generator is disconnected if there is a problem, to keep it safe from voltage that may feed back from the grid. The safety switches must be checked periodically to ensure that they are functioning correctly.

9.6 OVERSPEED SAFETY AND OVERLOAD CONTROLS TO PROTECT TOWERS

Wind turbines must have controls installed in the tower to protect the system from overspeed or overload conditions. Sensors continually monitor the speed of the blades to ensure that they do not overspeed. On larger wind turbines, the blade pitch and the nacelle yaw can be adjusted to ensure that the blades do not spin too fast in very high winds. If pitching the blades and adjusting the nacelle yaw do not keep the blade speed under control, a safety circuit to control overspeed will kick in and begin to apply the brakes on the rotor.

On small and mid-size wind turbines, the overspeed condition must be controlled to protect the blades from damaging the tower. The overspeed control on small and mid-size wind turbines will stop the blades and rotor and keep them stopped until the high winds subside. Once the wind speed returns to a safe level, the overspeed control will release the blades and allow them to begin to rotate again. You will need to check the data for each wind turbine you work on to determine the minimum speed and maximum speed at which the blades can operate, as well as the minimum and maximum wind speeds at which the wind turbine can operate.

Mechanically Blocking Blades for Safety Reasons

A wind turbine needs to have a service brake that can be set manually when maintenance is required. The regular brakes are used to bring the rotor to a stop, and then the service brake is set to lock the blade in position. If the blade has pitch control, it can be used at this time to slow the blade rotation prior to setting the brakes. Figure 9-57 shows the blades of a wind turbine pitched to where they will not harvest any wind. Notice that the narrow edge of the blade is pointing directly away from the wind and back toward the tower, so that the blade does not harvest the wind and try to turn, especially when the blades must be stopped for maintenance. The service brake for the blades and nacelle must be set before starting maintenance, and any other types of brakes must be applied to ensure that the blade does not turn. When maintenance has been completed, the brakes can be released so the blade can begin to turn again.

Magnetic-Release Safety Brakes on Blade Tips

Tip brakes or **safety brakes** are provided on the wind turbine blades of many small and medium-size wind turbines. One of the problems which can occur with all sizes of wind turbines is that they can overspeed during high winds and damage the blades or the tower. One way of preventing overspeed is to provide a safety brake or tip brake at the very tip of the wind turbine blade. Figure 9-58

FIGURE 9-57 The blades of this wind turbine have been adjusted so that they will not turn if the wind blows during maintenance procedures.

FIGURE 9-58 A magnetic tip brake is shown at the very end of the blade. When the blade begins to turn too quickly, the tip brake will slide into position and slow the blades down.

shows a magnetic tip brake. The magnetic tip brake is designed so that the braking mechanism will be kept in position by a magnetic field from a permanent magnet until the blade rotation begins to exceed its highest rated safe speed. When the wind exceeds safe rating and the blade begins to rotate faster than the rated safe speed, centrifugal force will cause the tip brakes to drop into position. This occurs because the centrifugal force exceeds the force of the magnet. When the tip brake drops into position, it creates an irregular shape for the blade and automatically causes that to began to unload wind energy and slow down. When the blade has slowed to a safe speed, the springs on the tip brake will cause them to snap back into the safe position, where they are again held by the magnet. The tip brakes can be used over and over to ensure that the blades never exceed their rated safe rpm.

Another type of brake provides a special mechanical tip at the very end of the blade that acts as a brake in a similar fashion to the one controlled by the magnet. This type of brake will stay in the stored position as long as the blade rpm is below the safe maximum. Any time the wind speed increases to a point that it causes the blades rpm to exceed the safe speed, the mechanical mechanism for the tip brake will cause the tip of the blade to rotate 90 degrees and cause an irregular surface to occur at the end of the blade. This irregular surface will cause the blade to unload the wind energy and slow down automatically. When the speed of the blades reaches a safe condition, springs cause the tip of the blade to return to its normal condition, which allows the blade to pick up wind energy and increase its speed again. If the speed should exceed safe speed again, the tip will again rotate and slow the blade down. This process can occur over and over to help ensure that the blades never exceed their safe operating speed.

It is important to inspect tip brake assemblies periodically to ensure that the mechanism that holds them in position will operate. You also need to test the tip brake to ensure that it works correctly and slides into position at higher speeds.

Tip brakes are designed specifically for use on the tip of a wind generator's blades so that the physical shape of the blade tip will be changed relative to the remainder of the blade and this will cause the rotor's speed to slow down. The tip of the blade can rotate approximately 90 degrees to its longitudinal axis, or the end segment of the blade that acts is released from its position when the centrifugal force exceeds the magnetic force. Centrifugal force increases as the speed of the blade increase. When the blade starts spinning too fast, the amount of centrifugal force will exceed the strength of the magnet that holds the blade tip in the "run" position, and it will cause the end of the blade to deploy and change the shape of the blade so that it begins to shed wind energy and naturally slows the blade down.

If the brake system uses a rotating tip mechanism, the blade tip will rotate with respect to the remainder of the blade when the blade is subjected to this excessive force due to excessive blade speed. Once the blade tip has been rotated, it will cause the blade to begin to shed wind energy, which causes the rotation of the rotor to slow. It will remain deployed until the wind slows or the rotor is slowed. When the rotor has slowed to a safe speed, a spring causes the tip brake to be reset. The tip brake will also be reset if the blade is stopped with a mechanical brake.

FIGURE 9-59 A single-bladed turbine teeters the blade to slow it down and stop it from rotating. (Courtesy of Powerhouse Wind Ltd., www.powerhousewind.co.nz.)

Another way to alter the blade so that it does not harvest wind is to teeter or feather the blade. Figure 9-59 shows a single-bladed wind turbine that teeters (feathers) its blade so that its pitch is changed to where it will not harvest wind energy. This action causes the blade pitch to change and the rotor to stop rotating.

Shut-Down Brakes

On a wind turbine, the control system and the brake system operate together to shut down the wind turbine blades and lock them in place to keep them from rotating. When these brakes are applied, they are called **shut-down brakes**. Any time the wind speed is below approximately 10 mph, the wind turbine generator will produce minimal electrical energy. (The minimum wind speed for some wind turbines to begin producing electrical energy may be as low as 6 mph.) When the wind speed is too low, the blades will be stopped in what is called a *shut* *down*. When the wind turbine is in a shut-down state, its brakes are set and the blades locked so they cannot rotate. The wind turbine may also need to have its blades stopped and the brakes applied to ensure that the blades do not rotate during maintenance procedures.

When the instruments on the wind turbine measure wind speed in excess of the minimum speed, a control signal is sent to the brakes that causes them to be released. The tip blade brakes and the dynamic brakes are released and the rotor begins to pick up speed and rotates until the high-speed shaft at the generator is turning at more than 1800 rpm. When the generator begins to receive exciter current and begins to produce voltage, the voltage will come up to the standard for the grid voltage in frequency, phase, and voltage level. When the voltage from the generator matches the voltage for the grid, a connector switch closes and allows the generator voltage to begin to flow into the grid.

If the wind speed increases above its safe level (typically, speeds above 50–60 mph), the controller sends the signal to shut down the turbine by activating it brakes. On some larger wind turbines, the control system begins to change the pitch of the of the blades to slow the rotor speed, and then changes the direction of the nacelle so that the blades are no longer pointed directly into the wind so that the rotor speed is slowed to a safe level rather than stopping the blades all together. This type of speed control makes the wind turbine more efficient while protecting the blade and tower from overspeed conditions.

The controller also protects the tower and the turbine blades when the wind speed is too slow. This condition is called underspeed or low-wind condition. When a low-wind condition occurs, the wind turbine blades are shut down and not started again until the wind speed rises above minimum speed again.

If the wind turbine generator begins to put out more power than it is rated for, the controller will put out a slow-down or shut-down command to slow the rotor speed, which causes the generator to produce less power until it gets back under control. If the mechanical friction brakes are used for extended periods of time, they will begin to overheat. If the brakes overheat while the rotor speed is being slowed, the turbine blades must be stopped completely so that the brakes can be allowed to cool down before the turbine is allowed to start again.

Braking Systems and Fail-Safe Systems

Braking systems are often part of the fail-safe system on a wind turbine. The term *fail-safe* means that if every part of the system, such as electrical, hydraulic, and mechanical energy sources, fail, the brakes will still actuate. When the wind turbine is originally designed and all of the systems are built, the fail-safe concept is designed in and built into many of the safety systems. It is important to test the fail-safe systems during periodic maintenance procedures.

Hydraulic Parking Brake

Larger wind turbines that have hydraulic systems may also have a hydraulic parking brake. The hydraulic parking brake receives its power and energy from hydraulic fluid. You learned about the hydraulic system in Chapter 5, where you learned that the size or diameter of the hydraulic cylinder and the amount of hydraulic pressure determines the maximum force the hydraulic brakes can apply. The hydraulic cylinder is large enough to provide sufficient force and energy to activate the parking brake system. Whenever maintenance is being accomplished on the wind turbine and its blades need to be locked in the park position, the hydraulic brake is applied.

9.7 BIRDS AND BIRD SAFETY AROUND TOWERS

The tip speed of wind turbine blades is fast enough to kill a bird if it comes into contact with it. Early versions of wind turbines tended to have very fast rotor speeds, and some smaller residential systems have very high rotor and blade speeds. The high-speed rotor made it possible for wind turbines to kill birds if they got into the path of the fast-moving blades. Also, many older wind turbines had external ladders that made it easy for birds to build nests on. When the young birds left the nest, some of them might fly into the blades.

Today, most of these issues have been addressed. The speed of the turbine blades is much slower and is usually less than 20 rpm, and the ladders on the outside of wind turbines have been moved inside or covered so that birds cannot easily build nests. As a result, the number of bird kills has been substantially reduced. The actual number of bird kills is not easily recorded.

When issues such as turbine bird kills are discussed, many times people do not put the entire issue in perspective. For example, if you research the major causes of bird kills in the United States, you will find that domestic cats that are allowed to run loose are the major cause of bird kills. Other causes include birds colliding with aircraft, birds killed on the highway by automobile and truck traffic, birds of prey such as owls and hawks that hunt smaller birds, and weather (blizzards and extreme cold weather) that reduce the availability of food, all of which contribute to more bird kills per year than do wind turbines. Even though bird kills are minimal, designers should do all that is possible to limit birds from using towers for habitat and preventing bird kill by wind turbines.

9.8 TOWER MAINTENANCE

Since wind turbine towers are subjected to the elements and extremes of the weather, the wind turbine tower must be inspected periodically for hardware that may come loose or corrosion that is taking place. On lattice-type towers, there is a wide variety of bolts and nuts as well as other types of fasteners that hold the tower together. These fasteners must be inspected periodically to ensure that they are not coming loose or corroding, which might allow the tower to fail. It is also important to remember that wind turbine blades may cause some amount of vibration when they rotate. The vibration may cause some of the hardware or fasteners to come loose over time. The wind turbine must be scheduled for tower maintenance periodically to ensure that the tower remains stable and reliable over time.

Larger monopole towers that are bolted together in sections must also be inspected to ensure that the nuts and bolts that hold the sections together remain tight and torqued to specification. It is also important to inspect the inside and outside surfaces of the monopole tower to ensure that it does not have any damage due to oxidation. Some towers may also require painting of external surfaces to protect them from the weather.

Questions

1. Describe what a monopole wind turbine tower looks like.
2. Describe what a lattice-type wind turbine tower looks like.
3. Explain what guy wires are and how they are installed.
4. Explain how the gin pole and tilting-type tower works to erect a wind turbine pole.
5. Identify the steps required to erect a lattice type wind turbine tower.
6. Explain how a large monopole steel or cast concrete tower is assembled.
7. Explain why a crane must be taller than the wind turbine tower it is assembling.
8. What is the role of the FAA in deciding where a wind turbine should be located?
9. Identify two types of safety equipment that wind turbine technicians use when they work above ground level.
10. List several way you could get to the top of the tower of a monopole wind turbine.

Multiple Choice

1. An external service platform for a wind turbine tower
   a. Allows technicians to ride in an electrical lift to the top
   b. Allows technicians to climb an external ladder to the top
   c. Allows technicians to climb an internal ladder to the top
   d. All the above
2. Which of the following types of wind turbine towers is easiest to climb?
   a. A vertical-axis wind turbine tower
   b. A lattice-type tower
   c. A monopole tower
   d. All of the above
3. Which federal government agency must be consulted about the height of a wind turbine?
   a. The EPA
   b. OSHA
   c. The FAA
   d. All of the above
4. Why is it important to use a template when pouring a concrete pad for a wind turbine tower with feet?
   a. The template ensures that the size of the pad is the exact size of the wind turbine tower.
   b. The template ensures that the bolts for the feet of the pad are in the correct locations.
   c. The template ensures that the thickness of the pad is the exact size for the tower.
   d All the above.
5. OSHA regulations cover
   a. Safety issues
   b. Height issues for wind turbine towers
   c. Environmental issues
6. ANS/ASSE Z359 covers
   a. Environmental issues
   b. Bird kill and environmental issues
   c. Fall-protection issues
7. Which of the following is safety gear used for working above the ground on a wind turbine?
   a. Full-body safety harness
   b. Shock-absorbing lanyard
   c. Safety line
   d. All the above
8. Guy wires
   a. Help reinforce the tower and keep it more rigid
   b. Help conduct electricity from the generator to the base of the tower
   c. Help reinforce the foundation of the concrete pad
   d. All the above
9. One of the problems with wind turbines located over oceans is
   a. Salt water causes wind turbine towers to corrode.
   b. It is difficult to secure the power cables to the ocean floor.
   c. Storms may cause damage to the wind turbine tower.
   d. All the above
10. The tilting-type tower uses a gin pole
    a. To erect a small wind turbine tower
    b. To make the quick evacuation of a tower easier and safer
    c. To make the lattice-type tower more rigid
    d. All the above

CHAPTER 10

Wind Turbine Installations and Wind Farms

OBJECTIVES

After reading this chapter, you will be able to:

- Explain the term *turnkey*.
- Identify four types of site issues that you may encounter when you are selecting a location for a wind turbine.
- Explain why a small residential wind turbine system may need a power electronic frequency converter (inverter).
- Explain two advantages to locating a wind turbine on agricultural land.
- List three types of commercial applications of wind turbines.
- Identify the five largest wind farms in the United States as of 2009.
- Identify the total amount of electrical energy produced by wind turbines worldwide as of 2008.
- Identify two advantages of locating wind turbines over water.
- Identify two wind farms that are located offshore in Europe.
- Identify two disadvantages of locating wind turbines over water.

KEY TERMS

Commercial wind turbine
Electronic inverter
Gigawatt
Megawatt
Offshore wind farm
Power electronic frequency converter
Project development plan
Residential wind turbine
Saltwater corrosion
Underground transmission lines
Underwater cable
Wind farm
Wind site assessment

OVERVIEW

This chapter will help you gain a better understanding of wind turbine installations, such as large wind farms, offshore wind farms, smaller installations of two or three wind turbines, and finally, single wind turbines connected to residential or commercial establishments. The chapter information includes issues that help determine where individual wind turbines should be located and where larger wind turbine farms should be located. These issues include project development, wind site assessment, issues found with individual sites, and other problems encountered when trying to determine the best location at which to place one or more wind turbines.

Information in this chapter will also include ways to determine the best location at which to place a small residential wind turbine, home-made wind turbine, commercial wind turbine, and finally, a wind turbine farm. You will also learn about where the largest wind farms are located in the United States and worldwide, and where the newest growth in wind farm installations will be. You will also learn about how some countries that are located on oceans have learned to develop offshore wind farms.

10.1 PROJECT DEVELOPMENT

To analyze information about the best location in which to place a wind turbine, you will need to create a **project development plan**. This project development plan will include things such as selecting the proper site for the wind turbine or wind turbine farm, identifying available wind resources and predicting energy output for various sized wind turbines, identifying landowners and developing landowner agreements, identifying grid interconnections if necessary and utility companies that control the grid in the area, identifying government agencies that control the country, state, county, or city regulations that will affect your selection, and finally, developing an environmental plan to ensure that the wind turbine conforms to all of these requirements.

A number of commercial companies will complete the project development process for you on a fee basis. The fee may be included in the cost of the wind turbine installation if you select their equipment. Other companies do not sell wind turbine equipment directly; rather, they will recommend one or more brand names and construction companies that will help with the installation and start-up process. Some of these projects fall under the category of *turnkey* projects, which means that all aspects of the project through start-up are taken care of, and then the project is turned over to the owner. A turnkey project may be more expensive, but all of the details identified in the project development plan will be taken care of for the owner of the project.

10.2 WIND SITE ASSESSMENT

The **wind site assessment** includes gathering data about wind speeds, wind direction, the number of hours per day the wind blows, and the number of days per month the wind blows. Data loggers are used in conjunction with anemometers and wind direction instruments to compile the information from a number of these instruments spread across a site and at different heights. Typically, wind site assessments include Global Positioning System (GPS) information to provide precise locations to connect with the wind data information. Other useful information can be provided by photographs or satellite images that show all the necessary information about the location and the surrounding environment.

Other important issues that must be taken into consideration include the geology of the land where larger wind turbines will be located. Problems may exist with soft soil, or wetlands that do not drain well, which will restrict the size and weight of the wind turbine towers and nacelles. Other issues such as height restrictions may limit the placing of wind turbines, such as in close proximity to airports or other areas where there are height restrictions. Some cities, such as in resort areas, may prohibit wind turbines altogether.

In some states, government agencies will help with the site assessment process, in order to entice more wind turbines to be located in the state. Some states have also provided financial incentives if wind turbines are placed in their states. In some cases you will find that other wind turbines have already been installed in nearby locations, which may make it simpler to complete the wind site assessment because information such as grid connection, utility company contacts, state forms, and other common information may already be available and completed from previous projects. In these instances the information from the original installations will be specific to the site, such as wind availability on that site and different amounts of wind availability at different heights.

10.3 SITE ISSUES

There are a number of site issues that you will become aware of as you begin to select locations for wind turbines. These site issues can be grouped into the following categories: environmental issues, noise issues, height issues, and electrical power and connection issues.

Environmental issues may include how well the wind turbine fits in with its environment. For example, if the wind turbine is located near a wildlife area, there has been concern about certain species of birds being killed at a higher rate than normal. There have also been discussions about placing offshore wind turbines in locations that migratory birds frequently fly through or use as resting areas, for fear that it may change these migration routes or that birds may be injured or killed by the wind turbines. Other environmental issues may include impact on the environment where roads and access must be created to move the wind turbine to the wind turbine site. Also there are concerns about possible environmental damage due to installing high-power lines and large power transmission towers near critical environmental locations.

Another issue that may affect the placement of wind turbines is called visual impact. Since wind turbines are normally among the tallest structures in the area, there are people who are concerned about these taller structures visually impacting areas such as mountain passes, large prairies, or near-shore developments. Since wind mills have been a fairly common sight on farms for the 150 years, these issues are not nearly so severe when the larger wind turbines are placed on farms.

Another concern about wind turbine placement is noise. The faster the blades on the wind turbine turn, the more noise is produced. When a wind turbine is running at low speeds, this is typically not an issue. When multiple wind turbines are running at high wind speeds, however, the noise will be carried with the wind to nearby residents, which may become a problem. Noise restrictions have been recently included in many zoning laws which may prohibit the location of wind turbines near residential locations. These restrictions may include how close a wind turbine can be located to the residential area, or it may restrict the number of decibels of noise that will be permitted near residential areas.

Other concerns when wind turbine sites are being planned include issues about maintaining the wind turbine site, or removing older equipment as it wears out. Many communities are concerned that companies will abandon older equipment and leave the equipment in place, where it may become an eyesore after it has served its useful purpose and is no longer maintained. Some regulations require companies to establish escrow accounts in which money will be put aside for the purpose of removing older equipment in the event these companies go bankrupt or abandon a wind turbine site.

10.4 VISUAL AND LANDSCAPE ASSESSMENT

Since wind turbines can extend over 300 ft into the air, they are visible for quite a distance. In some cases, such as agricultural or farmland, this is not an issue because the number of people in the immediate area is small. One method to combat this concern is to paint the wind turbines in neutral colors or colors that make the wind turbine blend into its environment. This in turn, however, may create issues with aircraft, and the aero industry, which would rather have the wind turbines painted in more vivid colors that are easier to identify at a distance or in low-visibility weather. Newer wind turbines tend to be larger and taller than the previous types, which tends to compound visual issues.

Another issue that may have visual and landscape concerns is the construction of larger electrical power transmission towers to get the electricity that the wind turbine produces into the grid. There are basically two ways to work around this problem. One way is to select a site for the wind turbine farm that is close to existing power lines, which will mean that a minimum number of new power lines will be needed to get power to the grid. Another way of working around this issue is to install as much of the power transmission cables underground as possible. **Underground transmission lines** are electrical power lines that are buried underground. These power lines may be capable of transmitting power several miles to a location where the construction of larger power transmission towers does not create an issue.

Other ways to head off problems with visual concerns is to hold a series of meetings with residents and businesses in the area where the wind turbine installation is planned. These meetings can get input from area residents and businesses and may include a thorough discussion of pros and cons about the site. In these meetings, residents can express concerns, which many times can be integrated into the plan to head off problems in the future.

10.5 SMALL RESIDENTIAL WIND TURBINE SYSTEMS

Small residential wind turbine systems are available for purchase and installation in a number of sizes from 5 to 25 kW. A **residential wind turbine** is one designed specifically for a home or apartment complex. The small residential wind turbine system is designed to offset the use of electricity from the power company. Most residential users consume between 1000 and 2000 kW per month, so the system may only be able to provide a portion of the total energy needed by the residence each week.

The total amount of electrical power from the wind turbine that will be available will depend on the amount of wind and the number of hours the wind blows each day. These small residential wind turbine generators can produce DC voltage which can be stored in batteries or used directly by moving DC power through an **electronic inverter**. The inverter is also known as a **power electronic frequency converter** (PEFC), and it takes AC electricity produced by a wind turbine that has a frequency of any value and converts it to DC electricity and then back to AC electricity with exactly 60 Hz frequency. The inverter can also be used to take DC voltage from a wind turbine or batteries and convert it to AC voltage with a frequency that is exactly 60 Hz so it can be integrated directly with the grid or used by the consumer.

The inverter allows the wind turbine generator to provide electricity with exactly 60 Hz frequency whenever the blades are turning, regardless of the speed. If the wind turbine produces DC power at night or when the residence does not need the power, it can be stored in batteries. If the generator produces AC power or the power is sent through an inverter, a net meter may be used to sell this excess power back to the grid. If storage batteries are used, their DC voltage can be sent back through the inverter into the residence and used as 60-Hz AC power at a later date.

Figure 10-1 shows an example of a residential wind turbine that is located in a suburban area. Some wind turbines used in suburban areas may have height restrictions and other zoning restrictions that may restrict the size and height of the system. These systems are also used for cabins or other remote areas where traditional grid electricity is not available. In these systems, the wind turbine may be used in addition to solar panels, and the total energy is stored in large banks of batteries. When the residence needs to use power, the circuits in the home are connected through the inverter, and the power put into the batteries from the solar energy and

FIGURE 10-1 A wind turbine in a residential setting. Zoning may limit the height and noise of the wind turbine. (Courtesy of Southwest Windpower.)

the wind turbine are combined to provide sufficient energy for the home. The inverter must be large enough to convert all the power the home needs at any given time.

Figure 10-2 shows a small wind turbine to supply power to a rural home. In most rural areas, zoning and other regulations are not as strict. Figure 10-3 shows a wind turbine installation in an agricultural area, where the regulations and zoning are usually more lax with regards to tower heights and noise pollution. Also, many states provide additional incentives to locate wind turbines on agricultural land. A wind turbine in an agricultural setting can be installed in a field near the residence, and the wiring from the wind turbine is run underground to the residence or farm buildings. The underground cable must be buried far enough below ground not to interfere with plowing or other agricultural functions that may disturb the earth to a depth of 2–3 ft. Some installations may need to integrate with field

FIGURE 10-2 A small wind turbine in a rural area. Height and noise limits may be less restrictive than in a suburban area. (Courtesy of Southwest Windpower.)

FIGURE 10-3 A wind turbine installed in an agricultural area, where height and noise restrictions generally do not limit the height or noise level of wind turbine towers. (Courtesy of John Fellhauer, jfellhauer@surenergy.us.)

drainage tile that is already installed in a field, so field maps may need to be studied prior to selecting the site and installing the wind turbine and any underground electrical cables.

10.6 HOME-MADE WIND TURBINE SYSTEMS

Another type of wind turbine system is the small home-made system. These systems usually have 2.5–10 kW output. The small home-made system may be built completely from scratch, including the turbine blades, which are usually constructed from either wood or fiberglass, or it can be built from a kit that requires some assembly. These kits usually require the turbine blades to be installed on the turbine and the entire assembly connected to the pole and lifted into place as a completed assembly. The remainder of the assembly and installation includes electrical wiring, an electrical disconnect switch, and electrical controls. In most cases, since single residents do not use electricity 24 hours a day, energy storage system such as batteries are also required. Since the wind turbine generator does not have any speed control, it will produce an AC voltage that does not have regulated frequency. If the system has a DC generator, the speed of the blades does not need to be controlled, since the DC electrical power can be directed to the batteries at nearly any voltage level. You should remember that the speed at which the DC generator turns will control the level of voltage that the generator produces.

When batteries are used to store the generated DC electrical power, two additional issues must be taken into consideration. First, an electronic control must be added to the batteries circuit to ensure that they do not become overcharged. If battery charging is not controlled, the wind turbine generator may continue to add electrical power to the batteries after they are fully charged, and this overcharged condition may cause the cells to dry out and become damaged. Another issue that must be controlled is when the wind turbine has not run for some period of time and the batteries are drained down due to usage of electrical power in the residence. If the wind turbine generator needs a small amount of voltage to excite the generator field to get it to begin to produce power, it is important that an electronic circuit control is provided to ensure that the batteries never drain down to 0 V. This electronic cut-out switch will disconnect the batteries before they are completely drained and then reconnect them when the wind turbine blades begin to turn again and the generator uses the electricity remaining in the battery to excite the generator field. In some cases the overcharge control and the low-voltage control are built into the electrical frequency converter (inverter).

A power electronic frequency converter or inverter will be needed to produce single-phase AC voltage with a controlled 60-Hz frequency and a regulated voltage level. The inverter can be connected directly to the generator output, or it can be connected to the battery circuit, where it is used to convert the stored DC voltage from the battery to AC. The batteries for these types of applications require significant inspection and maintenance so that they do not run dry or become overcharged or undercharged.

During the installation process, the wiring from the wind turbine generator to ground connections is attached securely to the tower so that it will not come loose in higher winds. Guy wires are also used to stabilize the tower and turbine so that it does not move in the wind. Figure 10-4 shows a small wind turbine located near a home. Figure 10-5 shows a close-up of a small wind turbine, in which you can see the smaller size of the turbine and blades and the guy wires for this installation.

FIGURE 10-4 Small wind turbine located near a home. (Courtesy of Southwest Windpower.)

FIGURE 10-5 Close up of the wind turbine in Figure 10-4. Notice the size of the turbine and blades, and the guy wires to stabilize the tower.

Home-made systems are constantly evolving so that people who want to install a wind system to provide electrical energy can do so with the help of a few people. These small systems will help offset the electric bill but will not replace the power company completely. These smaller systems typically require more maintenance than some of the larger systems designed for commercial use. This means that the person who installs this type of system needs to be able to perform the maintenance or have someone who will be willing to do the maintenance to keep the system operating. If you have to pay for the maintenance, the cost will typically consume the benefits gained by offsetting the electric bills. When you are deciding the feasibility and cost of installing and operating a small home system, you must include all of these things in your cost estimate.

10.7 COMMERCIAL WIND TURBINE SYSTEMS

Small and medium-size wind turbines that are sized from 10 to 100 kW are available to be installed right on the property of a commercial establishment. A **commercial wind turbine** application is one where a small business or small industry has a wind turbine and either uses the electrical energy it produces directly to offset the energy it purchases from the electric utility or sells the electricity back to the grid. Small to mid-size commercial establishments are finding that they generally have room on their

FIGURE 10-6 A wind turbine at a marine sales and service facility.

property to install a wind turbine and use the electrical energy it produces to offset some of power they previously purchased from an electric utility. In most cases these mid-size wind turbines are grid-tied and have a net meter so the owner can sell any excess electrical power back to the utility. Typically, these commercial applications do not use the same amount of electrical power at every hour during the day, so there are times when the wind is blowing and they are not using the power and it can be sold back to the electric utility. Figure 10-6 shows a wind turbine located at a marine sales and service facility. The wind turbine for this application works well because during the summer months the facility requires power approximately 12 hours per day, and any excess produced in the evening is sold to the grid. Also, during the winter months, when electrical use is lower, the electricity the wind turbine produces can be used for electric heating equipment to help heat the building, or it can be sold back to the utility through the grid if other forms of heating energy such as natural gas are less expensive. It is important to remember that any wind turbine system that is connected to the grid requires safety disconnect switches and transformers to ensure that the system connection meets the codes for the grid.

Figure 10-7 shows a wind turbine installed for a small commercial establishment that is primarily office space. The wind turbine in this application can produce the electrical power the building needs for lighting and ventilating. Any excess electrical power the wind turbine produces is sent back to the grid and sold to the electric utility. Another advantage of having a wind turbine produce power is that some commercial applications work with other commercial applications that will provide incentives in contracts for efforts to produce "green" or environmentally safe energy. Sometimes these incentives, plus original grants, will shorten the payback period for

FIGURE 10-7 A wind turbine installed at a small building for a small commercial establishment, which is primarily office space.

the system. Many times these types of systems are installed by companies that have an environmental or energy philosophy that requires them to utilize all possible technologies to limit pollution and help the environment.

Figure 10-8 shows a wind turbine at a railroad switching yard. The wind turbine in this application is large enough to provide the electrical power required to operate all the lighting in the yard, which operates 24 hours a day, 7 days a week. This type of wind turbine is perfect for this application, since the railroad yard has a nearly constant electrical load throughout the day and throughout the year. The wind turbine can be designed and specified larger than the electrical demand, and any excess electrical power that is generated during high winds can be sold to the electric utility through the grid. If the wind turbine is a little smaller than the demand, the project payback period is designed basically around the wind turbine providing electrical power to offset the electric bill on a monthly basis. Since this type of application has electricians and other maintenance personnel on site working on railroad engines and switchyard electrical applications, they can receive training and do the majority of work on the wind turbine installation. Large diesel engines that are used on the railroad to pull a train are

FIGURE 10-8 A wind turbine installed at a railroad switching yard.

basically a large generator driven by a diesel engine, so these personnel will be familiar with electrical generation and control.

10.8 WIND FARMS

The next type of installation that will be discussed two or more wind turbines placed in the same location. A **wind farm** is a group of wind turbines that may be two or more; some large wind farms have more than 100 wind turbines all placed close to each other so they can share the electrical grid connection, transformers, and maintenance equipment. In some cases, several hundred wind turbines are placed in the same location.

There are many advantages to grouping wind turbines in close proximity to each other. One of these advantages is a single interconnection to the grid, through either electrical switching or a substation. This also allows the multiple wind turbines to share larger transformers that are used to increase the voltage to the amount needed at the grid. Another advantage of placing multiple larger wind turbines on the same property is that they can all be serviced at nearly the same time, requiring paying only once for the transportation to bring the large crane onto the property that is needed for the maintenance of the nacelle and blades. Many times the cost of transportation to move the crane to the wind turbine property is over half the cost of the maintenance. In other applications, where larger numbers of wind turbines are located in close proximity, the crane can be purchased or leased and maintained at the property, so that it never has to move farther than between wind turbine towers. This also allows these applications to hire full-time maintenance crews that remain with these wind turbines throughout the year. If the wind farm is smaller and has only two or three wind turbines, the equipment to maintain them must be rented, and the personnel to do the maintenance are generally contracted for short periods of time.

Another advantage of wind farms is that it is less expensive to install additional wind turbines, because much of the work has already been completed for previous installations. Also, if any zoning restrictions were negotiated, they can be negotiated for all the wind turbines on the site, which is sometimes easier than negotiating for a dozen sites which may cross state, county, and other jurisdictional lines. Typically, complaints about visual appearance are minimized because the wind farm produces a tax base for the area and also provides additional electrical power, which is something most industrial regions and states are looking for to entice additional industry to their areas. In some cases there are large areas of unused land that are perfect sites for wind turbines, so these wind farms serve the purpose of producing energy while using land that would not be used for any other purpose.

The newest trend in wind turbine installations includes larger installations that may provide up to half the electrical power that a small coal-fired plant would provide. These wind farms produce sufficient power to make additional construction of other types of electrical power plants unnecessary. Some of the larger energy producers, such as BP, General Electric, and Siemens, are involved in large wind projects in the United States. Other energy companies and utilities have also proposed large wind farm projects throughout the United States and in some cases for offshore applications. Table 10-1 lists 11 of the largest wind farms in the United States as of 2008. The names on the list are not as important as the fact that a large number of wind turbine farm projects have recently been completed or are currently under construction. It is also important to note that at one time

TABLE 10-1 Largest Wind Farms in the United States as of 2008

Wind Farm	State	Size (kW)	Condition
Klondike Wind Farm	Oregon	400	Completed
Lone Star Wind Farm	Texas	400	Completed
Peetz Wind Farm	Colorado	400	Completed
Sweetwater Wind Farm	Texas	585	Under construction
San Gorgonio Pass Wind Farm	California	619	Completed
Capricorm Ridge Wind Farm	Texas	662	Completed
Tehachapi Pass Wind Farm	California	685	Completed
Horse Hollow Wind Energy Center	Texas	736	Completed
Fowler Ridge Wind Farm	Indiana	750	Completed
Sherbino Wind Farm	Texas	750	Under construction
Roscoe Wind Farm	Texas	781	Under construction

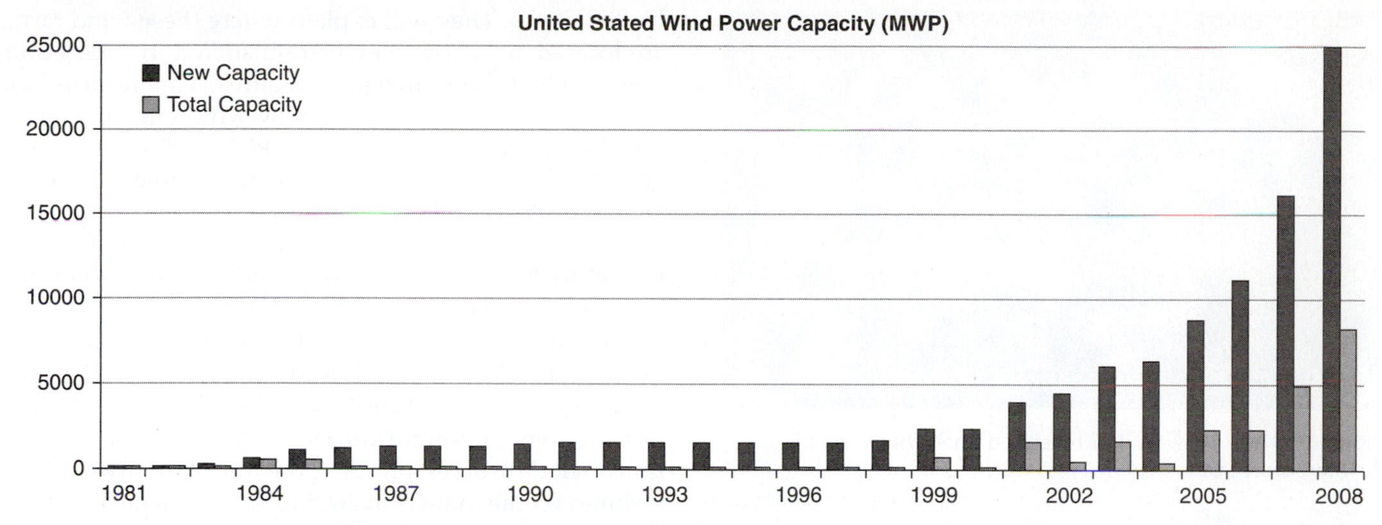

FIGURE 10-9 Electrical energy production and capacity from wind turbines in the United States. (Courtesy of U.S. Department of Energy.)

most of these larger wind farms were located in Texas and California, whereas today they are being placed in nearly every state where wind power production is sufficient to operate large wind turbines throughout the year.

Figure 10-9 shows the amount of electrical energy produced from wind turbines from 1981 through 2008. The left side of the chart shows the output energy in megawatts; recall that a **megawatt** is 1,000,000 watts (1 million watts). You can see that in 2008 we were reaching 25,000 MW of production in the United States alone. You can also see how the amount electrical energy produced by wind power has increased dramatically since 2000. Government incentives and more efficient wind turbines are the cause of this dramatic increase in the amount of electrical energy produced from wind turbines. It is important to understand that each megawatt produced from wind power offsets electrical power produced by coal, nuclear power, or oil, and represents a new installation of those types of energy production that was not needed. As government incentives increase and electricity produced by wind turbines becomes more efficient, the cost per unit of generated electricity is getting lower. However, it is important to understand that the cost of electrical energy produced by wind energy is by far the most expensive when compared to electricity generated by coal, nuclear, and hydroelectric facilities. The major reasons for making the investment in wind turbine generation is to provide a cleaner environment and to produce electrical power from renewable energy that does not have to be imported from other countries.

The amount of electrical energy produced by wind turbines from 1996 through 2008 increased steadily in the United States as well as worldwide, and the amount of energy produced has increased dramatically since the year 2000. Total world production of electricity from wind turbines is now over 120,000 MW.

Another way to review the increase of energy production from wind turbines is to compare the amount of energy produced by country. As of 2008, countries that produced the largest amount of electrical energy included (in order of the largest producers to the smallest) the United States, Germany, Spain, China, India, Italy, France, the United Kingdom, and Japan. It is important to understand that each of these countries has large numbers of projects under construction and are continually adding to the electrical generation capacity. It is possible that the leaders will quite frequently change positions on the list of who produces the most electrical energy from wind turbines.

In 2005, for instance, Germany was the leading producer of wind energy, followed by Spain, the United States, India, France, the United Kingdom, China, France, Italy, and Japan. In 2005, Germany produced almost twice as much wind energy as the next closest producers. Since 2005, the wind production and capacity in the United States has more than doubled to over 25,000 MW.

Small Community Wind Farm Projects

Many small communities have control over their electric utility company and have choices in where they purchase their electrical energy. In these communities, one of the options is to install wind turbines or lease land to companies that will install and operate wind turbines so that the electrical energy from them can be used by the community and offset electrical energy it would otherwise have to purchase off the grid. If the amount of electrical energy these wind turbines produces is larger than what can be used, it is sold back to the grid to offset times when more electrical energy must be purchased. Figure 10-10 shows an example of this type of installation. The picture shows four 1.8-MW, utility-scale wind turbines installed in Bowling Green, Ohio. Their electric utility system is part of Amp Ohio, which is a cooperative of utility companies for area communities which have joined together to produce and purchase energy at the lowest cost. The wind turbines provide approximately 7.2 million kWh of electricity annually, which is

FIGURE 10-10 Four wind turbines at a small wind farm in Bowling Green, Ohio.

enough to power approximately 3000 homes. These wind turbines are near a county landfill and are located in an area that is primarily agricultural, about 6 miles from the city. The four turbines are each 391 ft tall.

Mid-Sized Wind Farm Projects

In some parts of the country there is enough rural agricultural area that has sufficient wind that blows strong enough over most of the year. Some of the areas where mid-sized wind farms are located include Minnesota, Iowa, and Nebraska. Since this area is mainly agricultural and far enough from cities, large numbers of large-size wind turbines can be installed in one location. Since there is not a large demand for electricity in these areas, the majority of the electricity produced by these wind turbines is sent into the grid and distributed to cities that are farther away. Figure 10-11 shows the wind turbines on a mid-sized wind farm.

Large Wind Farm Projects

The following paragraphs will provide more detailed descriptions of some of the large wind turbine farms in the United States. They will explain where these wind farms are located, and how they were installed in stages until they reached their current capacities. This information will give you a better idea of how wind energy is being installed in the United States and how the capacity is growing very rapidly. You can refer to Table 10–1 to see all of the sites and their capacities.

KLONDIKE WIND FARM, OREGON Klondike wind farm is located just east of Wasco, Oregon, which is 110 miles east of Portland, Oregon. This wind farm originally came online in 2001 and consisted of 16 wind turbines that produced up to 24 MW of electrical power. The second phase of the project, brought online in 2005, added 50 additional wind turbines, bringing the total capacity of all the turbines on this project in 2005 to 99 MW, which was sufficient power for over 25,000 homes. The third phase of this wind farm consisted of an additional 122 wind turbines that produced 221 MW and was completed in 2007. This wind farm is currently the largest in the northwestern United States.

LONE STAR I AND II WIND FARM, TEXAS The Lone Star wind farms are located in Shackelford and Callahan Counties in Texas, just north of Abilene, Texas. This wind farm consists of two separate sections called Lone Star I and Lone Star II, and it is located on approximately 36,000 acres. Lone Star I was brought into production in December 2007 and has 88 Gamesa G83 2.0-MW turbines and 12 Gamesa G87 2.0-MW turbines. Lone Star I has an installed capacity of 200 MW, which can power over 60,000 homes. The second phase of the project, Lone Star II, has 100 Gamesa G87 2.0-MW turbines that produce 200 MW. It came online in 2008. Figure 10-12 shows the Lone Star I and II wind farm.

FIGURE 10-11 A mid-sized wind farm. (Courtesy of EWT International.)

FIGURE 10-12 Lone Star I and II Wind Farm is located in Shackelford and Callahan Counties, Texas. This wind farm is located north of Abilene, Texas. (Courtesy of Lone Star ©2008, Horizon Wind Energy LLC. Special thanks to Lone Star Wind Farm and Horizon Wind Energy LLC.)

PEETZ WIND FARM, COLORADO The Peetz Wind Farm, also called the Peetz Table Wind Energy Center, is a 400-MW wind farm located in Peetz, Colorado, which is in northeastern Colorado. Peetz is 150 miles northeast of Denver, Colorado. This wind farm uses 267 General Electric wind turbines, and the 400 MW they produce can power nearly 120,000 homes.

Peetz Table Wind Energy Center is located on Colorado's eastern plains, which is home to some of the best winds available for wind turbines in the United States. This project was developed in three stages, and the last stage was completed in 2007. The project also provided a 78-mile-long electrical transmission line to connect to the grid. The latest turbines for the third phase included 1.5-MW generators with 80-ft blades and 2.3-MW generators with 90-ft blades.

SWEETWATER WIND FARM, TEXAS Sweetwater Wind Farm is located in Nolan County, 40 miles west of Abilene, Texas. The Sweetwater Wind Farm is just off Interstate Highway 20. Sweetwater Wind Farm produces 585 MW and was created in five stages. The fourth stage of the wind farm included 135 Mitsubishi 1.0-MW wind turbines and 46 Siemens 2.3-MW wind turbines. The fifth stage of construction included 46 Siemens 2.3-MW turbines. The fifth stage of construction added another 80 MW of power and was completed and became fully operational in late 2007. Figure 10-13 shows the wind turbines at Sweetwater Wind Farm.

FIGURE 10-13 Sweetwater Wind Farm is located in Nolan County, west of Abilene, Texas. (Courtesy of iStockphoto.com)

SAN GORGONIO PASS WIND FARM, CALIFORNIA San Gorgonio Pass Wind Farm is located on the eastern slope of San Gorgonio Pass near White Water, California. White Water is east of Riverside, California. This location is one of the windiest places in southern California and is near the Coachella Valley. Two other large wind farms, the Altamont Pass Wind Farm and the Tehachapi Pass Wind Farm, are also located in this area. As of January 2008, the San Gorgonio Pass Wind Farm consisted of 3218 units delivering 615 MW. White Water, California, is just east of Riverside, California, which is a fairly large city.

CAPRICORN RIDGE WIND FARM, TEXAS Capricorn Ridge Wind Farm is a 662.5-MW wind farm located in Sterling and Coke Counties in Texas. Sterling and Coke Counties are about 100 miles southwest of Abilene, Texas, and the wind farm is located south of Interstate 20. The wind farm has 342 General Electric wind turbines, each of which produces 1.5 MW, and 65 Siemens wind turbines, each of which produces 2.3 MW. The wind turbines on this farm are capable of generating enough electricity for more than 220,000 homes. Figure 10-14 shows the wind turbines of Capricorn Ridge Wind Farm.

FIGURE 10-14 Capricorn Ridge Wind Farm is located in Sterling and Coke Counties, Texas, 100 miles southwest of Abilene and just south of Interstate 20. (Courtesy of Thinkstock)

FIGURE 10-15 Tehachapi Pass Wind Farm is located on the east and south areas of Tehachapi Pass and is one of California's larger wind farms. It is located east of Bakersfield, California. (Courtesy of iStockphoto.com)

FIGURE 10-16 Horse Hollow Wind Energy Farm in Texas is southwest of Abilene in west central Texas. (Courtesy of Thinkstock)

TEHACHAPI PASS WIND FARM, CALIFORNIA Tehachapi Pass Wind Farm is located on the east and south areas of the Tehachapi Pass and is one of California's larger wind farms. The wind farm is located east of Bakersfield, California, just off Highway 58. The turbines have been in place since the early 1980s and the original wind turbines were the largest in their day, but today they are much smaller than newer turbines on that wind farm. The Tehachapi Wind Farm has approximately 5000 wind turbines, and it is the second largest collection of wind generators in the world. Its collective electrical power output is 800 million kWh of electricity, which is enough to meet the needs of 350,000 residential customers every year. Figure 10-15 shows the wind turbines of Tehachapi Pass Wind Farm.

HORSE HOLLOW WIND ENERGY CENTER, TEXAS Horse Hollow Wind Energy Center is located in Taylor and Nolan Counties, Texas, which is southwest of Abilene, Texas. It has 735.5 MW of capacity, with 291 GE Energy 1.5-MW wind turbines and 130 Siemens 2.3-MW wind turbines spread over nearly 47,000 acres. The first phase of the Horse Hollow Wind Energy Center was completed in late 2005 and produces 213 MW. Phase two of the wind farm produces 223.5 MW and was brought online in the second quarter of 2006. Phase three produces 299 MW and was bought online at the end of 2006. Figure 10-16 shows the wind turbines of Horse Hollow Wind Energy Center in Texas.

FOWLER RIDGE WIND FARM, INDIANA The Fowler Ridge Wind Farm in Benton County, Indiana, generates 750 MW and was built in two phases. Benton County is about 90 miles northwest of Indianapolis. Dominion and BP Energy are partners for 650 MW, with BP retaining sole ownership of 100 MW. The total capacity of the first phase is 400 MW, and it was brought online at the end of 2008. The first phase consists of 222 wind turbines, of which 182 are Vestas V82 that produce 1.65 MW each and 40 of which are Clipper C-96 wind turbines that produce 2.5 MW each.

The second phase, Fowler Ridge II Wind Farm, will produce 350 MW. Construction began in early 2009 and it was expected to be online in the first quarter of 2010. The second phase will use133 GE SLE wind turbine generators, with each wind turbine rated at 1.5 MW. Figure 10-17 shows the wind turbines of Fowler Ridge II Wind Farm.

SHERBINO WIND FARM, TEXAS The Sherbino Wind Farm is located in Pecos County in west Texas. The wind farm is approximately 40 miles east of Fort Stockton, Texas, on approximately 10,000 acres. The first phase produces 150 MW from 50 Vestas V-90 wind turbine generators, each of which has a capacity of 3 MW. When stage two is complete, the overall project will have a potential capacity of 750 MW, which will generate enough electricity to power approximately 225,000 average U.S. homes. Figure 10-18 shows the wind turbines of the Sherbino Wind Farm in Texas.

ROSCOE WIND FARM, TEXAS The Roscoe Wind Farm is located near Roscoe, Texas, which is a city in Nolan County, Texas. Roscoe is 9 miles west of Sweetwater, Texas, and 50 miles west of Abilene, Texas. The wind farm has three phases completed and will be one of the world's largest wind farms with 627 wind turbines and a total capacity of 781.5 MW. The first and second phases of the wind farm came online in 2008 and produce 335.5 MW.

The third and fourth phases were expected to be on online in late 2009, and they will raise the total capacity of Roscoe Wind Farm to 781.5 MW, which will temporarily move Roscoe to the position of being the largest wind farm in the United States. As a comparison, the 627 wind turbines at Roscoe Wind Farm produce slightly less than the smaller nuclear power plants in the United States. The 781.5 MW will be enough energy to power more than 250,000 average homes.

FIGURE 10-17 The Fowler Ridge Wind Farm in Benton County, Indiana. Benton County is about 90 miles northwest of Indianapolis. (Photograph courtesy of BP p.l.c.)

FIGURE 10-18 The Sherbino Wind Farm is located 40 miles east of Fort Stockton in Pecos County in west Texas. (Courtesy of Vestas Wind Systems A/S.)

10.9 OFFSHORE INSTALLATIONS IN THE UNITED STATES

An **offshore wind farm** is one that is located away from the shoreline in the waters of a lake, river, or ocean. The largest offshore installation of wind turbines in the United States is slated for Gulf Coast waters approximately 10 miles off South Padre Island, Texas. The wind farms will consist of 500-ft-tall wind turbines. South Padre Island is about 25 miles south of Brownsville, Texas. These wind turbines are not a visual problem in this area, since there have been oil rigs in the area for years. Since the wind farm will be over 10 miles from the shore, the turbines will not be visible from shore, and any noise will not be noticeable. The biggest problem in putting wind farms far offshore is that the electrical power is sent to shore through underwater cables, and the farther the wind farm is offshore, the longer the cables must be, which may lead to voltage drops and cable losses. **Underwater cables** are electrical cables that are positioned under the surface of the water, usually along the bottom of the ocean or lake.

Off shore locations are desirable because the winds in these areas are stronger, more constant, and last longer through each day and through the year. Other offshore installation sites have been investigated on the East Coast at Martha's Vineyard in Massachusetts, and in the Pacific near Hawaii. These sites may take years to develop because of concerns about altering or destroying the beauty of the landscape or seascape.

One problem with placing wind farms in the Gulf of Mexico is that the area there is subject to hurricanes or strong tropical storms winds that may last for several days at a time. The winds in even a small hurricane may be over 80 mph and could seriously damage the wind turbines. A number of technical innovations are being investigated, including floating platforms that could be moved, or designs in which the blades can be stored so that only the tower silhouette remains exposed in strong winds.

Another location where wind turbines are being considered is on the Great Lakes. The winds over the water of the Great Lakes are stronger than at nearby locations on the land. The lake winds would provide a strong resource of electrical power from wind. The other advantage of placing wind turbines over freshwater areas rather than saltwater is that corrosion will be less of a problem in freshwater locations. **Saltwater corrosion** is corrosion that is caused by the chemical makeup of the salt, which attacks metal and causes it to rust quickly if the metal is not painted or treated. Corrosion in freshwater areas is still a problem due to the constant splashing of the water on the metal parts of the towers that are exposed or not coated or painted. Another reason that Lake Erie and Lake Ontario are being considered is that they are fairly shallow—and there are locations where the water depth is less than 30 ft and the lake bottom is solid enough to support large towers. This is very important, since the depths of the ocean mean that turbine towers located there may need to be well over 50 ft to reach bottom.

10.10 LARGE OFFSHORE WIND FARMS IN EUROPE

Great Britain (England and Scotland), Denmark, and Germany have large numbers of offshore wind farms that have been in production for over 20 years. Newer and larger wind turbines are being installed on offshore wind farms to provide additional electrical energy for these countries. This section provides more information about these offshore wind farms.

Offshore wind turbine location is generally considered to be at least 10 km (6 miles) or more from land. The offshore wind turbines have several advantages over those located on inland areas in Europe. For example, they are less obtrusive at sea than turbines on land, as their apparent size and noise is limited because of the distance they are located from shore. The average wind speed at sea is usually higher and the winds last longer than over land. Wind turbines at sea can be floating or anchored deep in the sea floor. Offshore installation is typically more expensive than onshore installation. Offshore towers are can generally be taller than onshore towers because there are no height restrictions as there are on land. Power transmission from offshore turbines is through underwater cables, often using high-voltage DC if the distance between the wind turbine and shores is large. Offshore saltwater causes substantial corrosion, which raises the overall cost of maintenance since corrosion of the towers and welds is continuous. Repairs and maintenance are usually more costly than for onshore turbines, and this leads to installing the largest wind turbines possible to produce the most electrical energy with the fewest number of wind turbines.

Another advantage that offshore wind turbines have over onshore units is that the tower, nacelle, and blades are limited in size because of transportation issues when they need to be transported by trucks though the often narrow roads of Europe. When a wind turbine is located offshore, its size is limited only by the size of the ship or barge used to move it out to sea.

As of 2008, Europe led the world in development of offshore wind power, due to strong wind resources and shallow water in the North Sea and the Baltic Sea. Denmark installed the first offshore wind farms, and for years was the world leader in offshore wind power. More recently, the United Kingdom took the lead in late 2008.

In 2009, the first deep-water floating wind turbines were anchored off the coast of Stavanger, Norway, in water that is over 200 m deep. The wind turbines are mounted on towers that are over 120 m tall, and each produces 2.3 MW. The following paragraphs provide information about several of these offshore wind farms.

Lynn and Inner Dowsing Wind Farm, England

Lynn and Inner Dowsing Wind Farm is located in the shallow waters off the coast of Lincolnshire, England. This wind farm has 54 Siemens 3.6-MW wind turbines that have a total generating capacity of 194 MW. Power from the wind turbines is transmitted through undersea cables that run between the wind turbines and shore. Once the power reaches the shore, it is transmitted by underground cables to a new substation at Middlemarsh, Skegness.

The wind turbine towers in this wind farm are located in water that is over 59 ft deep, and the tubes for the towers are pushed into the seabed up to 80 ft. The remainder of the wind turbine tower is mounted on the foundation in two pieces. Figure 10-19 shows the wind turbines of Lynn and Inner Dowsing Wind Farm.

Plans to create the world's largest offshore wind farm off the coast of Britain include a massive array of 341 giant wind turbines in the Thames Estuary, with construction starting in 2009. When this wind farm is completed, the turbines will generate up to 1 **gigawatt** (1 GW) of electricity, which is 1,000,000,000 watts (1 billion watts), which is equivalent to 1,000 MW or 1 million kW.

FIGURE 10-19 Burbo Bay Wind Farm off the coast of England. (Courtesy of Siemens Wind Energy.)

Nysted Wind Farm, Denmark

The Nysted Wind Farm is a joint Danish–Swedish wind farm that was installed in 2003 near Lolland, Denmark, with 72 Siemens wind turbines, each producing 2.3 MW. The 72 wind turbines give a total generating capacity of 166 MW. Lolland is the fourth-largest island of Denmark and is located in the Baltic Sea. Figure 10-20 shows the wind turbines of Nysted Wind Farm. The personnel who work on these wind turbines are ferried to and from them in boats.

FIGURE 10-20 Wind turbines at the Nysted Wind Farm near Lolland, Denmark. (Courtesy of Siemens Wind Energy.)

Questions

1. What does the term *turnkey* mean?
2. Identify four types of sight issues that you may encounter when you are selecting a location for a wind turbine.
3. Explain why a small residential wind turbine system may need a power electronic frequency converter (inverter).
4. Name two advantages of locating a wind turbine on agricultural land.
5. List three types of commercial applications of wind turbines.
6. Identify the five largest wind farms in the United States as of 2009.
7. Identify the total amount of electrical energy produced by wind turbines in the United States as of 2008.
8. Name two advantages of locating wind turbines over water.
9. Identify two wind farms that are located offshore in Europe.
10. Name two disadvantages of locating wind turbines over water.

Multiple Choice

1. Four categories of site issues are
   a. Environmental issues, noise issues, height issues, and connection issues
   b. Environmental issues, time issues, noise issues, and height issues
   c. Environmental issues, height issues, connection issues, and rules issues.
   d. All the above
2. Small residential wind turbine systems may need an electronic inverter if
   a. The wind turbine generator is used to charge batteries in the home, and all the loads in the home require AC voltage.
   b. The wind turbine generator produces DC voltage, which needs to be converted to AC voltage.
   c. The wind turbine blade speed is usually not controlled and can turn the generator shaft at any speed; an inverter can control the frequency of the AC voltage output.
   d. All the above.
3. The three largest wind farms in the United States as of 2009 are
   a. Horse Hollow Wind Energy Center, Roscoe Wind Farm, and Fowler Ridge Wind Farm
   b. Roscoe Wind Farm, Sherbino Wind Farm, and Fowler Ridge Wind Farm
   c. Sherbino Wind Farm, Roscoe Wind Farm, and Tehachapi Pass Wind Farm
   d. All the above
4. Where is the largest offshore wind farm in the United States planned to be built?
   a. Off the coast on the Atlantic Ocean
   b. Off the coast on the Gulf of Mexico
   c. Off the coast on the Pacific Ocean
   d. All the above
5. Identify the biggest advantages of locating wind turbines offshore.
   a. The wind turbine towers can be taller because there are no height restrictions.
   b. Larger wind turbines with larger blades can be used because barges carry the wind turbine parts to the final location and there are no restrictions due to narrow roads.
   c. The wind turbines do not corrode as badly when located over saltwater.
   d. All the above.
   e. Only a and b.
6. Where is the world's largest offshore wind farm planned to be located in 2009?
   a. Off the shores of Great Britain
   b. Off the shores of Sweden
   c. Off the shores of Germany
   d. Off the shores of Japan

7. What state in the Midwestern United States has the largest wind farm?
   a. Ohio
   b. Indiana
   c. Illinois
   d. Michigan
8. How is electrical power transmitted from offshore wind farms to shore?
   a. Underwater cables
   b. Underground cables
   c. Large overhead power transmission towers that are anchored to pylons in the ocean floor
   d. Large transformers mounted on barges
9. Which state has the most number of the 10 largest wind farms?
   a. Texas
   b. California
   c. Indiana
   d. Colorado
10. What equipment is used during a wind site assessment to determine the exact location for a wind turbine?
    a. A city or county map.
    b. A compass
    c. A GPS (Global Positioning System) unit
    d. All the above

CHAPTER 11

Installation, Troubleshooting, and Maintenance of Wind Generating Systems

OBJECTIVES

After reading this chapter, you will be able to:

- Explain the fundamentals of troubleshooting.
- Explain the difference between a symptom and a problem.
- Develop a procedure for locating a loss of voltage in an electrical circuit.
- Explain the troubleshooting process that is used to check wind turbines.
- Troubleshoot the electrical, hydraulic, and mechanical systems of a wind turbine.
- Perform periodic maintenance on the electrical, hydraulic, and mechanical systems of a wind turbine.
- Explain the steps in installing a wind turbine.
- Explain the start-up procedures for a wind turbine.

KEY TERMS

Continuity test
Megger
Overhaul
Periodic maintenance
Problem
Site preparation
Symptom
Troubleshooting
Troubleshooting matrix

OVERVIEW

This chapter will provide information about installation and troubleshooting wind turbine generators. Extensive information about the installation process will be provided, which will take you through the steps of installing a wind turbine and tower. You will also learn the basic steps involved in the troubleshooting process, regardless of which system you are troubleshooting. Some manufacturers of wind turbines provide tables and matrixes that you can use in troubleshooting the wind turbine or any subsystem. You will also learn that one of the main aspects of troubleshooting is learning how to tell the difference between a symptom and a problem. Many times technicians find themselves troubleshooting symptoms and never get to the root cause of the problem.

Detailed troubleshooting will be provided on each subsystem of the wind turbine, including the blades, the gearbox or transmission, and the generator. Material will also be provided about how to troubleshoot electrical problems inside the nacelle and on the ground inside the electrical power panels and any other disconnect equipment. Information will also be provided on how to troubleshoot electrical problems from the generator, through all the distribution cabling, and the connections to the grid. You will also learn how to troubleshoot the hydraulic system and the mechanical systems used to control the pitch of the blades, the braking system, as well as troubleshooting problems in other mechanical systems and the tower.

The final sections of this chapter will cover periodic maintenance of a wind turbine system and major overhaul of a wind turbine system. When you have completed this chapter you will have a very strong understanding of the knowledge necessary to troubleshoot and maintain a wind turbine system.

11.1 STEPS IN THE INSTALLATION OF A WIND TURBINE

Installation of a wind turbine consists of several steps that are completed in sequence. The entire process begins when a customer expresses an interest in obtaining electricity from wind power. This generally results in a first meeting in which the customer is interested in determining what is involved in putting a wind turbine on his site. He will also be interested in the initial cost and the ongoing operating and maintenance costs. He will want to know about any subsidies in the form of grants or write-offs that are available from the federal government or from his state or local government. He may also want some basic information about how the wind turbine operates and functions. During this process you will want to find out what the customer intends to do with the electricity: simply supply electricity for a process such as a home or industry, or provide electricity for a grid-tied system.

The next step in the process is to identify the amount of wind that is available at the location where the customer would like the wind turbine located. From this information you will be able to determine the size of the wind turbine that will provide the amount of electricity the customer wants, and an estimate of how much electricity the wind turbine will provide on a daily and yearly basis. After this information has been identified, you will be able to show the customer one or more wind turbines that will meet the specifications, and will be able to provide a cost for the complete system and a tentative schedule of when the installation work can be completed and the generator can begin producing electricity.

Once contracts have been written and signed, the wind turbine system, which consists of the wind turbine, the tower, the electrical controls, and the concrete pad, can be ordered. The availability of each of these items will help determine the final schedule for installation. When the size of the wind turbine and its tower has been determined, the remainder of the equipment designed for it can be ordered and installation can begin. For example, once the size of the turbine and the tower has been determined, site preparation can begin. **Site preparation** includes preparing the concrete pad and its footings as well as preparing all underground conduits for the electricity portion of the system, which must be located under the concrete pad. The exact locations of the mounting bolts for the tower, and the locations for electrical conduits and panels, must be precisely determined during this process. Once the concrete is poured, it will be very difficult to make any changes. An excavating company will be involved in digging the footer and holes for the pad. The excavators will also determine the size of the hole and the amount of concrete that will be required to mount the tower securely so that it can withstand the highest design wind speeds. When some larger wind turbines are installed on wind farms in remote areas, access roads must be created to allow trucks to deliver the large blades and towers to the site.

Site Preparation

Site preparation includes preparation of the pad and mounting for the wind turbine as well as providing access to the area for service vehicles and cranes that may be needed during the installation or at a later date for maintenance on the system. In some applications, the turbine will be mounted on a site that already has improvements such as paved roads or gravel roads and electrical service from the local power company. In other applications, the location of the wind turbine may be so remote that roads and electrical service must be brought from existing locations to the site where the turbine will be located. If the turbine and tower are large, the access road to the site must be capable of supporting heavy loads as well as large enough access areas to allow the transport vehicles to get in and out safely. In some cases where multiple wind turbines are being installed on a wind farm, this may involve creating a subbase surface for the roadway that must be built from scratch, or it may involve just upgrading an existing access road for larger loads. In some cases, soil samples may need to be taken to determine if the soil can handle a highway surface that can accommodate the heavy loads for the access vehicles to bring the wind turbine and the tower to the site. Soil samples may also be needed to determine how large and how deep the foundation for the concrete pad must be to support the wind turbine for the strongest winds for which it is designed. Once the foundation has been excavated, the dirt that has been removed must be hauled away or leveled off around the wind turbine site. Care must be taken to ensure that water drainage after heavy rain or snow is able to move excess water away from the concrete pad. Site preparation also includes determining the closest point to tie into the grid.

The next step in the process of creating the concrete pad includes placing reinforced steel that is used to strengthen the concrete pad. Figure 11-1 shows a typical concrete pad for a medium-sized 50-kW wind turbine. The

FIGURE 11-1 A pad for a wind turbine is readied for installation. Notice that the mounting bolts for the tower are precisely placed when the concrete is poured. The underground electrical conduit is also put into position before the pad is poured, and the electrical cabinet is mounted before the tower is set. (Courtesy of John Fellhauer, jfellhauer@surenergyohio.com.)

FIGURE 11-2 Notice the smaller size of the concrete pad for a monopole tower for a wind turbine. The concrete pad is deep enough to support the wind turbine in the strongest winds for which it is designed. (Courtesy of John Fellhauer, jfellhauer@surenergyohio.com.)

concrete pad in this figure is about 20 ft by 20 ft in area and probably about 6–10 ft deep. Companies that specialize in pouring concrete foundations and pads will be able to help you determine the proper size from a soil sample where the pad is to be located. Figure 11-2 shows a much smaller concrete pad that is needed for a monopole tower. The depth of the concrete base is determined by the size of the wind turbine and its designed maximum speed. In some applications, as shown in Figure 11-1, the electrical panel is mounted directly above the conduits that have been positioned in the concrete pad. In other applications, such as the monopole tower in Figure 11-2, the condits and electrical panels are not mounted on the concrete pad.

Permits and Other Documents Needed to Install a Wind Turbine

Another step in the process of installing a wind turbine is applying for and obtaining the construction permits and other permissions required by federal, state, and local government agencies. These may also include permission from local zoning boards that have control over any construction in their areas. Some states and countries have streamlined this process to ensure that permits are handled in a timely fashion. Some of these locations have created a process to entice the maximum number of wind turbines to be located in their areas. If the wind turbine is going into a residential area, other safety and environmental conditions must be met, such as view obstruction and noise abatement. Some zoning rules requires that wind turbines be located at some distance from homes, buildings, or highways, so that wind turbine blades do not become a hazard if they come loose or come off the rotor hub. In some cases the permits and other documents must all be obtained prior to the deal been agreed on. In other cases these documents are applied for and obtained during the planning process. This process becomes much easier when multiple wind turbines are placed in the same area.

Mounting the Wind Turbine Blades to the Rotor and on to the Nacelle

When the wind turbine is being installed, the turbine blades may be connected to the rotor hub while it is still on the ground. On other wind turbines, the blades and rotor are mounted to the nacelle once the tower is erected, and the nacelle is placed on the tower. Figure 11-3 shows several nacelles waiting for blades to be attached. Figure 11-4 shows a wind turbine blade ready to be shipped from the factory. The nacelle and blades in these figures are for larger types of wind turbines. Basically, the size and weight of the turbine blades determine whether they

FIGURE 11-3 Nacelles ready for blades to be installed on their rotor prior to mounting on the tower. (Courtesy of John Fellhauer, jfellhauer@surenergyohio.com.)

FIGURE 11-4 Turbine blades ready to be shipped to site. (Courtesy of EWT International.)

FIGURE 11-5 The nacelle of a large wind turbine is being installed on the pole after it has been erected and secured. (Courtesy of Northern Power Systems Inc.)

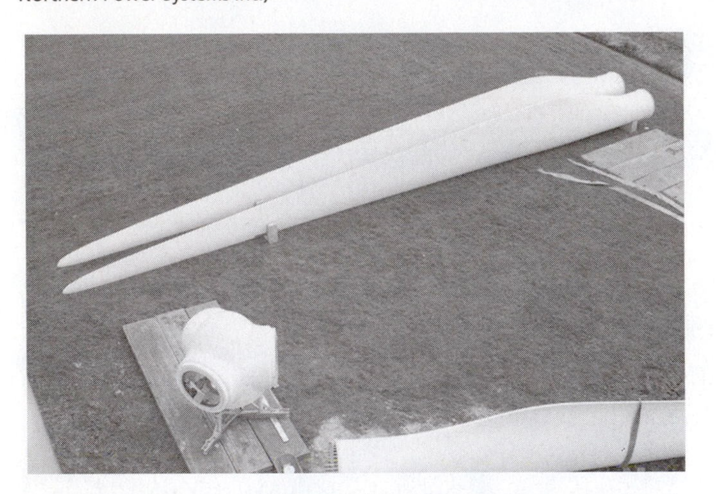

FIGURE 11-6 A wind turbine blade is bolted to the rotor while it is on the ground. The rotor with all blades attached is then mounted on the nacelle. (Courtesy of EWT International.)

are mounted to the wind turbine while it is on the ground or after it is in the air. For some smaller wind turbines, the blades can be attached to the rotor and the rotor can be connected to the nacelle while the tower is on the ground. Then a large crane or several cranes lift the complete assembled wind turbine into place, with the blades already attached to the hub and the hub connected to the nacelle.

Another way to get the blades onto the nacelle is to connect the blades to the rotor hub while the rotor hub is on the ground. Then a large crane lifts the rotor hub, with the blades attached, up to the nacelle, and the connections are made in the air. For some larger systems or where the wind turbine systems are located over water, a helicopter is used to lift the blades and hub as an assembly up to the nacelle, where the final connection is made in the air. The reason for this is that the blades and rotor would make the nacelle and tower too heavy to pick up if they were all assembled while it was on the ground. Figure 11-5 shows the nacelle being placed on the pole by a large crane.

Figure 11-6 shows wind turbine blades ready to be mounted on the rotor of a wind turbine. A large number of bolts are used to mount the wind turbine blade safely

FIGURE 11-7 Rotor and blades being mounted onto the nacelle of a wind turbine. (Courtesy of Fotolia.)

to the rotor. The next step in the installation process is to lift the blades and rotor assembly up to the nacelle and secure it where it is connected to the front of the nacelle. The rotor assembly is actually connected directly to the low-speed shaft that extends through the front of the nacelle. Figure 11-7 shows the wind turbine blades and rotor assembly being lifted into position where they will be connected to the low-speed shaft that protrudes out the front of the nacelle. In the installation of larger wind turbines, the component parts are lifted with a large crane if the wind turbine is located on land, and with a helicopter if it is a wind turbine based over water or at sea.

Mounting the Nacelle or Wind Turbine to the Tower

When mid-size wind turbines are installed, the nacelle can be attached to the tower while the tower is still on the ground. In some cases the turbine blades can also be mounted to the nacelle while it is still on the ground. Since a mid-size wind turbine is not too heavy, the entire assembly can be lifted into place by a large crane. Figure 11-8 shows technicians connecting the nacelle of a mid-size wind turbine to its tower. Figure 11-9 shows the technicians tightening the nuts and bolts that hold the nacelle to the tower. This process would be much more difficult to complete if the nacelle was attached to the tower after the tower was erected. When the tower is still on the ground, the technicians can get the amount of force and torque required to tighten the nuts to the bolts.

Figure 11-10 shows turbine blades and the rotor assembly being connected to the nacelle. The tower and nacelle must be supported high enough off the ground so that the rotor and blades can be mounted without them touching the ground. A wooden support or metal support

FIGURE 11-8 The nacelle of a mid-sized wind turbine is mounted to the tower while it is on the ground. (Courtesy of John Fellhauer, jfellhauer@surenergyohio.com.)

FIGURE 11-9 The Nacelle is secured to the tower on the mid-sized wind turbine using nuts and bolts. The nuts are tightened on the bolts while the tower is still on the ground. (Courtesy of John Fellhauer, jfellhauer@surenergyohio.com.)

FIGURE 11-10 After the nacelle is attached to the tower, the wind turbine blades and rotor are mounted to the shaft that protrudes through the front of the nacelle. Notice the wooden structure used to keep the tower off the ground during this work. (Courtesy of John Fellhauer, jfellhauer@surenergyohio.com.)

FIGURE 11-11 In this installation, the turbine blades are a little larger and heavier, and they must be supported while they are mounted to the rotor. Notice that the nacelle and tower are supported off the ground with a wooden structure to provide enough space to attach the blades to the rotor without the blades touching the ground. (Courtesy of John Fellhauer, jfellhauer@surenergyohio.com.)

can be designed to be adjustable to ensure that the blades are clear of the ground when the parts are assembled to the tower and nacelle. Figure 11-11 shows the larger blades of a wind turbine being supported while they are mounted to the rotor.

Installing the Electrical Cables and Making Electrical Connections

The next step in the installation process is routing the electrical cables from the nacelle down through the tower so it can be connected to the electrical equipment on the ground. In most small and mid-size wind turbines, electrical cables can be routed through the tower while it is on the ground. Figure 11-12 shows how the electrical cables are routed through the tower. Figure 11-13

FIGURE 11-12 Electrical cables are routed from the generator in the nacelle through the tower while the tower is on the ground. (Courtesy of John Fellhauer, jfellhauer@surenergyohio.com.)

FIGURE 11-13 The ends of the electrical cable that are routed to the ground have molded plug ends so that they cannot be connected incorrectly. (Courtesy of John Fellhauer, jfellhauer@surenergyohio.com.)

shows the ends of the electrical cable that are routed to the ground. These cables have molded plug ends so that they cannot be connected incorrectly. Another cable that is mounted to the electrical panel on the ground also has molded ends that will mate with the cable from the nacelle when they are connected. This type of electrical connection allows for varying cable lengths and ensures that the cables are connected correctly in the field.

When the electrical cable is routed through the tower, it needs to be connected to the wiring that is connected to the generator at the top of the tower. Figure 11-14 shows a technician making the electrical connections in a weatherproof box near the top of the tower. After the electrical connections are made at the top, the technician moves to the bottom of the cable and tests the cable for continuity. When the electrical connection at the top end of the cable has been completed, the technician will secure the cable to the tower at several points to ensure that it does not swing freely in the wind, which would eventually damage the cable.

FIGURE 11-14 A technician connects the electrical cable to the cable that is connected to the generator. The electrical box is weatherproof to keep rain and moisture out of the box when the cover is screwed in place. (Courtesy of John Fellhauer, jfellhauer@surenergyohio.com.)

FIGURE 11-15 Technicians are placing a lifting sling around the nacelle so the tower can be lifted into place. The other end of the sling is attached to the cable from the crane. (Courtesy of John Fellhauer, jfellhauer@surenergyohio.com.)

After the blades and rotor are mounted onto the tower and the electrical connections are made, a lifting sling is placed securely around the nacelle. The other end of the sling is attached to the cable from the crane. It is very important that the sling is secured correctly around the nacelle and blades so they do not slip out of the sling or are damaged when the tower is lifted into place. Figure 11-15 shows technicians placing a sling around the nacelle and rotor blades and connecting the cable from the crane.

During the time when the wind turbine is being lifted, it is important to secure the turbine blades to the tower so that they do not rotate freely. If the turbine blades are allowed to rotate, they may come into contact with the ground and be damaged, or they may swing into one of the technicians, who could be injured. Figure 11-16 shows technicians securing the turbine blades to the tower so that they cannot rotate when the tower is being lifted into place. After the tower is erected and secured, a

FIGURE 11-16 Technicians secure the turbine blades to the tower during the lifting process to keep them from swinging freely and possibly becoming damaged or hitting a technician during the lifting process. (Courtesy of John Fellhauer, jfellhauer@surenergyohio.com.)

FIGURE 11-17 A typical crane that is used to lift large wind turbine towers into place.

FIGURE 11-18 The crane begins to take tension out of the cable and sling and test the rigging to ensure that everything is secure. (Courtesy of John Fellhauer, jfellhauer@surenergyohio.com.)

technician will need to climb the tower and remove the cables that have been used to secure the blades to the tower during the lift.

A typical crane is shown in Figure 11-17. The crane used to lift the wind turbine must be able to lift a load that is larger than the actual weight of the wind turbine. For example, a lattice-type tower may weigh as much as 7000 lb, and the turbine blades, rotor, and drive train may weigh another 5000 lb, so the crane must be able to lift over 12,000 lb. The actual weight of these components will be listed in the technical data for the tower, rotor, and drive train.

When the crane begins to lift the tower and wind turbine, a safe practice is to allow the crane to lift the tower a few inches and test the balance and to ensure the sling and cable will not slip. Since the tower is moved only a few inches at this point, the tower can easily be lowered quickly to make any adjustments. Once the technicians are sure that the sling and cables are secure and operating correctly, the tower is lifted into place by the crane. Figure 11-18 shows the crane taking up the tension in its cable and the sling and beginning to lift the tower. Figure 11-19 shows a turbine and tower lifted well off the ground as the crane lifts the tower into the vertical position. Figure 11-20 shows the tower settled down on its mounting bolts. After the tower is in the vertical position and is set on its bolts, technicians quickly put nuts on the bolts and tighten them to secure the tower in position.

FIGURE 11-19 The crane has the tower, drive train, and blades lifted approximately halfway up to the vertical position. (Courtesy of John Fellhauer, jfellhauer@surenergyohio.com.)

FIGURE 11-20 Technicians quickly place nuts on the mounting bolts after the tower is set vertical and placed on the bolts. (Courtesy of John Fellhauer, jfellhauer@surenergyohio.com.)

FIGURE 11-21 The electric cable from the wind turbine is plugged into the cable that is coming up from the cabinet. A small loop is placed in the cable to relieve stress on the cable.

Making Electrical Connections to the Wind Turbine Electrical Panel

The next step in the installation process is to connect electrical wiring from the tower to the electrical panel for the wind turbine. When the concrete pad is poured, a large metal conduit is installed that is used to pull the main wiring into the electrical cabin. The technicians slide a pull tape into the conduit at the end where it enters the electrical cabinet. They then push on the pull tape, forcing it completely through the conduit so that its end is hanging out of the conduit at the pad. The electrical cables, which have molded plug end connections, are connected to the pull tape so that the free ends of the cable can be pulled into the electrical cabinet and leave the molded plug ends out of the conduit at the concrete pad end, so they can be connected with the cables that were fed down from the generator through the tower,. When the two plug ends are connected, a small loop is made in the cable to provide strain relief so that the cable does not become stressed by being too tight. Figure 11-21 shows the cable molded plug ends connected and the stain relief loop in the cable.

The other free ends of this cable are pulled through the conduit with the pull tape. The cable is trimmed to length and the free ends are connected to the electrical cabinet. Figure 11-22 shows a diagram that is mounted on the front door of the electrical cabinet. A two-wire cable comes down the tower from the generator to the electrical panel, where it connects into the main circuit breaker.

Figure 11-23 shows the electrical panel and the large black wires coming in at the bottom. After the cable from the wind turbine is connected to the electrical panel, the wires that will connect the wind turbine to the grid are connected to the cabinet. The electrical cabinet has a PLC controller or another type of computer controller that will ensure the voltage the wind turbine produces gets to the grid.

Figure 11-24 shows an example of the electrical service entrance for the property where the wind turbine is being installed. You can see the overhead cables from the utility power company. These cables come from a set of transformers that provide 240 V for this business. The wires come down the mast directly to an electric meter

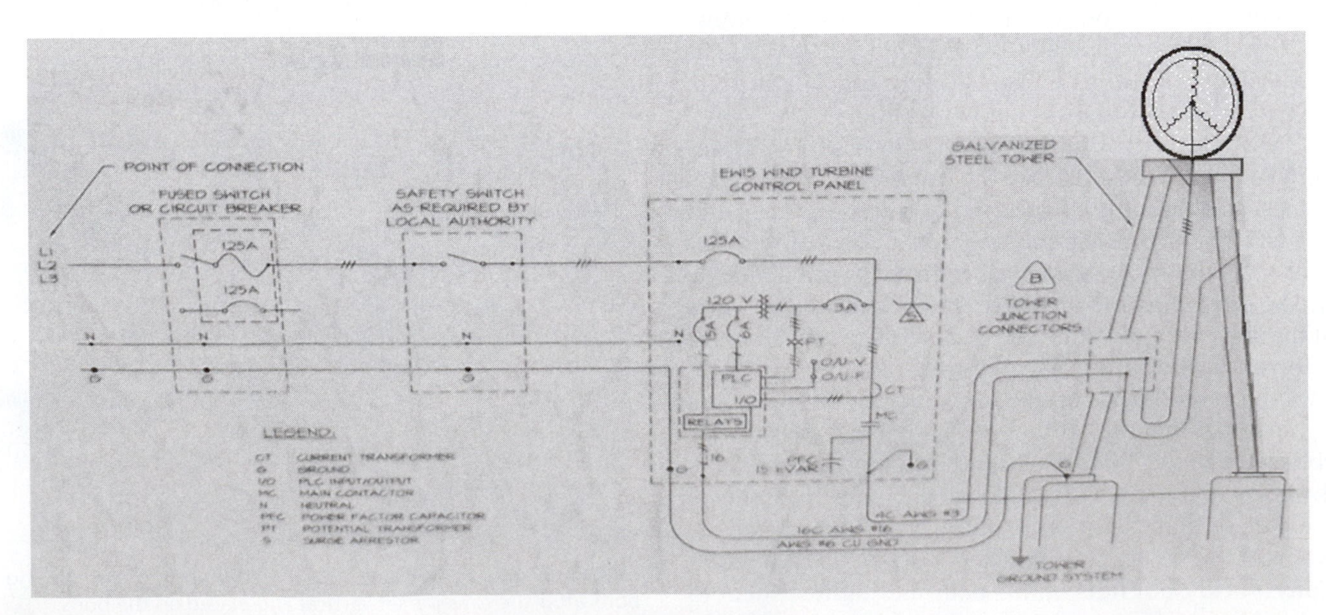

FIGURE 11-22 An electrical diagram is mounted on the door of the electric panel.

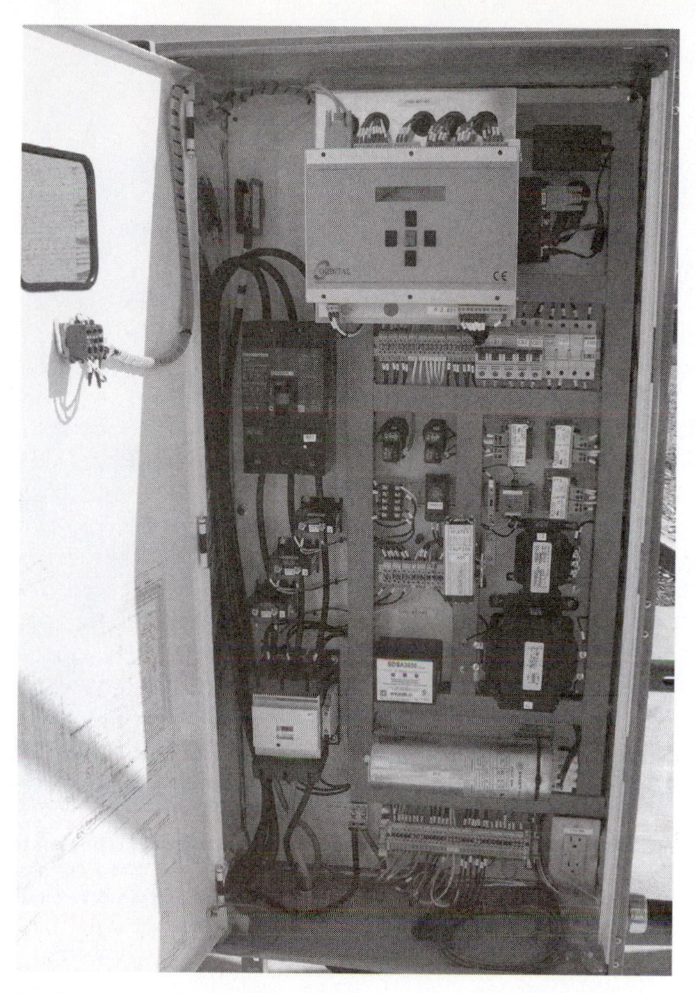

FIGURE 11-23 Electrical cabinet for a wind turbine. The wires from the wind turbine generator come into the cabinet on the left side and are connected to the circuit breaker.

FIGURE 11-24 The overhead electrical service entrance from the utility and electrical meter are mounted where the voltage comes into the building. Notice the electrical meter on the far left. The two boxes in front of this panel are for the wind turbine, and they provide a transformer and the required disconnect safety switch. Figure 11-25 shows the transformer and disconnect more clearly.

panel that measures all the power coming from the electrical utility to the company. A disconnect switch is also connected in line with the meter. The disconnect switch allows all the voltage from the power utility to be disconnected from the building if electrical maintenance must be completed inside the building.

When the wind turbine is installed, this meter will be replaced with a net meter, which can measure voltage coming from the power company or voltage that is created by the wind turbine. Typically, any excess voltage that is produced by the wind turbine and not used immediately by the customer can be put back into the grid.

Figure 11-25 shows the disconnect switch mounted on top of the transformer for the wind turbine. The National Electrical Code (NEC) requires that a safety disconnect switch must be connected between the wind turbine and the existing wiring. The disconnect switch can be manually switched to the open position any time the voltage from the wind turbine needs to be turned off or isolated. The electrical wiring that comes from the disconnect switch and transformer for the wind turbine is connected to the wiring at the meter. This is called the *tie point* for the system, and it is the place where the wind turbine is connected to the grid.

Testing the Wind Turbine for Its Initial Run

After all the electrical connections have been completed and tested, a technician can climb the tower and remove the ropes that were used to keep the blades from rotating. When technicians need to climb the tower or do work off the ground, they must wear personal protective equipment (PPE) such as a safety harness, as prescribed by OSHA. Figure 11-26 shows a technician climbing a tower while wearing a safety harness. The technician uses a lanyard to tie off the safety harness to the tower so that he is protected from falling if he loses his grip or slips. After the technician gets to the top of the tower, he can remove the ropes from the blade. It is important that the technician stays clear of the blades and works on the wind turbine from inside the tower. In some wind turbines, a mechanical brake is used to keep the blade from turning while technicians are on the tower or working near the blades and rotor.

FIGURE 11-25 The electrical disconnect switch and transformer that is connected to the cables from the wind turbines.

FIGURE 11-26 A technician wears a safety harness and is tied off to the tower as he climbs the tower to remove the ropes that were used to tie the blades off to the tower when the tower and wind turbine were erected. (Courtesy of John Fellhauer jfellhauer@surenergyohio.com.)

When the technician has inspected everything on the wind turbine and is sure that the system is safe to start up, he can climb down the tower. When the wind speed reaches the start-up threshold, the turbine blades are released and allowed to begin turning and generating electricity. During this period, the speed of the blades and the amount of voltage that is generated is closely monitored. Also, the mechanical parts of the wind turbine, such as the blades, rotor, low-speed shaft, gearbox or transmission, high-speed shaft, and generator, are monitored for signs of vibration, misalignment, or other mechanical problems. After the wind turbine has produced voltage for several hours, technicians stop the turbine blades and then go back over all of the nuts and bolts that were installed when the wind turbine was assembled, and also check any other mechanical systems that might have become loose or not be functioning correctly. The technicians also look for signs of wear or excess heat that could damage the system. These small problems are easy to detect and repair before the system is put online full-time but can become major problems or damage the wind turbine if they are not detected. It is important to remember that once a wind turbine is placed into production, it will run without close supervision, so it may run for long periods without anyone being on site with it.

11.2 OVERVIEW OF TROUBLESHOOTING

Many technicians believe that **troubleshooting** is a process to determine what is broken when they are checking out a system. This section will show you that it is actually easier to identify what is working correctly on a system and, through a process of elimination, focus only on the part of the system that is faulty. This section will help you understand all the details that you must understand to become an efficient and effective troubleshooter.

You will learn to use a consistent procedure each time you are troubleshooting and to go through a methodical system to test components. Some technicians fail at troubleshooting because they basically try to remember what was broken on the system the last time and work from there, as though the problem is simply repeating itself. Others fail because they do not use a time-tested procedure; instead, they randomly start testing different parts or sections of the system with no concept of what they are testing or why. In this section, you will learn to develop a procedure that you can use every time you troubleshoot a wind turbine system.

The first step in troubleshooting a wind turbine system is to determine what the system was doing before it had the problem. Sometimes, simply reviewing the data collection system for the wind turbine can do this. You can look at the data and think about a series of questions to determine if the wind turbine system was operating correctly before this malfunction occurred, or if the problem has been occurring for some time. After you check out all the wind turbine systems, you may try and operate different parts of the wind turbine system. Sometimes you will find that the wind turbine system can operate partially, or that one part of the wind turbine system, such as the blades and rotor, can operate, whereas another part, such as the generator, cannot. This will allow you to focus on just the part of the system that is not functioning correctly.

When you first begin to troubleshoot a wind turbine system, you may feel that it is a large job that is very complex. As you work with more experienced technicians, you will find that they use a number of methods to determine what is wrong with the system. For example, they may spend a few minutes just looking at the complete system to get an idea of what components are running or turned on. They may also try to identify what power systems the wind turbine system has, such as the mechanical components, electrical components, or other subsystems such as hydraulics, and see whether any or all of these are operating correctly. Another important part of this process is to identify what types and how many subsystems the wind turbine system has. You also want to determine what type of wind turbine you are working on: Is it an upwind or a downwind system? Does it use an electric motor or hydraulics system to change the yaw direction or the pitch of the blades, or are these changed by the wind or centrifugal force? Does the system have a gearbox or transmission, or is it a direct-drive system, and what type of generator and what type of controls are used in the system?

Finally, you may take a minute to determine whether the wind turbine system uses regular hard-wired controls or a computer or PLC. If the system has a computer or PLC, is there any data that indicates what the wind turbine has been doing the last few hours or days, or are there fault codes or status lights indicating problems?

A second step in the process of troubleshooting includes using your other senses, beyond vision. For example, you may want to listen for noise from the rotor, shafts, and gearboxes to determine if they are under an extreme load. Are there any gears, bearings, or other mechanical parts that are squeaking or groaning, which indicates that they are running improperly and creating friction? Another sound that you should listen for is for motors such as yaw motors that hum but do not run. These clues will help you later when you are trying to determine what is wrong with the wind turbine system. Another sense that is usually overlooked is your sense of smell. In some cases you will be able to smell a generator, electrical component, or bearing that is overheating before you are able to determine from some other test that it has a problem.

Using Company Resources Such as Manuals, Bulletins, Safety Alerts, and Help Desks

Most wind turbine manufacturers have websites or telephone centers that will provide you with an additional source of information regarding problems that may occur with their systems. A manufacturer may have a website that posts bulletins, or it may send bulletins to its dealers to make them aware of problems that have been encountered. The manufacturer may also provide detailed technical manuals or a technical help desk that you can email or talk to directly about conditions and problems you are encountering. The manufacturer may also want you to report the kinds of problems you are finding in the field, so it can add them to its database. The manufacturer's website may also have a page called FAQ (Frequently Asked Questions), which consists of questions and answers about common areas of its system. Many times the answers on the FAQ page will provide the information that you need to complete a test, or it will provide information that will lead you to another source.

Sometimes you can use the Internet to search for the problem you are encountering to get help, or you might join an online group or forum for wind turbine technicians who will respond to questions that you post on the forum. When you use the Internet to search for similar problems, your search may provide detailed studies of similar problems, or you may locate a bulletin or report that you did not know exists.

Troubleshooting from Most Likely to Least Likely

Another way to look at troubleshooting is to look for things that are most likely to occur and check them out first, and then move on to things that are less likely to occur. For example, when a wind turbine has been running for many months without any problems, you might want to check things that are most likely to fail, such as parts that need lubrication, parts that might come loose over time, and parts that may be subject to more wear and tear as they run longer. You probably don't want to start by looking for things that may only happen once in 10 years. As you become more familiar with wind turbine systems, you will learn which components are more prone to failure and what their failure mode is. You will also find that certain problems tend to occur with specific brands or models.

Swapping Parts

When you have tested the wind turbine system and have identified a faulty component, you may want to swap a part to ensure that the part you suspect is faulty is actually bad. Swapping parts may involve going to the service vehicle or wherever spare parts are stored and locating a good part, then swapping it with the suspected part. You need to be aware that this procedure has both advantages and drawbacks. One advantage of swapping a part is that if the part is expensive or has a two- or three-day shipping time, you can ensure that you have actually determined the part is faulty before you order a new one and wait for it to be shipped. One problem is that some technicians defer to the component swap method before thoroughly testing the part; they basically guess which part is bad and therefore may waste a lot of time changing parts that are not bad.

The basic rule of thumb to follow when swapping parts is to ensure that you have completely tested the part while it is in its natural operating state and you have completed an alternative test with power off. If you have completed these tests and are sure the part is faulty, you can safely change or swap the part to verify your suspicions.

Warranty Work

At times, you will be requested to work on a new wind turbine installation that may be under warranty. This means that the company that installed or manufactured the system has guaranteed that it will back some or all of the parts and a portion of the labor for the work. If you suspect or have records that show the wind turbine system is new and is within the warranty period for parts or labor, you will need to verify some of the things you propose to do, such as changing parts. You may have to contact the company that manufactured or installed the wind turbine to locate the proper repair part, and you may have to send the part that is faulty back to the company so it can check it to see what caused the failure. You will find that when you are changing warranty parts, the company will ask for more detailed information about the entire system, such as installation date, number of hours the system has run, and what caused the problem you are working on, such as complete failure caused other parts failing, or that this an isolated problem. The company will also want the complete model and serial number of any parts you are sending in and replacing, so that it can determine if this is an original part that was installed when the system was new, or that the part has been previously replaced in the field. Sometimes you will be called to work on a system that has the same problem occur more than once within the warranty period. If this is the case, the company may want proof that the first time the part was changed out, it was replaced properly and that the second failure was not due to the part being installed improperly or that service was not carried out correctly.

Keeping Accurate Records and a Maintenance Log for the Wind Turbine

When you begin working on wind turbine systems, it is important that each system has its own maintenance logbook, which is used to record any and all maintenance that is done on the system, from periodic maintenance to failures and parts that are replaced. Accurate dates and times and the number of hours the system has run are important data to enter into the logbook. Some systems create a logbook record electronically through the PLC or computer system. Other systems may use a software program that monitors the wind turbine continually to log specific data that you can look up and use when you are troubleshooting. The type of data gathered by these systems may include wind speed, direction of the wind, time of day, kilowatts the generator is producing, and what percentage of full power the system is operating at. This data may be provided in table or graph form. This type of information will be a very valuable record when you are trying to troubleshoot the system.

11.3 UNDERSTANDING THE TROUBLESHOOTING PROCESS

When you are requested to track down a problem on a wind turbine, you will need to invoke a troubleshooting process. When technicians first learn to troubleshoot, they think the process involves finding the problem as quickly as possible. Instead, you will learn that the best way to find a problem is to determine what parts of the system are working correctly and then eliminate them from the process. If the system has multiple energy sources such as mechanical (wind turbine blades and turning the rotor), electrical (generator and transmission cables), power transmission (shafts, gearboxes, and transmissions), and a hydraulics system including brakes and blade trimming, you will need to check out each system and determine which ones are working correctly and which ones are not.

Another way to understand this concept is to think about troubleshooting a wind turbine system the way you would try to locate a specific card in a deck of playing cards. Each card in the deck represents a different problem. Suppose I pull one card from the deck and then ask you to guess what the card is. This would be similar to the wind turbine having a problem in one of its systems and you are trying to troubleshoot the system to identify what system has the problem. As you know, there are 52 cards in a standard deck of cards, which has hearts and diamonds that are red cards, and clubs and spades that are black cards. If you just start guessing, without a method, it may take you 52 guesses to hit the correct card. This section will show you a method that you can use when you are troubleshooting a wind turbine system.

For this example, let's say the card I have pulled from the deck is the 9 of hearts. Using our process, your first question is to ask whether the card is a black card. The answer is no, so we can deduce that the card that was pulled must be a red card. The important thing is that this one question eliminates 50% of the deck. We now know the card must be a heart or diamond, so we can stop thinking about all of the spades and clubs in the deck.

The next question will be whether the card is a heart. By now you can see that it does not matter whether

I ask if the card is a heart or if I ask if it is a diamond, because the answer will indicate which one it is. In this case the answer is yes, the card is a heart, so I can stop thinking about all of the diamonds. With two questions, we have eliminated 75% of the cards.

Now that I know the card is a heart, I will continue to ask questions that eliminate the remaining cards 50% at a time. Since I know there are 13 hearts ranging from a high of ace to a low of 2, my next question will be whether the card is lower than an 8. The answer is no, so I have eliminated the 2, 3, 4, 5, 6, and 7. The remaining cards are 8, 9, 10, jack, queen, king, and ace. Since the queen or jack is the midpoint, I next ask whether the card is lower than the queen. The answer is yes, so now I know the card must be a jack, 10, or 9. The next question is whether the card is lower than the 10. Since the answer is yes, I now know the card is the 9 of hearts.

You may be wondering what the example of finding a card in a deck of cards has to do with troubleshooting. The four suits of cards—diamonds, spades, hearts, and clubs—represent the subsystems on most wind turbine systems: blades and rotor, electrical (generator and transmission cables), power transmission (shafts, gearboxes, and transmissions), and the hydraulics system including the brakes and blade trimming. When you come upon a wind turbine that is not generating electricity, your first questions should determine which of these systems are working and which are inoperative. You need to develop a series of questions and tests that will eliminate the most possibilities.

Let's say that you have found that the turbine blades are turning but the system is not producing any voltage out of the system when you measure it at the electrical cabinet. You can see that when the wind turbine does not produce voltage at the electrical cabinet, any number of parts of the system could be faulty, such as the turbine blades, rotor, low-speed shaft, gearbox, high-speed shaft, generator, electrical cables, disconnect switch, and circuit breakers. Using the troubleshooting process, you could start by asking whether the generator shaft is turning. If the answer is yes, the turbine blades, rotor, low-speed shaft, gearbox, and high-speed shaft are all working and can be eliminated from the process. You can then concentrate your troubleshooting tests on the generator, its controls, and the wiring down to the electrical panel.

If the answer to the question about the generating shaft turning is that the generator shaft is not turning, then you would eliminate the generator and its cables and concentrate on the parts of the system that make the high-speed shaft turn. You can see that using a troubleshooting procedure like this, you will be able to find the problems much more quickly, by eliminating all the parts of the system that are working correctly. The key to using this process is to properly identify the subsections of the wind turbine system and identify which sections send power to the other sections.

11.4 WHAT IS THE DIFFERENCE BETWEEN A SYMPTOM AND A PROBLEM?

When you first start to learn to troubleshoot, you may find conditions that make the wind turbine inoperable and you make the repair only to find that the condition reoccurs. For example, if you are called to work on a wind turbine that has a generator that does not produce voltage, and you find a circuit breaker that has tripped, you will reset the circuit breaker and the generator will begin to produce voltage again, only to trip the same circuit breaker later in the day. What has occurred is that you found a **symptom** rather than the **problem**. The tripped circuit breaker opens an electrical circuit that caused the generator to stop producing voltage, but you have not found the actual problem that is causing the circuit breaker to trip. When you do a more though investigation of the condition, you ultimately find the generator has a loose wire that is causing the circuit breaker to trip. When you tighten the nut on the wire terminal, you have fixed the problem, and the circuit breaker will not trip again. This process is called finding the problem rather than working on the symptom. When you find the actual problem, it is called finding the root cause of the problem. When you are troubleshooting and find something that is causing a condition that makes the wind turbine inoperable, you will need to ask yourself if you have found a symptom or the problem.

11.5 USING TROUBLESHOOTING TABLES AND A TROUBLESHOOTING MATRIX

When you are troubleshooting a wind turbine, a good source of information is the manual that comes with the system. Some larger wind turbines provide one or more operating, maintenance and troubleshooting manuals. These manuals may have one or more tables that show a step-by-step procedure to check out the subsystems. Figure 11-27 shows an example of this type of troubleshooting table. The table has three columns: symptoms, possible cause, and corrective action. The column of symptoms outlines the types of problems you may encounter. The middle column shows one or more possible causes for this symptom. The third column shows the things you should consider to correct the problem. You should remember that a symptom is not necessarily the problem, so you may need to review the symptoms in detail in order to determine the actual problem. For example, if the symptom is the loss of voltage, you can correct this symptom by replacing a fuse or resetting a circuit breaker, but you must look deeper into the system to find out what is the problem that is causing the fuse to blow. You should also understand that this type of table is designed to be used as a guide, and it may not be detailed enough to help you find every problem. Some manufacturers also provide a **troubleshooting matrix** or troubleshooting table that shows the sequence of operation. This table shows what happens in step 1 and progresses through the sequence of operations for the

Troubleshooting Chart for Small Wind Turbine Electrical Problems

Symptoms	Possible Cause	Corrective Action
1. Propeller turns too slowly even in strong wind, will not start.	a. Brakes on (Run Switch on turned on). b. Battery dead or low voltage. c. Shorted diode in controller rectifier. d. Bad generator brushes. e. If new installation, generator wired incorrectly.	a. Turn Run Switch to on position. b. Charge battery or replace if defective. c. Check diodes and replace bad ones. d. Check brushes and replace worn brushes. e. Check generator wiring and change as needed.
2. Propeller runs too fast, no mechanical noise, no output.	a. Controller disconnected generator from battery or grid. b. One or more generator windings open or wires between generator and battery open. c. Controller diodes open.	a. Check controller fault for disconnect condition b. Test generator windings for opens and replace generator if needed. c. Check diodes for open, and replace bad ones.
3. Propeller runs too fast, no mechanical noise.	a. Battery over charging, no load for generator. b. If new installation, generator wired incorrectly.	a. Check battery voltage regulator fault, and check for open wires between generator and load. b. Check generator wiring and change as needed.
4. Propeller runs too fast output less than 50%, mechanical noise.	a. Wires open between generator and battery or load. b. Open diode in controller. c. Bad brushes or slip rings on generator. d. If new installation, generator wired incorrectly.	a. Test for open wires between generator and battery or load. Replace bad wires. b. Check for open diodes and replace bad ones. c. Check for worn or bad brushes and replace as needed. d. Check generator wiring and change as needed.
5. Propeller runs too slow, output less than 50%, no mechanical noise.	a. Battery voltage low or dead battery. b. If new installation, generator wired incorrectly.	a. Charge battery or replace bad battery. b. Check generator wiring and change as needed.

FIGURE 11-27 A troubleshooting chart for a wind turbine.

system. This type of troubleshooting information will help you determine what part of the system is working and at what point the system stopped.

11.6 TROUBLESHOOTING WIND TURBINE GENERATION AND TRANSMISSION PROBLEMS

Some problems that you have to deal with will involve the wind turbine blades, rotor, low-speed shaft, transmission, high-speed shaft, or the generator. The problems that you will encounter with these components will generally involve the transmission of power from the blades through all the shafts and transmission on to the generator. Typical problems will be loose inter connections between parts such as the rotor and low-speed shaft, the low-speed shaft and the gearbox, the gearbox and the high-speed shaft, or the high-speed shaft and the generator. Each of these connections involves nuts, bolts, and other connection hardware that will need to be checked to ensure they remain tight throughout the life of the wind turbine. Each of these connection points also has the chance to become misaligned, so they must all be checked for alignment and vibration. A complete vibration analysis is important to ensure that power is transferred smoothly from the turbine blades all the way to the generator. The alignment must be checked as part of the preventative maintenance program on a scheduled basis at periodic times throughout the life of the turbine.

Also, the alignment of all of the shafts and gearbox must be checked any time there is any suspicion of vibration. Vibrations from a misaligned shaft will be amplified at each point in the system and eventually cause major damage to bearings and other parts of the system that maintain the shaft mounting. If any connection point in this system becomes loose, it will introduce misalignment and cause vibrations through the shafts. These vibrations will eventually be delivered to the bearings, where severe damage will occur.

Another part of the mechanical system that must be checked periodically or troubleshooted when a problem occurs is the pitch control system for the blades. The pitch control system provides control of the blade pitch. On some systems this is an overt control system that uses mechanical parts or hydraulics to move the blades. This system has two parts for the control. One is the sensors that tell the PLC or computer where the blade pitch should be, and the second part is the output control of the hydraulics or mechanical system that actually moves the blades. Basically, the sensor tells the computer where to move the blades and then checks to see if the blades moved to that position. When you are troubleshooting this part of the system, you should start at the sensor to determine where it is indicating the blade should be. The signal from this sensor will be analog in nature and will be between 0 and 100% of full value. The next step in the troubleshooting process is to see how the PLC or computer is measuring this input sensor. The control circuit in the PLC will indicate to the output control for the pitch travel how far the pitch should be moved. When this part of the system is tested, it is important that the turbine is not under a full wind load. You should also be able to put this circuit under manual control and force a change in value from 0 to 100% to cause the blade to move through its full range of movement. When the blade pitch control is suspected of having a problem, you should put the system in manual control and test it through its full range. If it can move through its full range of motion when it is in manual control, you should then put it in automatic control and allow the sensor to dictate the pitch for the blade.

If the blade control is functioning correctly, and the blade is moving the rotor when the wind is blowing, then the next step in troubleshooting is to follow the energy flow through the low-speed shaft to the gearbox. You will need to troubleshoot the gearbox to see whether the ratio of the input speed and output speed is as rated. Also, the gearbox needs to be tested for excessive vibration and any problems in creating excess heat due to the gearbox failing. The parts of the gearbox that can fail are the meshing of the gears and the lubrication of the gears. If the gears begin to wear, are overloaded, or suffer from a lack of lubrication, the gearbox will begin to overheat and cause damage. If you suspect the gearbox of having problems, you may do the work yourself, or some companies call in a mechanical specialist to check the system in more detail and change out the gearbox as a unit if necessary.

The other part of the low-speed shaft and the high-speed shaft that needs to be checked is the universal joint. The system may have a universal joint on each end of both shafts to help even out the misalignment and vibration that may work its way into the shafts. The universal joint must be checked for lubrication and to be sure that it has not become loose. If there are any problems with the universal joint, it generally must be replaced rather than repaired.

11.7 TROUBLESHOOTING MECHANICAL PROBLEMS AND TOWER PROBLEMS

Another part of the mechanical system that will need to be testing and troubleshooting is the yaw control. The yaw control system works the same way as the pitch control in that a sensor indicates where the wind turbine yaw should be positioned. This system should also be checked out under manual control, where a value from 0 to 100% can be entered into the computer or PLC and its control program will send a value to the yaw control motor to cause it to rotate clockwise or counterclockwise. When the system is under manual control, the technician needs to be aware of the yaw position so it does not get out of control. If the yaw control system responds to manual control signals that cause it to rotate in each direction, then the system can be put under automatic control and the sensors will tell the PLC where to move the yaw motor.

The tower must also be inspected periodically to ensure that it remains securely anchored and is not showing any signs of corrosion or other damage. The tower must also be inspected to ensure that it remains in the same position that it was anchored in when it was erected. If the tower has guy wires, they must be checked to ensure they have not started to wear and that their tension is still correct.

The nacelle needs to be inspected where it is attached to the tower to ensure that the connecting nuts and bolts remain secure. It is also important to check the nacelle to see that its seals do not become worn and that any access doors fit securely. Generally, the tower and nacelle do not create problems that need troubleshooting; rather, they simply need to be inspected periodically to ensure that they remain intact and as they were on the day of installation. Most wind turbine manufacturers provide a detailed list of inspection points for this operation. It is important that the time schedule is followed as closely as possible. The time schedule is generally specify in hours of running, or strictly by a calendar date.

11.8 TROUBLESHOOTING ELECTRICAL PROBLEMS

When you encounter electrical problems in a wind turbine system, you will need to be able to read electrical schematics (ladder) diagrams and wiring diagrams as well as electronic diagrams to troubleshoot the systems. You also need to know how to make proper measurements using a voltmeter, ohmmeter, ammeter, or Megger. You will also learn when it is appropriate to use each type of meter for these measurements. If you need more information about basic electricity, refer the material provided in Chapter 12 of this text.

When to Use a Voltmeter to Troubleshoot

The voltmeter is the meter of choice for most technicians. The voltmeter allows technicians to test many electrical circuits with power applied. At first it may sound dangerous to test voltage on circuits to which power is still applied. You will learn that it is safe to do these tests as long as you remain aware of the potential for electrical shock if you touch a terminal that has electrical voltage applied. Personal protective equipment (PPE) is available to ensure that you are able to work safely around electrical circuits that still have power. This includes leather and rubber gloves, boots with heavy rubber soles, and thick rubber mats to isolate you from becoming grounded. It is important that you learn to make these voltage measurements safely and read the meter accurately. You will find that it is a very important part of troubleshooting to determine the exact point where voltage is interrupted in an electrical circuit when a failure has occurred. By being able to determine where voltage is interrupted, you will be able to identify the point where the electrical problem has occurred.

Being able to measure voltage on a live circuit also allows technicians to make a large number of preliminary tests without disconnecting any wiring or components. If you are called to troubleshoot a wind turbine system that has stopped generating voltage completely, it more than likely has an open circuit, where voltage has been interrupted. The first test that you should make with the voltmeter is to measure the source voltage in the electrical cabinet for the system where it comes in from the electric utility. You should always set the voltmeter to the highest voltage range until you have determined the actual voltage. After measuring the source voltage, you will need to use the electrical diagram provided for the wind turbine system and measure voltage throughout the circuit wherever you suspect the voltage has stopped. When you find the loss of voltage, you can turn off the power to the system, lock it out and tag it out, disconnect the wire or component that you suspect is faulty, and test it with an ohmmeter.

This troubleshooting may include checking the coil or contacts of relays, as well as the voltage coming out of electronic circuits. Most electrical diagrams indicate the amount of voltage that you should measure at different points in the circuit. You should be able to use a voltmeter to locate the point where voltage has been lost in a circuit. You should remember that the schematic (ladder) diagram indicates the sequence of operations of the wind turbine electrical system. The wiring diagram indicates the location of wire terminals on the components where you would find the components mounted if you were looking in the electrical cabinet. The wiring diagram looks like you took a picture of the electrical cabinet.

You also should be aware of some problems that you might encounter when measuring voltage. The first problem is voltage feeding backwards through a circuit. This usually occurs in circuits that have one blown fuse or an open wire somewhere in the circuit. If you have one blown fuse in a circuit that has voltage supplied by two lines, it is possible for voltage from the remaining supply line, which has a good fuse, to feed through a transformer winding or motor winding and flow backwards up to the bottom of the blown fuse. When you test the output side of each fuse, your meter will measure voltage and it will appear that all of the fuses are good. If you are aware that this can happen, you can use a fuse puller and pull all of the fuses from their holders and test each fuse for continuity with an ohmmeter. It is important to remember that you must remove the fuse from the fuse holder and ensure that no power is applied if you intend to use in ohmmeter to make the test for continuity to determine whether the fuse is good or bad.

You should also remember that if a circuit has one blown fuse, you cannot have any current flow in the circuit, so you could check for current flow too if a circuit is under power. You will also get feedback voltage through a any motor when either the run or the start winding is open if it is a single-phase motor and if any one of the three-phase windings of a three-phase motor is open and the other windings are good. All you need to do is be aware that feedback voltage does occur, so you can double-check the source of voltage if you suspect that the voltage you are reading may be coming from a feedback source. Another way to ensure that feedback is not occurring is to remove a terminal or wire from any component such as a motor, relay coil, or indicator lamp that could possibly be a path for feedback voltage.

Another electrical problem you need to be aware of is that the electronic controls on a wind turbine system may cause odd values of voltage to show up in a circuit. If a circuit has an inverter, or if it uses variable-frequency drives, they can produce voltages that are more or less than 60 Hz. Since most digital and analog voltmeters are designed for measuring voltages that are exactly 60 Hz, the voltage reading on the meter could be slightly more or less than it actually is due to a change in frequency. This may make the technician think that the circuit is faulty because the voltage is higher or lower than the voltage indicated on the print or specification sheets, when in fact it is operating correctly. If you use a true RMS-reading voltmeter, you will always get an accurate reading in these types of circuits regardless of the frequency. As long as you are aware that the voltage may be higher or lower in an electronic circuit and everything is still working correctly, you can take voltage measurements with either type of voltmeter. It is also important to be aware that if you have the meter set for an AC voltage reading and the voltage you are actually measuring is a DC voltage, the measured voltage that is indicated on the meter will not be correct. This will also be true if you have the meter set to measure DC voltage and the voltage you are measuring is AC voltage. Another way to get around this problem is to use a scope meter, which is actually a digital oscilloscope, which will show the waveform of the voltage you are measuring as well as display the amount of voltage.

It is also important to remember to use the voltmeter to measure a known quantity of voltage from either a small battery that you carry in your toolbox or a known 110-V AC power source to ensure that the voltmeter and your meter leads are working correctly when you first take it out to make measurements. It is important to ensure that your voltmeter is able to measure a test voltage correctly before you use it to measure voltage in a live circuit. If the meter is broken or if the meter leads are faulty, the meter will indicate the voltage is zero even though the circuit is fully powered. Technicians have encountered problems when they touched a circuit that had electrical power applied, after they checked the voltage check and their meter indicated 0 V. For this reason you can see why it is important that you use a test voltage of a known value such as a small battery to ensure that the meter is working correctly before using the meter to troubleshoot a live circuit.

Some technicians also carry a secondary meter that simply detects the presence of any voltage. This type of simple meter provides an audible tone or ticking noise that indicates voltage is present and live in a circuit. The meter is an excellent addition to a voltmeter to make sure the circuit is safe to work on. A good rule to use when working on an electrical circuit is to assume that every circuit has voltage applied and is live, and treat it that way.

When to Use an Ohmmeter to Troubleshoot

It is important to understand when to use an ohmmeter. You must have all power to a circuit turned off before you use an ohmmeter to take measurements, since the ohmmeter presents very low impedance (resistance) to the circuit. When impedance is low, it will create a short-circuit path if you place the meter leads across terminals that have voltage present. This will cause the large current to create sparks and possibly cause molten metal from copper wire or allow the meter to explode in your face. Putting the meter terminals of an ohmmeter on the terminals where voltage is applied will certainly damage the meter.

In most cases, you will use the ohmmeter to make a **continuity test** with power off, after you have used a voltmeter to locate an open fuse, faulty switch, open motor winding, or open wire. The continuity test measures the amount of resistance. Basically, the test indicates one of two states, very low resistance that indicates a fuse is good or a switch is closed, or very high resistance that indicates a fuse or switch is open. When you are making the continuity test with an analog motor, you should zero the meter by touching the meter leads together and use the lowest ohm setting. If you do not get the meter to indicate some amount of small resistance or zero resistance when you touch the leads together during this test, the meter itself might have a blown fuse, or there might be a problem in the leads. The lowest resistance setting will detect the slightest amount of resistance in the switch or wire. Resistance that is very high is called infinite and indicates that an open circuit is present in the contacts or the wire. If you are using a digital voltmeter, be sure to check what the meter uses to indicate an infinite resistance measurement, such as a blinking number on the display or a very high number such as 9999. You can test the meter to see what value it displays when it is measuring infinite by turning the meter on and setting it to the resistance measurement. As long as the tips of the meter leads are not touching, this setting simulates an infinite reading. Check the meter display and notice what it displays as an infinite reading. Next, touch the tips of the meter leads together, which is a low-resistance measurement. Check the meter display when the tips of the meter leads are touching and notice that the display should show 000 ohms.

If you are testing a wire and get a reading that indicates infinity, this test indicates that the wire is faulty (open) and must be replaced. If you are testing a switch, be sure you are on the correct terminals that make a contact set, and then change the switch from its open to closed position several times to ensure that you are checking it correctly. If the meter reading remains at infinity, the switch contacts are faulty (open), and the switch must be replaced.

Another use for an ohmmeter is to measure the actual number of ohms in the windings of a motor, transformer, or coil. When you are testing a single-phase motor, you should measure the resistance in each winding accurately so that you can determine that the winding with the highest resistance reading is the *start winding* and the winding with the lowest resistance reading is the *run winding*. If you are measuring the resistance of a three-phase motor, each of the windings should have the same value of resistance. If you are troubleshooting electronic circuits, you will also need to measure the amount of resistance accurately. For example, when you are trying to identify the terminals of a diode, transistor, silicon-controlled rectifier (SCR), or triac, you will need to measure the resistance and determine the polarity of the terminals. Be sure to switch the meter to the setting that will provide the most accurate resistance reading for the electronic device. Since the ohmmeter has its own battery for its power source, it may have a specific setting that should be used for measuring electronic components.

Testing Components and Wiring with a Megger

A **Megger** is a specialized ohmmeter that uses approximately 600–2500 V to check cables, transformer windings, and motor windings for leaking current through their insulation. Megger is a registered trademark of AVO International (formerly the Biddle Co.), but the term is widely used for all similar testers regardless of manufacturer. To use a Megger, the voltage to the circuit must be turned off, and these wires and components must be isolated during this test. The Megger puts the voltage into the motor winding or cable for a short time, usually less than several minutes, and tests for leaking current, which indicates a breakdown of the resistance in the winding or cable. The results of the leaking current through the insulation are displayed on the Megger in units of resistance (megohms or millions of ohms). A low reading (less than about 100 megohms) indicates possible winding or cable problems. You are looking for very high resistance readings, the higher the better. A high resistance reading on a winding indicates that the insulation is in good working condition and the winding is not likely to short out. Sometimes moisture will get into the motor winding or transformer winding and cause the resistance of the winding to drop below what it normally would be. The megohm test is usually used on motors, transformers, or cables that you suspect are starting to fail but have not completely failed. A Megger test allows you to detect these components before they fail completely and cause damage to additional components in the system. You should also test suspected windings before you decide to remove them and replace a component in the system, since changing out major components requires additional downtime for the system. If you are fairly sure that a winding is shorted or is starting to create a short circuit, the Megger test will verify your suspicions. A regular ohmmeter can find major shorts and opens in a wire or winding, but it cannot measure the higher values of resistance the way a Megger can.

When to Use a Clamp-on Ammeter to Troubleshoot

A clamp-on ammeter is a valuable troubleshooting instrument that can indicate the presence of current in a circuit. For example, if you suspect that a motor with three-phase supply voltage has one blown fuse, you can quickly determine which leg of the three phases has the blown fuse by measuring the current in each of the three legs with the clamp-on ammeter when power is applied. The leg that has the blown fuse will indicate 0 A, and the two legs where the fuses are good will indicate current that is higher than rated for the circuit, because the motor will be drawing locked-rotor current since it will not be able to start. This test is more effective than the voltage-drop test for a blown fuse, which is subject to feedback voltages. Be aware that you can only allow the power to be applied for a few seconds for this type of test, or the remaining fuses will be subjected to very high currents until one of them opens.

The clamp-on ammeter is also useful for determining the current load of any circuit wire. This is especially useful in determining how close the circuit wire is to overloading, which will cause a fuse to blow. You can also determine if a motor is beginning to wear out and has bad bearings or lack of lubrication. If the motor current is higher than its rating, you can begin to suspect that it is overloaded. If the load is within tolerance, you can suspect that the motor needs lubrication or the bearing is bad. The clamp-on ammeter is also useful in measuring the output current of the turbine generator. Many wind turbine systems have dedicated current meters mounted in their electrical panels. These ammeters use a donut transformer called a current transformer that looks like a circle and has the main wire from the generator running through its open center. The transformer samples the amount of current in the wire, and an electronic circuit converts the small amount of induced current to an accurate current measurement for the system. Figure 11-28 shows three current transformers in the electrical cabinet for a wind turbine. Each of the three main power lines from the turbine generator passes through a current transformer. These three current transformers enable the system to measure and indicate the correct amount of current for each line and send it to the SCADA system so it can be monitored continually or recorded for future use. The amount of current the wind turbine generator produces at any instant is an indication of how much power it is converting from the wind.

FIGURE 11-28 Current transformers used to measure current in the main electrical cables.

11.9 TROUBLESHOOTING HYDRAULIC PROBLEMS

Some wind turbines use hydraulic systems to control turbine blade pitch, the yaw brake system, the high-speed shaft brake system, the rotor and hub lock system, and a braking system called the aerodynamic brake that helps slow down turbine blade rotation during high winds. Figure 11-29 shows a basic hydraulic system used in a wind turbine. The hydraulic system consists of a hydraulic pump that is turned by an electric motor, a reservoir that holds the fluid, a pressure-relief valve, directional valves, flow control valves, and actuators such as cylinders or hydraulic motors. The hydraulic system has a volume of hydraulic fluid that is stored in a reservoir, which is cleaned as it passes through one or more filters or strainers as the pump moves it.

The pressure-relief valve, directional valves, and flow control valves can be controlled manually or by an electrical solenoid, which is basically a coil of wire that becomes a strong magnet when electrical current passes through it. The solenoid valve allows the hydraulic system to be controlled by the PLC or computer control system. The solenoid valves can be controlled in an on/off manner by the controller or, on larger wind turbine systems, these valves can be analog, which means they can control the valve from 0 to 100%. These types of hydraulic valves are also called proportional valves. Having a feedback sensor placed inside them, which makes them a servo valve or servo system, more accurately controls some of the valves.

When you are troubleshooting a hydraulic system, the problem is usually expressed as a condition in which the actuator is not properly functioning, such as a cylinder rod that does not extend or retract. The troubleshooting complaint may also be that an actuator is not moving fast enough. The hydraulic system will have gauges that

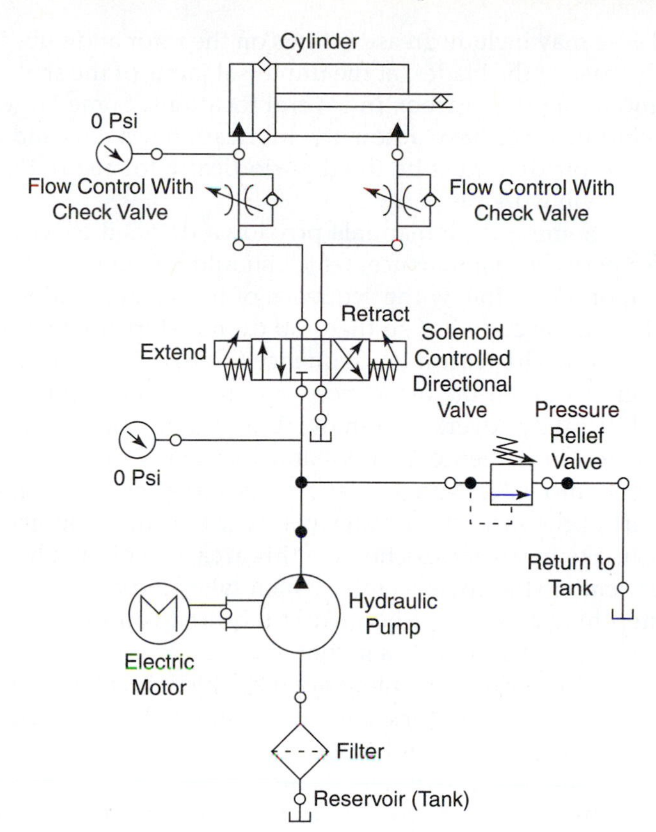

FIGURE 11-29 Basic hydraulic diagram for extending or retracting a hydraulic cylinder that can activate any of the brake systems on the wind turbine.

indicate the amount of pressure at various points in each hydraulic circuit. These may include readouts that are displayed by the PLC or computer to the SCADA system, where technicians can easily check them. When you are troubleshooting a hydraulic system, you will be able to use the pressure readings and the hydraulic diagram to determine what is causing the problem.

The first test in the hydraulic system is to determine whether the system has a sufficient volume of fluid and whether the pump is moving the fluid to create flow and pressure. It is important to remember that pressure will exist in a system only when flow has filled the volume of the hoses, piping, valves, and actuators. The amount of pressure a system will have depends on the specification to which the pressure-relief valve is set. If the gauges indicate that the pressure of the system is zero, you would start your tests by ensuring that the electric motor is turning the pump and there is sufficient fluid. It is possible that the electric motor shaft is turning, but the coupling between the motor and hydraulic pump is damaged. It is also possible that the pump is damaged and not capable of moving the volume of hydraulic fluid necessary to create pressure and cause the actuators to move correctly.

If the gauges indicate the proper pressure, you can assume that the pump and motor are working correctly. The next step in the troubleshooting process is to try and cause the actuator to move. If the actuator is a hydraulic cylinder, it should be able to extend or retract when the electrical command is sent to the directional valve solenoid. If the hydraulic cylinder does not move when the solenoid receives an electrical signal, you can check to see whether the coil of the solenoid has become magnetic, by placing a small metal screwdriver near it. If the solenoid has become magnetic, the metal screwdriver will be attracted to it. If it is not, the problem will be that the solenoid coil has an open, or the electrical signal did not reach the solenoid coil. You can test whether the solenoid coil has an open by removing the electric wires connected to the coil and using an ohmmeter to test the coil for continuity. You can also check for voltage at the two wires that connect to the coil to see whether they are receiving an electrical signal.

If the directional valve solenoid coil is receiving electrical signal and is magnetic, you can check whether the valve has changed position by placing a small metal screwdriver against the end of the spool to see if changes position. If it changes position, the hydraulic fluid should be directed to the hydraulic cylinder to make it extend. If the hydraulic cylinder does not extend when you have full pressure on it, you may suspect that the cylinder rod is jammed somewhere. You can see that it is easy to troubleshoot the hydraulic system when you use a systematic approach.

11.10 PERIODIC MAINTENANCE OF WIND GENERATION SYSTEMS

Periodic maintenance on a wind turbine consists of a series of visual inspections, small tests, and checks that are carried out on a schedule. The schedule for when the inspection should occur will be defined by a number of operating hours, or by a calendar date, whichever comes first. You will be able to find the date of the previous inspection or the number of hours the wind turbine has operated since the previous inspection by checking the maintenance log and comparing the hour meter to the number of hours needed to trigger the next test. Some wind turbines provide this information through their SCADA system so that the system can be scheduled ahead of time to be taken out of production for a time for inspection.

Most wind turbine manufacturers include a section in their service manual that outlines the parts of the system that need to be tested and visually inspected during the periodic maintenance procedure. A table may be provided that includes the number of hours and calendar dates that will trigger specific periodic maintenance procedures. For example, you may need to check the oil in the gearbox at a certain number of hours, and then change the oil after a certain longer number of hours. Most wind turbine manufacturers provide a complete procedure that includes a checklist of points that should be inspected and tests that should be completed.

The reason periodic maintenance is conducted on the basis of specific operating hours or calendar dates is that manufacturers of basic parts that make up the wind turbine system have run tests and have a history of when these parts may fail, and the types of wear and tear they will receive throughout their life. For example, the gears and bearings in a gearbox have been thoroughly tested, and the manufacturers have learned how long the parts in the gearbox can run before they begin to wear out. They also know the number of hours that the lubricants can be used before they begin to deteriorate. The history on these parts help manufacturers determine what needs to be inspected, and what needs to be changed at specific dates throughout the life of the parts. The theory of periodic maintenance is that it is less expensive to take a wind turbine out of service for periodic inspection before it has a major failure. If the wind turbine has a major failure, it must be taken out of commission until the cause of the failure is determined and the repairs have been made. If this includes purchasing and shipping of parts, it may add several days to the downtime, which causes the wind turbine to be out of commission and not producing electricity for a longer period of time.

When a wind turbine is scheduled for periodic maintenance, any parts that need to be changed can be ordered and shipped a week or two ahead, so they will be on hand and ready to be changed out when the turbine is being inspected. This results in less downtime and a more controlled procedure. It is also much less expensive to provide proper lubrication of mechanical parts, bearings, and gears on a scheduled basis than it is to allow them to run dry and fail completely. If mechanical systems, shafts, and gearboxes are properly maintained, inspected, and lubricated, they may last for 5 or 10 years.

One of the major concerns during periodic maintenance is to ensure the safety of the technical personnel completing the test, and the safety of the wind turbine. OSHA documents prescribe a number of procedures that must be followed when maintenance is being performed on wind turbines. For example, the blades must be stalled and brakes applied or secured in some manner when personnel are inspecting them. Any time that technicians must climb a tower or leave the ground during inspections, they must have the proper personal protective equipment, including safety harnesses and lanyards. Most wind turbine manufacturers have identified a number of safety procedures that must be followed when periodic maintenance is performed. It is important that you become aware of the safety issues, and conduct a safety meeting with other personnel who will be working with you when you do a periodic maintenance inspection.

Whether you are performing periodic maintenance on large, medium-size, or small wind turbines, you will be requested to use proper lubrication as specified in the technical service manuals. You must obtain the proper lubrication before the date the maintenance is to be performed, so you will be able to change any lubrication that is required, such as oil or hydraulic fluid, and you will be able to add oil or grease to lubrication points throughout the system. It is important that you obtain the lubrication that is specified in the technical manuals, as the manufacturer has determined that these brands and weights of lubricants will make the system perform at its peak for a longer period of time.

Many manufacturers provide detailed diagrams or pictures of the lubrication points on the wind turbine. These may include grease fittings on the rotor and hub, at the base of the blades, at the universal joints of the shafts, and along the gearbox in several locations. Some larger wind turbines have automatic lubrication systems and a reservoir that must be filled periodically to ensure that they will work properly.

Some service manuals provide a detailed checklist for periodic maintenance, which should be followed. It is important to follow the sequence of tests that are identified on the checklist, so that you do not skip any important test. The reason the sequence is provided is so that you can do all the inspections and test of a certain area while safety covers are removed. If you do not complete the test in sequence, it is possible that you will remove a safety or inspection cover and replace it after doing only part of the inspection. Later you will find that you need to make further inspections in this area, so you will have to remove the covers again, which takes more time, or a step may be omitted because it requires you to remove and replace the covers a second time.

The periodic maintenance inspection may be broken into several parts. For example, there may be a complete list of inspections and tests that must be done on the tower and any guy wires that are providing support for the tower. Another series of tests may involve the electrical system that is in the nacelle, and on the ground. A final series of tests may include all the mechanical and rotating parts of the system, such as the yaw control system, the turbine blades, the blade pitch system, the rotor and hub, the low-speed shaft, the gearbox, the high-speed shaft, and the generator. These tests will also include checking the brake system that is used on any of these parts.

Maintenance on the tower consists of visually inspecting any hardware that is used to hold the parts of the tower together, such as nuts and bolts or rivets. This inspection also involves checking to ensure that there are no cracks in any integral parts. A further inspection for corrosion of these parts is also necessary. If any of these problems exist, the service manual will outline the type of repair that needs to be performed. If the tower is embedded in a concrete pad or attached to concrete pad, the concrete pad should be inspected to ensure that it is not deteriorating, and that the bolts that hold the tower in place are not coming loose. If the wind turbine is installed over fresh water or salt water, additional inspections may be required to look for corrosion that occurs in these environments.

Visual inspection of the turbine blades may be outlined in the periodic maintenance inspection. This includes checking the blades for any damage due to bird strikes, lightning strikes, or ice damage. The technician must also check the blades for any deterioration due to aging in sunlight.

Some periodic maintenance on the electrical system must be performed on the system when power is turned off, locked out, and tagged out. After the system is locked out and tagged out, you must make one additional test with the voltmeter to verify that the power is turned off. At this point the periodic maintenance inspection procedure may request that you physically test each electrical connection in the panel and ensure that they are tight and secure. This may require that you use a screwdriver or wrench to physically test each connection to ensure it is tight. Other maintenance procedures may require that you turn the power on and take a number of voltage or current measurements to ensure that the system is operating correctly. If the procedure is to be performed with voltage applied, you must be sure you are using the correct safety procedures while taking the readings.

Periodic maintenance on the hydraulic system may include checking the hydraulic oil level, checking the condition of the hydraulic oil, and replacing hydraulic oil filters. After the turbine has run for a number of years, you may be asked to take an oil sample to be sent to a lab for analysis to determine whether any metal is showing up. The laboratory test can determine the source of the metal particles in the oil, which will indicate a bearing or a moving part in a pump or valve is beginning to fail. If the results of this test indicate that a pump or other part of the system is failing, you will be required to order that part and replace it as soon as possible. You should be aware that you may not be requested to do the same tests and inspections during each periodic maintenance that is scheduled. This means that at some time intervals, you will only be asked to inspect a fluid level, and at longer intervals you will be requested to change that fluid.

The major point to remember about periodic maintenance is that when it is scheduled properly and performed on schedule, it will save wear and tear on the system and help keep downtime to a minimum. Some technicians may feel it is permissible to extend the period of time to perform the periodic maintenance beyond the manufacturer's specifications, or skip portions of the maintenance procedure and consider this action as saving time and money. In reality, when periodic maintenance is not performed on schedule, the cost and downtime in the end will be considerably more.

11.11 MAJOR OVERHAUL OF A WIND TURBINE

At times a wind turbine may sustain a major failure, which requires a major **overhaul** of the system. Also, some wind turbines have an operating life that is determined and specified as a number of years or operating hours, and as the wind turbine runs down toward the end of this life span, certain parts of the wind turbine may need to be changed out in what is called a major overhaul. The major overhaul may include changing blades, or the shafts and drive train for the system. The philosophy of this type of overhaul is based on the fact that it may be less expensive to change the major parts of the system that move and wear as they produce electricity, rather than allowing the wind turbine to fail completely, so that it needs to be replaced. Since the tower does not receive the same amount of wear and tear as the moving parts, it may have a life span that is double that of any of the moving parts, so it may be cost-effective to change the drive train and generator at some point rather than take the wind turbine down and start over.

Individual subsystems of the wind turbine may also be identified for major overhaul when the system comes to the end of its operating life. For example, the generator may be one of the parts of the wind turbine that wears faster than other parts. It is possible to remove and replace the electrical generator as part of a major overhaul. If the generator is removed and replaced, the wind turbine will be able to operate for a longer number of years than it was originally designed for. Another feature of a major overhaul of a wind turbine is that all the major parts can be certified or warranted, which gives the owner basically a new system. This becomes a cost-effective means to keep the wind turbine in service for a longer period of time.

In some cases the complete nacelle for a wind turbine can be removed and taken to a shop where it is overhauled and refurbished. Since the overhaul process takes several weeks or months, a new nacelle or one that has been refurbished can be put on the pole immediately. This type of overhaul is called swapping, as it allows the major parts of the wind turbine to be swapped out and new or refurbished parts installed in their place. This allows the wind turbine to be put back into production with a minimum amount of downtime. It also allows the overhaul team to take their time and test and change out parts at a slower pace in a controlled environment in a shop. This process also allows for older systems to be refurbished at a much lower cost and may provide one or more replacement nacelles to be stored where they will be ready to be used as replacements. Many overhaul facilities have a test system available to turn the main shaft in the nacelle without the blades being connected to the rotor. This allows the complete drive train and generator to be tested in a controlled environment in a shop, which allows technicians to certify the operation prior to putting the nacelle on a pole and attaching blades. If the technicians detect any problem, they can fix them more easily while the nacelle is on the ground in the shop. It is also possible to load the generator with an electrical load bank or by connecting it to the grid and allowing the generator to become fully loaded. This allows the technicians to check the drive train, gearbox, and generator under full load and make any corrections to shaft alignments and vibration checks to ensure the system is able to run at full power when it is installed back on the tower. If this overhaul process is not completed in a controlled environment, it could take much longer and not be as effective.

Overhaul of the wind turbine can be a planned activity that extends the life of the system, or it can be an activity that occurs when a major problem develops, such as a lighting strike or a fire, that causes a large part of the system to fail The process can also be used to develop a system of storing backup parts and repair parts for wind turbines. Replacement parts can be assembled, tested, and put into storage, where they can wait to be used when a complete overhaul is called for.

Questions

1. List the steps in completing a site preparation for a new wind turbine installation.
2. List the basic parts of the hydraulic system that you would need to test if the gauges are showing zero system pressure.
3. Identify the parts of the drive train system for a wind turbine, from the rotor to the generator.
4. Explain the advantages of a major overhaul of a wind turbine system.
5. Explained why maintaining a strict periodic maintenance schedule is important to extend the life of a wind turbine.
6. Identify the things that you should do during the installation procedure when a wind turbine is still on the ground.
7. Explain the things you should be aware of if wind turbine blades are attached to the rotor while the wind turbine is on the ground.
8. Explain why the nacelle may have to be installed on the wind turbine pole after the pole is erected.
9. Explain why it is important to create tests that indicate what sections of the wind turbine system are working correctly when you are troubleshooting a problem in a wind turbine system.
10. Explain when you would use a voltmeter, an ohmmeter, and a clamp-on ammeter when troubleshooting the electrical system of a wind turbine.

Multiple Choice

1. Site preparation for a wind turbine includes
   a. Ensuring that access roads are available to get the wind turbine and its parts to the site
   b. Excavating and pouring concrete for the pad
   c. Determining the closest point to tie in to the grid
   d. All of the above
   e. Only a and c.
2. Wind turbine blades
   a. Can be mounted to the rotor when the tower is on the ground for small and mid-size units
   b. Can be mounted to the rotor after the tower has been erected and the nacelle is installed for large systems
   c. Need to be handled with extreme care during the installation process
   d. All the above
   e. Only a and c
3. If the wind turbine blades are mounted to the rotor before the tower is erected,
   a. The blades must be secured to the tower while the tower is erected.
   b. The blades must be allowed to rotate freely when the towers erected.
   c. Only one of the blades should be mounted to the rotor before erecting the tower.
   d. Turbine blades should never be mounted to the rotor before erecting the tower.
4. In small and mid-size wind turbines, the electrical cables from the generator
   a. Should be routed through the tower while the tower is on the ground.
   b. Should not be installed until after the tower is erected.
   c. Small and mid-size wind turbines do not need electrical cables between the generator and the electrical panel on the ground.
   d. Should be pulled from the ground to the top of the tower with a rope.
5. If the wind turbine blades are installed while the tower is on the ground,
   a. The tower can be rolled to one side when the blades are connected.
   b. The tower does not need to be supported, as the blades can touch the ground without damaging them.
   c. The tower will need to be supported high enough off the ground so the blades do not touch the ground when they are installed.
   d. Wind turbine blades should never be installed on the tower while it is still on the ground.
6. A current transformers are used in the electrical panel
   a. To sense the amount of current the system is producing
   b. To sense the amount voltage the system is producing
   c. To sense the amount of resistance the system has
   d. The sense the amount of torque on the turbine blades
7. The best way to troubleshoot the components in a wind turbine is to
   a. Check every single part to ensure that it is working.
   b. Perform tests that will eliminate large portions of the system as operating correctly.
   c. Turn off all power and visually inspect all the parts.
   d. Concentrate on parts that have been a problem in the past.
8. The difference between a symptom and a problem is that
   a. If you fix a symptom, you have actually repaired the system.
   b. A symptom and a problem are always the same thing.
   c. A symptom such as a blown fuse may cause the system to stop, but it does not indicate what is causing the problem.
   d. You only need to look for symptoms because they are usually the same as the problem.
9. Overhauling a wind turbine system
   a. Is nothing but advanced periodic maintenance
   b. Can save a considerable amount of money and extend the life of the wind turbine
   c. Is more expensive than installing a complete new system
   d. Something that should be done when the system is first installed, before start-up.
10. When you are troubleshooting the electrical system,
    a. You must always turn off and lock out all electrical power.
    b. There are times when you must have the power applied to the system so you can find where power is lost due to faulty components.
    c. The best way to find all problems is to check for continuity throughout the system with the power off.
    d. All the above.

CHAPTER 12

Electrical and Electronic Fundamentals for Wind Generators

OBJECTIVES

After reading this chapter, you will be able to:

- Explain the term *voltage.*
- Explain the term *current.*
- Explain the term *resistance.*
- Understand Ohm's law and how it is used to calculate volts, ohms, and current.
- Explain the function of an electrical wiring diagram.
- Explain the function of an electrical ladder diagram.
- Explain why the voltage at each branch of a parallel circuit is the same as the supply voltage.
- Explain the symbols M, k, m, and μ as used in units of measure, and provide an example of how each is used.
- Discuss the advantage of connecting switches in series with a load.
- Explain the term *infinite resistance.*
- List two things you should be aware of when making resistance measurements with an ohmmeter.
- Explain why you should set the voltmeter and ammeter to the highest setting when you first make a measurement of an unknown voltage or current.
- Explain which two materials are combined to make P-type and N-type materials.
- Explain the operation of a diode (PN) junction and show the input AC waveform and the output DC waveform.
- Identify a four-diode and a six-diode full-wave bridge rectifier.
- Explain the operation of a power electronic frequency converter (inverter).

KEY TERMS

Closed switch
Coil
Contactor
Diode
Fuse
Fuse disconnect
Inverter
Ladder diagram
Light-emitting diode (LED)
Motor starter
N-type material
Normally closed
Normally open
Open circuit
Open fuse
Open switch
Overload
Parallel circuit
P-type material
Power circuit
Rectifier
Relay
Schematic (ladder) diagram
Series circuit
Series-parallel circuit
Switch
Transformer
Transistor
Voltage drop
Wiring diagram

OVERVIEW

As a technician, you must thoroughly understand electricity so that you will be able to troubleshoot the electrical components of systems, such as generators, switches, and controls. This chapter will provide you with the basics of electricity, which will be the building blocks for more complex circuits that you will encounter in the field. You do not need any previous knowledge of electricity to understand this material. The simplest form of electricity to understand is direct-current (DC) electricity, so many of the examples in this chapter use DC electricity. It is important to understand, however, that the majority of electrical components and circuits in field equipment use alternating-current (AC) electricity, so some of the examples use AC electricity. This chapter begins by defining basic terminology such as voltage, current, and resistance and showing simple relationships between them.

12.1 BASIC ELECTRICITY AND A SIMPLE ELECTRICAL CIRCUIT

It is easier to understand the basic terms of electricity if they are connected with a working circuit with which you can identify. As an example, we will use the motor that moves a conveyor belt. The motor must have a source of voltage and current. *Current* is a flow of electrons, and *voltage* is the force that makes the electrons flow. This is just an introduction; much more detail about the terms will be provided later in this chapter.

This circuit is shown in Figure 12-1. It shows that the source of voltage is 110 V, which is identified by L1 and N. The **switch** is a manually actuated switch for this circuit, so the switch provides a way to turn the motor on and off. When the switch is moved to the on position, or closed, voltage and current move through the wires to the motor. When voltage and current reach the motor, its shaft begins to turn. The wires in the circuit are called *conductors*. The conductors provide a path for voltage and current, and they provide a small amount of opposition to the current flow. This opposition is called *resistance*.

After the motor has run for a while, the switch is opened, the voltage and current are stopped at the switch, and the motor is turned off. When the switch is closed again, the voltage and current again reach the motor. Now that you have an idea of the basic terms *voltage*, *current*, and *resistance*, you are ready to learn more about them.

The source of energy that makes the shaft in the motor turn is called electricity. *Electricity* is defined as a flow of electrons. You should remember from the previous discussion that the definition of *current* is also a flow of electrons. In many cases the terms *electricity* and *current* are used interchangeably. *Electrons* are the negative parts of an atom, and *atoms* are the basic building blocks of all matter. All matter can be broken into one of 118 materials called *elements*. Two of the elements that are commonly used in electrical components and circuits are copper, which is used for wire, and iron, which is used to make many of the parts, such as motors.

To understand electricity at this point, we need to understand only that the atom has three parts: electrons, which have a negative charge; protons, which have a positive charge; and neutrons, which have no charge (also called a neutral charge). Figure 12-2 is a diagram of the simplest atom, which has one electron, one proton, and one neutron. From this diagram you can see that the proton and neutron are located in the center of the atom. The center of the atom is called the *nucleus*. The electron moves around the nucleus in a path called an *orbit*, in much the same way the earth moves around the sun.

Because the electron is negatively charged and the proton is positively charged, an attraction between these two charges holds the electron in orbit around the nucleus until sufficient energy, such as heat, light, or magnetism, breaks it loose so that it can become free to flow. Any time an electron breaks out of its orbit around the proton and becomes free to flow, it is called electrical current. Electrical current can occur only when the electron is free from the atom so it can flow. The energy to create electricity can come from burning coal, nuclear power reactions, or hydroelectric dams.

The atom shown in Figure 12-2 is a hydrogen atom; it has only one electron. In the circuit presented in Figure 12-1, the copper wire is used to provide a path for the electricity to reach the conveyor motor. Because copper is the element most often used for the conductors in circuits, it is important to study the copper atom. The copper atom is different from the hydrogen atom in that the

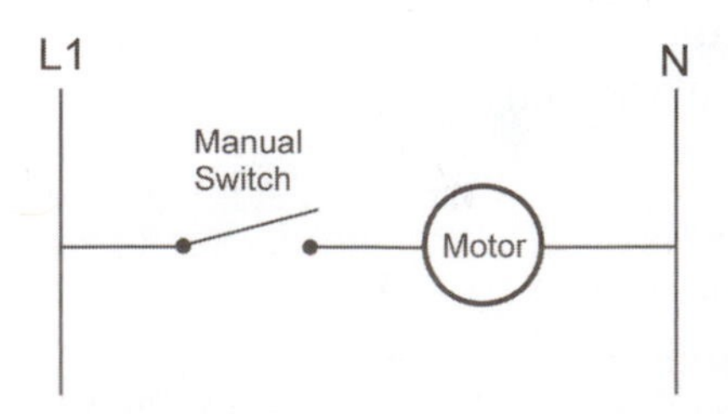

FIGURE 12-1 Electrical diagram of a manually operated switch controlling a motor.

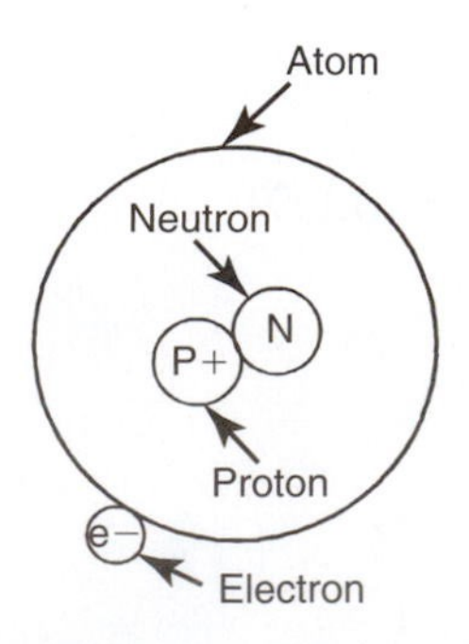

FIGURE 12-2 An atom with its electron, proton, and neutron identified.

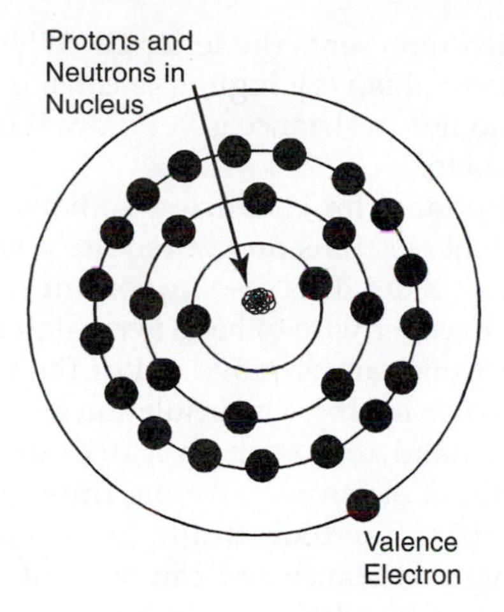

FIGURE 12-3 A copper atom has 29 electrons. This atom has one electron in its valence shell that can break free to create current flow.

copper atom has 29 electrons, whereas the hydrogen atom has only 1. The copper wire that provides a path for the electricity for the motor contains millions of copper atoms.

Figure 12-3 shows a diagram of the copper atom with its 29 electrons, which move about the nucleus in a series of orbits. Because the atom has so many electrons, they will not all fit in the same orbit. Instead, the copper atom has four orbits. These orbits are referred to as *shells.* There are also 29 protons in the nucleus of the copper atom, so the positive charges of the protons will attract the negative charges of the electrons and keep the electrons orbiting in their proper shells. The attraction for the electrons in the three shells closest to the nucleus is so strong that these electrons cannot break free to become current flow. The only electrons that can break free from the atom are the electrons in the outer shell. The outer shell is called the *valence shell,* and the electrons in the valence shell are called *valence electrons.* Notice that the copper atom has only one electron in the valence shell. All good conductors have a low number of valence electrons. For example, gold—which is one of the best conductors—has only one valence electron, and silver—which is also a good conductor—has three valence electrons.

In the motor circuit we studied earlier, a large amount of electrical energy, called *voltage,* is applied to the copper atoms in the copper wire. The voltage provides a force that causes a valence electron in each copper atom to break loose and begin to move freely. The free electrons actually move from one atom to the next. This movement is called *electron flow;* more precisely, it is called *electrical current.* Because the copper wire has millions of copper atoms, it will be easier to understand if we study the electron flow among three of them. Figure 12-4 is a diagram of the electrons breaking free from three separate copper atoms.

When the electron breaks loose from one atom, it leaves a space, called a *hole,* where the electron was located. This hole will attract a free electron from the next nearest atom. As voltage causes the electrons to move, each electron moves only as far as the hole that was created in the adjacent atom. This means that electrons move through the conductor by moving from hole to hole in each atom. The movement will begin to look like cars that are bumper to bumper on a freeway.

Example of Voltage in a Circuit

Voltage is the force that moves electrons through a circuit. Figure 12-5 shows the effects of voltage in a circuit in which two identical light bulbs are connected to separate voltage supplies. Notice that the bulb connected to the 12-V battery is much brighter than the bulb connected to the 6-V battery. The bulb connected to the 12-V battery is brighter because 12 V can exert more force on the electrons in the circuit than 6 V, and higher voltage will also cause more electrons to flow.

In the original diagram of the motor, the amount of voltage is 110 V. The energy that produces this force comes from a generator. The shaft of the generator is turned by a source of energy such as steam. Steam is created from heating water to its boiling point by burning coal or from a nuclear reaction. The energy to turn the generator shaft can also come from water moving past a dam at a hydroelectric facility. In the light-bulb example, the voltage comes from a battery. The voltage in a battery was previously stored in a chemical form when the battery was originally

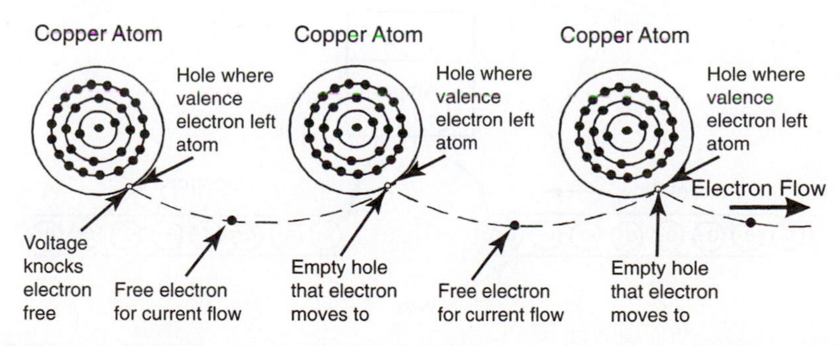

FIGURE 12-4 Voltage is the force that knocks one electron out of its orbit. In this diagram the valence electron in the atom on the far right breaks free from its orbit and leaves a hole. The electrons in the atoms to the left of the first atom move from hole to hole to create current flow.

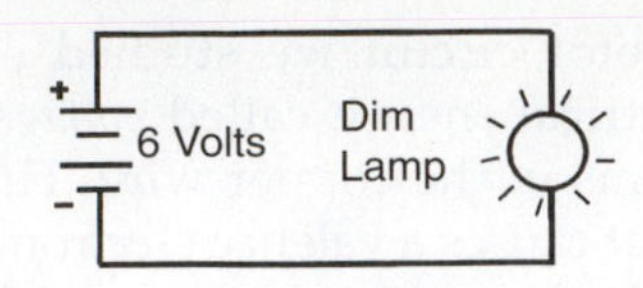

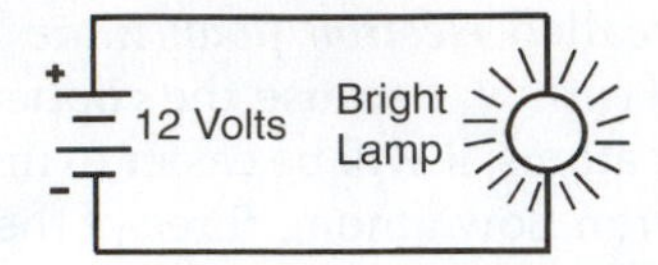

FIGURE 12-5 Two identical light bulbs with different voltages applied to them. The light bulb with 12 V applied to it is brighter than the one with 6 V applied to it.

charged. The larger the amount of voltage in a circuit, the larger the force that can be exerted on the electrons, which makes more of them flow. The scientific name for voltage is *electromotive force* (EMF).

Example of Current in a Circuit

The definition of *current* is a flow of electrons. The actual number of electrons that flow at any time in a circuit can be counted, and their number is very large. The unit of current flow is the *ampere* (A). The word *ampere* is usually shortened to amp. One ampere is equal to 6.24×10^{18} electrons flowing past a point in 1 second (s). Figure 12-6 shows an example of the number of electrons flowing that equals 1 A. In this diagram you can see that each electron that is flowing is counted; when 6,240,000,000,000,000,000 electrons pass a point in 1 s, it is called an *ampere* (A). The ampere is used in all electrical equipment, such as motors, pumps, and switches, as the unit for electrical current.

Example of Resistance in a Circuit

Resistance is defined as the opposition to current flow. Resistance is present in the wire that is used for conductors and in the insulation that covers the wires. The unit of resistance is the ohm, or Ω (the capital Greek letter omega, which represents the letter O). When the resistance of a material is very high, it is considered to be an insulator, and if its resistance is very low, it is considered to be a conductor.

It is important for conductors to have very low resistance so that electrons do not require a lot of force to move through them. It is also important for insulation that is used to cover wire to have very high resistance so that the electrons cannot move out of the wire into another wire that is nearby. The insulation also ensures that a wire on a hand tool such as a drill or saw can be touched without electrons traveling through it to shock the person using the tool. Examples of materials that have very high resistance and can be used as insulators are rubber, plastic, and air.

The amount of resistance in an electrical circuit can also be adjusted to provide a useful function. For example, the heating element of a furnace is manufactured to have a specific amount of resistance so that the electrons will heat it up when they try to flow through it. The same concept is used in determining the amount of resistance in the filament of a light bulb. The light bulb filament of a 100-W light bulb has approximately 100 Ω of resistance; 100 Ω is sufficient resistance to cause the electrons to work harder as they move through the filament, which causes the filament to heat up to a point where it glows. When the amount of resistance in a circuit is designed to convert energy, such as a heating element or motor, it is referred to as a *load*.

Identifying the Basic Parts of a Circuit

It is important to understand that each electrical circuit must have a *supply of voltage, conductors* to provide a path for the electrons to flow, and at least one *load*. The circuit may also have one or more controls. Figure 12-7 shows an example of a typical electrical circuit with the voltage supply, conductors, control, and load identified. In this circuit, the voltage source is supplied through terminals L1 and N, the load is a motor, and the two wires connecting the motor to the voltage source are the conductors. The manual switch is a control, because it controls the current

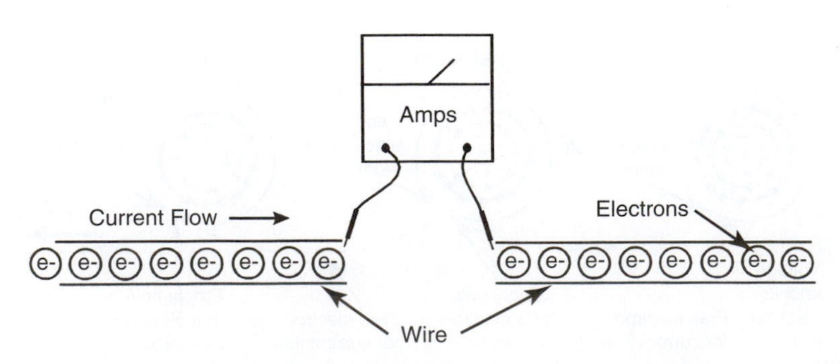

FIGURE 12-6 Electrons passing a point where they are measured. When the number reaches 6,250,000,000,000,000,000 in 1 s, 1 A of current has been measured.

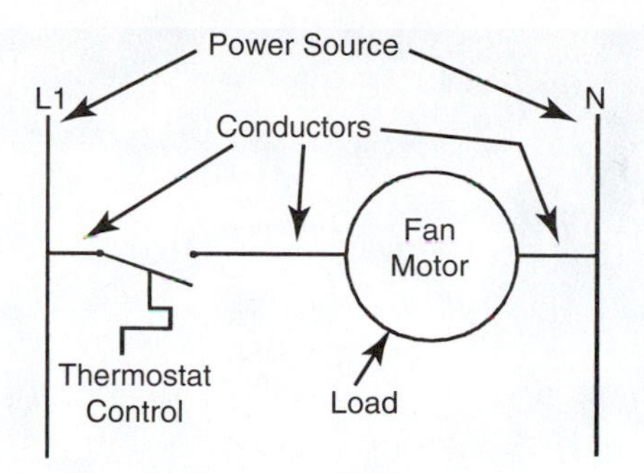

FIGURE 12-7 The basic parts of a circuit: the power source, conductors, control, and load.

flow to the motor. In a large wind turbine, it is possible to have more than one load. For example, the system may have multiple hydraulic pumps. You will learn more about these components later.

Wiring Diagrams and Ladder Diagrams

There are two basic ways to diagram electrical circuits. The components can be shown in a wiring diagram or a ladder diagram. The **wiring diagram** shows the components of a circuit as you would see them if you took a picture of the circuit. The **ladder diagram** shows the sequence of operation of all the components. Figure 12-8 illustrates a wiring diagram with two switches and a light. The switches and wires are shown where you would find them in a picture. The wiring diagram helps you know where the components are located in an electrical cabinet and where the terminals are located. This is important when you are installing components in a circuit. Figure 12-9 shows the same circuit, but this time drawn as a ladder diagram. The ladder diagram clearly shows that the two switches are wired in parallel with each other so that either can turn on the light. This is called *showing the sequence of operation*. In the wiring

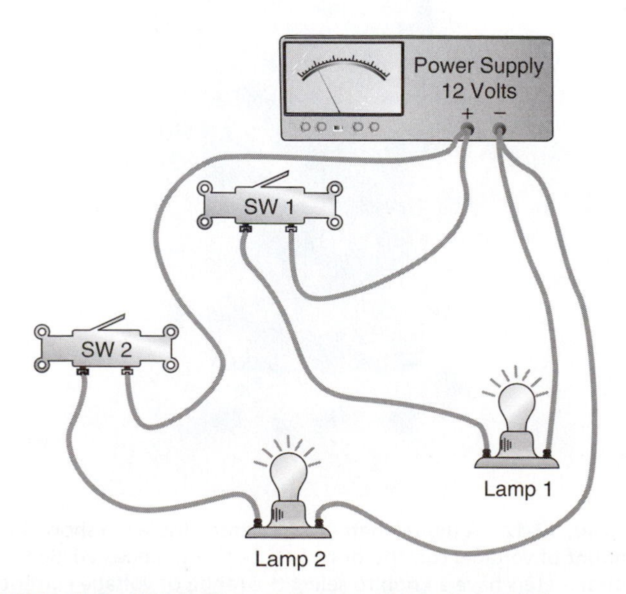

FIGURE 12-8 Wiring diagram for two circuits in parallel.

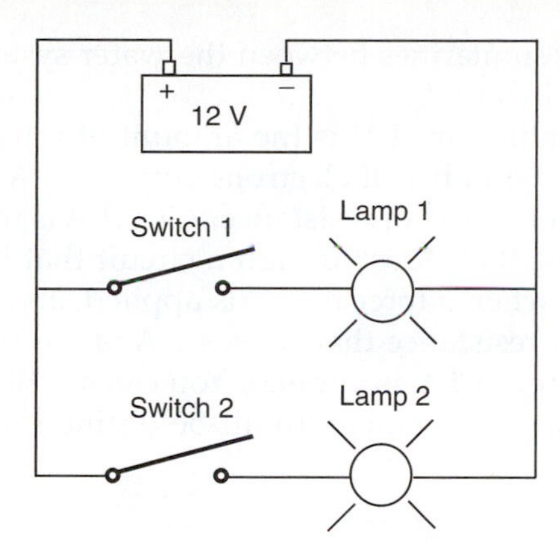

FIGURE 12-9 Ladder diagram of two circuits in parallel, each with one switch and one lamp.

diagram it may be difficult to determine if the switches are connected in series or parallel, whereas in the ladder diagram it is very easy to see how the switches operate. You will learn that one of these diagrams is not necessarily better than the other; rather, there is an appropriate time to use each. You will also see that it is easy to convert from either type of diagram to the other.

Equating Electricity to a Water System

It may be easier to understand electricity if you think of it as a system such as a water system, with which you may be more familiar. If you examine water flowing through a hose, for example, the flow of water would be equivalent to the flow of current (amps), and the water pressure would be equal to voltage. If you stepped on the hose or bent the end over, it would create a resistance that would slow the flow, just as resistance in an electrical circuit slows the flow of current. Figure 12-10 is a diagram that

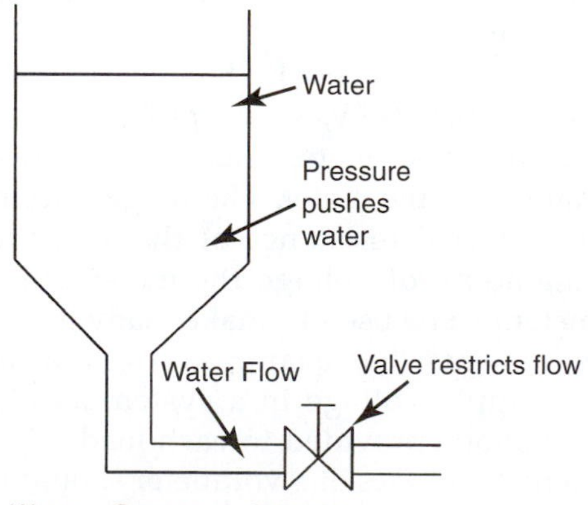

FIGURE 12-10 A diagram showing similarities between a water system and an electrical system. The flow of water is similar to the flow of current (amps), the pressure on the water is similar to voltage, and the resistance from the valve causes the flow of water to slow just as resistance in an electrical system causes current flow to slow.

shows the similarities between the water system and an electrical circuit.

By definition, 1 V is the amount of force required to push the number of electrons equal to 1 A through a circuit that has 1 Ω of resistance. Also, 1 A is the number of electrons that flow through a circuit that has 1 Ω of resistance when a force of 1 V is applied, and 1 Ω is the amount of resistance that causes 1 A of current to flow when a force of 1 V is applied. You can see that voltage, current, and resistance can all be defined in terms of each other.

FIGURE 12-11 An analog meter has a needle that is moved by a magnetic field. The face of the meter indicates the amount being measured and the knob on the front is used to determine whether volts, amps, or ohms are being measured and the range of values the meter can measure. (Courtesy of Triplett Corporation.)

12.2 MEASURING VOLTS, AMPS, AND OHMS

It is important to be able to measure the amount of voltage (volts), current (amperes), and resistance (ohms) in a circuit in a wind turbine system to be able to determine when it is working correctly or something is faulty. If you are checking a tire on your automobile, you can see if it is flat by simply looking at it. Because the flow of electrons through a wire is invisible, it is not possible to simply look at the wire and determine if it has voltage applied to it or if it has current flowing through it, so you must use a meter to make voltage, current, and resistance measurements.

Measuring Voltage

The meter used to measure voltage is called a voltmeter. Figures 12-11 and 12-12 show two types of voltmeters. The meter in Figure 12-11 is called an *analog-type voltmeter* because it uses a needle and a scale to indicate the amount of voltage being measured. The meter in Figure 12-12 is called a *digital-type voltmeter* because it displays numbers to indicate the amount of voltage being measured.

The voltmeter has very high internal resistance, approximately 20,000 Ω/V, so its probes can be safely placed directly across the terminals of the power source without damaging the meter. The range selector switch adjusts the internal resistance of the meter to set the maximum amount of voltage the meter can measure. The voltmeter can be used to make many different voltage measurements. For example, it can measure the amount of supply voltage in a system as well as the amount of voltage provided to each load. Figure 12-13 shows where the probes of a voltmeter should be placed to safely make voltage measurements in a circuit. In Figure 12-13a, the voltmeter probes are placed on the terminals of a battery to measure the amount of battery voltage supplied. Figure 12-13b shows the proper location for placing the voltmeter probes to measure the voltage available to lamp 1, and Figure 12-13c shows the proper location for placing the voltmeter probes to measure the voltage available to lamp 2. In each case the voltmeter probes are placed across the terminals where the voltage is being measured.

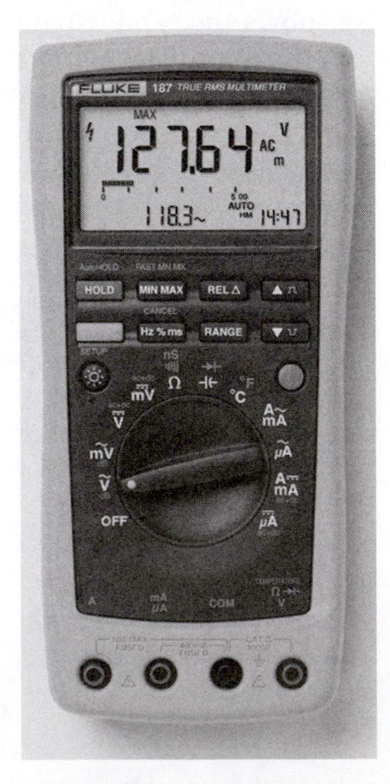

FIGURE 12-12 A digital meter has a digital display to show the amount of voltage, current, or resistance being measured. Some digital meters have a knob to select the range of voltage current and resistance, while others are self-ranging. (Reproduced with permission of Fluke Corporation.)

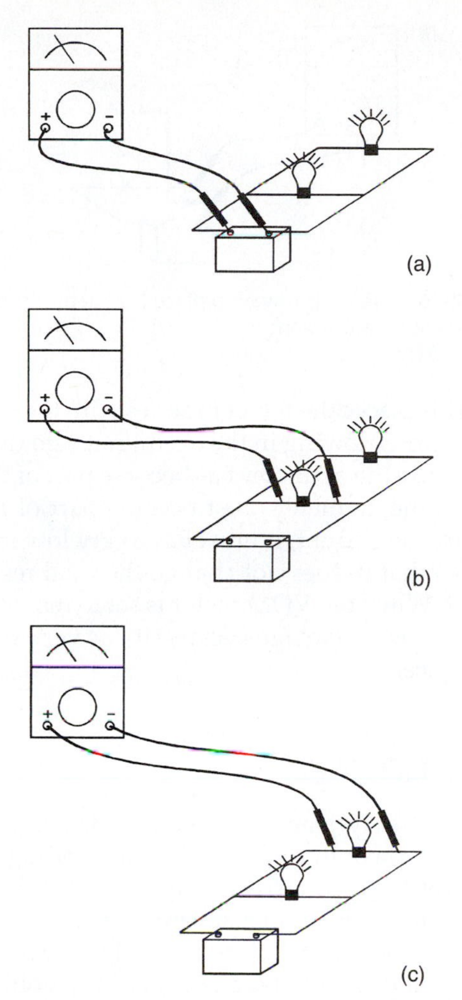

FIGURE 12-13 (a) The proper way to place the probes of a voltmeter to measure the amount of voltage available at the battery. (b) The proper way to place the voltmeter probes to measure the voltage available at lamp 1. (c) The proper way to place the voltmeter probes to measure the voltage available at lamp 2.

SAFETY NOTICE

It is important to remember that the voltage to a circuit must be turned on to make a voltage measurement. This presents an electrical shock hazard. You must take extreme caution when you are working around a circuit that has voltage applied, so that you do not come into contact with exposed terminals or wires when you are making voltage measurements.

Measuring Large and Small Currents

The meter shown in Figure 12-14 is called a clamp-on ammeter, and it is used to measure current in larger AC circuits (some clamp-on ammeters can measure DC current, too). From the figure you can see that this type of meter has two *claws*, or jaws, at the top that form a circle when they are closed. A button on the side of the meter is depressed to cause the jaws to open so that they can fit around a wire without having to disconnect the wire at

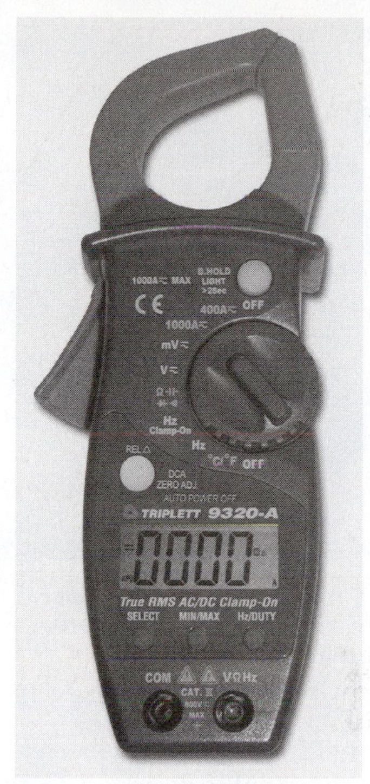

FIGURE 12-14 A digital clamp-on ammeter that is used to read high currents. (Courtesy of Triplett Corporation.)

one end. The jaws of the clamp-on ammeter are actually a **transformer** that measures the amount of current flowing through a wire by *induction*. When current flows through a wire, it creates a magnetic field that consists of flux lines. Since AC voltage reverses polarity once every 1/60 s, the flux lines collapse and cause a small current to flow through the transformer in the claw of the ammeter. This small current is measured by the meter movement, and the display indicates the actual current flowing in the conductor clamped by the meter. For this reason, the clamp-on ammeter can only measure current in one conducting wire at a time. If the clamp-on ammeter was connected around more than one wire at a time, the current in one wire would cancel the current measured in the other wire, which would cause the ammeter to measure 0 A. The clamp-on ammeter for measuring AC current has several ranges so it can measure lower currents (less than 10 A) and larger currents, that is, currents larger than 100 A. If you are trying to measure a very small amount of AC current, you can wrap the wire two or three times around a claw and it will cause the amount of current to multiply by 2 if you use two coils, or to multiply by 3 if you use three coils around the claw. The *multiplier* allows the ammeter to measure smaller currents more accurately.

There is also a version of the clamp-on ammeter that can measure DC current. This type of meter must be specifically designed for measuring DC current.

An ammeter may be analog or digital, and it is used to measure electrical current. A digital ammeter is used to measure current that is less than 10 A. This type of ammeter is built into a special type of meter that is usually called a VOM (volt-ohm-milliammeter) because it can measure

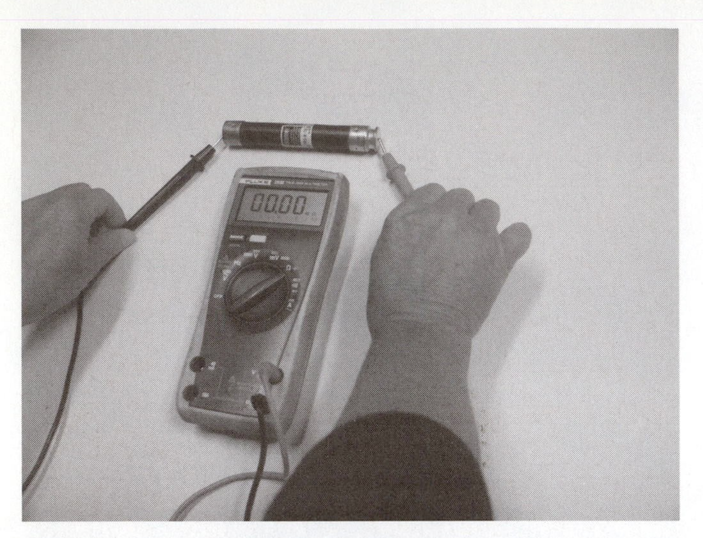

FIGURE 12-15 A true RMS digital VOM that can read voltage, resistance, current from milliamps through 10 A, and also measures temperature and capacitance.

volts, ohms, or milliamps. When the meter is used to measure current, it is actually designed to read very small amounts of current, such as 1/1000 A, called a *milliamp* and also written as 0.001 A. (Recall that the prefix *milli-* means one-thousandth.) A VOM is shown in Figure 12-15 measuring the resistance of a fuse. The analog VOM meter that was shown in Figure 12-11 has a single meter movement that causes a needle to deflect when it senses current flow. This means that when a voltage is applied to a VOM meter, it must be routed through resistors that convert it to current that actually flows through the meter movement. When the VOM meter is set to measure current or resistance, the same meter movement is used. The only part of the meter that changes as it is switched from being a voltmeter to an ammeter or an ohmmeter is the arrangement of resistors to limit the amount of current flow through the meter movement.

Measuring Milliamps

At times you will need to measure the current in a circuit in milliamps, since the amount of current is less than 1 A. For example, you may need to measure the amount of current used by a hydraulic solenoid valve to ensure that it is working correctly. You will also need to make milliamp measurements on the solid-state control boards found in wind turbine control systems. The most important part about reading milliamps is that the ammeter must be placed in series with the current it is reading. For the meter to be put in series with the current it is reading, an open must be created in the circuit. Figure 12-16 shows that an open is made in the circuit when a milliamp measurement is made, and the terminals of the ammeter should be placed on the wires where the hole has been made in the circuit. The open is created by removing one of the wires from the battery terminal. One of the meter probes is placed on the battery terminal, and the other is placed on the end of the wire that was removed from the

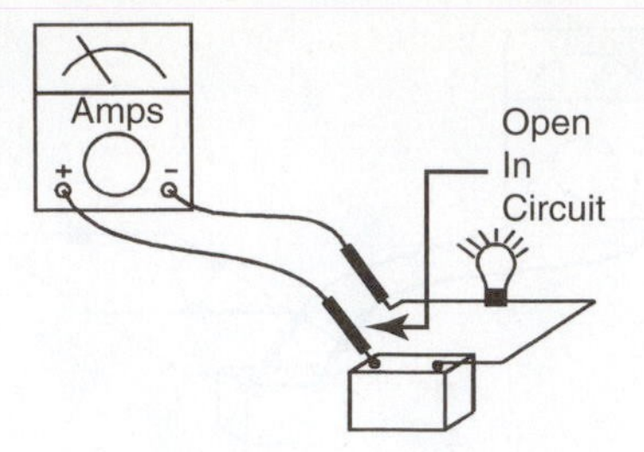

FIGURE 12-16 The proper way to place the meter probes to make a current measurement.

battery. This places the meter in series with the circuit, and all the electrons flowing in the circuit must go through the meter because the ammeter has become part of the circuit.

Since the ammeter must become part of the circuit to measure the current, it must have very low internal resistance so that it does not change the total resistance of the circuit. When the VOM meter is set to read milliamps, it is vulnerable to damage because it has very small internal resistance.

SAFETY NOTICE

If you mistakenly place the meter leads across a power source as you would to take a voltage reading when the meter is set to read current, the meter will be severely damaged. You can also be severely burned because the meter can explode, since it is a very-low-resistance component when it is set to read current. This would be similar to taking a piece of bare wire and bending it so it could be inserted into the two holes of an electrical outlet. You must always be sure to check how you have the meter set when you are making voltage, current, or resistance measurements.

As a technician you will be expected to use both a milliamp meter and a clamp-on ammeter. Milliamp meters are used to measure small amounts of current, and clamp-on meters are used to measure larger currents such as the amount of current a motor is using. If a motor is using (pulling) too much current, you will be able to determine that it is overloaded and that it will soon fail.

Digital VOM meters and some analog VOM meters have circuits to measure both DC and AC current. These types of meters have several ranges for measuring very small current in the range of milliamps up to 10 A.

Making Measurements with Digital VOM Meters

When you make a measurement with a digital VOM meter, you should follow the same rules as when you use an analog VOM meter. The major difference is that the measurement is presented as a digital number on the meter's display. All you need to do is observe the range selector switch to determine the units that the meter is measuring and use the value that is displayed as the measurement. Be sure to observe polarity (+/−) if the voltage or current is DC. Some digital meters also provide an audible signal for resistance measurements, which is useful when you are locating wires in multiconductor cables

Measuring Resistance and Continuity with a Digital VOM

You can use a digital VOM to measure resistance for several purposes. For example, you can use the digital meter to measure continuity, which is a test in which you are looking for zero resistance, or infinite resistance in a component such as a fuse or switch. If a fuse has zero resistance, it is considered good; and if it has infinite resistance, it is considered open. When you are using a digital voltmeter to measure continuity, you should begin the test by first checking to see if the meter leads and terminals are good. You can start by touching the two meter leads together; this represents zero resistance to the meter, and the meter display should show zero or near zero ohms. If the meter displays zero ohms when you touch the two leads together, it means the meter is good and the leads are good. When you hold the two leads apart, the meter is reading infinite resistance, and you should see the indication for infinite resistance. On some digital meters the display will show OL, and on other digital meters it will display a large number such as 9999. You can test your digital meter for reading zero and infinite by setting it to one of the ohms positions, touching the leads together, and observing the reading for zero ohms (continuity), then allowing the leads not to touch and observing the reading of the display for infinite. Figure 12-17 shows a digital meter measuring infinite resistance, which indicates that the fuse is open. Notice that the display is showing OL. Figure 12-18 shows a digital meter measuring zero resistance, which indicates that the fuse is good. Notice that the display is showing 0.0.

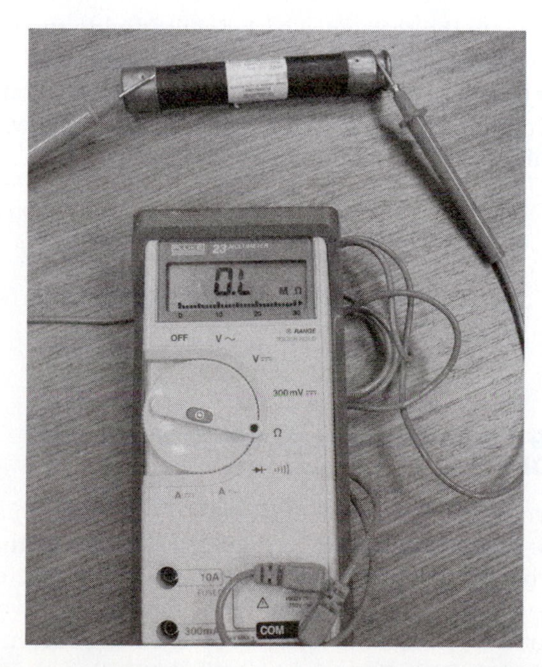

FIGURE 12-17 A digital meter showing high resistance (infinite resistance) in an open fuse.

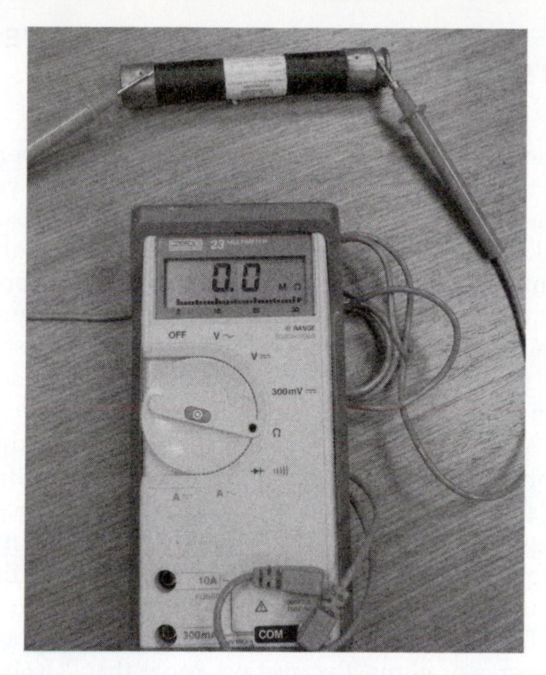

FIGURE 12-18 Meter showing low resistance when the fuse is good.

Measuring continuity means that you are looking for zero resistance or infinite resistance. The continuity test is good for measuring the resistance through a switch, views, or piece of wire to determine if it has low resistance, or infinite resistance. A component is said to have *continuity* if its resistance is low. If the resistance is high (infinite) the wire, switch, or fuse is said to be *open*. Some digital meters have an audible beep that can be selected with the continuity test, to provide an audible signal when the resistance is low and you have continuity. This is quite useful when testing fuses, switches, or wires, and it allows you to make the test very quickly because you are not looking at the display for a number, rather you are listening for the audible beep.

The third type of measurement that you will make with the ohmmeter is to measure actual resistance in a motor winding, or the resistance of a variable resistor. In this case you are actually trying to identify the specific amount of resistance, so you will need to read the meter accurately as its display will show the number of ohms the meter is measuring. For example, you may be trying to compare the windings and a transformer, or the windings of a motor, so you will need to know the exact number of ohms of resistance in the winding. When you make this resistance measurement, you must be sure that no power is applied to the circuit, and that the component you are measuring has its leads isolated so it does not have interference from nearby resistance. When you touch the meter leads to the two ends of the coil, you can then look at the meter; the amount of resistance shown in the display is the amount of resistance the meter is measuring.

12.3 USING OHM'S LAW TO CALCULATE VOLTS, AMPERES, AND OHMS

There are times when you are working in a circuit that you need to calculate or estimate the volts, amperes, or ohms the circuit should have. In most cases when you are troubleshooting an electrical problem, you will also need to make measurements of the volts, amperes, and ohms to determine if the circuit is operating correctly. The problem with taking a measurement is that you will not be sure if the values are too high or too low or if they are just about right. A calculation can tell you if the measurements are within the estimated values the circuit should have.

A relationship exists between the volts, amperes, and ohms in every DC circuit that can be identified by Ohm's law. *Ohm's law* states simply that the amount of voltage in a DC circuit is always equal to the amperes multiplied by the resistance in that circuit. If a circuit has 2 A of current and 4 Ω of resistance, the total voltage is 8 V.

The formula can be changed to determine the amount of amperes or ohms. Because we know that $2 \times 4 = 8$, it stands to reason that $8 \div 2 = 4$, and $8 \div 4 = 2$. Thus, if you measure the volts in a circuit and find that you have 8 V, and if you measure the amperes in the circuit and find that you have 2 A, you can determine that you have 4 Ω by dividing 8 by 2. You can also determine the number of amperes you have in the circuit by measuring the volts and ohms and calculating: $8\text{ V} \div 4\ \Omega = 2\text{ A}$.

Ohm's Law Formulas

When scientists, engineers, and technicians do calculations, they use letters of the alphabet to represent units such as volts, amps, and ohms. The letters are accepted by everyone, so when a calculation is passed from one person to another, everyone uses the same standard abbreviations.

In Ohm's law formulas, the letter E is used to represent voltage. The E is derived from first letter of the word *electromotive force.* The letter R is used to represent *resistance.* The letter I is used to represent current (amperes); it is derived from the first letter of the word *intensity.*

Ohm's law is represented by the formula $E = IR$. When you use Ohm's law, the basic rule of thumb is that you should use E, I, and R to represent the unknown values in a formula when you begin calculations, and you should use V for volts, A for amps, and Ω for ohms as units of measure when you have determined an answer or whenever the values are known or measured.

Using Ohm's Law to Calculate Voltage

When you are solving for voltage, you need to know the amount of current (amperes) and the amount of resistance (ohms). If you do not know these two values, you cannot calculate voltage. For example, if you are asked to calculate the amount of voltage when the current is 20 A and the resistance is 40 Ω, you can multiply 20 by 40 to get an answer of 800 V. To solve a problem, start with the formula and substitute the values of I and R. Then you can continue the calculation. Notice that the units A (for amperes) and Ω (for ohms) are added when the values are known.

$$E = IR$$

$$E = 20\text{ A} \times 40\ \Omega$$

$$E = 800\text{ V}$$

Using Ohm's Law to Calculate Current

The formula for calculating an unknown amount of current in a circuit is

$$I = \frac{E}{R}.$$

To calculate the current, divide the voltage by the resistance. For example, if you have measured the circuit voltage and found that it is 50 V, and you have measured the resistance and found that it is 10 Ω, the current is found by dividing 50 V by 10 Ω, which equals 5 A.

$$I = \frac{E}{R}$$

$$I = \frac{50\text{ V}}{10\ \Omega}$$

$$I = 5\text{ A}$$

Using Ohm's Law to Calculate Resistance

The formula for calculating an unknown amount of resistance in a circuit is

$$R = \frac{E}{I}.$$

To calculate the resistance of a circuit, divide the voltage by the current. For example, if you have measured voltage of 60 V and current of 12 A, the resistance is found by dividing the voltage by the current. Again, start with the formula and substitute values to get an answer:

$$R = \frac{E}{I}$$

$$R = \frac{60\text{ V}}{12\text{ A}}$$

$$R = 5\ \Omega$$

Using the Ohm's Law Wheel to Remember the Ohm's Law Formulas

A learning aid has been developed to help you remember all the Ohm's law formulas. Figure 12-19 shows this aid, and you can see that it is in the shape of a wheel, or pie. The top of the pie has the letter E, representing voltage; the bottom left has the letter I, representing current (amperes); and the bottom right has the letter R, representing resistance (ohms).

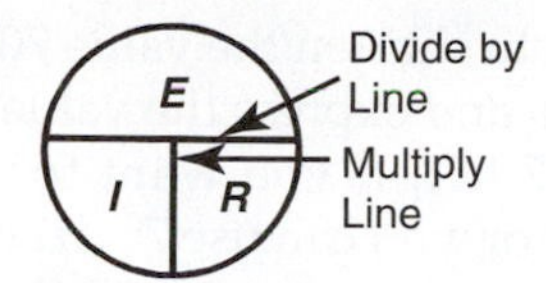

FIGURE 2-19 Ohm's law pie, a memory tool for remembering all the Ohm's law formulas.

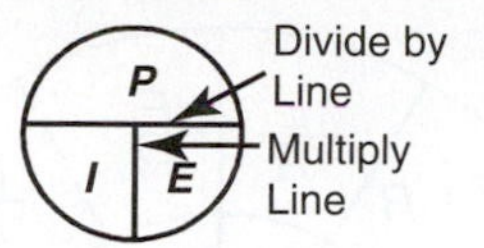

FIGURE 12-21 Formulas for calculating power, current, and voltage.

Figure 12-20 shows how to apply the formulas. If you want to remember the formula for voltage, put your finger over the E: The letters I and R show up side by side. The vertical line separating the I and R is called the *multiply line*, because that is the math function you use with I and R. This represents the formula $E = IR$.

If you want to know the formula for current (I), put your finger over the I: The remaining letters are E over R. The horizontal line that separates the E and R is called the *divide-by line*, because to determine I, you divide E by R. This represents the formula

$$I = \frac{E}{R}.$$

If you want to know the formula for resistance (R), put your finger over the R: The remaining letters are E over I. This represents the formula

$$R = \frac{E}{I}.$$

Calculating Electrical Power

Electrical power (P) is work that electricity can do. The units for electrical power are *watts* (W). Watts are determined by multiplying amps times volts ($P = IE$). Most electrical loads that are resistive in nature, such as heating elements, are rated in watts. For example, if a heating element is rated for 5 A and 110 V, it uses 550 W of power. You can calculate the amount of power by using Watt's law, $P = IE$: 5 A × 110 V = 550 W.

This basic formula can be rearranged in the same way as the Ohm's law formula. You can calculate current by dividing the power (550 W) by the voltage (110 V), and you can calculate voltage by dividing the power (550 W) by the current (5 A). Figure 12-21 shows a pie that is a memory aid for remembering all of the formulas for Watt's law. In this figure, the original formula, $P = IE$, can be found by placing your finger over the P. The formula for solving for current,

$$\left(I = \frac{P}{E}\right)$$

can be found by placing your finger over the I, and the formula to solve for voltage,

$$\left(E = \frac{P}{I}\right)$$

can be found by placing your finger over the E.

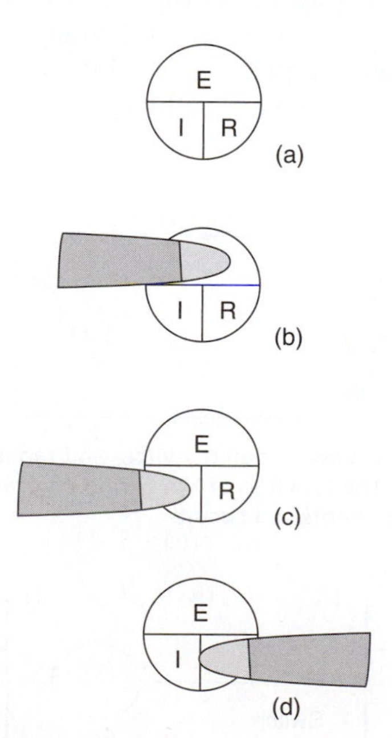

FIGURE 12-20 (a) The Ohm's law pie is a memory tool for remembering all of the Ohm's law formulas. (b) Put your finger over the E if you want the formula to find voltage. (c) Put your finger over the I if you want the formula to find current. (d) Put your finger over the R if you want the formula to find resistance.

Presenting All the Formulas

When you are troubleshooting, you may be able to get a voltage reading easily. If you can find a data plate that provides the wattage for a heating element or other load, you will be able to use one of the formulas that was presented to calculate (estimate) the current or the resistance. This estimate will give you an idea of what the values should be so that you can determine whether the system is working correctly. When you are in the field making these measurements, it may be difficult to remember all the formulas, so a chart has been developed that provides all the formulas. In Figure 12-22 you can see that the chart is in the shape of a wheel with spokes emanating from an inner ring. A cross is shown in the middle of the inner wheel, and it separates the inner wheel into four sections. These sections are represented by the letters P (power, in watts), I (amps), E (volts), and R (ohms).

The formulas in each section of the outer ring correspond to P, I, E, or R. The formulas for wattage are shown emanating from the inner section marked with P. All the formulas for current are shown emanating from the inner section marked with I. All the formulas

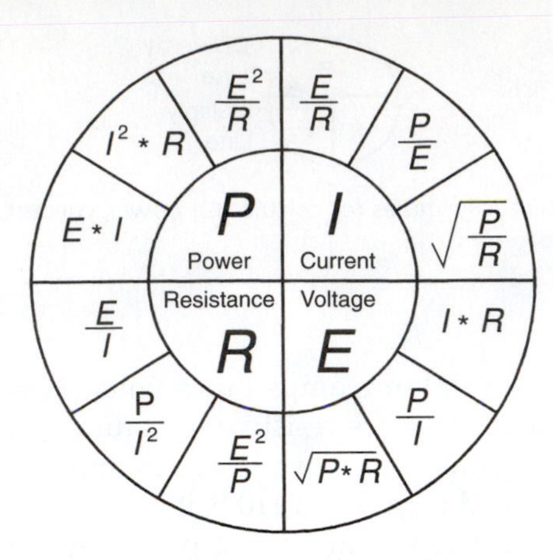

FIGURE 12-22 All formulas for calculating power, current, voltage, and resistance.

for voltage are shown emanating from the inner section marked with *E*, and all the formulas for resistance are shown emanating from the inner section marked with *R*.

If you are trying to calculate the power in a circuit or the power used by any individual component, you could use any of the three formulas shown on the wheel emanating from the section identified by the letter *P*. Thus, you could use $P = EI$, $P = I^2R$, or $P = E^2/R$. Your choice of formula will depend on the two values that are given.

Using Prefixes and Exponents with Numbers

At times you have seen numbers that need a lot of room to print—such as 5,000,000 or 0.0000003. When you need to write a long number on the face of a meter or in a calculation, the number may need to be written in a form that does not take up so much space. Several standards have been adopted throughout the fields of mathematics and electricity for this purpose. Table 12-1 shows several ways to write numbers using less space. For example, if you want to shorten the number that represents the value for 5,000,000 ohms, you could use the prefix *mega* and express the value as 5 megaohms or, using symbols, 5 MΩ. If you want to work with this number in a calculator, you can use its exponential form, which is 5^{e6} Ω. The exponent may be expressed verbally as "5 times 10 to the 6th power."

TABLE 12-1 Prefixes, Symbols, and Exponents Commonly Used in Electricity and on Electrical Equipment

Prefix	Symbol	Number	Exponent
Mega	M	1,000,000	10^6
Kilo	k	1,000	10^3
Milli	m	0.001	10^{-3}
Micro	μ	0.000001	10^{-6}

If you need to shorten the value 7000 ohms, you can use the prefix *kilo* and express the value as 7 kiloohms or, using symbols, 7 kΩ. If you want to use the exponent form in a calculator, you can use 7^{e3} Ω. When you express the exponent verbally, you say "7 times 10 to the 3rd power."

If you wanted to short the value 0.005 A, you can use the prefix *milli* and express the value as 5 milliamps or 5 mA. To use the exponent in a calculation, you can use 5^{e-3} A. To express the exponent verbally, you say "5 times 10 to the negative 3rd power." If you want to shorten the value 0.000002 A, you can use the prefix *micro* and express the value as 2 microamps or 5 μA, where μ is the Greek letter mu. To use the exponent in a calculator, you can use 5^{e-6} A. To express the exponent verbally, you say "5 times 10 to the negative 6th power."

12.4 FUNDAMENTALS OF ELECTRICAL CIRCUITS

A basic circuit consists of a power source, conductors (wires), one or more switches (controls) that can close to allow current to flow through or open to stop current flow, and a load such as a motor. The basic diagram can be drawn in two ways. The diagram in Figure 12-23 shows these components drawn as a wiring diagram, and the diagram in Figure 12-24 shows the components in a ladder diagram. The wiring diagram is used to show the location of components in a circuit, whereas the **schematic (ladder) diagram** is used to show the sequence of operation. The reason the diagram in Figure 12-24 is called a ladder diagram is that as additional loads are added, they are shown on additional "rungs," which makes the overall appearance of this diagram look like a wooden ladder. The load in the first line of the diagram

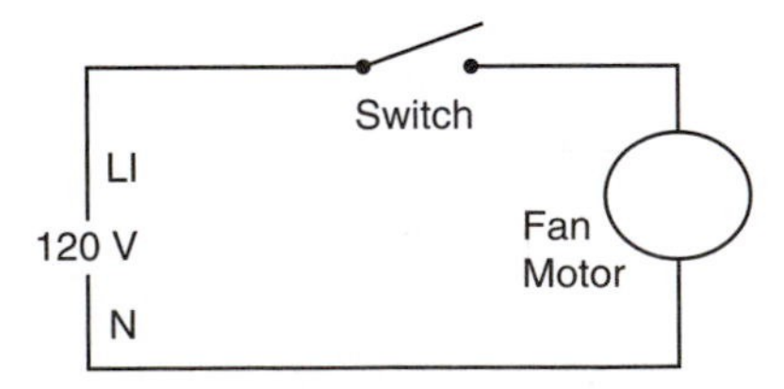

FIGURE 12-23 A power source, switch, and fan motor shown in a wiring diagram. The function of the wiring diagram is to show the location of components in a circuit.

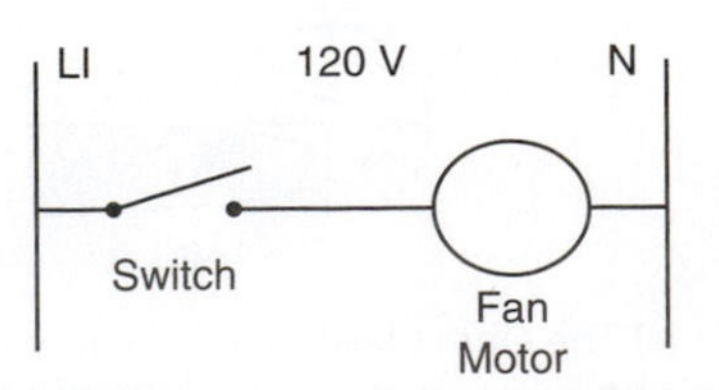

FIGURE 12-24 A power source, switch, and fan motor shown in a schematic (ladder) diagram. Notice that there is only one rung on this ladder.

will energize before the load in the second or third line (rung). Equipment manufacturers provide a schematic (ladder) diagram or a wiring diagram or both to help you install and troubleshoot a system.

Some basic terms are provided here to help you understand how electricity works. For example, a switch can have two basic conditions, open or closed. When a switch in a circuit is open, it is called an **open switch**. Electrons cannot flow past the open contacts, so all current in the circuit stops and the load is turned off. For example, if the switch is controlling a motor, when the contacts are opened, current cannot flow through the switch and the motor is off. When the switch is closed, it is called a **closed switch**, and the current is allowed to flow and the motor will run.

Simple circuits consisting of resistors will be used to explain the relationships among voltage, current, and resistance in a series circuit. Resistors are small components that are common in modern electronic circuit boards used to control systems. The motors and other loads in systems act in some part as resistances, so it is important to understand the relationships among multiple resistances in series circuits. Understanding resistances in series circuits and in parallel circuits is also essential for understanding the solid-state components that are commonly used in solid-state controls in control systems.

As a technician you must understand the effects of switches in series circuits and electrical loads in series circuits. You will learn about the relationships among switches and loads that are connected in series. Later we will explain the effects of electrical switches and loads that are connected in parallel.

The relationships among voltage, current, and resistance in series-parallel circuits will also be introduced. It will be easier to understand these relationships when a known value of resistance is used. A component called a resistor is used to provide the resistance for these simple series circuits. Each resistor has a color code consisting of colored bands that are painted on the resistor to identify its resistance value. Ohm's law will also be used to help you understand how voltage and current should react in all types of series circuits. The basic rules discussed in this chapter will help you troubleshoot larger circuits that are commonly found in control systems.

12.5 EXAMPLES OF SERIES CIRCUITS

All electrical circuits have at least a power source, one load, and conductors to provide a path for current. As circuits become more complex, switches are added for control. These switches can provide a variety of functions, depending on the way they are connected in the circuit. We have already shown examples of a simple circuit that includes a power source, a switch for control, and a motor as the load. This circuit was called a **series circuit** because the current has only one path to travel to the load and back to the power source.

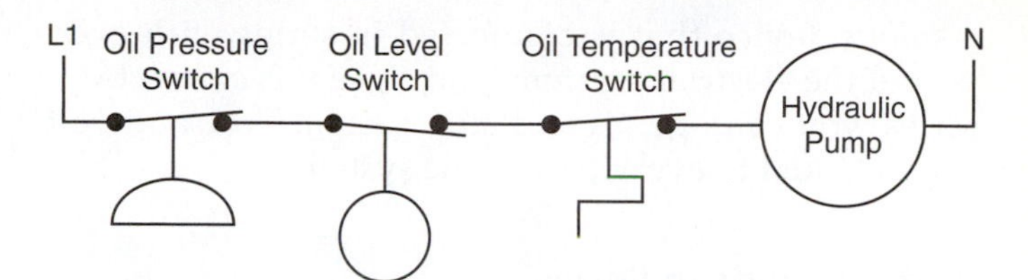

FIGURE 12-25 Series circuit with oil pressure switch, oil level switch, and oil temperature switch controlling a hydraulic pump.

When additional switches are added to the circuit to control the load, they can be connected so that all current continues to have only one path by which to travel around the circuit. Figure 12-25 is also a series circuit, because the current has only one path by which to travel from L1 through the three switches to the load and back to N. If any of the three switches is opened, current is interrupted and the motor will be turned off. When all the switches are closed, the motor will run.

Since all three of these switches are connected in series, each is capable of opening and causing current to stop flowing to the motor. If additional operational or safety switches need to be added to this circuit, they will be connected in series with the original switches. It is important to remember that when switches are connected in series, any one of them can open and stop current flow in the entire circuit.

Connecting Switches in Series

Switches are connected in series to control motors and pumps. The basic rule about series circuits is that any time a switch that is connected in series is open, all current stops and any loads that need the current will be deenergized. It is important to understand that all current in a series circuit stops whenever there is an open switch, open fuse, or open conductor.

Other Components Connected in Series

A **fuse** is another electrical component that is connected in series with all wiring, motors, and controls. Fuses are designed to protect the circuit for the maximum amount of current the wire can handle. If for some reason the current exceeds the normal amount, the fuse will overheat and its fusible link will melt, creating an **open fuse**. Since the fuse is in series with all loads and wires in the control system, it will cause the power to the system to be turned off. Having the fuses wired in series makes them able to open and protect the entire circuit, because an open anywhere in the series circuit will cause current to stop in all parts of the circuit.

A **fuse disconnect** is a switch that is combined with the fuses. The disconnect is connected in series with the fuses so that all power to the unit can be turned off when you need to change out an electrical component in the system. In this way, the disconnect switch is a safety device that is connected in series with every part of the electrical system, and since it is connected in series, you can open it and stop current flow and voltage potential to every part of the system.

Adding Loads in Series

In electrical systems such as those found in wind turbines, loads such as motors are not connected in series with other loads; instead, they are connected in parallel with one another so that they all receive the same voltage. Some heating elements for electric heating are connected in series. You will also learn about how resistors are connected in series to create useful voltage drop circuits for electronic circuit boards and components used in wind turbine electronic controls.

Loads are not generally connected in series because they would split the amount of applied voltage. For example, if you connect two light bulbs that have the same amount of resistance in their filaments in series with each other and supply the circuit with 120 V, each light bulb will receive 60 V. Each bulb will glow only half as brightly as it would if it received the entire 120 V.

In some cases, loads are connected in series, such as in electric heating elements. If the amount of resistance for each heater is equal, the supply voltage will be split equally among them. For example, if the supply voltage to the heaters is 240 V, each heater will receive 120 V. This means that each heater must be rated for 120 V. This type of circuit allows smaller heaters to be used on larger voltage sources.

Another problem with connecting loads in series is that if one of the loads has a defect and develops an **open circuit**, the current to all the loads in the circuit will be interrupted. An example of this type of circuit is a string of lights that is used to decorate a Christmas tree. If all the lights are connected in series, they will all go out if one of the lights burns out. The advantage of connecting 50 lights in series is that each of the light bulbs will receive approximately 2.4 V when the circuit is plugged into a 120-V power source.

Using Ohm's Law to Calculate Ohms, Volts, and Amps for Resistors in Series

In electronic circuits, some of the loads are resistors. These resistors are sized so that the supply voltage is dropped to smaller increments. These types of circuits are widely used in electronic circuits for wind turbine equipment. Several such circuits will be used to explain how voltage and current are affected by changes in resistance. Figure 12-26 shows an example of three resistors connected in series with a voltage source (battery). Since the resistors are connected in series, there is only one path for current. The resistors are numbered R_1, R_2, and R_3, and the supply voltage is identified as E_T. The formula for calculating total resistance in a series circuit

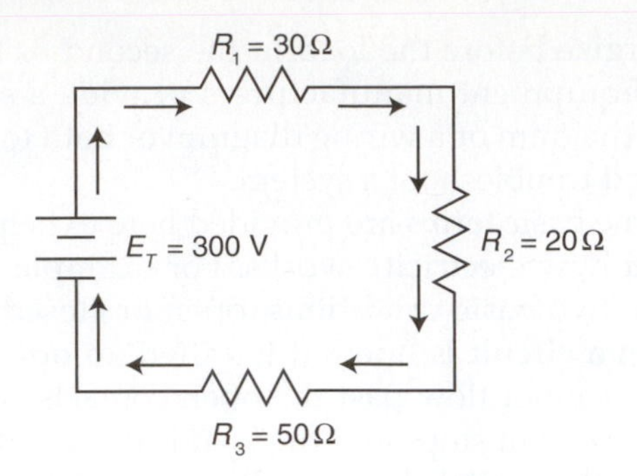

FIGURE 12-26 A series circuit with three resistors connected in series to create one path for current.

is $R_T = R_1 + R_2 + R_3$. If more than three resistors are used in the circuit, the additional resistors are added to the first three to obtain the total resistance. The arrows show conventional current flow. [Electron flow would show the flow of electrons (current) moving from the negative battery terminal to the positive terminal. This text will use conventional current flow in its explanations and diagrams.] If you would rather use electron-flow theory, the current would flow in the direction opposite to the arrows.

The series circuit in Figure 12-26 has three resistors: R_1 is 30 Ω, R_2 is 20 Ω, and R_3 is 50 Ω. The total resistance for this circuit can be calculated by the formula

$$R_T = R_1 + R_2 + R_3$$
$$R_T = 30\,\Omega + 20\,\Omega + 50\,\Omega$$
$$R_T = 100\,\Omega$$

Solving for Current in a Series Circuit

The total current can be calculated by the Ohm's law formula $I = \frac{E_T}{R_T}$, $I = \frac{300\text{ V}}{100\,\Omega}$, or $I = 3$ A. The total current in any series circuit is the same everywhere in the circuit because there is only one path for the current. This means that once you find the total resistance of a series circuit, you can divide total voltage by total resistance and find the total current, and you also have determined the current that flows through each resistor. Since the current is the same everywhere in the series circuit, it may be identified as I_T or I_1 if the current is known at resistor R_1.

Calculating the Voltage Drop Across Each Resistor

The voltage drop across each resistor can be calculated by the Ohm's law formula ($E = IR$). It is important to remember that the current in the series circuit is the same in all places, so if I_T is equal to 3 A, then I_1, I_2, and I_3 are also equal to 3 A. In this case the voltage drop across resistor R_1 is calculated by the formula

$$E_1 = I_1R_1 \quad E_1 = 3\text{ A} \times 30\,\Omega \quad E_1 = 90\text{ V}$$

If you place the probes of a voltmeter on each side of resistor R_1, you will measure 90 V. This is the actual voltage that the resistor is causing to *drop* when current is flowing through it. The **voltage drop** across resistor R_2 is calculated by the formula

$$E_2 = I_2R_2 \quad E_2 = 3\text{ A} \times 20\ \Omega \quad E_2 = 60\text{ V}$$

The voltage drop across resistor R_3 is calculated by the formula

$$E_3 = I_3R_3 \quad E_3 = 3\text{ A} \times 50\ \Omega \quad E_3 = 150\text{ V}$$

From these calculations you can see that the voltage drop across R_1 is 90 V, across R_2 it is 60 V, and across R_3 it is 150 V. If you add all these voltage drops, you will find that they are equal to the supply voltage. This means that $E_1 + E_2 + E_3 = E_T$ (90 V + 60 V + 150 V = 300 V).

EXAMPLE 12-1

This circuit has three resistors connected in series and has a power source of 450 V. Resistor R_1 is 40 Ω, R_2 is 30 Ω, and R_3 is 20 Ω. Calculate the total resistance for this circuit. After you have calculated the total resistance, divide the total voltage by the total resistance to calculate the total current. When you have determined the total current for this circuit, calculate the voltage drop that would be measured across each resistor.

SOLUTION

R_T is calculated by the formula $R_T = R_1 + R_2 + R_3$.

$$R_T = 40\ \Omega + 30\ \Omega + 20\ \Omega \quad R_T = 90\ \Omega$$

I_T is calculated by the formula $I_T = \frac{E_T}{R_T}$.

$$I_T = \frac{450\text{ V}}{90\ \Omega} \quad I_T = 5\text{ A}$$

Since current is the same in all parts of the circuit,

$$I_T = 5\text{ A} \quad I_1 = 5\text{ A} \quad I_2 = 5\text{ A} \quad I_3 = 5\text{ A}$$

The voltage that is dropped across each resistor from the current flowing through it is calculated by the formulas

$$E_1 = R_1 \times I_1 \quad E_1 = 40\ \Omega \times 5\text{ A} \quad E_1 = 200\text{ V}$$
$$E_2 = R_2 \times I_2 \quad E_2 = 30\ \Omega \times 5\text{ A} \quad E_2 = 150\text{ V}$$
$$R_3 = R_3 \times I_3 \quad E_3 = 20\ \Omega \times 5\text{ A} \quad E_3 = 100\text{ V}$$

You can check your answers by adding all the drops to see that they equal the total supply voltage:

$$E_T = E_1 + E_2 + E_3 \quad E_T = 200\text{ V} + 150\text{ V} + 100\text{ V}$$
$$E_T = 450\text{ V}$$

Calculating the Power Consumption of Each Resistor

The power consumed by each resistor or the power consumed by the total circuit can be calculated by any of the three formulas for power (wattage): $P = EI$, $P = I^2R$, and $P = E^2/R$. We previously calculated the voltage, current, and resistance for each resistor in the circuit. This means that we can use any of the three formulas to calculate the power for any resistor. For this example we will use all three formulas to show that they all give the same result. We will use the values for R_1: $E_{R1} = 200$ V, $I_{R1} = 5$ A, and $R_1 = 40\ \Omega$.

$$P = EI \quad P = 200\ V * 5\text{ A} \quad P = 1000\text{ W}$$
$$P = I^2R \quad P = (5\text{ A})^2 * 40\ \Omega \quad P = 1000\text{ W}$$
$$P = E^2/R \quad P = (200\text{ V})^2/40\ \Omega \quad P = 1000\text{ W}$$

Since all three formulas are equivalent, you can use any of the three that you choose. The major factor in deciding which formula to use will be the values that you have been given or that you can determine by measuring.

Calculating the Power Consumption of an Electrical Heating Element

The power formula can also be used to calculate the power consumption of any other type of resistance used in a circuit. The electric heating elements used in an electric heating system are a *large resistance load*. If you know the amount of resistance in the element and the amount of voltage applied to the circuit, you can calculate the current the element uses and the amount of power it consumes. For example, if the heating element has 10 Ω of resistance and it is connected to 240 V, it will draw 24 A, and it will consume 5760 W.

EXAMPLE 12-2

A heating element in an electric heating system has a resistance of 2.5 Ω. Calculate the current and the power consumption of this heating element if the heating system is connected to 240 V AC.

SOLUTION

The amount of current is calculated as 240 V/2.5 Ω = 96 A. The amount of power can be calculated by the formula $P = I2R$ or $P = IE$. Using $P = I2R$, we have (96)2 × 2.5 Ω = 23,040 W.

Using $P = IE$, we obtain 96 × 240 = 23,040 W.

(Note: This answer can also be expressed as 23.040 kW.)

12.6 PARALLEL CIRCUITS

Parallel circuits are used frequently in electrical systems. The difference between a series circuit and a **parallel circuit** is that all the loads in a parallel circuit have the same voltage supplied, whereas in a series circuit each load has a different amount of voltage depending on its resistance. The load components in electrical systems, such as pump motors, must all be provided with the same voltage, so they must be connected in parallel when they are in the same circuit. Each point where a resistor is connected in parallel in this circuit is called a *branch circuit*. A parallel circuit can have any number of branch circuits. A parallel circuit allows current to return to the power supply by more than one path. These paths are identified by arrows in the wiring diagram.

The current in a parallel circuit is additive, and the formula to calculate total resistance is $I_T = I_1 + I_2 + I_3 + \cdots$ (the . . . means that additional currents can be added to the total). Another point to understand is that the current in a parallel circuit gets larger as more loads are added to the circuit. The parallel circuit also provides a means for disconnecting any load from the power source while still supplying voltage to the remaining loads. This is accomplished by placing a switch in each branch circuit just ahead of each resistor.

When a switch is placed in the branch circuit, it becomes a **series-parallel circuit**, in which some parts of the circuit are series in nature and other parts are parallel. For example, the fuse and disconnect for the circuit must be able to interrupt all power that is supplied to the loads in a circuit, so the fuse and disconnect must be connected in series with the power supply. The important point to remember when working with series-parallel circuits is that you use the formulas for series circuits for the part of the circuit that is in series, and the formulas for parallel circuits for the part of the circuit that is in parallel. It is also possible to combine parts of the circuit to make it a simplified series or simplified parallel circuit.

Calculating Voltage, Current, and Resistance in a Parallel Circuit

Voltage, current, and resistance can be calculated in a parallel circuit just as in a series circuit by using Ohm's law. Figure 12-27 shows an example circuit with the voltage, current, and resistance calculated at each point in the circuit. From this diagram you can see that the supply voltage is 300 V. Since the supply voltage is 300 V, you can determine that the voltage of each branch circuit across each resistor is also 300 V, because the voltage at each branch circuit in a parallel circuit is the same as the supply voltage.

The current that is flowing through each resistor can be calculated by using the Ohm's law formula for current: $I = E/R$. The current in each branch circuit is calculated from this formula and is placed in the diagram beside each resistor.

$$
\begin{aligned}
I_1 &= E/R_1 & I_1 &= 300\ \text{V}/30\ \Omega & I_1 &= 10\ \text{A}\\
I_2 &= E/R_2 & I_2 &= 300\ \text{V}/20\ \Omega & I_2 &= 15\ \text{A}\\
I_3 &= E/R_3 & I_3 &= 300\ \text{V}/60\ \Omega & I_3 &= 5\ \text{A}\\
I_T &= E_T/R_T & I_T &= 300\ \text{V}/10\ \Omega & I_T &= 30\ \text{A}\\
I_T &= I_1 + I_2 + I_3 & I_T &= 10\ \text{A} + 15\ \text{A} + 5\ \text{A} & I_T &= 30\ \text{A}
\end{aligned}
$$

If the voltage and current are given and the resistance needs to be calculated, the Ohm's law formula for resistance can be used: $R = E/I$.

$$
\begin{aligned}
R_1 &= E/I_1 & R_1 &= 300\ \text{V}/10\ \text{A} & R_1 &= 30\ \Omega\\
R_2 &= E/I_2 & R_2 &= 300\ \text{V}/15\ \text{A} & R_2 &= 20\ \Omega\\
R_3 &= E/I_3 & R_3 &= 300\ \text{V}/5\ \text{A} & R_3 &= 60\ \Omega\\
R_T &= E/I_T & R_T &= 300\ \text{V}/30\ \text{A} & R_T &= 10\ \Omega
\end{aligned}
$$

If the total resistance and the total current in a circuit are known, you can calculate the voltage for the circuit. If you know the current and resistance at any branch circuit, you can also calculate the voltage using the Ohm's law formula: $E = IR$. The nice part about calculating voltage is that once you determine the voltage at any branch circuit, the same amount of voltage is at every other branch circuit and at the supply. The same is true if you calculate the voltage at the supply; you do not need to calculate the voltage at any other branch, because it will be the same.

$$
\begin{aligned}
E_T &= I_T R_T & E_T &= 30\ \text{A} \times 10\ \Omega & E_T &= 300\ \text{V}\\
E_1 &= I_1 R_1 & E_1 &= 10\ \text{A} \times 30\ \Omega & E_1 &= 300\ \text{V}\\
E_2 &= I_2 R_2 & E_2 &= 15\ \text{A} \times 20\ \Omega & E_2 &= 300\ \text{V}\\
E_3 &= I_3 R_3 & E_3 &= 5\ \text{A} \times 60\ \Omega & E_3 &= 300\ \text{V}
\end{aligned}
$$

EXAMPLE 12-3

Use the circuit in Figure 12-28 to calculate the individual branch currents, total current, and total resistance.

SOLUTION

Use the following formulas to calculate the current for each branch:

$$
\begin{aligned}
I_1 &= E/R_1 & I_1 &= 240\ \text{V}/60\ \Omega & I_1 &= 4\ \text{A}\\
I_2 &= E/R_2 & I_2 &= 240\ \text{V}/30\ \Omega & I_2 &= 8\ \text{A}\\
I_3 &= E/R_3 & I_3 &= 240\ \text{V}/15\ \Omega & I_3 &= 16\ \text{A}
\end{aligned}
$$

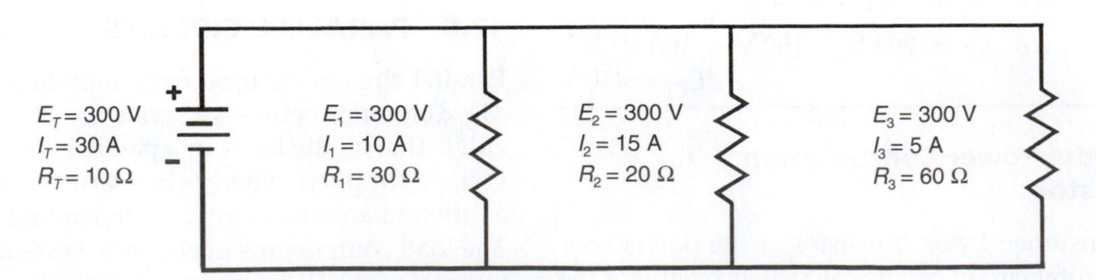

FIGURE 12-27 A parallel circuit with the voltage resistance and current shown at each load.

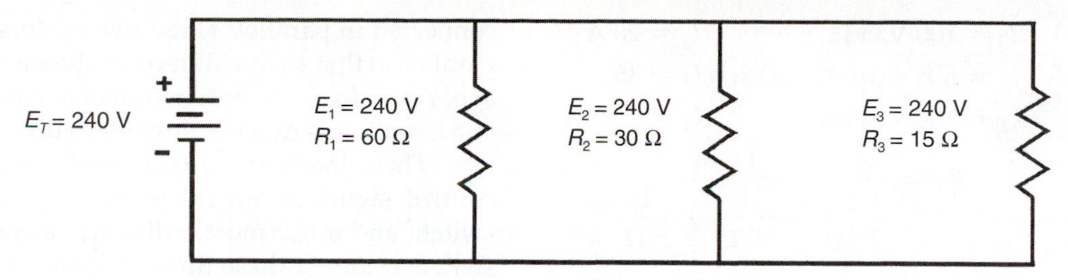

FIGURE 12-28 Parallel circuit showing voltage and resistance for the circuit in Example 12-3.

Use the following formula to calculate the total current I_T:

$$I_T = I_1 + I_2 + I_3 \quad I_T = 4\text{ A} + 8\text{ A} + 16\text{ A} \quad I_T = 28\text{ A}$$

Use the following formula to calculate total resistance R_T:

$$R_T = E_T/I_T \quad R_T = 240\text{ V}/28\text{ A} \quad R_T = 8.57\ \Omega$$

At times you will need to calculate the total resistance of a parallel circuit when only the branch resistance and supply voltage are provided. You can calculate the individual currents at each branch circuit and then calculate the total with the formula $I_T = I_1 + I_2 + I_3$. After you have determined the total current, you can use the formula $R_T = E_T/I_T$.

Another method, called *the product over the sum method*, can be used, which requires you to use the formula for calculating total resistance in a parallel circuit. This method is called the product over the sum method because you multiply the two resistors to get the product, and you add the two resistors to get the sum. The third step in the calculation includes dividing the product by the sum (product over sum). The formula is written as follows:

$$R_T = \frac{R_1 \times R_2}{R_1 + R_2}$$

You should notice that with this formula you can calculate the total resistance of only two resistors at a time. If the circuit has three resistors, you will need to find the total of the first two resistors in the branch and then use their total with the third resistor in the formula again to find the *grand* total resistance. We will use this method to calculate the total resistance for these resistors that are connected in parallel: $R_1 = 30\ \Omega$, $R_2 = 20\ \Omega$, $R_3 = 60\ \Omega$.

Since we can only use two resistances at a time in this method, in the first step we will find the resistance of R_1 and R_2. The amount of the resistance of R_1 and R_2 in parallel will be called *equivalent resistance*. Since this is the first equivalent resistance, we will call this resistance R_{eq1}. Next we will find the parallel resistance of R_{eq1} and R_3. Since we have three resistors, the amount of the resistance of R_{eq1} and R_3 will be the total resistance (R_T) for the circuit.

$$R_{eq_1} = \frac{R_1 \times R_2}{R_1 + R_2} \quad R_{eq_1} = \frac{30\ \Omega \times 20\ \Omega}{30\ \Omega + 20\ \Omega} \quad R_{eq_1} = 12\ \Omega$$

$$R_T = \frac{R_{eq_1} \times R_3}{R_{eq_1} + R_3} \quad R_T = \frac{12\ \Omega \times 60\ \Omega}{12\ \Omega + 60\ \Omega} \quad R_T = 10\ \Omega$$

From these calculations you can see that the total parallel resistance is 10 Ω. *It is important to understand that in all parallel circuits, the total resistance will always be smaller than the smallest branch circuit resistance.* You can see that in this circuit, the smallest resistance in the branch circuits is 20 Ω, and the total resistance is 10 Ω.

The next calculation shows the total resistance calculated from the formula

$$R_T = \frac{1}{\frac{1}{R_1} + \frac{1}{R_2} + \frac{1}{R_3}} \text{ or } \frac{1}{R_T} = \frac{1}{R_1} + \frac{1}{R_2} + \frac{1}{R_3}$$

This formula is designed to be used with a calculator. If you do not have a calculator, it is recommended that you use the previous method.

EXAMPLE 12-4

Use the parallel circuit in Figure 12-29 to calculate the current through each resistor, the total current, and the total resistance for this circuit. The supply voltage is 100 V.

SOLUTION

$$I_1 = E/R_1 \quad I_1 = 100\text{ V}/20\ \Omega \quad I_1 = 5\text{ A}$$

$$I_2 = E/R_2 \quad I_2 = 100\text{ V}/10\ \Omega \quad I_2 = 10\text{ A}$$

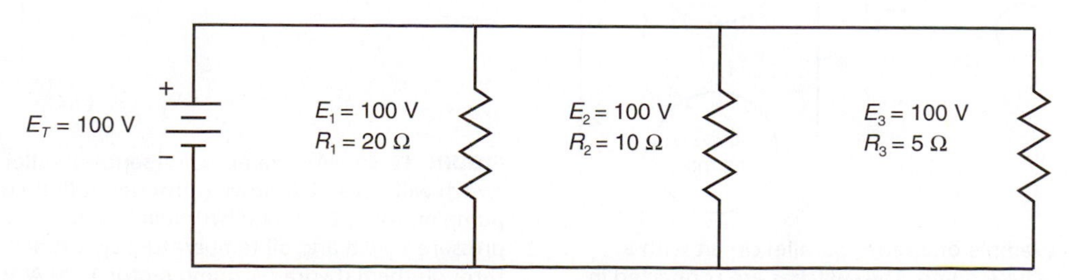

FIGURE 12-29 Parallel circuit for Example 12-4.

$I_3 = E/R_3 \quad I_3 = 100\text{ V}/5\ \Omega \quad I_3 = 20\text{ A}$

$I_T = I_1 + I_2 + I_3 \quad I_T = 5\text{ A} + 10\text{ A} + 20\text{ A} \quad I_T = 35\text{ A}$

$R_T = E_T/I_T \quad R_T = 100\text{ V}/35\text{ A} \quad R_T = 2.86\ \Omega$

$$R_T = \frac{1}{\frac{1}{R_T} + \frac{1}{R_2} + \frac{1}{R_3}} \quad R_T = \frac{1}{\frac{1}{20\ \Omega} + \frac{1}{10\ \Omega} + \frac{1}{5\ \Omega}}$$

$$R_T = 2.86\ \Omega$$

Calculating Power in a Parallel Circuit

The formula for calculating power in a parallel circuit is the same as the formula for a series circuit: $P_T = P_1 + P_2 + P_3$. The formula for power at each individual resistor is found from the original Watt's law formula $P = IE$. This means that you must calculate the power consumed by each branch resistor and then add them all together to get the total power consumed. For example, in the parallel circuit shown earlier in Figure 12-27, the voltage at R_1 is 300 V and the current is 10 A, the voltage at R_2 is 300 V and the current is 15 A, and the voltage at R_3 is 300 V and the current is 5 A. The following calculations are used to determine the power consumed by each individual resistor and the total power used by the whole circuit.

$P_1 = I_1E_1 \quad P_1 = 10\text{ A} \times 300\text{ V} \quad P_1 = 3000\text{ W}$

$P_2 = I_2E_2 \quad P_2 = 15\text{ A} \times 300\text{ V} \quad P_2 = 4500\text{ W}$

$P_3 = I_3E_3 \quad P_3 = 5\text{ A} \times 300\text{ V} \quad P_3 = 1500\text{ W}$

$P_T = P_1 + P_2 + P_3 \quad P_T = 3000\text{ W} + 4500\text{ W} + 1500\text{ W}$

$P_T = 9000\text{ W}$

You could also calculate the total power consumed by this circuit by using the formula

$P_T = I_TE_T \quad P_T = 30\text{ A} \times 300\text{ V} \quad P_T = 9000\text{ W}$

12.7 SERIES-PARALLEL CIRCUITS

As a technician you will work on circuits for electrical systems. These systems will have a number of switches and loads that are connected in series and parallel. A typical series parallel circuit is shown in Figure 12-30. It has a hydraulic pump motor and a lube pump motor that are connected in parallel. These two motors are connected in parallel so that each will receive the same amount of voltage. When loads such as motors are connected in parallel in a circuit, it is called a **power circuit.**

These loads are also connected in series with three control switches: an oil pressure switch, an oil level switch, and a thermostat that acts as an oil temperature switch. If any of these three switches is open, the current to both loads will be stopped. The switches are connected in series for this reason. Since this circuit has some switches connected in series, and two loads connected in parallel, it is called a series-parallel circuit.

The parts of the circuit that are connected in series will follow all the rules for series circuits, and the parts of the circuit that are connected in parallel will follow all the rules for parallel circuits. This series-parallel circuit will provide the best of series circuits and the best of parallel circuits to supply voltage to the loads and control their operation. This type of series-parallel circuit is very common in field equipment that you will encounter.

Series-Parallel Circuits Used in Wind Turbines

Series-parallel circuits are used in a variety of wind turbine equipment. You can use the basic information that you have learned about series-parallel circuits to make some simple observations about the size of wires and switches based on the voltage and current for which they will need to be rated. Figure 12-31 shows an example of a series-parallel circuit for an electrical system. In this diagram you can see that the circuit has some switches in series with motors, and the motors are connected in parallel with one another. For example, if a disconnect switch was connected at the top of the circuit at L1, it would be connected

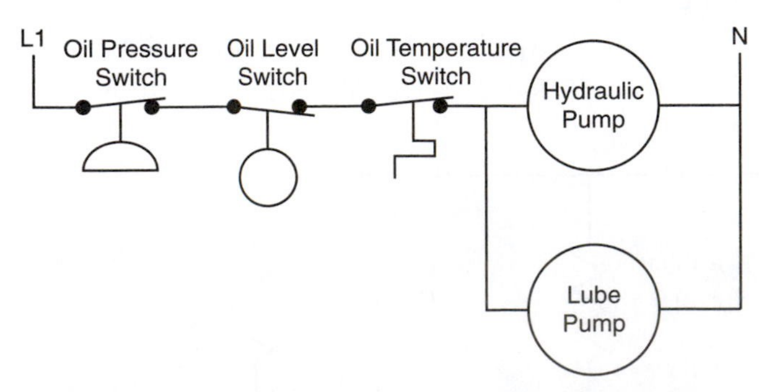

FIGURE 12-30 An example of a series-parallel circuit with a hydraulic pump and a lube pump. The switches are connected in series, and the two motors are connected in parallel.

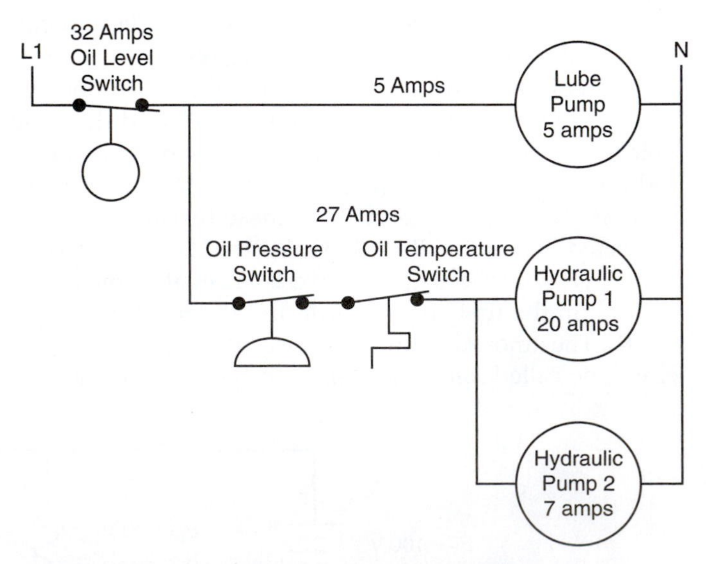

FIGURE 12-31 An example of a series-parallel circuit. The oil level switch will have 32 A flowing through it (lube pump, 5 A; hydraulic pump motor 1, 20 A; and hydraulic pump motor 2, 7 A). The oil pressure switch and oil temperature switch have only 27 A flowing through them (hydraulic pump motor 1, 20 A; hydraulic pump motor 2, 7 A).

in series with the entire circuit. This means that if you open the disconnect switch, the entire circuit is disconnected.

The oil pressure switch and oil temperature switch are connected in series with the hydraulic pump motor 1 and hydraulic pump motor 2. The lube motor draws 5 A, hydraulic pump motor 1 draws 20 A, and hydraulic pump motor 2 draws 7 A; therefore, the oil level must be rated for at least 32 A, because all of the current will flow through it. If you were going to select a fuse to put in L1 to protect everything in the circuit, you would need to select a fuse that is larger than 32 A.

The oil pressure switch and oil temperature switch are connected in series with hydraulic pump motor 1 and hydraulic pump motor 2, so they must be rated to carry the 27 A used by these two motors. This also means that the current to these motors will be interrupted whenever either of these switches is opened. All three motors are connected in parallel with one another, so they will all receive the same amount of voltage, but the amount of current each will draw will be different unless their load characteristics are identical. The amount of total current that is needed to supply the three motors can be calculated by the formula $I_T = I_1 + I_2 + I_3$.

12.8 CAPACITORS AND CAPACITIVE REACTANCE

A *capacitor* is an electrical device that has two terminals and has the capability of storing a charge. The capacitor is made of two conducting plates that are separated by an insulator called a dielectric. If the capacitor is an electrolytic capacitor, one of the terminals is negative and the other is positive. In all other capacitors, the two terminals do not have polarity. The negative terminal is identified by a minus sign, and its terminal looks like a half-circle. The positive terminal is identified by a plus sign.

The operation of the capacitor can best be described by applying a DC voltage to it and allowing it to charge and then to discharge. At first, since the capacitor is not charged, electrons (current) begin to flow from the negative terminal of the battery to the negative terminal of the capacitor when the circuit is complete. It is important to understand that current does not flow through the capacitor, since its two plates are separated by a dielectric, which is a good insulator. Electrons continue to flow until the voltage (charge) on the capacitor plate is equal to the battery voltage. At this point, there is no longer a difference of potential between the capacitor and the battery, so current flow is stopped. Since the dielectric between the positive and negative plate of the capacitor is a good insulator, theoretically it will remain charged indefinitely when the switch is opened and the capacitor is disconnected from the battery.

Since there is a large potential difference between the positive plate and the negative plate of the capacitor, the electrons will flow from the negative plate to the positive plate until the charge on each plate is equal, when the capacitor is considered discharged. At this point, if the switch is moved to position A again, the capacitor will charge again. When the capacitor is placed in a circuit with DC voltage, the capacitor will become charged and remain charged until the circuit is changed to allow the capacitor to discharge. If the capacitor is placed in a circuit with AC voltage, it will charge and discharge at the frequency of the voltage.

Resistance and Capacitance in an AC Circuit

In a previous section we learned that when capacitors or inductors are used in an AC circuit, they create an opposition that is similar to resistance. The opposition caused by a capacitor is called *capacitive reactance*, and the opposition caused by an inductor is called *inductive reactance*. The combined opposition caused by capacitive reactance, inductive reactance, and resistance in an AC circuit is called *impedance*. The main difference between capacitive reactance, inductive reactance, and resistance is that a phase shift occurs between the voltage waveform and current waveforms. Figure 12-32a is a diagram of a capacitor in an

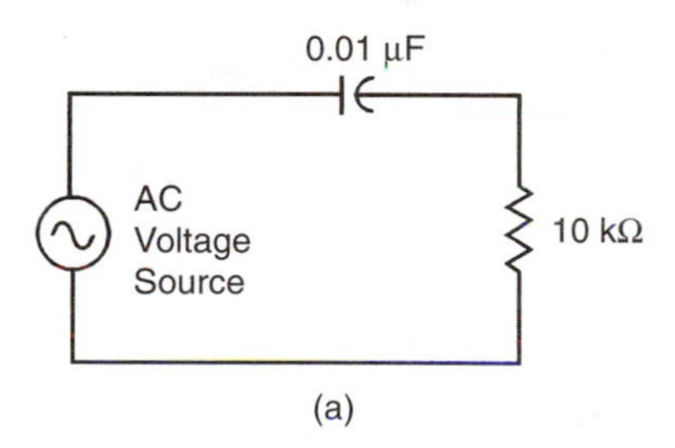

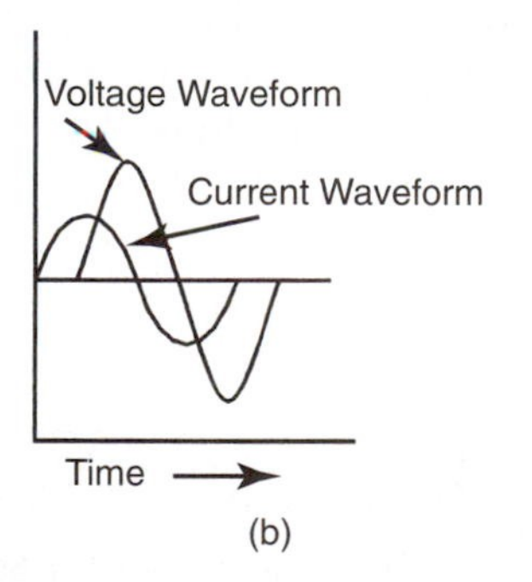

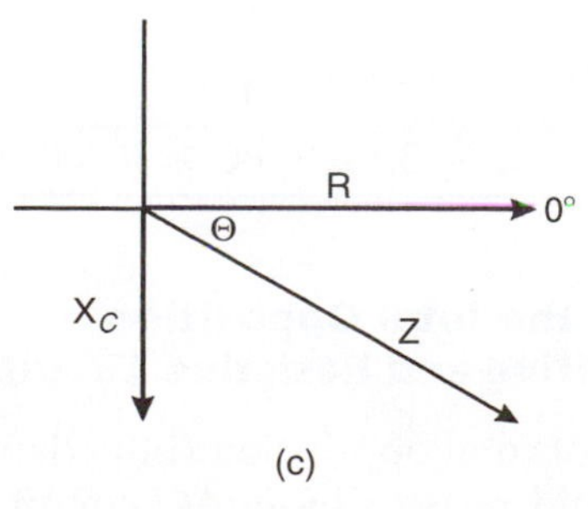

FIGURE 12-32 (a) Capacitor and resistor in an AC circuit. (b) Waveform of voltage and current for the AC circuit. Notice that the current waveform leads the voltage waveform. (c) Vector diagram showing the relationship between voltage across the resistor and the voltage across the capacitor.

AC circuit with a resistor. Figure 12-32b shows the voltage and current waveforms for this circuit, and Figure 12-32c shows the vector diagram that is used to calculate the amount of phase shift for this diagram. A vector diagram is a graph derived from a trigonometry calculation used to determine phase angle.

When voltage encounters a capacitor in a circuit, it will take time to charge up the capacitor and then discharge it. This causes the reactance, which becomes an opposition to the voltage waveform. The capacitor does not affect the current waveform.

Calculating Capacitive Reactance

The total amount of opposition caused by the capacitor is called *capacitive reactance* and is indicated in symbols by X_C. Even though you may never need to calculate capacitive reactance, you should understand the effects of changes in capacitance and frequency on the amount of capacitive reactance. Capacitive reactance can be calculated by the following formula:

$$X_C = \frac{1}{2\pi FC}$$

where $\pi = 3.14$

F = frequency

C = capacitance in microfarads

From this formula you can see that you must know the value of the capacitor and the frequency of the AC voltage. For example, if a 40-μF capacitor is used in a 60-Hz circuit, the amount of opposition (capacitive reactance) is 66.35 Ω. Notice that because the capacitive reactance is an opposition, its units are ohms. This calculation is as follows:

$$X_C = \frac{1}{2\pi FC} = \frac{1}{2 \times 3.14 \times 60 \times 0.000040} = 66.35\ \Omega$$

EXAMPLE 12-5

Calculate the capacitive reactance of a 60-Hz AC circuit that has a 90-μF capacitor.

SOLUTION

$$X_C = \frac{1}{2\pi FC} - \frac{1}{2 \times 3.14 \times 60 \times 0.000090} = 29.49\ \Omega$$

Calculating the Total Opposition for a Capacitive and Resistive Circuit

The amount of total opposition (impedance) caused by a capacitor and resistor in an AC circuit can be calculated. For example, if a circuit has 70 Ω of resistance and 40 Ω due to capacitive reactance, the total impedance must be calculated using the distance formula, because the voltage and current in the circuit are out of phase. The formula for calculating impedance of this circuit is

$$Z = \sqrt{R^2 + X_c^2}$$

$$Z = \sqrt{70^2 + 40^2} \qquad Z = 80.62\ \Omega$$

12.9 RESISTANCE AND INDUCTANCE IN AN AC CIRCUIT

When inductors (coils of wire) are used in an AC circuit, they create an opposition that is similar to resistance. The opposition caused by an inductor is called *inductive reactance.* The main difference between capacitive reactance, inductive reactance, and resistance is that a phase shift between the voltage and current waveforms occurs. Figure 12-33a is a diagram of an inductor in an AC circuit with a resistor. Figure 12-33b shows the voltage and current waveforms for this circuit, and Figure 12-33c shows the vector diagram that is used to calculate the amount of phase shift for the circuit.

When current encounters an inductor in a circuit, it will take time to charge up the inductor and then discharge

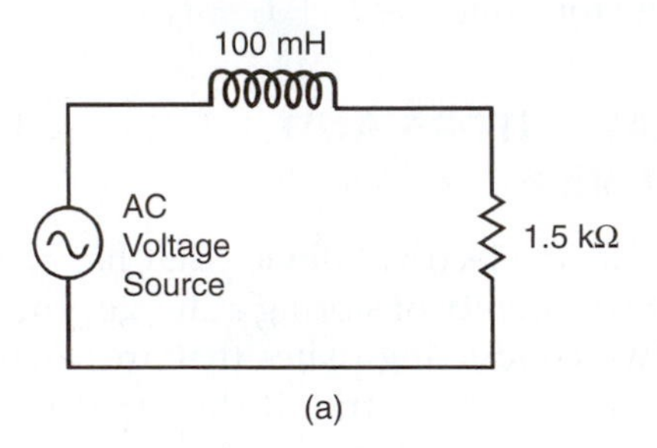

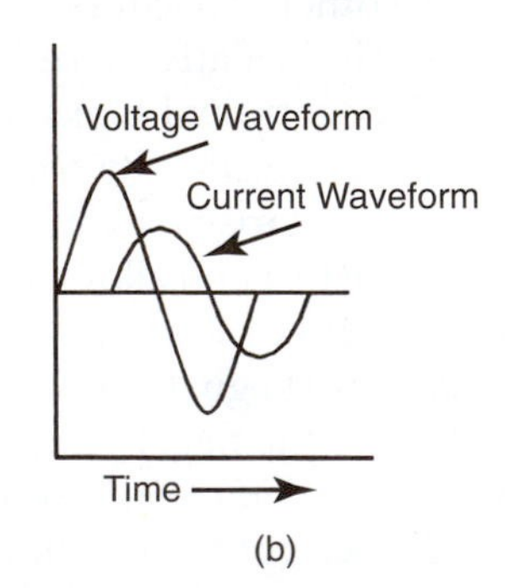

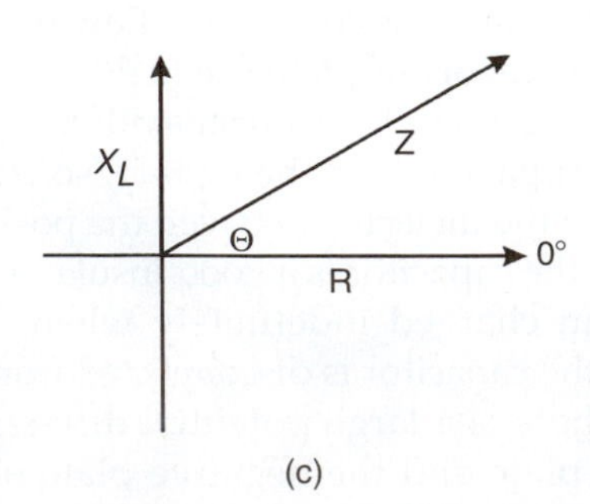

FIGURE 12-33 (a) An inductor and a resistor in an AC circuit. (b) Waveforms for the voltage and current in the inductive and resistive circuit. (c) Vector diagram showing voltage across the inductor leading the voltage across the resistor.

it. This causes the reactance, which becomes an opposition to the current waveform. The inductor does not affect the voltage waveform. In Figure 12-33b you can see that the current waveform *lags,* or starts later than, the voltage waveform.

Calculating Inductive Reactance

The total amount of opposition caused by the inductor is called *inductive reactance,* indicated in symbols by X_L. As in the case of capacitive reactance, inductive reactance can be calculated, and you should remember that even though you do not calculate reactance in the field when you are troubleshooting, it is important that you understand the effect that changing the size of the inductor or changing the frequency has on the total amount of inductive reactance. Inductive reactance can be calculated by the following formula:

$$X_L = 2\pi FL$$

where $\pi = 3.14$
$F =$ frequency
$L =$ inductance in henries (H)

From this formula you can see that you must know the value of the inductor and the frequency of the AC voltage. For example, if an 80-H inductor is used in a 60-Hz circuit, the amount of opposition (inductive reactance) is 30,144 Ω (30.144 kΩ). Notice that because the inductive reactance is an opposition, its units are ohms. This calculation is as follows:

$$X_L = 2\pi FL = 2 \times 3.14 \times 60\text{ Hz} \times 80\text{ H} = 30{,}144\ \Omega$$

EXAMPLE 12-6

Calculate the inductive reactance of a 60-Hz AC circuit that has a 20-H inductor.

SOLUTION

$$X_L = 2\pi FL = 2 \times 3.14 \times 60 \times 20 = 7{,}536\ \Omega$$

12.10 IMPEDANCE; CALCULATING THE TOTAL OPPOSITION FOR AN INDUCTIVE AND RESISTIVE CIRCUIT

The amount of total opposition (impedance) caused by an inductor and resistor in an AC circuit can be calculated. For example, if a circuit has 70 Ω of resistance and 50 Ω due to inductive reactance, the total impedance must be calculated with a vector diagram, because the voltage and current in this circuit are out of phase. The formula for calculating impedance of this circuit is

$$Z = \sqrt{R^2 + X_L^2}$$
$$Z = \sqrt{70^2 + 50^2}$$
$$Z = 86.02\ \Omega$$

12.11 TRUE POWER AND APPARENT POWER IN AN AC CIRCUIT

In a DC circuit you could simply multiply the voltage times the current and determine the total power of the circuit. In an AC circuit you must account for the current caused by any resistors and calculate it separately from the power caused by resistors and capacitors or resistors and inductors. The reason for this is that when current is caused by a resistor, it is called *true power* (TP), and current caused by capacitive reactance and inductive reactance is called *apparent power* (AP). Apparent power does not take into account the phase shift caused by the capacitor or inductor. The power that occurs when current flows through a resistor is called true power because the resistor does not cause a phase shift between the voltage waveform and the current waveform.

The main point to remember is that true power can be used to determine the heating potential in an AC circuit. Thus, if you have 1000 W due to true power, you can determine that you can get 1000 W of heating power from this circuit.

12.12 CALCULATING THE POWER FACTOR

The amount of true power and the amount of apparent power in an AC circuit combine to become a ratio called the *power factor* (PF). The formula for power factor is

$$\text{PF} = \frac{\text{TP}}{\text{AP}}$$

EXAMPLE 12-7

Calculate the power factor for a circuit that has 600 V-A of apparent power and 500 W of true power. Note that when volts are multiplied by amperes in a reactive circuit, the units for apparent power are V-A, which stands for volt-amperes.

SOLUTION

$$\text{PF} = \frac{\text{TP}}{\text{AP}} = \frac{500\text{ W}}{600\text{ VA}} = 0.83, \quad \text{or} \quad 83\%$$

The true power in a circuit is always smaller than the apparent power, so the power factor is always less than 1. In a pure resistance circuit, the true power and the apparent power are the same, so the power factor is 1. When the power factor becomes too low, the power company adds a penalty to the electric bill that increases the bill substantially. When a factory has an electric bill of more than $20,000 per month, this penalty becomes important. In these types of applications, the power factor can be corrected by adding extra capacitance to an inductive circuit (having large or multiple motors) or by adding extra inductance to a capacitive circuit. Commercial power factor-correction systems are available for these applications.

12.13 HOW TO CHANGE POWER FACTOR WITH INDUCTORS OR CAPACITORS

When you are working with a power factor that is caused by induction, capacitors can be added to bring the power factor back to 1, which is also called unity. If the power factor is caused by excessive capacitance, you can add inductance coils to correct the power factor. When wind turbines are connected to the grid, the power factor must be corrected before the connection is made. The power factor for most AC voltage generated by wind turbines needs to be corrected by adding capacitance. Capacitor banks are used to make the correction.

12.14 VOLT-AMPERE REACTANCE (VAR)

The units of apparent power are VAR, which stands for volt-ampere reactance. When the VAR is very large, it is measured in thousands of VAR and the units kVAR are used. The units of VAR are the same whether the apparent power is caused by capacitance or by inductance.

12.15 THREE-PHASE TRANSFORMERS

Figure 12-34 shows examples of three-phase transformers that you will encounter on the job. The transformer may be mounted near the wind turbine equipment, or it may be located in a transformer vault (a special room where transformers are mounted). Sometimes the transformers are mounted on a utility pole just outside the commercial site. Figure 12-35 shows a three-phase transformer with its cover removed so you can see three separate transformer windings. The three-phase transformer operates exactly like a single-phase transformer in that AC voltage is applied to the primary side of the transformer and induction causes voltage to be created in the secondary winding. In the case of the three-phase transformer, the 120-degree phase shift of the three-phase voltage applied to the primary windings will be maintained in the secondary side of the transformer.

FIGURE 12-34 Typical three-phase transformers with their covers in place. (Courtesy of Pearson Education/PH College)

FIGURE 12-35 A typical three-phase transformer with its cover removed, showing the three independent transformers and connection terminals.

When you are working on a commercial system, you may need to make connections on the secondary side of the transformer to the disconnect box, or between the secondary side of the transformer and a load center. You may also need to make voltage tests at the terminals of the transformer. You can use the diagram that is provided on the side of the transformer to identify the correct terminals to use. In most cases, you will not be required to make any electrical connections at the primary side of the transformer, since the primary voltage may be rather high (12,470 V AC or higher). The electrical technicians for the electric utility company or technicians from a high-voltage service company will make the primary voltage connections for the transformer. You will be expected to make the connections for the wind turbine equipment and at the various switchgear.

The Wye-Connected Three-Phase Transformer

Figure 12-36 is a diagram of a *wye-connected three-phase transformer*. Notice that the physical shape of the transformer windings looks like the letter Y, which gives this configuration its name. Also notice that this diagram shows only the secondary coils of the transformer. It is traditional

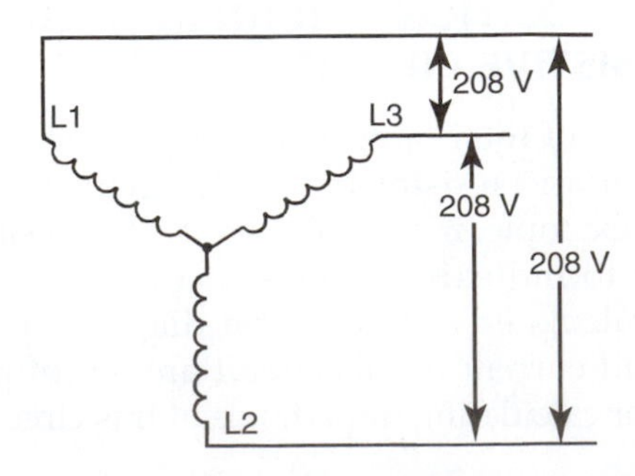

FIGURE 12-36 A diagram showing the secondary windings of a three-phase transformer connected in a wye configuration. The voltages available at L1–L2, L2–L3, and L3–L1 are all 208 V for this transformer.

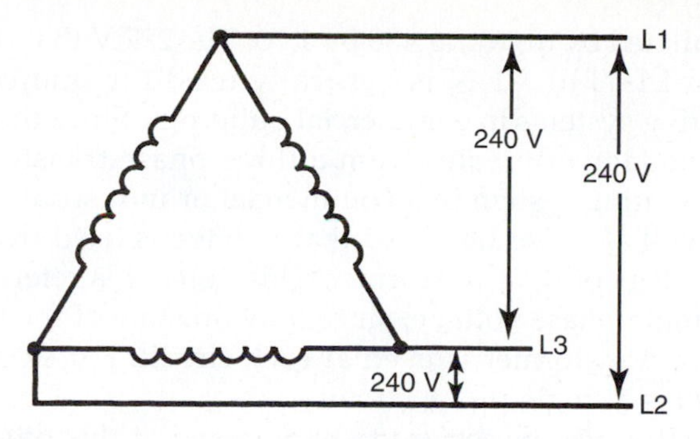

FIGURE 12-37 Electrical diagram of the secondary windings of a transformer connected in a delta configuration. Notice that the voltage at L1–L2, L2–L3, and L3–L1 is 240 V.

to show only the primary-side connections or only the secondary-side connections when discussing a transformer power distribution system, since showing both the primary and secondary connections in the same diagram tends to become confusing.

The amount of voltage measured at the L1–L2, the L2–L3, and the L3–L1 terminals on the secondary winding will be 208 V for the wye-connected transformer if its turns ratio is set for low voltage. If the turns ratio of the transformers is set for high voltage, the amount of voltage between each terminal will be 480 V. The voltage indicated between the windings for the transformer shown in the diagram is 208 V, which indicates that the turns ratio for this transformer will provide the lower voltage. If 480 V were needed, a transformer with a higher turns ratio would be used. The primary and secondary voltage for each transformer is provided on its data plate and can be specified when the transformer is purchased and installed.

The Delta-Connected Three-Phase Transformer

Figure 12-37 is a diagram of a *delta-connected three-phase transformer*. Notice that the shape of the transformer windings in this diagram looks like a triangle (Δ). This shape is the Greek letter D, which is named *delta*. This diagram also shows only the secondary side of the three-phase transformer.

The amount of voltage measured at L1–L2, L2–L3, and L3–L1 is 240 V for the delta-connected transformer if it is wired for its lower voltage. If the delta-connected transformer is wired for its higher voltage, the voltage at L1–L2, L2–L3, and L3–L1 is 480 V. If the delta-connected transformer is wired for its lower voltage, it is very easy to differentiate it from a wye-connected transformer. If the delta-connected transformer and wye-connected transformer are both connected for their higher voltage, you cannot tell them apart, since both will provide 480 V between their terminals.

If you need to know whether the transformer windings are connected as delta or wye, you can check the physical connections or you can make an additional voltage measurement between each line and the neutral terminal of the transformer if one is provided. The next section will explain these measurements.

Delta- and Wye-Connected Transformers with a Neutral Terminal

Figure 12-38 shows diagrams of a wye-connected transformer and a delta-connected transformer, each with a neutral terminal. The neutral terminal on the wye-connected transformer is at the point where the three ends of the individual windings are connected together. This point is called the *wye point*.

Note that the neutral point for the delta-connected transformer is actually the midpoint of the secondary winding between L1 and L3. This point is essentially the center tap of one of the transformer windings. Traditionally, it is the winding that is connected between L1 and L3 if a neutral connection is used with the three-phase transformer.

Figure 12-39a shows the amount of voltage that you would measure between terminals L1 and L2 of a wye-connected transformer and between terminals L1 and N, and L2 and N. Notice that the voltage between L1 and L2 is 208 V, so this transformer is wired for the lower voltage. The voltage between L1 and N and between L2 and N is shown as 120 V. Figure 12-39b shows the amount of voltage that you would measure between terminals L1 and L3 and between L1 and N, and L3 and N for a delta-connected transformer. The voltage between L1 and L3 is 240 V, so this transformer is connected for its

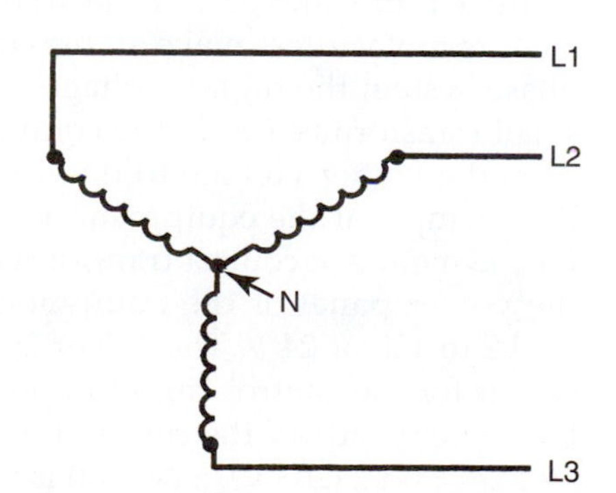

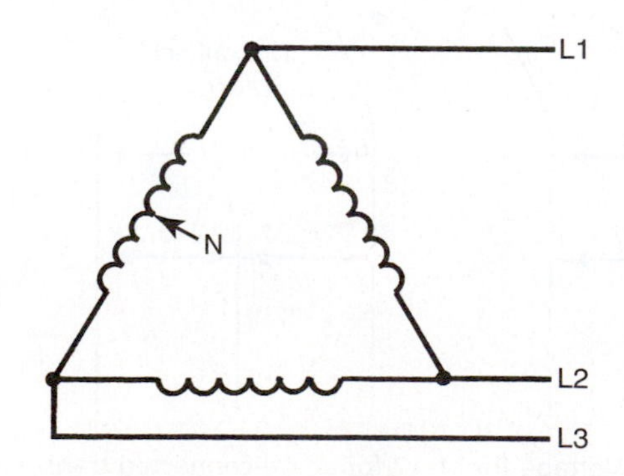

FIGURE 12-38 Electrical diagram of a wye-connected transformer with a neutral point and a delta-connected transformer with a neutral point.

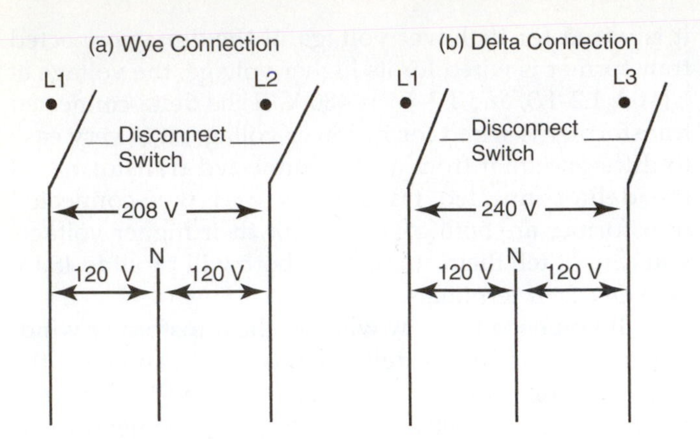

FIGURE 12-39 (a) Voltage for L1–L2 for a wye-connected transformer is 208 V and for L1–N or L2–N is 1120 V. (b) Voltage for L1–L3 for a delta-connected transformer is 240 V and for L1–N or L1–L3 is 120 V.

lower voltage. The voltage between L1 and N and between L3 and N is shown as 120 V. You should notice that since the neutral point for the delta-connected transformer is exactly halfway on the transformer winding, the voltage L1–N and L3–N will always be exactly half of the voltage between L1 and L3.

This is the main difference between a wye-connected transformer and a delta-connected transformer. The voltage at L1–N and at L3–N for any delta-connected transformer is always exactly half, and the L1–N or L2–N voltage for a wye-connected transformer will always be more than half the voltage at L1–L2. The exact amount of voltage L1–N can be calculated by dividing the voltage between L1 and L2 by 1.73, which is the square root of 3($\sqrt{3} = 1.73$). The square root of 3 is used because of the relationship among the phase shifts of the three phases. Note that 208 divided by 1.73 is 120 V.

The same relationship of voltage between L1 and L2 and L1 and N exists when the transformers have turns ratios for their higher voltage. Figure 12-40a shows a wye-connected voltage between L1 and L2 of 480 V, and between L1 and N of 277 V. The 277 V can be calculated by dividing 480 by 1.73. The 277 V that comes from L1–N or L2–N is generally used for fluorescent lighting systems in commercial buildings. Since the supply voltage originates from a three-phase transformer, the lighting system in a commercial or industrial building will also use L3–N, so that voltage is used from all three legs of the transformer. This voltage is referred to as single-phase voltage, since only one line of the three-phase transformer is used at each circuit. For example, L1–N is a single-phase circuit.

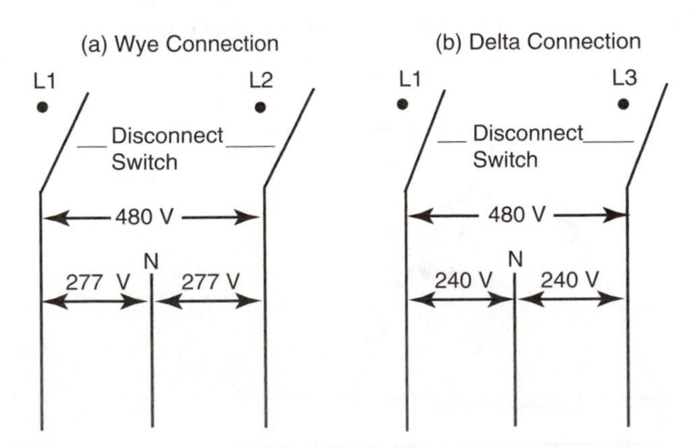

FIGURE 12-40 (a) Voltage for L1–L2 for a wye-connected transformer is 480 V and for L1–N is 277 V. (b) Voltage for L1–L3 for a delta-connected transformer is 480 V and for L1–N or L3–N is 240 V.

It is also important to understand at this time that L1–L2, L2–L3, or L3–L1 could each be used to supply complete power for an electrical system. Even though this type of power supply uses two legs of the transformer, it is still called a single-phase power supply because only one phase of voltage is used at any instant in time. For example, if the system is powered with voltage from L1–L2, during any given half-cycle of AC voltage, the power source for the system will come from L1, and then during the next half-cycle it will come from L2. The source of power will continue to oscillate between L1 and L2, but only one phase is in use during any instant in time.

Figure 12-40b shows that the higher voltage for the delta system between L1 and L3 is also 480 V, but this time the voltage between L1 and N or L3 and N is 240 V. Again the L1–N or L3–N voltage is exactly half of the supply voltage, since the neutral point on the transformer is the center tap of one of the transformers. It is important to understand that it is very easy to distinguish between a wye-connected power source and a delta-connected power source by measuring the voltages L1–L3 and L1–N. If the L1–N voltage is half of the L1–L3 voltage, the system is a delta-connected system. If the L1–N voltage is more than half, it is a wye-connected system. You should remember that the L1–N wye voltage can always be calculated by dividing the L1–L2 voltage by 1.73.

The High-Leg Delta System

When a three-phase transformer system is used for the power source for an electrical system, it may have a neutral tap. It is important to remember that a three-phase system does not need to have a neutral to operate correctly. The neutral is added only if the lower voltage (120 V) is needed for some part of the system. Generally, equipment manufacturers make all the components in a three-phase system the higher voltage, or they may supply a small transformer inside the equipment power panel to drop the higher voltage to the necessary voltage level. For example, if the equipment needs 208 V three phase for the motors, a control transformer can be provided in the power panel of the equipment to drop the 208 V L1–L2 to 120 or 24 V. The 120 or 24 V is used to provide power for the control circuit to power relay coils. Since the primary side of the control transformer can be powered by L1–L2 (208 V), a neutral is not needed in this system. If the control transformer or motors require 120 V for power, a neutral tap, called a neutral, is needed, and

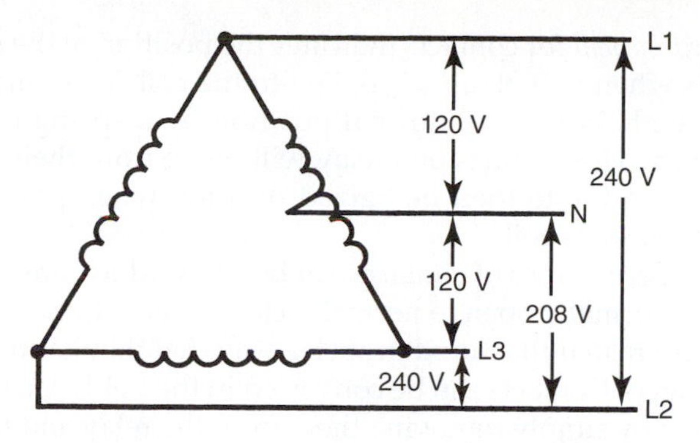

FIGURE 12-41 A high-leg delta voltage occurs between L2 and N because the center tap to produce the neutral is the center tap of the L1–L3 winding.

these components are connected between L1 and N, L2 and N, or L3 and N.

If the transformer is connected as a delta connection, it is important to understand that a different voltage becomes available between L2 and N. Figure 12-41 shows the diagram for this voltage. From this diagram you can see that the secondary winding of this three-phase transformer is connected as a delta transformer. The voltage for L1–N and for L3–N is 120 V, so the voltage between L1 and L2, L2 and L3, or L3 and L1 is 240 V.

A different voltage occurs between L2 and N. Since the winding between L1 and L3 has a center tap, it stands to reason that the voltage between L1 and N or between L3 and N will be exactly half of the voltage between L1 and L3. Since the center tap is not between L2 and L1, the voltage for L2–N must come from one complete phase (240 V) and half of the next phase (120 V). Since this voltage uses two phases, the 240 V and the 120 V are out of phase, and the resulting voltage from them is 208 V.

Since the 208 V between L2 and N comes from two phases, the transformer will overheat if it is used to power any components that require 208 V. For this reason the voltage is called the *high-leg delta voltage,* to indicate that it is derived from L2–N of a delta-connected transformer and that it should not be used to power 208-V components. It is very important to understand that the L2 leg of the transformer is perfectly usable when it is used with L1–L2 or L2–L3 as part of the three-phase system or the 240-V single-phase system. The only problem occurs when the L2 terminal is used in conjunction with the neutral, which creates the L2–N voltage of 208 V.

The L2 terminal in any power distribution box for a delta-wired system should always be marked with an orange wire to identify it as the high-leg delta. In some areas, the high-leg delta is also called the *wild leg, red leg,* or *stinger leg*.

12.16 THEORY AND OPERATION OF A RELAY AND CONTACTOR

A **relay** is a magnetically controlled switch that is the main control component in an electrical system. Figure 12-42 shows a typical relay, and Figure 12-43 shows

FIGURE 12-42 A typical large relay.

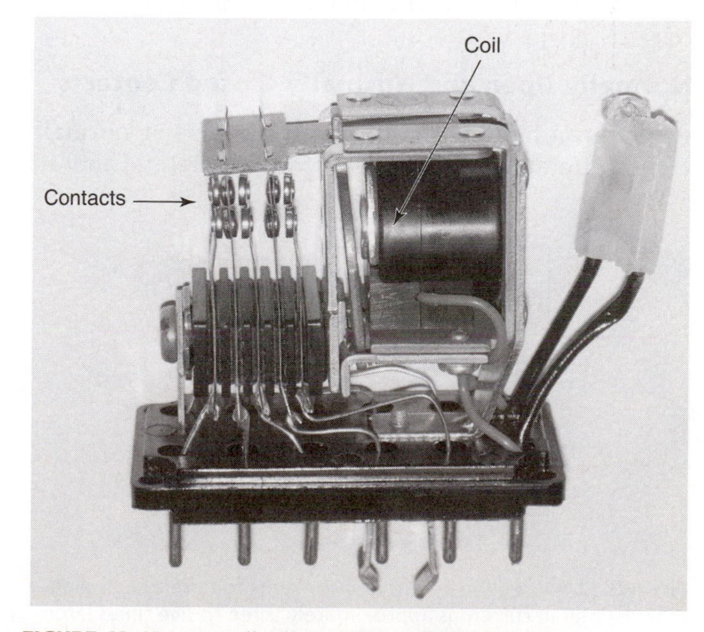

FIGURE 12-43 A small relay with its coil and contacts shown.

a cutaway view of a smaller plug-in type relay with its contacts and coil identified. The relay can consist of a single **coil** and a number of sets of contacts. The coil becomes an electromagnet when it is energized, and its magnetic field causes each set of normally open contacts to close and each set of normally closed contacts to open. The contacts are basically switches that are operated by magnetic force. The part of the relay that moves and causes the contacts to move is called the *armature*. Power is applied to the coil of the relay first, and the magnetic flux causes the armature to move and causes the contacts to change position. The coil is part of the control circuit, and the contacts are part of the load circuit.

You need to think of the relay as two separate pieces, the coil and the contacts, even though they are mounted near each other and operate almost simultaneously. It is important to understand that the coil must be energized first, and a split second later the magnetic field built up in the coil will cause the contacts to move.

Pull-in and Hold-in Current

When voltage is first applied to the coil of a relay, it draws excessive current because the coil of wire presents only resistance to the circuit when current first starts to flow. As the flow of current increases in the coil, inductive reactance begins to build, which causes the current to become lower. The current creates a strong magnetic field around the coil, which causes the armature to move. The armature movement causes the induction in the magnetic coil to change, so that less current is required to maintain the position of the armature.

Figure 12-44 is a diagram showing the *pull-in current* and the *hold-in current*. The pull-in current is also called the *inrush current*, and the hold-in current is also called the *seal-in current*. The pull-in current is typically three to five times larger than the hold-in current.

Normally Open and Normally Closed Contacts

A relay can have **normally open** contacts or **normally closed** contacts. It is important to understand that the word *normal* for contacts indicates the position of the contacts when no voltage is applied to the coil. The contacts can be held in their normal position by a spring or by gravity. The contacts of a relay will move from their normal position to their energized position when power is applied to its coil.

Some types of contacts can be changed or converted from normally open to normally closed in the field. Other types are manufactured in such a way that they cannot be changed. Contacts can be converted in the field by a technician by simply removing them from the relay and turning them upside down. When a set of normally open contacts is inverted, the contacts become normally closed, and when normally closed contacts are inverted, they become normally open contacts. This means that a technician in the field can change the contacts in a relay to get the exact number of normally open or normally closed contacts needed for the application.

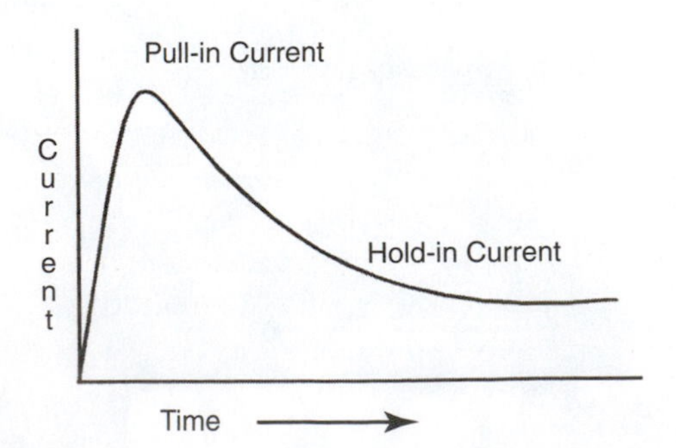

FIGURE 12-44 Pull-in and hold-in currents for a relay coil. Notice that the pull-in current is approximately three to five times larger than the hold-in current.

Ratings for Relay Contacts and Relay Coils

When you change a relay that is worn or broken, you must ensure that the voltage rating of the coil of the new relay matches the voltage of the control circuit exactly. This means that if the voltage for the control circuit is 24 V AC, the coil must be rated for 24 V. If the control voltage is 120 V AC, the coil must be rated for 120 V. The voltage rating for a relay coil is stamped directly on the coil. The coil may also have a color code. If the coil is rated for 24 V AC, it will be color-coded black. If the coil is rated for 120 V AC, it should also be color-coded red or have a red-colored stamp on the coil. If the relay coil is rated for 208 or 240 V AC, it will be color-coded green or identified with a green stamp or green printing on the coil. DC coils are color-coded blue. It is important to understand that the current rating for a relay coil is seldom listed on the component. If it is important to know the current rating for the coil, you can look for it in the catalog or on the specification sheet that is shipped with the new relay. If you change a relay, you must also make sure that the rating for the contacts meets or exceeds the current rating and the voltage rating of the load to which it will be connected.

Contact ratings are grouped by voltage and by current. The voltage ratings are generally broken into two groups, of 300 V and 600 V. This means that if you are using the contacts to control 240 or 208 V AC, you should use contacts that have a 300-V rating. If you are using the contacts to control a 480-V AC motor, you will need to use 600-V–rated contacts.

The current rating of contacts is listed in amperes or horsepower. The current rating or horsepower rating must exceed the amount of current the relay is controlling. This means that if the relay is controlling 12 A, the contacts need to be rated for more than 12 A of current. The current rating of the contacts and the voltage rating for the contacts are printed directly on the contacts or on the side of the relay.

Identifying Relays by the Arrangement of Their Contacts

Some types of contact arrangements for relays have become standardized so that they are easy to recognize when they are ordered for replacement or when you are trying to troubleshoot them. Figure 12-45 shows examples of some standard types of relay arrangements. Figure 12-45a shows a relay with a set of normally open contacts. This type of relay can also have a single set of normally closed contacts instead of normally open contacts. Since this relay has only one contact and it can only close or open this type of relay, it is called a single-pole, single-throw (SPST) relay. The word *pole* here refers to the number of contacts, and the word *throw* refers to the number of terminals to which the input contacts can be switched. Since the contact in this relay has one input and it can be switched to only a single output terminal, it is said to have a single throw.

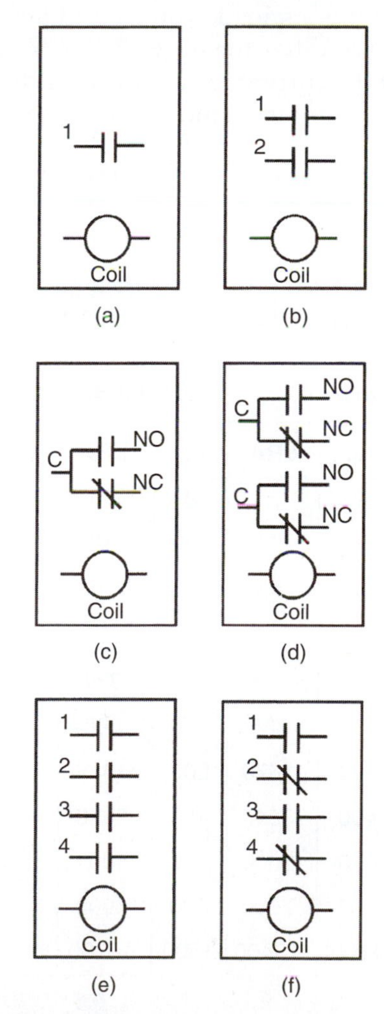

FIGURE 12-45 (a) A relay with a single set of normally open contacts. This type of relay is called a single-pole, single-throw (SPST) relay. (b) A relay with two individual sets of contacts. This relay is called a double-pole, single-throw (DPST) relay. (c) A relay with two sets of contacts that are connected at one side at a point called the common (C). The output terminals are identified as normally open (NO) and normally closed (NC). This type of relay is called single-pole, double-throw (SPDT). (d) A relay with two sets of SPDT contacts. This relay is called a double-pole, double-throw (DPDT) relay. (e) A relay with multiple sets of normally open contacts. (f) A relay with a combination of normally open and normally closed contacts.

Figure 12-45b shows a relay with two sets of normally open contacts. Since this relay has two sets of single contacts, it is called a double-pole, single-throw (DPST) relay. The *double-pole* part of the name refers to the two individual sets of normally open contacts, and the *single-throw* part of the name refers to the one output terminal for each contact. When the coil is energized, both sets of contacts move from their normally open position to the normally closed position.

Figure 12-45c shows a relay with a set of normally open and a set of normally closed contacts that are connected on the left side. The point where this connection is made is called the *common point* and is identified with the letter C. When the relay coil is energized, the normally open part of the contacts will close, and the normally closed part of the contacts will open. Since these contacts have a common point as the input terminal, and two output terminals [one normally open (NO) and one normally closed (NC)], the relay is called a single-pole, double-throw (SPDT) relay. The most important feature of this relay is that the contacts have two terminals on the output side, so it is called a double-throw relay. The single-pole, double-throw relay is used where two exclusive conditions exist and you do not want them ever to occur at the same time.

Figure 12-45d shows two sets of single-pole, double-throw contacts, so it is called a double-pole, double-throw (DPDT) relay. In this case the term *double-throw* is used because two sets of normally open/normally closed contacts are provided. Each set has a common point on its left side (input side) and a terminal that is connected to the normally open (NO) set and a terminal that is connected to the normally closed (NC) set on its right side. This type of relay is used where the exclusion is needed and the loads are 208 or 240 V AC where they need power from both L1 and L2. In this type of application, L1 will be connected to the common terminal (C) of one set of contacts, and L2 will be connected to the common terminal (C) of the other set. This will cause L1 and L2 to be switched the same way in both conditions.

Figure 12-45e shows multiple sets of normally open contacts. This type of relay can have any number of sets of normally open contacts. The additional sets of contacts can be added to the original contacts in some types of relays. If the original relay is manufactured with this provision, you can purchase the additional contacts and add them to the original relay by placing them on top of the original relay and tightening the mounting screws to make the additional contacts operate with the relay armature. The contacts for this type of relay can all be normally closed if the application requires it. The main feature of this type of relay is that it can have any number of contacts.

Figure 12-45f shows a relay with multiple sets of individual normally open and normally closed contacts. This type of relay is similar to the one shown in

Figure 12-45e except that in this type of relay the contacts can be any combination of normally open or normally closed sets. In most cases, the contacts in this type of relay are convertible in the field. As a technician, you can add sets of contacts and change them from normally open to normally closed or vice versa as needed. In most installations, the relays are provided in the original equipment and you need only to identify them for installation and troubleshooting purposes.

The Difference Between a Relay and a Contactor

A **contactor** is similar to a relay in that it has a coil and a number of contacts. The contactor, however, is larger, and its contacts can carry more current. A *relay* is generally defined as magnetically controlled contacts that carry a current of less than 15 A. A *contactor* is defined as having contacts that are rated for 15 A or more. Some manufacturers do not follow the 15-A rating, so sometimes you will find a relay that has a current rating for its contacts in excess of 15 A, and you also may find contactors with contact ratings of less than 15 A. In general, the main difference is that a contactor is specifically designed so that its contacts can carry a larger amount of current, up to 2250 A. Contactors are rated by the National Electrical Manufacturers Association (NEMA), and their sizes range from size 00 to size 9. They may also be rated using horsepower (hp).

In most systems, motors are controlled by contactor rather than relays, because most motors draw more than 15 A of current. When you look at a contactor, you can see that it looks like a very large relay.

NEMA Ratings for Contactors

Figure 12-46 shows a table of NEMA ratings for three-phase magnetic contactors and motor starters, showing that a size 00 contactor is rated to safely carry up to 9 A for a continuous load. You should also notice that the contacts are rated for up to 575 V. You should remember that the current rating for contacts depends on the size of the contacts, and the voltage rating of the

Table 2-4-1

HORSEPOWER (HP) AND LOCKED-ROTOR CURRENT (LRA) RATINGS FOR THREE-PHASE, SINGLE-SPEED FULL-VOLTAGE MAGNETIC CONTROLLERS FOR LIMITED PLUGGING AND JOGGING-DUTY

Size of Controller	Continuous Current Rating (Amperes)	At 200V 60 Hz		At 230V 60 Hz		At 380V 50 Hz		At 460V 60 Hz		At 575V 60 Hz		Service-Limit Current Rating* (Amperes)
		Hp	LRA	Hp	LRA	Hp	LRA	Hp	LRA	Hp	LRA	
00	9	1.5	46	1.5	40	1.5	30	2	25	2	20	11
0	18	3	74	3	70	5	64	5	53	5	42	21
1	27	7.5	151	7.5	140	10	107	10	88	10	70	32
2	45	10	255	15	255	25	255	25	210	25	168	52
3	90	25	500	30	500	50	500	50	418	50	334	104
4	135	40	835	50	835	75	835	100	835	100	668	156
5	270	75	1670	100	1670	150	1670	200	1670	200	1334	311
6	540	150	3340	200	3340	300	3340	400	3340	400	2670	621
7	810	...	...	300	5000	...	...	600	5000	600	4000	932
8	1215	...	...	450	7500	...	...	900	7500	900	6000	1400
9	2250	...	...	800	13400	...	...	1600	13400	1600	10700	2590

*See clause 4.1.2

4.1.2 Service-Limit Current Ratings

The service-limit current ratings represent the maximum rms current, in amperes, which the controller shall be permitted to carry for protracted periods in normal service. At service-limit current ratings, temperature rises shall be permitted to exceed those obtained by testing the controller at its continuous current rating. The current rating of overload relays or trip current of other motor protective devices used shall not exceed the service-limit current rating of the controller.

FIGURE 12-46 NEMA ratings for three-phase contactors.

(Courtesy of National Electrical Manufacture Association, NEMA.)

contacts depends on the way the relay is manufactured so that arcs do not jump between different terminals. This means that contacts that are rated for a higher voltage have more plastic or insulating material between the sets of contacts so that arcs do not jump from the contacts to other parts of the relay or to other sets of contacts. This table also identifies the load as a maximum horsepower rating. This means that you may identify the load from its current rating or its horsepower rating and select the proper contactor size to safely control the load.

You should notice from this table that the next larger size of contactor is a size 0. This contactor is rated for up to 18 A on a continuous basis. The size 9 contactor is the largest, and its current rating is 2250 A.

12.17 WHY MOTOR STARTERS ARE USED

Motor starters are magnetically controlled devices that are usually used in larger commercial applications. The motor starter is a larger version of a relay or contactor and is used to control larger motors. The motor starter also has *overcurrent protection* for motors built into it, and relays and contactors do not. This overcurrent protection is called an *overload* and is sized to trip if the amount of current drawn by the motor exceeds the designated limit.

Relays and contactors that are used to start motors in electrical systems are designed to close their contacts and provide current to the motors in the system. Their main function is to turn on and off to provide current to these motors. The motors in these applications are protected against overcurrent by fuses or internal overloads that are built into the winding. These motors are protected by fuses or circuit breakers in the disconnect for short-circuit protection, and by the overloads in motor starters to protect against slow overcurrents. If an open motor has a problem such as loss of lubrication or overload, it will draw extra current, which will damage the motor if the overload is allowed to continue for any length of time. The overloads in the motor starter sense the excess current and trip the motor starter so that its contacts open and stop all current flow to the motor until a technician resets the overloads manually.

The Basic Parts of a Motor Starter

Figure 12-47 shows a typical motor starter with all its parts identified. This is a three-pole starter that is used in three-phase circuits. The incoming voltage is connected at the top at the terminals identified as L1, L2, and L3, and the motor leads are connected to the bottom terminals, which are identified as T1, T2, and T3. The three major sets of contacts are located in the top part of the motor starter, and the overload assembly is mounted in the lower part of the motor starter. The coil is located in the middle of the motor starter, and it has an indicator that shows the word *ON* when the coil is energized, and *OFF* when the coil is de-energized.

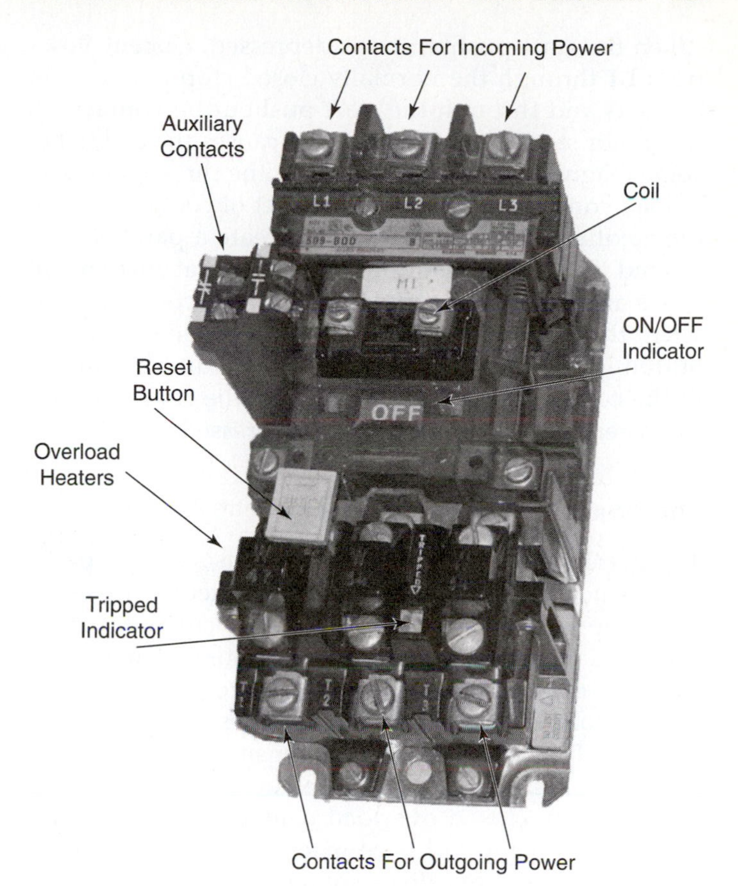

FIGURE 12-47 A typical motor starter.

The motor starter has three sets of contacts that are in series with a heater. This ensures that all the current that flows to the motor must pass through a heater. If the current flowing to the motor is normal, the heater does not provide sufficient heat to cause the overload to trip. If the motor draws excess current, that current flowing through the heater will cause it to create excess heat that will trip the normally closed overload contacts. Since the normally closed overload contacts are connected in series with the motor starter coil, current to the motor starter coil will be interrupted when the overload contacts open. A reset button on the motor starter must be manually reset to set the overload contacts back to their normally closed position.

The control circuit is shown as a ladder diagram at the bottom of the figure. You can see that the motor starter coil is energized when the start pushbutton is depressed. A manual pushbutton is used for this example because it is easier to understand, but most motor starters in electrical applications are controlled by a manual switch or some other type of control.

The motor starter also has one or more additional sets of normally open contacts called *auxiliary contacts*. The auxiliary contacts are connected in parallel with the start pushbutton. These contacts serve as a seal-in circuit after the motor starter coil is energized. The start pushbutton is a momentary-type switch, which means that it is spring-loaded in the normally open position.

When the start pushbutton is depressed, current flows from L1 through the normally closed stop pushbutton contacts and through the start pushbutton contacts to the motor starter coil. This current causes the coil to become magnetized so that it pulls in the three major sets of load contacts and the auxiliary set of contacts. When the auxiliary contacts close, they create a parallel path around the start pus-button contacts so that current still flows around the start pushbutton contacts to the coil when the pushbutton is released. Since the stop pushbutton is connected in series in this circuit, the current to the coil is de-energized and all of the contacts drop out when the stop pushbutton is depressed.

The Operation of the Overload

The **overload** for a motor starter consists of two parts. The heater is the element that is connected in series with the motor, and all the motor current passes through it. The heater is actually a heating element that converts electrical current to heat. The second part of the overload device is the trip mechanism and overload contacts. The trip mechanism is sensitive to heat, and if it detects excess heat from the heater, it trips and causes the normally closed overload contacts to open. Since the motor starter coil is connected in series with the normally closed overload contacts, all current to the coil is interrupted and the coil becomes de-energized when the overload contacts are tripped to their open position. When the coil becomes de-energized, the motor starter contacts return to their open position, and all current to the motor is interrupted. When the overload contacts open, they remain open until the overload is reset manually. This ensures that the overloaded motor stops running and cools down until someone investigates the problem and resets the overloads.

Figure 12-48 shows a typical heating element for a motor starter overload device, which is called the heater. In the cutaway view you can see that the heater is actually a heating element that converts electrical current into heat as it passes through the heating element. You should also notice the knob protruding from the bottom of the heater. This knob has a shaft that is held in

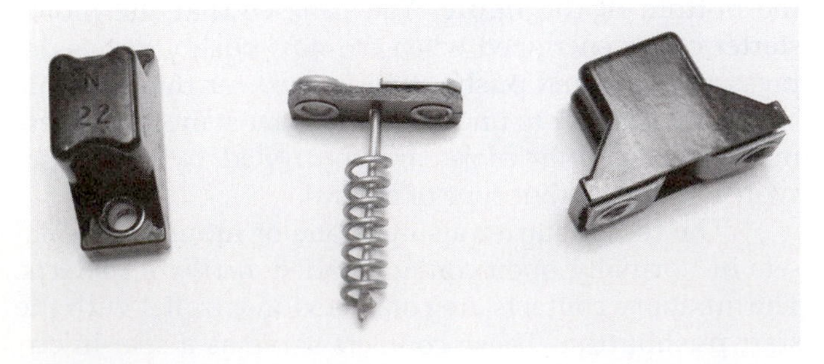

FIGURE 12-48 A typical heater assembly. Notice the ratchet that protrudes from the bottom of the heater.

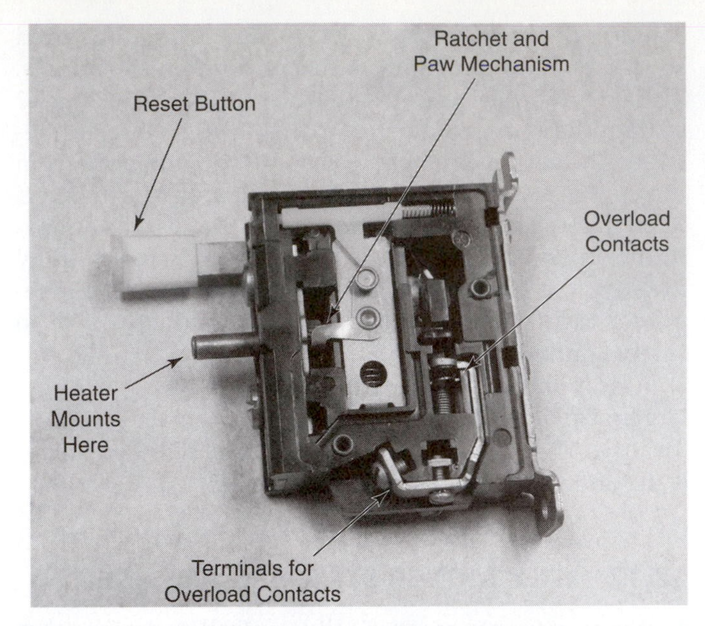

FIGURE 12-49 Overload with all its parts identified.

position inside the heater, and teeth machined into the part that protrudes from the heater. The teeth are called the *ratchet mechanism*.

Figure 12-49 shows all of the parts of the overload assembly, including the reset button, the shaft on which the heating element is mounted, the ratchet and paw mechanism, the overload contacts, and the terminals that connect to the contacts. The trip mechanism consists of the ratchet from the heater and a paw. The ratchet is actually the toothed knob that protrudes from the bottom of the heater. The paw has spring pressure that tries to rotate the ratchet. Because the heater holds the shaft of the ratchet tight with solder, the paw cannot move. When the heater becomes overheated, it melts the solder that holds the ratchet in place and allows it to spin freely. When the ratchet spins, it allows the paw to move past it, which in turn allows the normally closed contacts to move to their open position.

After an overload condition has occurred and the overload contacts have opened, the motor starter is de-energized and the motor stops running. When the motor stops running, the heating element is allowed to cool down. After the heating element cools down for several seconds, the reset button can be depressed, which moves the paw back to its original position, and the normally closed overload contacts move back to their closed position. If the motor continues to draw excess current when it is restarted, the excess current will cause the heater to trip the overload mechanism again. If the motor current is within specification, the heaters do not produce enough heat to cause the overload mechanism to trip.

Since the overcurrent condition must last for several minutes to cause the overload mechanism to trip, the overloads allow the motor to draw high locked-rotor

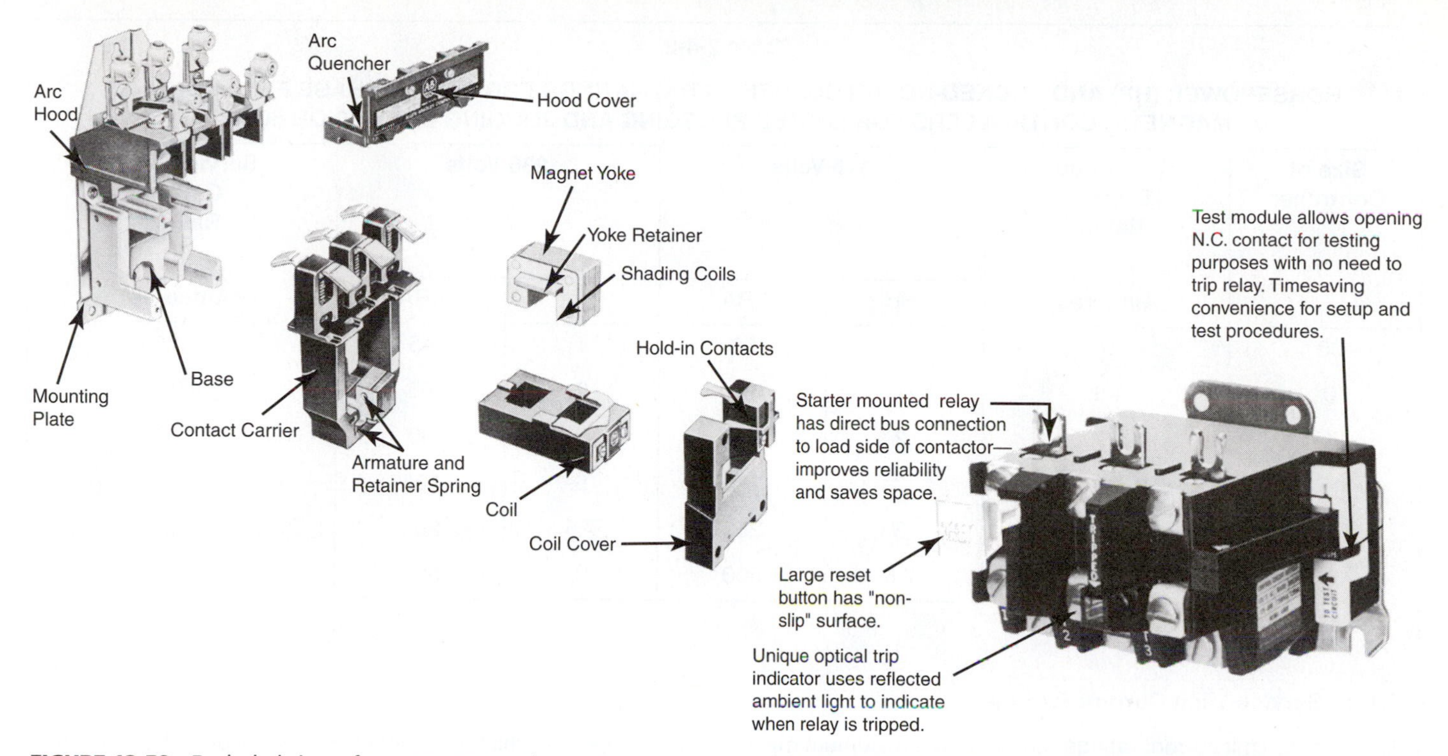

FIGURE 12-50 Exploded view of a motor starter. The stationary contacts are shown at the far left, and the movable contacts are shown to the right of the stationary contacts. The coil and magnetic yoke are shown in the middle. The overload mechanism is shown at the far right side.

amperage (LRA) during the few seconds the motor is trying to start, without tripping. If the motor has an overcurrent condition while it is running, the overloads allow the condition to last several minutes before the motor is deenergized. If the problem continues, the overloads will sense the overcurrent and trip to protect the motor.

Exploded View of a Motor Starter

It is easier to see all the parts of the motor starter in an exploded view, as shown in Figure 12-50. At the far left in this figure, you can see that the main contacts of the motor starter are much larger than those in a traditional relay. The coil is shown in the middle of the figure. The coil has two square holes in it that allow the magnetic yoke to be mounted through it. The magnetic yoke and coil are mounted in the movable contact carrier. When the coil is energized, it pulls the magnetic yoke upward, which causes the movable contacts to move upward until they make contact with the stationary contacts that are shown at the far left. The overload mechanism is shown at the far right in this figure. It is mounted at the lower part of the motor starter, and all current that flows through the contacts must also flow through the overload mechanism.

Sizing Motor Starters

At times you will need to select the proper size motor starter for an application. The size of motor starters is determined by the National Electrical Manufacturers Association (NEMA). The ratings refer to the amount of current the motor starter contact can safely handle. The sizes are shown in the table reproduced in Figure 12-51, where you can see that the smallest size starter is a size 00, which is rated for 9 A and is sufficient for a 2-hp motor connected to 480 V three phase or for a 1-hp single-phase motor connected to 240 V. You can see that the size 1 motor starter is rated for 27 A, which is sufficient for a 10-hp three-phase motor connected to 480 V or a 7½-hp single-phase motor connected to 240 V. A size 00 motor starter is about 4 in. high, and a size 2 motor starter is approximately 8 in. high. You will typically use up to a size 3 or 4 motor starter to protect motors for electrical systems. The size 4 starter will protect motors up to 100 hp. It is important to understand that the overload heaters for the motor starter can also be purchased for a specific current rating. This means that each motor starter can have a heater that is rated specifically to the amount of current the motor that is connected to it draws.

12.18 FUSES

Fuses perform a function similar to that of an overload, except a fuse uses an element that is destroyed when the overcurrent occurs. A fuse provides short-circuit protection, and the overload is not designed to protect the motor when short-circuit current occurs. The fuse has a thermal sensing element that is capable of carrying current. When

Table 2-4-2

HORSEPOWER (HP) AND LOCKED-ROTOR CURRENT (LRA) RATINGS FOR SINGLE-PHASE FULL-VOLTAGE MAGNETIC CONTROLLERS FOR LIMITED PLUGGING AND JOGGING DUTY, 50 OR 60 HZ

Size of Controller	Continuous Current Rating (Amperes)	115 Volts		230 Volts		Service Limit Current Rating* (Amperes)
		Hp	LRA	Hp	LRA	
00	9	1/3	50	1	45	11
0	18	1	80	2	65	21
1	27	2	130	3	90	32
1P	36	3	140	5	135	42
2	45	3	250	7.5	250	52
3	90	7.5	500	15	500	104

*See clause 4.1.2

4.1.2 Service-Limit Current Ratings

The service-limit current ratings represent the maximum rms current, in amperes, which the controller shall be permitted to carry for protracted periods in normal service. At service-limit current ratings, temperature rises shall be permitted to exceed those obtained by testing the controller at its continuous current rating. The current rating of overload relays or trip current of other motor protective devices used shall not exceed the service-limit current rating of the controller.

FIGURE 12-51 NEMA sizes for single-phase full-voltage magnetic controllers. Notice that the smallest motor starter is a size 00, and the largest is a size 3.

(Reprinted from NEMA ICS 2-2000,®2005.)

the current becomes excessive, the heat that is generated is sensed by the fuse element, which melts when the temperature is high enough.

Fuses are available in a variety of sizes and shapes for different applications. Figure 12-52 shows various types of cartridge-type fuses. Each fuse is sized for the amount of current it will limit. When the amount of current is exceeded, the fuse link melts and opens the fuse. The single-element fuse provides protection at a single level. This type of fuse is generally used for noninductive loads such as heating elements or lighting applications.

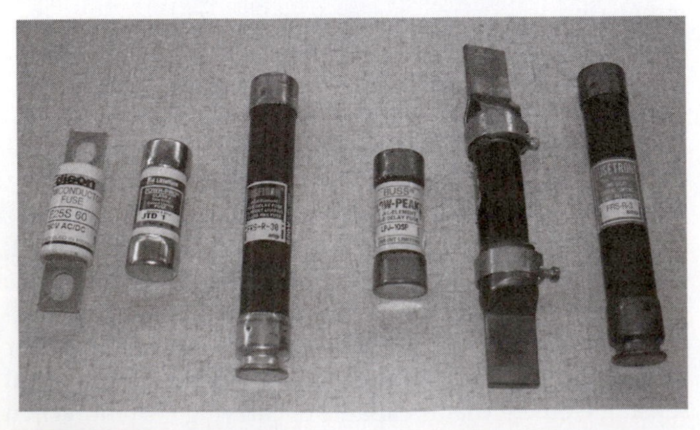

FIGURE 12-52 Various types of fuses.

The dual-element fuse provides two levels of protection. The first level is called slow overcurrent (overload) protection, and it consists of a fusible link that is soldered to a contact point and attached to a spring. When a motor is started, it draws locked-rotor amperage (LRA) for several seconds. This excess current causes heat to build in the fuse, and this heat is absorbed in the slow-overcurrent link. If the motor starts and the current drops to the full-load amperage level, the link will cool off, and the fuse will not open. If the LRA current continues for 30 s, the amount of heat generated will cause the solder that holds the slow-overcurrent link to melt. When the solder melts, the spring pulls the link open and interrupts all current flowing through the fuse. The slow-overcurrent link allows the fuse to sustain overcurrent conditions for short periods of time, and if the condition clears, the fuse will not open. If the condition continues to exist, the fuse will open.

The second type of element is called the short-circuit element. This element opens immediately when the amount of current exceeds the level of current the link is designed to handle. In the dual-element fuse, the short-circuit link is sized to be approximately 5 times the rating of the fuse. A short circuit is by definition any current that exceeds the full-load current rating by 5 times. (Some manufactures use the rating of 10 times.)

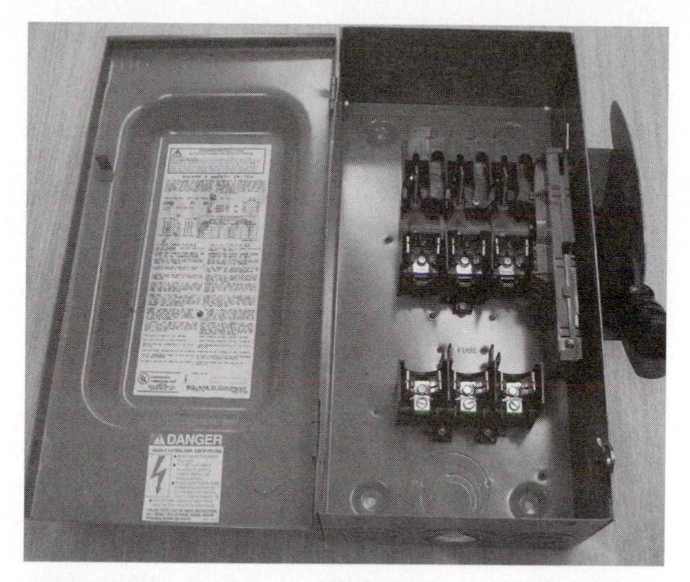

FIGURE 12-53 A fuse disconnect has the fuses in series with the remainder of the circuit.

The single-element fuse has only a short-circuit element in it. This type of fuse is generally not used in circuits to start motors, since the motor draws locked-rotor current. If a single-element fuse is used to protect a motor circuit, it must be sized large enough to allow the motor to start, and then it is generally too large to protect the motor against an overcurrent condition of 20%, which will eventually damage the motor if it is allowed to occur for several hours.

Fuse Disconnect Panels

The cartridge-type and screw-base fuses are generally mounted in a panel called a *fuse disconnect*. The fuse disconnect is normally mounted near the equipment it is protecting, and it serves two purposes: It provides a location to mount the fuses, andit provides a means of disconnecting the electrical supply voltage to a circuit. A three-phase disconnect is shown in Figure 12-53. The fuse disconnect has a switch handle that is used to disconnect main power from the circuit and the fuses so that the fuses can be safely removed and replaced or tested.

SAFETY NOTICE

Always use plastic fuse pullers to remove fuses from and replace them in a disconnect to protect yourself from electrical shock even when the fuse disconnect switch is in the off position. It is important to remember that even though the switch handle is open, line voltage is still present at the top terminal lugs in the disconnect. Never use metal pliers or screwdrivers to remove fuses.

Circuit Breakers and Load Centers

A *circuit breaker* is similar to a thermal overload in that it is an electromechanical device that senses both overcurrent and excess heat. Some circuit breakers also sense magnetic forces. Circuit breakers are mounted in electrical panels called *load centers*. The load center can be designed for three-phase circuits or for single-phase circuits. The single-phase panel provides 240-V AC and 120-V AC circuits. Figure 12-54 shows a typical load center without any circuit breakers mounted in it.

Circuit breakers are manufactured in three basic configurations, for single-phase 120-V AC applications that require one supply wire, 240-V AC single-phase applications that require two supply wires, and three-phase applications that require three wires. The three-phase circuit breakers can be mounted only in a load center that is specifically designed for three-phase circuits. Two-pole and single-pole breakers can be mounted in a single-phase or a three-phase load center.

The operation of the circuit breaker is similar to that of the thermal overload in that it senses excessive current that will trip its circuit after a specific amount of time. The main difference between the circuit breaker and the thermal overload is that the circuit breaker is mounted in the load center to protect both the circuit wires as well as the load. The circuit breakers are sized for the total current rating of the wire and all the loads that are connected to the wire. In some cases this means that the circuit breaker is sized for the current flowing to the motors. In these circuits it may be necessary to use a circuit breaker to protect the entire circuit, with

FIGURE 12-54 A load center on which circuit breakers are mounted. The load center is sometimes called a circuit breaker panel.

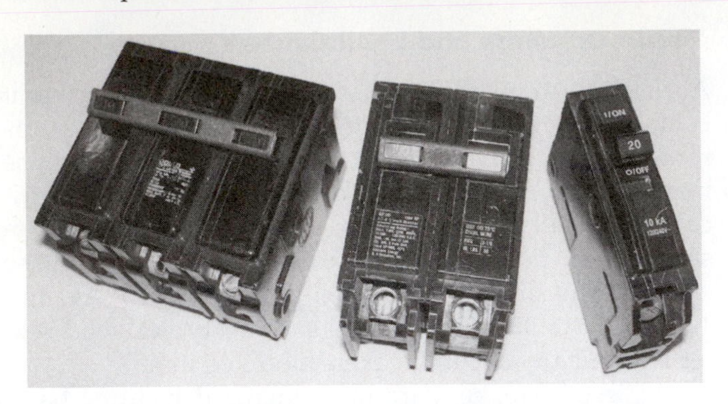

FIGURE 12-55 Single-pole, two-pole, and three-pole circuit breakers for load centers.

overloads at each motor to protect them individually from overheating. This means that the main job of the circuit breaker is to protect the circuit against short-circuit conditions and to protect the entire circuit against overcurrent conditions rather than to protect individual motors against overcurrent. This is why many circuits have a combination of circuit breakers and overload devices. Figure 12-55 shows examples of single-phase and three-phase circuit breakers.

12.19 ELECTRONIC COMPONENTS USED IN INVERTERS AND CIRCUITS

Electronics are used in many parts of wind turbine control systems, such as the power electronic frequency converter (inverter) that converts DC voltage to AC voltage, in circuit boards to control proportional hydraulic valves, and in variable-frequency drives that control motor speeds. Electronics make systems more efficient and more reliable. In this section you will find that electronic components provide functions that are similar to electrical components and systems you are already familiar with. The information in this section will provide names for all components and explain the operation of electronic devices such as diodes, which are the basic component for power supplies and inverters, transistors, and silicon-controlled rectifiers (SCRs) that have become commonplace in electrical systems because they provide better control than electromechanical devices and are less expensive to manufacture. Electronic components are also called solid-state components. You will also gain an understanding of P-type materials and N-type materials, which are the building blocks of all electronic components. After you have a basic understanding of P-type materials and N-type materials, you will be introduced to diodes, transistors, SCRs, and triacs, and you will see application circuits of each of these types of components. The theory of operation and troubleshooting techniques for each type of device will also be presented so that you will have a good understanding of how to determine whether a device or circuit board is working correctly. At first, most wind technicians think that it is difficult to troubleshoot electronic devices because they cannot see their operation. After you fully understand the theory of operation of each component and basic circuits, you will find it easy to determine whether components are working correctly or are faulty and must be replaced. You will also learn in this section that if you determine a component or circuit is faulty, you will change out the complete device or board rather than try to repair an individual component. You will also begin to understand that every component has a specific test that can prove that an electronic device is either good or faulty.

Conductors, Insulators, and Semiconductors

Early in this chapter you learned that atoms have protons and neutrons in their nuclei and electrons that move around the nucleus in regions of space called shells. The number of electrons in the atom is different for each element. For example, copper has 29 electrons, three of them located in the outermost shell. The outermost shell is called the valence shell, and the electrons in that shell are called valence electrons. The atoms of the most stable materials have eight valence electrons, which are found as four pairs. This means that an atom may have five, six, or seven atoms, and it will take less energy to add electrons to get a full shell (eight), or it may have one, two, or three electrons, and it will take less energy to give up these electrons to achieve the stable configuration of a full shell with eight electrons.

A *conductor* is a material that allows electrons (electrical current) to flow easily through it, and an *insulator* does not allow current to flow through it. An example of a conductor is copper, which is used for electrical wiring. An example of an insulator is rubber or plastic. The atomic structure of a conductor makes it easier for electrons to flow through it, and the atomic structure of an insulator makes it nearly impossible for any electrons to flow through it.

Figure 12-56 shows a simplified atomic structure for a conductor. Atoms of conductors can have one, two, or three valence electrons. The atom in this example has one valence electron. Since all atoms will gain or lose sufficient electrons to achieve a configuration of eight electrons (four pairs) in their valence shell, it takes less

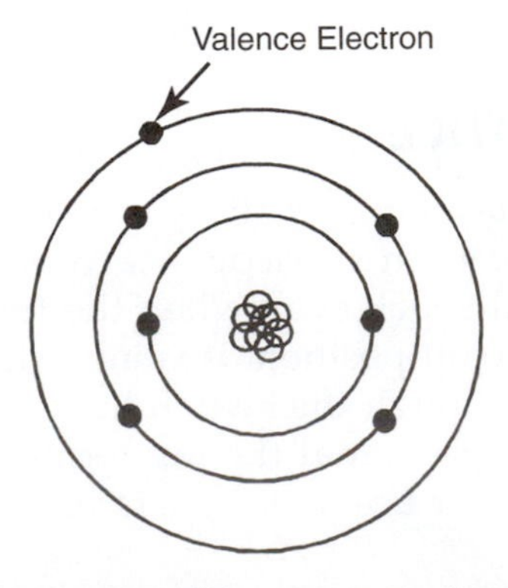

FIGURE 12-56 Simplified atomic structure of a conductor.

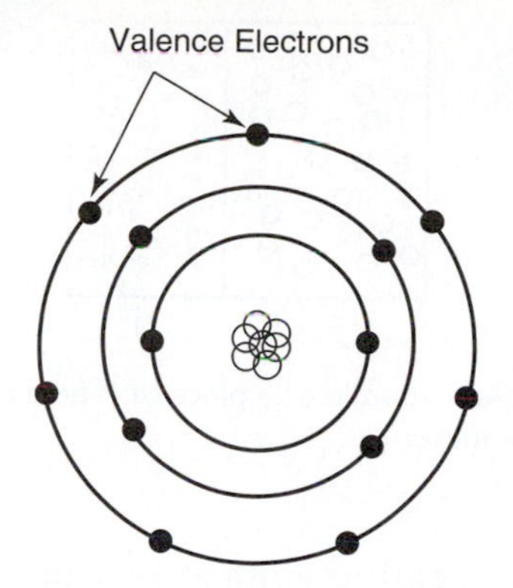

FIGURE 12-57 Simplified structure of an insulator.

energy for conductors to give up these electrons (one, two, or three) so that the valence shell becomes empty. At this point the atom becomes stable because, the previous shell becomes the new valence shell and it has eight electrons. The electrons that are given up are free to move as current flow.

Figure 12-57 shows a simplified atomic structure of an insulator. Insulators have five, six, or seven valence electrons. In this example, the atom has seven valence electrons. This structure makes it easy for insulators to take on extra electrons to get eight valence electrons. The electrons that are captured to fill the valence shell are electrons that would normally be free to flow as current.

Semiconductors are materials whose atoms have exactly four valence electrons. Since these atoms have exactly four valence electrons, they can take on four new valence electrons like an insulator to get a full valence shell, or they can give up four valence electrons like a conductor to get an empty outer shell. Then the previous shell that has eight electrons becomes the valence shell. Figure 12-58 shows a simplified atomic structure of a semiconductor material.

Combining Atoms

When solid-state material or other material is manufactured, large numbers of atoms are placed together. The structure that becomes most stable at this point is called a

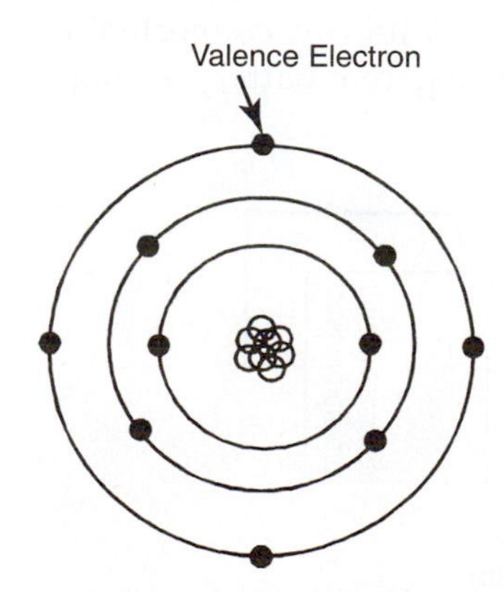

FIGURE 12-58 Simplified atomic structure of a semiconductor material.

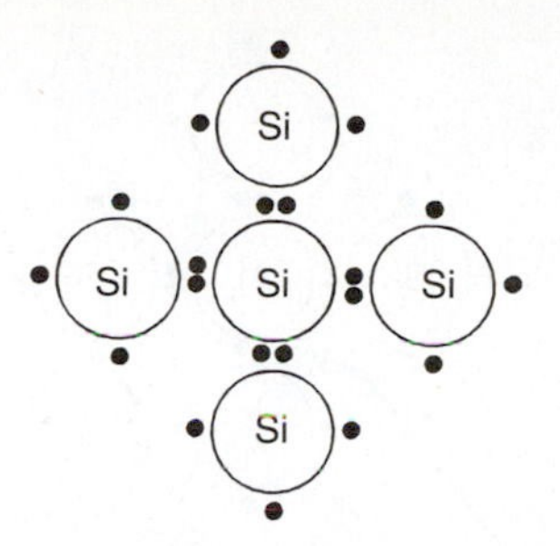

FIGURE 12-59 Atoms of silicon semiconductor material combine to create a lattice structure.

lattice structure. Figure 12-59 shows an example of the lattice structure that occurs when atoms are combined. In this diagram you can see that atoms of silicon (Si), which is a semiconductor material with four valence electrons, combine so that one valence electron from each of the neighbor atoms is shared, so that all atoms look and act as if they have eight valence electrons each.

Combining Arsenic and Silicon to Make N-Type Material

Other types of atoms can be combined with semiconductor atoms to create the special materials that are used in solid-state **transistors** and diodes. Figure 12-60 is a diagram of four silicon atoms combined with one atom of arsenic. In this figure you can see that arsenic has five valence electrons, and when the silicon atoms are combined with it, they create a very strong lattice structure. Each silicon atom donates one of its valence electrons to pair up with each of the valence electrons of the arsenic atom. Since the arsenic atom has five valence electrons, one of the electrons will not be paired up and will become displaced from the atom. This electron is called a *free electron*, and it can go into conduction with very little energy. Since this new material has a free (*negatively* charged) electron, it is called **N-type material**.

Combining Aluminum and Silicon to Make P-Type Material

An atom of aluminum can also be combined with semiconductor atoms to create the special material called P-type

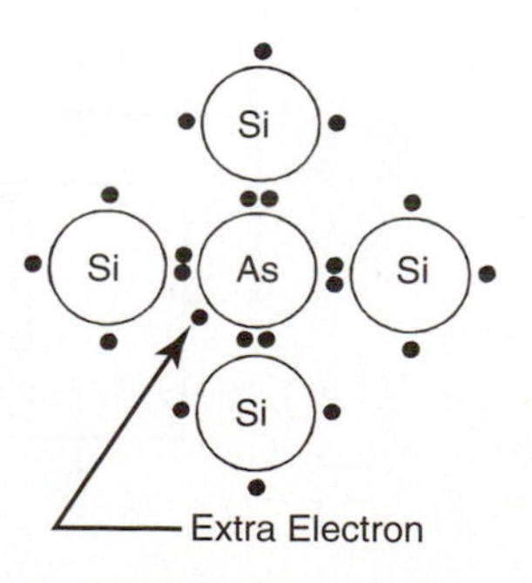

FIGURE 12-60 N-type material formed by combining four silicon atoms with a single arsenic atom.

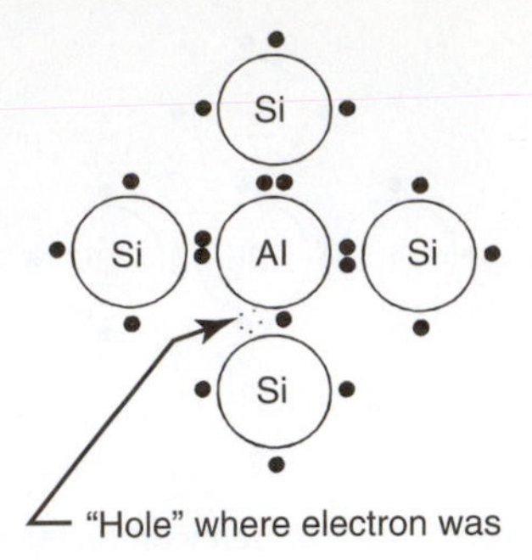

FIGURE 12-61 P-type material formed by combining four silicon atoms with a single aluminum atom.

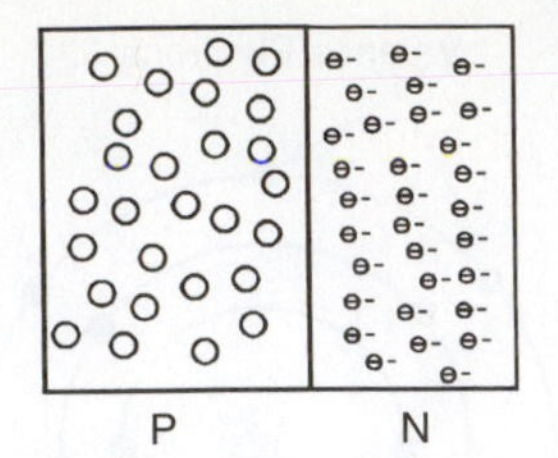

FIGURE 12-62 An example of a piece of P-type material joined to a piece of N-type material.

material. Figure 12-61 is a diagram of four silicon atoms combined with one atom of aluminum. In this figure you can see that aluminum has three valence electrons, and when the silicon atoms combine with it, they create a very strong lattice structure. Each silicon atom donates one of its valence electrons to pair up with each of the valence electrons of the aluminum atom. Since the aluminum atom has three valence electrons, one of the four aluminum electrons will not be paired up and there will be a space where any free electron can move into it to combine with the single electron. This free space is called a *hole,* and it is considered to have a positive charge since it is not occupied by a negatively charged electron. Since this new material has an excess hole that has a *positive* charge, it is called **P-type material**.

The PN Junction

One piece of P-type material can be combined with one piece of N-type material to make a PN junction. Figure 12-62 shows a typical *PN junction.* The PN junction creates an electronic component called a **diode**. The diode is the simplest electronic device. When DC voltage is applied to the PN junction with the proper polarity, it causes the junction to become a very good conductor. Conversely, if the polarity of the voltage is reversed, the PN junction becomes a good insulator.

Forward Biasing the PN Junction

When DC battery voltage is applied to the PN junction so that positive voltage is connected to the P-type material and negative voltage is connected to the N-type material, the junction is *forward-biased.* Figure 12-63a shows a battery connected to a PN junction so that it is forward-biased. In this figure you can see that the positive battery voltage causes the majority of holes in the P-type material to be repelled, so that the free holes move toward the junction, where they will come into contact with the N-type material. At the same time, the negative battery voltage also repels the free electrons in the N-type material toward the junction. Since the holes and free electrons come into contact at the junction, the electrons recombine with the holes to cause a low-resistance junction, which allows current to flow freely through it. When a PN junction has low resistance, it will allow current to pass just as if the junction were a closed switch.

It is important to understand that up to this point in this text, all current flow has been described in terms of *conventional current flow,* which is based on a theory that electrical current flows from a positive source to a negative return terminal. Now, however, you can see that this theory will not support current flow through electronic devices. For this reason, *electron current flow theory* must be used when discussing electronic devices. In electron current flow theory, *current* is the flow of electrons, and it flows from the negative terminal to the positive terminal in any electronic circuit.

Reverse Biasing the PN Junction

Figure 12-63b shows a battery connected to the PN junction so that it is *reverse-biased.* In this diagram, the positive battery voltage is connected to the N-type material, and the negative battery voltage is connected to

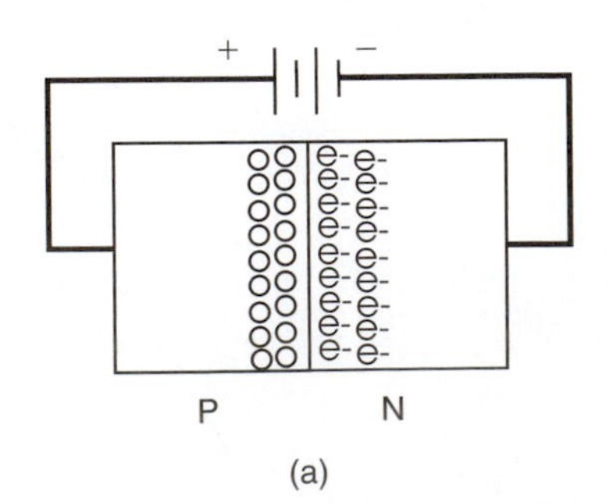

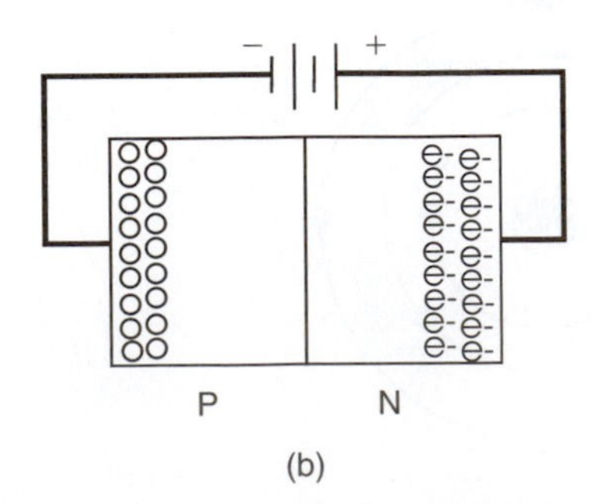

FIGURE 12-63 (a) A forward-biased PN junction. (b) A reverse-biased PN junction.

the P-type material. The negative voltage on the P-type material attracts the majority of holes in the P-type material, so they move away from the junction. Thus, they cannot come into contact with the N-type material. At the same time, the positive battery voltage that is connected to the N-type material attracts the free electrons away from the junction. Since the holes and free electrons are both attracted away from the junction, no electrons can recombine with any holes. Thus, a high-resistance junction is formed that will not allow any current flow. When the PN junction has high resistance, it will not allow any current to pass, just as if the junction were an open switch.

12.20 USING A DIODE FOR RECTIFICATION

In electronic circuits it is difficult to manipulate AC voltage. With the advent of electronics, DC voltage and current can be manipulated (changed) to control a variety of components in an electrical system such as inverters, proportional hydraulic valves, and other frequency-control systems. Since most equipment is supplied with AC voltage, electronic components are required to change the voltage into DC voltage so it can easily be changed. One or more diodes are used to change AC voltage to DC voltage. An electronic diode is a simple PN junction. Figure 12-64 shows the symbol for a diode. You can see that the symbol for the diode looks like an arrowhead that is pointing against a line. The part of the symbol that is the arrowhead is called the *anode,* and it is also the P-type material of the PN junction. The other terminal of the diode is called the *cathode,* and it is the N-type material. Since the anode is made of positive P-type material, it is identified with a + sign. The cathode is identified with a − sign, since it is made of N-type material. The AC power source produces a sine wave that has a positive half-cycle followed by a negative half-cycle. The diode converts the AC sine wave to half-wave DC voltage by allowing current to pass when the AC voltage provides a forward bias to the PN junction, and it blocks current when the AC voltage provides a reverse bias to the PN junction. The forward-bias condition occurs when the AC voltage sine wave provides a positive voltage to the anode and a negative voltage to the cathode. During this part of the AC cycle, the diode is forward-biased and has very low resistance, so current can flow. When the other half of the AC sine wave occurs, the diode becomes reverse-biased, with negative voltage applied to the anode and positive voltage applied to the cathode. During the time the diode is reverse-biased, a high-resistance junction is created, and no current will flow through it.

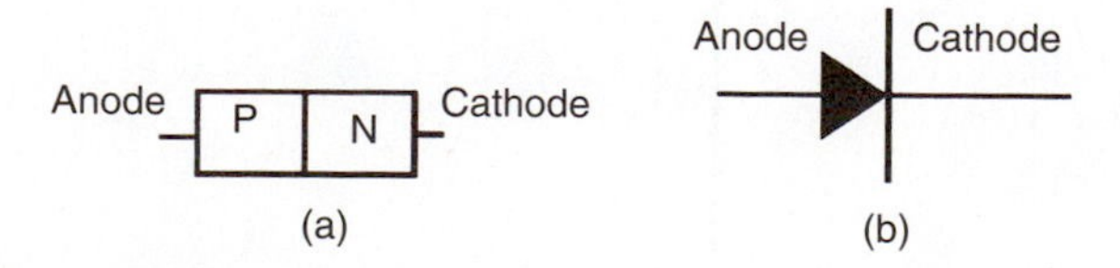

FIGURE 12-64 (a) PN junction for a diode. (b) Electronic symbol for a diode. The anode is the arrowhead part of the symbol, and the cathode is the other terminal.

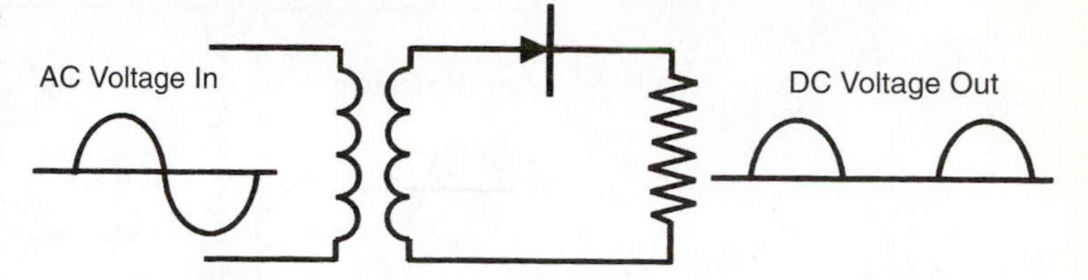

FIGURE 12-65 A single diode used in a circuit to convert AC voltage to DC voltage.

A **rectifier** is used to change AC voltage to DC voltage. (The process is called *rectification.*) One of the main jobs of the diode is to convert AC voltage to DC voltage. Most electronic circuits used in equipment need some DC voltage to operate. Since the equipment is connected to AC voltage, a power supply is required to provide regulated DC voltage for the solid-state circuits, and the diode is part of the power supply that rectifies the AC voltage to DC.

Half-Wave and Full-Wave Rectifiers

When one diode is used in a circuit to convert AC voltage to DC voltage, it is called a *half-wave rectifier,* since only the positive half of the AC voltage is allowed to pass through the diode, while the negative half is blocked. The rectifier shown in Figure 12-65 is a half-wave rectifier. The half-wave rectifier is not very efficient, since half of the AC sine wave is wasted.

If four diodes are used in the circuit, they can convert both the positive half-wave and the negative half-wave of the AC sine wave. Figure 12-66 shows a circuit with four diodes that is used to rectify AC voltage to DC voltage. Since the four diodes can convert both the positive half and the negative half of the AC sine wave, this type of rectifier is called a *full-wave rectifier.* It is important to understand that the full-wave bridge consists of four diodes. Some circuit boards provide four individual diodes on the board, while other boards package the four diodes into an integrated circuit (IC) that has four leads. Two of the leads are marked AC, and this is where the AC voltage is supplied to the bridge. The other two leads are identified as DC+ and DC−, and these leads provide DC voltage out of the bridge. If only one of the diodes in a bridge rectifier goes bad, the bridge will provide approximately half its rated voltage. If two or more diodes in the bridge rectifier go bad, the DC output voltage is usually zero.

Three-Phase Rectifiers

The speed of three-phase AC motors can be changed so that they run more efficiently by changing the frequency of the voltage supplied to them. In these applications, six diodes are used to convert three-phase AC voltage to DC

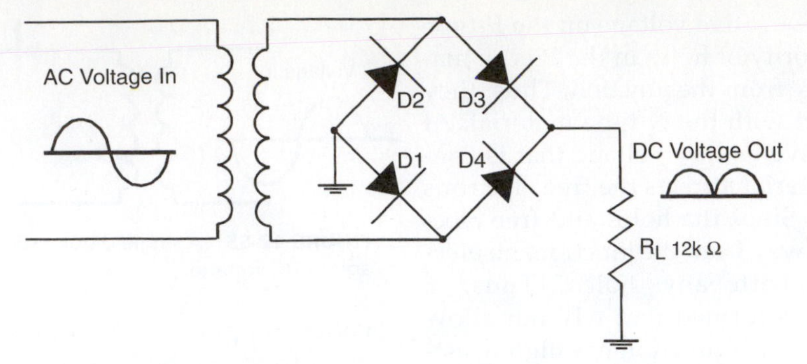

FIGURE 12-66 A four-diode, full-wave rectifier. This type of rectifier is often called a full-wave bridge rectifier, since the diodes are connected in a bridge circuit.

voltage, and then a microprocessor-controlled circuit converts the DC voltage back to three-phase AC voltage. The frequency of this voltage can be adjusted to change the speed of the motors. Figure 12-67 shows a six-diode three-phase rectifier. Notice that the supply voltage to the diodes is three-phase AC voltage, and the output voltage from the rectifier consists of six positive half-waves.

Most power supplies for battery-charging applications and motor drives use three-phase AC voltage rather than single-phase voltage. This means that the rectifier for these circuits must use a three-phase bridge, which has six diodes to provide full-wave rectification (two diodes for each line of the three phases). Figure 12-68 shows the electrical diagram for a three-phase bridge rectifier. In this figure you can see that the secondary winding of the three-phase transformer is connected to the diode rectifier. Phase A of the three-phase voltage from the transformer is connected to the point where the cathode of diode 1D is connected to the anode of diode 2D. Phase B is connected to the point where the cathode of diode 3D is connected to the anode of diode 4D, and phase C is connected to the point where the cathode of diode 5D is connected to the anode of diode 6D. The anodes of diodes 1D, 3D, and 5D are connected together to provide a common point for the DC negative terminal of the output power. The cathodes of diodes 2D, 4D, and 6D are connected to provide a common point for the DC positive terminal of the output power.

A good rule of thumb for determining the connections on diode rectifiers is that the AC input voltage is connected to the bridge where the anode and cathode of any two diodes are joined. Since this occurs at two points in the bridge, in a four-diode bridge the two AC lines are connected there without respect to polarity, since the incoming AC voltage does not have a specific polarity. The positive terminal for the power supply is connected to the bridge where the two cathodes of the diodes are joined, and the negative terminal is connected to the bridge where the two anodes of the diodes are joined.

Figure 12-68 also shows the waveforms for the three-phase sine waves that supply power to the bridge, and for the six half-waves of the output pulsing DC voltage. You should notice that since the six half-waves

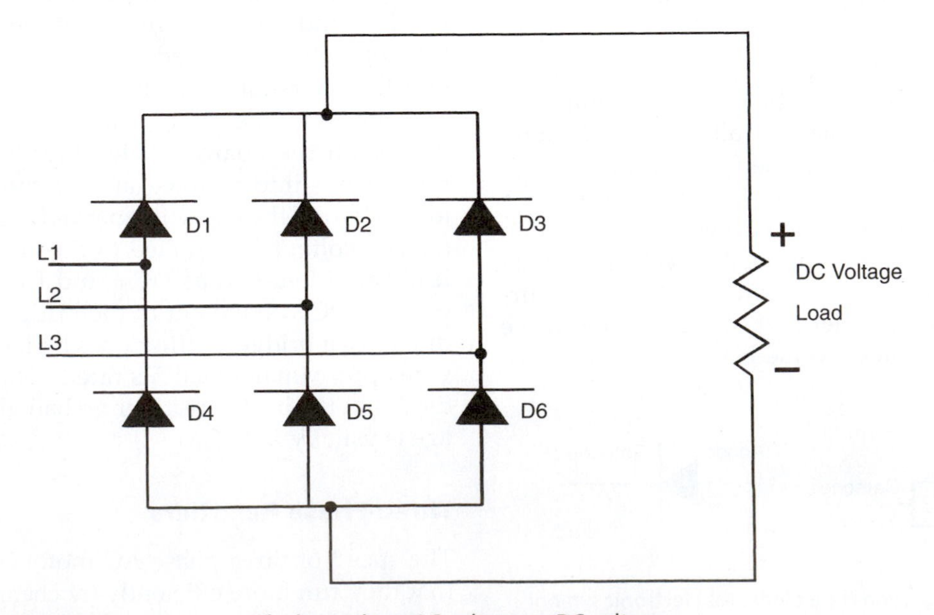

FIGURE 12-67 A six-diode bridge used to rectify three-phase AC voltage to DC voltage.

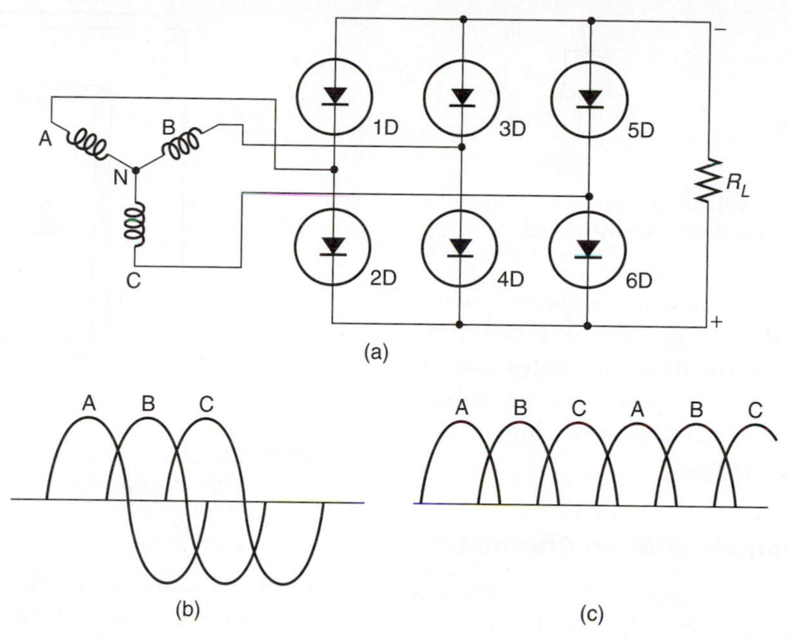

FIGURE 12-68 (a) Electrical diagram of a three-phase bridge rectifier connected to the secondary winding of a three-phase transformer. (b) Three-phase input sine waves. (c) Six half-waves for the DC output.

overlap, the DC voltage does not have a chance to get to the zero voltage point; thus, the average DC output voltage is very high.

A three-phase full-wave bridge rectifier is used where the required amount of DC power is high and the transformer efficiency must be high. Since the output waveforms of the half-waves overlap, they provide a *low ripple percentage.* In this circuit, the output ripple is six times the input frequency. Since the ripple percentage is low, the output DC voltage can be used without much filtering. This type of rectifier is compatible with transformers that are wye- or delta-connected.

Testing Diodes

One of the tasks you will perform as a technician is to test diodes to see whether they are operating correctly. One way to do this is to apply AC voltage to the input of the diode circuit and test for DC voltage at the output of the diode circuit. If the DC voltage at the power supply is half of what it is rated for in a four-diode bridge rectifier circuit, you can suspect that one of the diode pairs has one or both diodes opened. If this occurs, you can turn off all voltage to the diodes and use an ohmmeter to test each diode to determine which one is faulty.

When you are testing the diodes with an ohmmeter, it is important that all power to the diode circuit be turned off. You should remember from earlier chapters that the ohmmeter uses an internal battery as a DC voltage source. Since you know that the diode can be tested for forward bias and reverse bias with a DC voltage source, you can use the ohmmeter as the voltage source and the meter to test for high resistance and low resistance through the diode junction. Figure 12-69a shows an example of putting the positive ohmmeter terminal on the anode of the diode and the negative ohmmeter terminal on the cathode of the diode to cause the diode to go into forward bias. During this test the diode is forward-biased, and the ohmmeter should measure low resistance. When the ohmmeter leads are reversed as in Figure 12-69b, so that the negative meter lead is connected to the anode of the diode and the positive meter lead is connected to the anode of the diode, the diode is reverse-biased. When the diode is reverse-biased, the ohmmeter should measure infinite

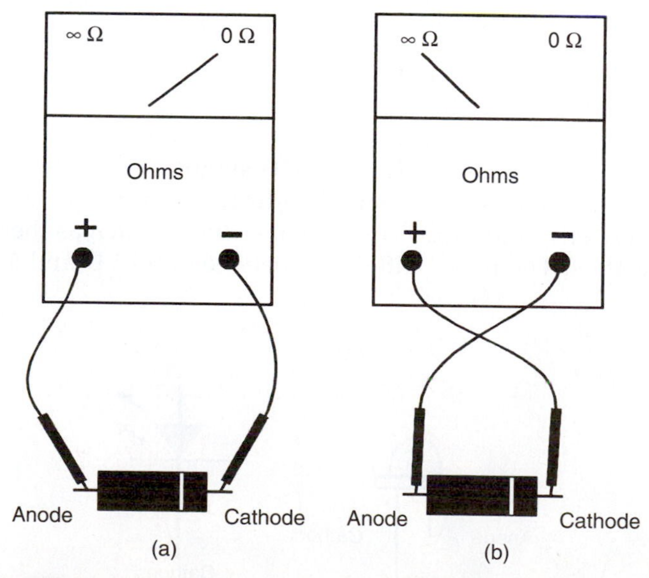

FIGURE 12-69 (a) Using the battery in an ohmmeter to forward bias a diode. The diode should have low resistance during this test. (b) Using the battery in the ohmmeter to reverse bias a diode. The diode should have high resistance during this test.

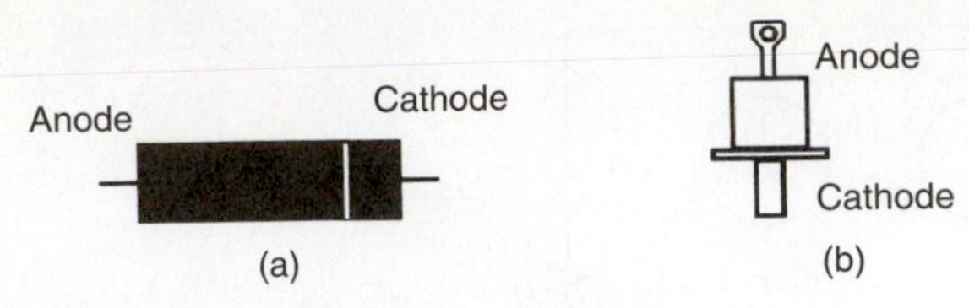

FIGURE 12-70 (a) Typical diode with anode and cathode identified. (b) Power diode with anode and cathode identified.

(∞) resistance. If the diode indicates high resistance when it is reversed-biased and low resistance when it is forward-biased, it is good. If the diode indicates low resistance during both the forward and the reverse bias tests, the diode is shorted. If the diode shows high resistance during both tests, it is opened.

Identifying Diode Terminals with an Ohmmeter

Since the ohmmeter can be used to determine if a diode is good or faulty, the same test can be used to determine which lead of a diode is the anode and which lead is the cathode. When you use the ohmmeter to test the diode for forward and reverse bias, you should notice that the ohmmeter indicates high resistance when the diode is reverse-biased and low resistance when the diode is forward-biased. When the meter indicates low resistance, you know the diode is forward-biased, so the positive lead is touching the anode and the negative lead is touching the cathode. This method will work when you are testing any diode. If the diode has markings, you can identify the cathode end of the diode because it has a strip around it. Figure 12-70 shows two types of diodes, with the anode and cathode identified in each.

12.21 LIGHT-EMITTING DIODES

A **light-emitting diode (LED)** is a special diode that is used as an indicator because it gives off light when current flows through it. Figure 12-71a shows a typical LED, and Figure 12-71b shows its symbol. In this figure the LED looks like a small indicator lamp. You will likely encounter LEDs on various controls such as thermostats. The major difference between an LED and an incandescent lamp is that the LED does not have a filament, so it can provide thousands of hours of operation without failing.

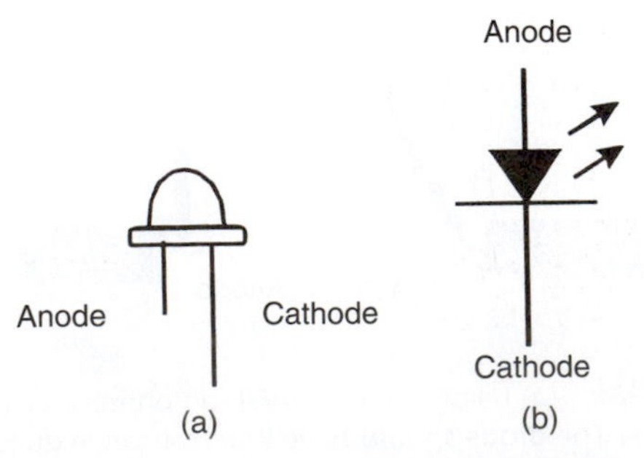

FIGURE 12-71 (a) A typical light-emitting diode (LED). Notice that the LED looks like a small indicator lamp. (b) The symbol for an LED.

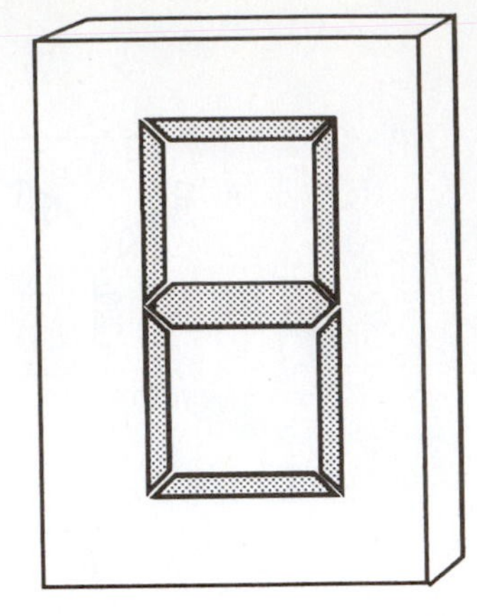

FIGURE 12-72 LEDs used in seven-segment displays. The seven-segment display can display numbers 0–9.

The LED must be connected in a circuit in forward bias. Since the typical LED requires approximately 20 mA to illuminate, it is usually connected in series with a 600-Ω to 800-Ω resistor that limits the current. If the voltage is higher, the size of the resistance will be larger. Figure 12-72 shows a set of seven LEDs connected to provide a seven-segment display. The seven-segment display has the capability to display all numbers 0–9. Seven-segment displays are used to display numbers on thermostats and other electronic devices.

LEDs are also used in opto isolation circuits where larger field voltages are isolated from smaller computer signals. The LED is encapsulated with a phototransistor. When the input signal is generated, it causes current to flow through the LED, and light from the LED shines on the phototransistor, which goes into conduction and passes the signal on to the computer.

12.22 PNP AND NPN TRANSISTORS

Two pieces of N-type material can be joined with a single piece of P-type material to form an NPN transistor. A PNP transistor can be formed by joining two pieces of P-type material with a single piece of N-type material. Figure 12-73 shows the electronic symbol and the material for both the PNP and the NPN transistors. The terminals of the transistor are identified as the *emitter, collector,* and *base.* The base is the middle terminal, and the emitter is the terminal identified by the arrowhead. In the transistor, the electrons are emitted from the emitter and collected by the collector, and the amount of electron flow between them is controlled by the voltage applied to the base.

Operation of a Transistor

A transistor can be connected in a circuit to perform a wide variety of functions. The simplest function that a

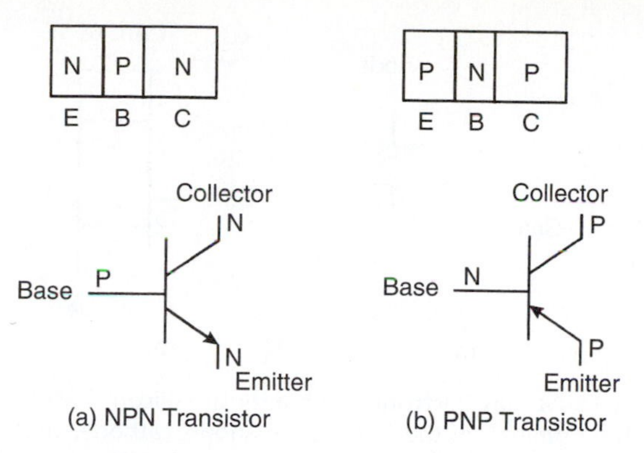

FIGURE 12-73 (a) Electronic symbol and diagram for an NPN transistor. (b) Electronic symbol and diagram for a PNP transistor.

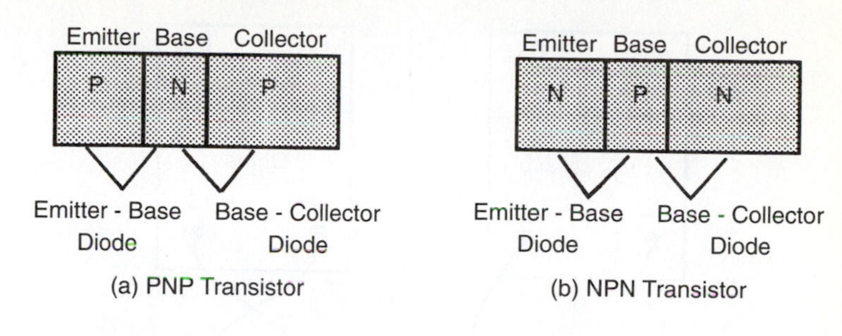

FIGURE 12-75 (a) A PNP transistor shown as its equivalent PN junctions. Each PN junction can be tested for forward bias and reverse bias. (b) An NPN transistor shown as its equivalent PN junctions.

transistor can provide is to act as an electronic switch. Figure 12-74 shows a transistor used as an electronic switch. In this type of application, the base terminal of the transistor acts like the coil of a relay, and the emitter–collector circuit acts like the contacts of a relay. When the proper amount and polarity of DC voltage is applied to the base of the transistor, the resistance between the collector and emitter is relatively low, which allows the maximum amount of circuit current to flow through the emitter–collector circuit. The transistor at this time acts like a relay with its coil energized.

When the polarity of the voltage on the base of the transistor is reversed, the emitter–collector circuit is changed to a high-resistance circuit, which acts like a relay when the coil is de-energized. The major advantage of the transistor is that a very small amount of voltage or current on the base can switch the transistor from high resistance to low resistance. Since the base current is very small and the current flowing through the collector is very large, the transistor is called an *amplifier*. Transistors are used in a variety of applications, including motor protection circuits, inverters, and variable-frequency motor drives.

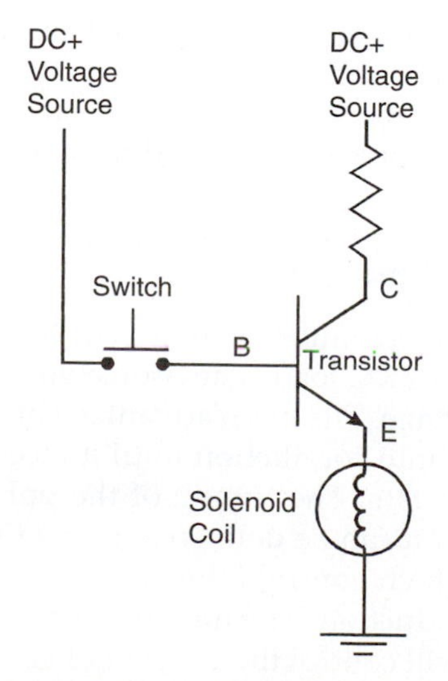

FIGURE 12-74 A transistor used as an electronic switch.

Figure 12-75 shows a PNP transistor and an NPN transistor as two PN diode circuits. The equivalent diode circuits are shown with each transistor to give you an idea of how the two junctions work together inside each transistor. Since each transistor is made from two PN junctions, each junction can be tested just like a single-junction diode for forward bias (low resistance) and reverse bias (high resistance). If you must work on a number of systems that have electronic circuits, you may purchase a commercial-type transistor tester that allows you to test the transistor either in the circuit or out of the circuit.

Typical Transistors

You will be able to identify transistors by their shape. Small transistors are used for switching control circuits, and larger transistors are mounted to heat sinks so that they can easily transfer heat. Figure 12-76 shows examples of types of transistors.

Troubleshooting Transistors

Transistors can be tested by checking each P and N junction for front-to-back resistance. Figure 12-77 shows these tests. You can see that each time the battery in the ohmmeter forward biases a PN junction, the resistance is low, and when the battery reverse biases the junction, the meter indicates high resistance. You can test a transistor in

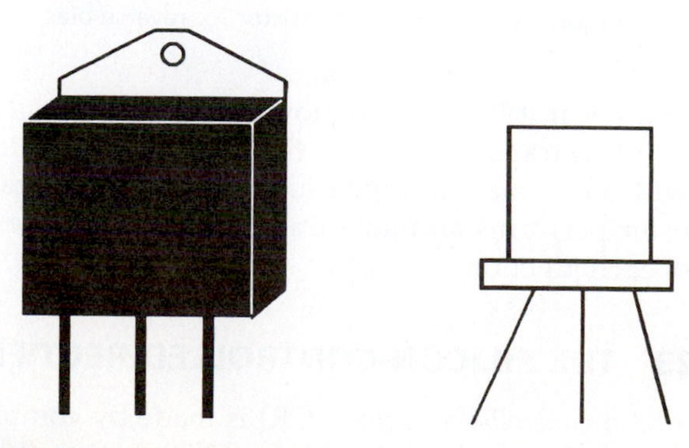

FIGURE 12-76 Typical transistors used for power control and switching.

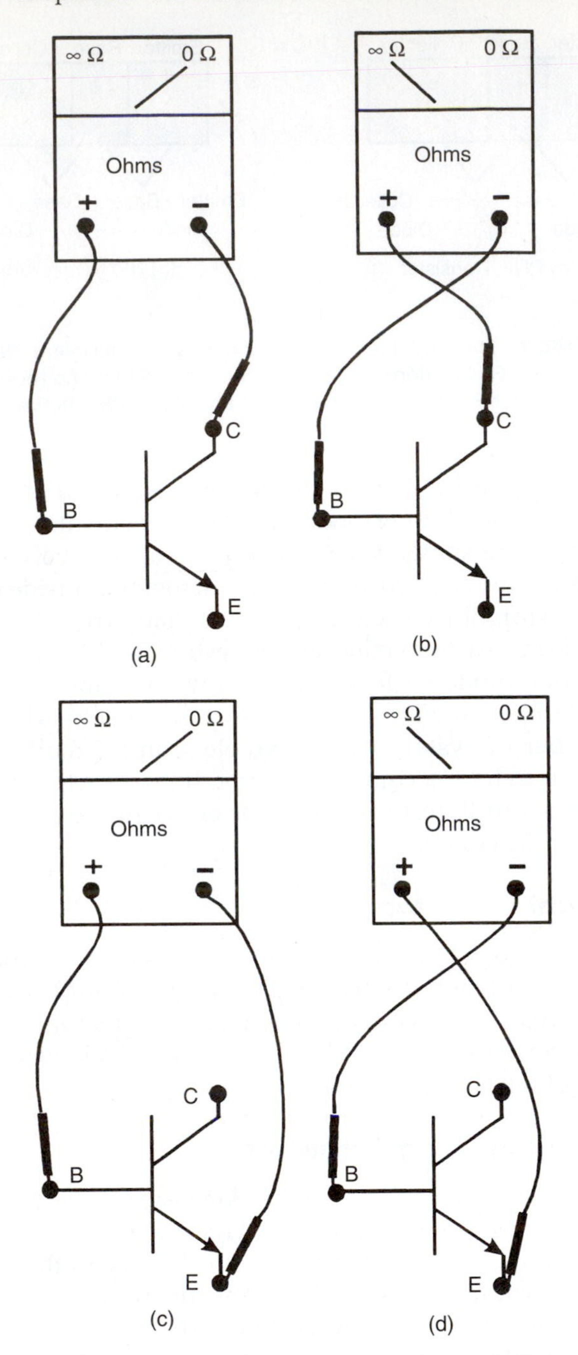

FIGURE 12-77 (a) Testing the base–collector junction of an NPN transistor for forward bias. (b) Testing the base–collector junction of an NPN transistor for reverse bias. (c) Testing the base–emitter junction of an NPN transistor for forward bias. (d) Testing the base–emitter junction of an NPN transistor for reverse bias.

this manner if it has been removed from the circuit. You can also test transistors while they are connected in circuit with a commercial-type transistor tester. The transistor tester performs similar front-to-back resistance tests across each junction.

12.23 THE SILICON-CONTROLLED RECTIFIER

The *silicon-controlled rectifier* (SCR) is made by combining four PN sections of material. Figure 12-78a shows the PN-material for the SCR, and Figure 12-78b shows its electronic

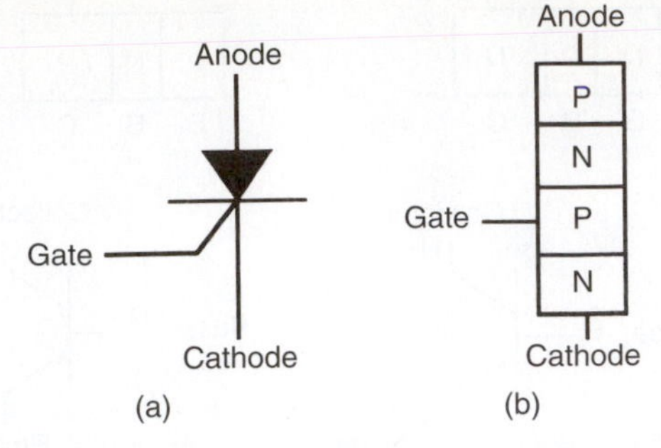

FIGURE 12-78 (a) Electronic symbol for the silicon-controlled rectifier. The terminals of the SCR are the anode, cathode, and gate. (b) P-type material and N-type material in an SCR. The P-type material and N-type material are combined to make a PNPN junction.

symbol. The terminals on the SCR are identified as anode, cathode, and gate. Since the SCR is basically a diode that is controlled by a gate, its symbol uses the arrow from the rectifier diode that you studied earlier in this chapter. When the SCR is turned on, it can conduct large amounts of DC voltage and current (over 1000 V and 1000 A) through its anode and cathode. The major difference between the SCR and the junction diode is that the junction diode is always able to pass current in one direction when the diode is forward-biased. The SCR is forward-biased by applying positive voltage to its anode and negative voltage to its cathode. At this point the SCR still has high resistance at its anode–cathode junction. If a positive voltage pulse is applied to the SCR gate, the SCR's anode–cathode junction will have low resistance, and the SCR will be in conduction. When the pulse is removed from the gate of the SCR, it will remain in conduction because positive current that comes through the anode will replace the voltage the gate provided. The only way to turn the SCR off is to provide reverse bias voltage to the anode–cathode or to reduce the current flowing through the anode–cathode to zero. You should remember that the AC sine wave has zero voltage right before it provides the negative half of its waveform. This means that if the SCR is powered with AC voltage, the SCR will be turned off when the AC waveform goes through 0 V and then to its negative half-cycle. When the AC voltage waveform goes positive again, a gate pulse can be provided and the SCR can go into conduction again. The gate provides a pulse that is used to cause the SCR to go into conduction.

Operation of the SCR

Figure 12-79 shows an SCR connected in a circuit to control voltage to a DC load. The source voltage for this circuit is AC voltage. The main advantage of the SCR is that it will not go into conduction until it receives a pulse of voltage to its gate. The timing of the pulse can be controlled so that it can be delivered at any time during the half-cycle, which controls the amount of time the SCR will be in conduction. The amount of time the SCR is in conduction will control the amount of current that flows through the SCR to its load. If the SCR is turned on

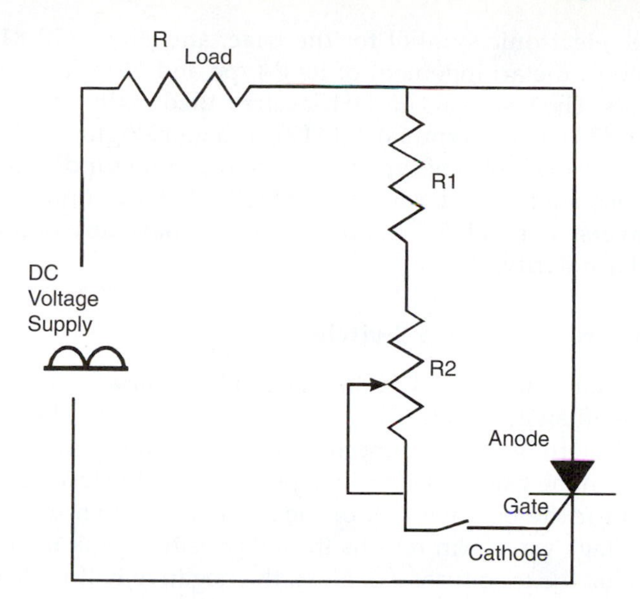

FIGURE 12-79 An SCR in a circuit controlling DC voltage and current to a load.

immediately during each half-cycle, it will conduct all the half-wave DC voltage just like a normal diode rectifier. If the gate delays the point at which the SCR turns on and goes into conduction at the 45-degree point of the half-wave, the amount of voltage and current the SCR conducts will be 50% of the fully applied voltage.

The other important feature of the SCR is that it will go into conduction only when its anode and cathode are forward-biased. This means that if the supply voltage is AC, the SCR can go into conduction only during the positive half-cycle of the AC voltage. When the negative half-cycle occurs, the anode and cathode will be reverse-biased, and no current will flow. This means that the SCR will automatically be turned off when the negative half of the AC sine wave occurs. Since the positive half of the AC sine wave occurs for 180 degrees, the SCR can provide control of only 0–180 degrees of the total 360-degree AC sine wave.

It is also important to understand that since the turn-off point of the SCR is fixed to the point where the sine wave begins to go negative, the SCR can be controlled only by adjusting the point where it is turned on. The point where the SCR is turned on and goes into conduction is called the *firing angle*. If the SCR is turned on at the 10-degree point in the AC sine wave, its firing angle is 10 degrees. If the SCR is turned on at the 45-degree point, its firing angle is 45 degrees. The number of degrees the SCR remains in conduction is called the *conduction angle*. If the SCR is turned on at the 10-degree point, its conduction angle will be 190 degrees, which is the remainder of the 180 degrees of the positive half of the AC sine wave.

Controlling the SCR

Figure 12-80 shows an SCR in a circuit with a unijunction transistor (UJT, a transistor with only one junction)

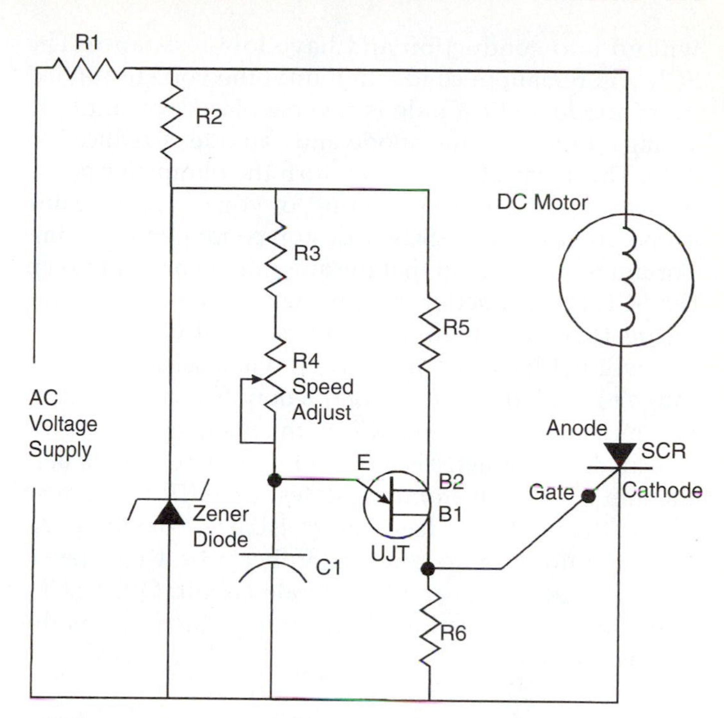

FIGURE 12-80 An SCR used to control a DC motor. A UJT is connected to the gate of the SCR to control its firing angle.

connected to its gate. The load in this circuit is a DC motor. This type of DC motor is often used as a small control motor. The circuit is powered by AC voltage, and the variable resistor in the oscillator (capacitor–resistor) circuit sets the timing for the pulse that is used to energize the gate of the SCR. You should notice that a diode rectifier provides pulsing DC voltage for the capacitor, which charges to set the timing for the pulse that comes from the UJT. Since this DC voltage comes from the original AC supply voltage, it will have the same timing relationship as the original sine wave. This means adjusting the pulse from the UJT to turn on the SCR gate at just the right time to control the firing angle of the SCR from 0 to 180 degrees. In reality, the firing angle is usually controlled from 0 to 90 degrees, which gives sufficient range of control to adjust the output DC voltage that is sent to the DC motor. You should remember that the speed of the DC motor can be controlled by adjusting the voltage sent to the armature and field. The load in this circuit could also be any other DC-powered load.

Testing the SCR

You will probably need to test an SCR to determine whether it is good or faulty. Since the SCR is made of PN junctions, you can use forward-bias and reverse-bias tests to determine whether it is good or faulty. In this test you should put the positive probe on the anode and the negative probe on the cathode. At this point the ohmmeter will still indicate that the SCR has high resistance. If you use a jumper wire and connect positive voltage from the anode to the gate, you should notice that the SCR will go into conduction and have low resistance. The SCR will remain in conduction until the voltage applied to its anode and cathode is reverse-biased, or until the voltage applied to the anode and cathode is reduced to zero. This means that you can turn the ohmmeter polarity switch to the opposite setting, or you can remove one of the probes and the SCR will stop conducting. It is important to understand that the amount of current to keep the SCR in conduction is approximately 4–6 mA. This means that some high-impedance digital volt/ohmmeters will not have enough current when set to the ohms range to keep the SCR in conduction. If this is the case, you may need to test the SCR with an analog ohmmeter. The analog ohmmeter is a type of ohmmeter with a needle and scale. You should also test the SCR for reverse bias to ensure that it has high resistance. Sometimes an SCR will not go into conduction because it has developed an open in its anode–cathode circuit. Other SCRs may stay in conduction at all times, which means the SCR is shorted.

12.24 THE TRIAC

A *triac* is basically two SCRs that have been connected back to back in parallel so that one of the SCRs will conduct the positive part of an AC signal and the other will conduct the negative part of an AC signal. As you know, the SCR can control voltage and current in only one direction, which means that it is limited to DC circuits when it is used by itself. Since the triac acts like two SCRs that are connected inverse parallel, one section of the triac can control the positive half of the AC voltage and the other section of the triac can control the negative half of the AC voltage. Figure 12-81a shows the electronic symbol for the triac, and Figure 12-81b shows the arrangement of its P-type and N-type materials. The terminals of the triac are called main terminal 1 (MT1), main terminal 2 (MT2), and gate. Figure 12-81c shows examples of typical triac semiconductor devices. Since the triac is basically two SCRs that are connected inverse parallel, MT1 and MT2 do not have any particular polarity.

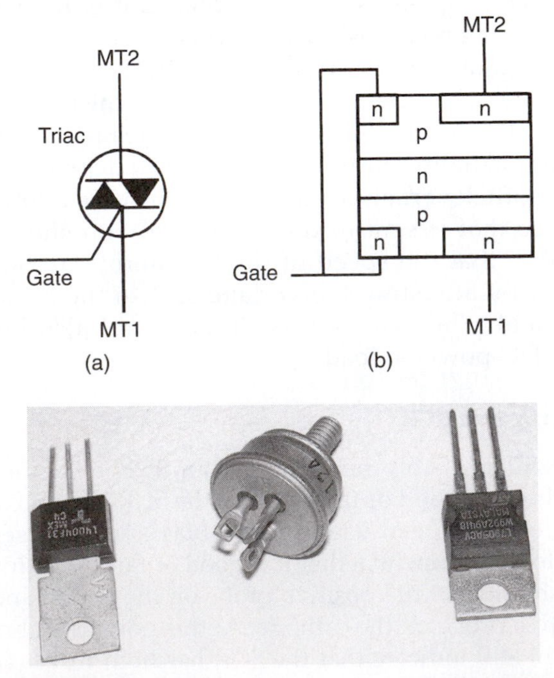

FIGURE 12-81 (a) Electronic symbol for the triac. (b) P-type and N-type materials in the triac. (c) Typical triac semiconductor devices.

Using a Triac as a Switch

A triac can be used in an electrical circuit as a simple on–off switch. In this type of application, the MT1 and MT2 terminals are connected in series with the AC load. When the gate gets a positive pulse signal, the triac turns on for the positive half of the AC cycle. When the AC voltage waveform returns from its positive peak to zero volts, the triac turns off. Next, the negative half-cycle of the AC voltage waveform reaches the triac, and it receives a negative pulse on its gate and goes into conduction again.

This means that the triac looks as though it turns on and stays on when AC voltage is applied. The load connected to the triac will receive the full AC sine wave just as if it were connected to a simple single-pole switch. The major difference is that the triac switch can be used for millions of on–off switching cycles. The other advantage of the triac switch is that the gate pulse can be a very small amount of voltage and current. This allows the triac to be used in temperature control, where the temperature-sensing part of a thermostat can be a small solid-state sensing element called a *thermistor*. The sensing element can also be a very narrow strip of mercury in a glass bulb that is very accurate but can carry only a small amount of voltage or current.

Another useful switching application for a triac is the solid-state relay. Figure 12-82 shows a thermostat used to control a triac which energizes the oil cooler solenoid coil to ensure that the hydraulic oil stays at the correct temperature. The temperature-sensing element is connected to the gate terminal of the triac.

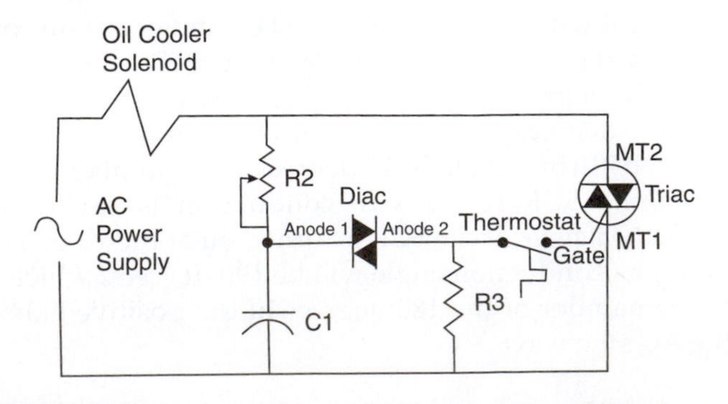

FIGURE 12-82 A triac used as a switch to turn on voltage to an oil cooler solenoid to ensure that the hydraulic oil stays cool. The triac receives its gate signal from a small amount of voltage that moves through the temperature-sensing element.

When the temperature increases, the thermostat sends a signal to the triac gate. The small amount of voltage flowing through the gate is sufficient to cause the triac to go into conduction and provide voltage to the solenoid valve.

Using a Triac for Variable Voltage Control

A triac can also be used in variable voltage control circuits, since it can be turned on anytime during the positive or negative half-cycle, in much the same way that the SCR is controlled for DC circuit applications. In this type of application, a diac is a device that is used to provide a positive and a negative pulse that can be delayed from 0 to 180 degrees to control the amount of current flowing through the triac. This type of circuit can be used to control the amount of current and voltage supplied to electric heating elements. This allows the amount of current and voltage to be controlled from zero to maximum by adjusting R2, which in turn allows the temperature to be controlled very accurately. A resistive-type temperature sensor can be used instead of the fixed R2 resistor, and the system will be able to hold the temperature for the nacelle or possibly the oil temperature control for a hydraulic system. Figure 12-83 is a diagram of a triac used to control an electric heating element powered by an AC voltage source. Notice that a variable resistor is connected with a capacitor to provide an oscillator pulse to the diac. Since the resistor and capacitor are connected to an AC voltage source, the pulse will be both positive and negative as the AC sine wave changes polarity. The triac is connected to the same AC voltage source so that the timing of the pulse from the diac to its gate will always be synchronized with the polarity of the voltage arriving at the main terminals of the triac.

Testing a Triac

Since a triac is made from P-type material and N-type material, it can be tested like other junction devices. The only point to remember is that since the triac is essentially two SCRs mounted inverse parallel to each other, some of the ohmmeter tests will not be affected by the polarity of the ohmmeter leads. In the first test of a triac, an ohmmeter should be used to test the continuity between MT1 and MT2. When no gate pulse is present, the resistance between these terminals should be infinite regardless of which ohmmeter probe is placed on each terminal. Figure 12-84 shows how the ohmmeter should be

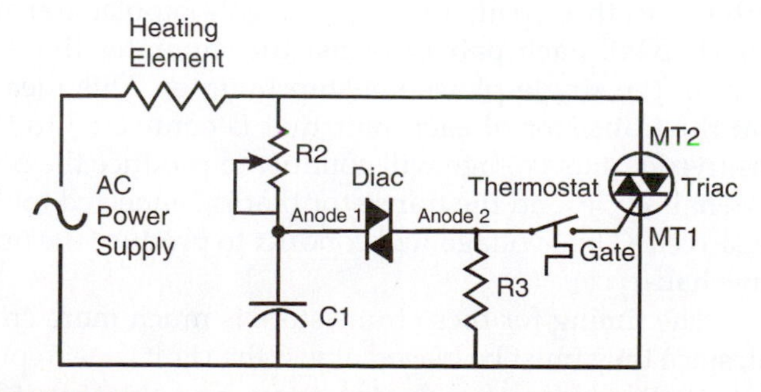

FIGURE 12-83 A triac used to control variable voltage to an AC electric heating element. Notice that a diac is used to provide a positive and a negative pulse to the triac gate.

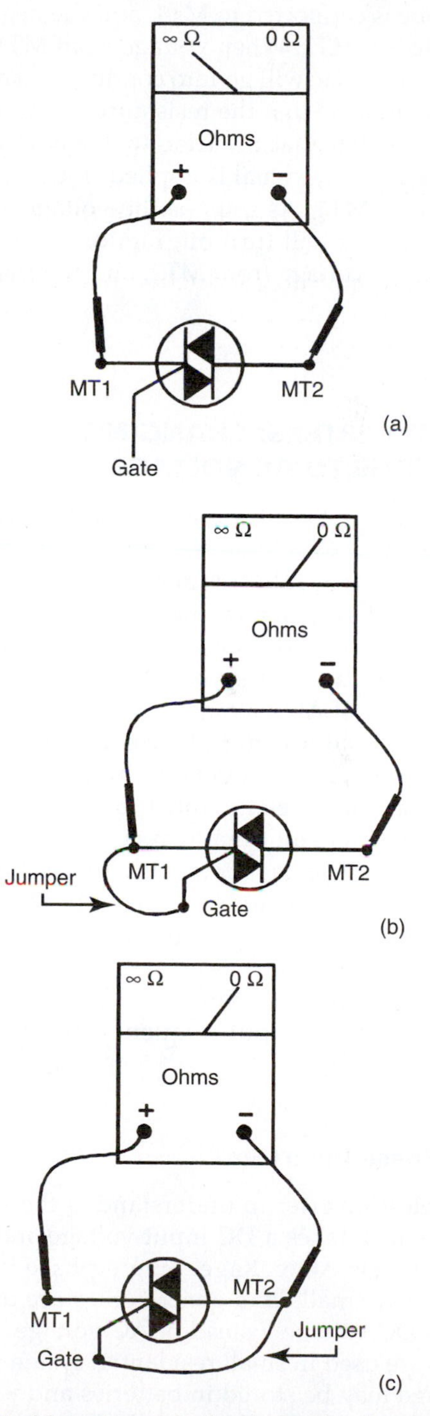

FIGURE 12-84 (a) Testing a triac by placing the ohmmeter positive probe on MT2 and the negative probe on MT1. When gate voltage is applied from the MT2 probe, the triac goes into conduction. (b) Testing the triac by placing the ohmmeter positive probe on MT1 and the negative probe on MT2. When gate voltage is applied from the MT2 probe, the triac goes into conduction. (c) The jumper is placed between MT2 and the gate.

connected to the triac. In Figure 12-84a you can see that the positive ohmmeter probe is connected to MT1 and the negative probe is connected to MT2. Since no voltage is applied to the gate, the resistance should be infinite. When voltage from MT2 is jumped to the gate, the triac will go into conduction and the ohmmeter will indicate low resistance.

In Figure 12-84b you can see that the positive ohmmeter probe is connected to MT1, and the negative probe is connected to MT2. When voltage from MT1 is jumped to the gate, the triac will go into conduction and the ohmmeter will indicate that the resistance is low. It is important to remember that the triac will stay in conduction only while the gate signal is applied from the same voltage source as MT1. As soon as the voltage source is removed, the triac will turn off. Figure 12-84c shows the gate receiving voltage from MT2, and the triac is in conduction.

12.25 INVERTERS: CHANGING DC VOLTAGE TO AC VOLTAGE

An **inverter** is a circuit designed specifically to change DC voltage to AC voltage. As you know, systems such as inverters, variable-frequency motor drives, and power control systems for small wind turbines convert AC power to DC and then convert the DC back to AC. This may sound like a strange way to provide an AC output voltage if AC voltage is the original supply, but in the case of the inverter, the frequency of the supply voltage may be any frequency, since the wind turbine turns at various speeds, and the output AC voltage needs to be a frequency that is very close to 60 Hz. In the case of power supplies for small wind turbines, the AC supply voltage needs to be changed to DC so it can be stored in a battery for use if the power supply is interrupted. Since the voltage is changed to DC and is stored in a battery, it must be changed back to AC to be usable in a home, where output frequency needs to be a constant 60 Hz.

Single-Phase Inverters

The simplest inverter to understand is the single-phase inverter, which takes a DC input voltage and converts it to single-phase AC voltage. Single-phase inverters are used in many small wind generators, since the generator produces DC voltage instead of AC voltage. These small inverters are used in small residential systems, where the DC voltage may be stored in batteries and used at times when the wind does not blow, or the batteries can be used as an interim storage component like a buffer to absorb short bursts of energy when a wind gust occurs or when the wind blows stronger than the power can be used at that time. The inverter simply ensures that the AC output voltage is exactly 60 Hz.

Using Transistors for a Six-Step Inverter

Figure 12-85 shows the electrical diagram of a single-phase inverter that uses four transistors. Since the transistors can be biased to any voltage between saturation and zero, the waveform of this type of inverter can be more complex than the traditional AC sine wave. The waveform shown in this figure is a *six-step* AC sine wave. Two of the transistors are used to produce the top (positive) part of the sine wave, and the remaining two transistors are used to produce the bottom (negative) part of the sine wave.

When the positive part of the sine wave is being produced, the transistors connected to the positive DC bus voltage are biased in three distinct steps. During the first step, the transistors are biased to approximately half-voltage for one-third of the period of the positive half-cycle. Then these transistors are biased to full voltage for the second third of the period of the positive half-cycle. The transistors are again biased at the half-voltage for the remaining third of the period. This sequence is repeated for the negative half-cycle. This means that the transistors that are connected to the negative DC bus are energized in three steps that are identical to the steps used to make the positive half-cycle.

Since six steps are required to make the positive and negative half-cycles of the AC sine wave, this type of inverter is called a *six-step inverter.* The AC voltage for this inverter is available at the terminals marked M1 and M2. Even though the AC sine wave from this inverter is developed from six steps, the motor or other loads see this voltage and react to it as though it was a traditional smooth AC sine wave. The timing for each sine wave is set so that the period of each is 16 ms, which means it will have a frequency of 60 Hz. The frequency can be adjusted by adjusting the period for each group of six steps.

Three-Phase Inverters

Three-phase inverters are much more efficient for industrial applications in which large amounts of voltage and current are required. The basic circuits and theory of operation are similar to those for the single-phase transistor inverter. Figure 12-86 is a diagram for a three-phase inverter with three pairs of insulated-gate bipolar transistors (IGBTs). Each pair of transistors operates like the pairs in the single-phase six-step inverter. This means that the transistor of each pair that is connected to the positive DC bus voltage will conduct to produce the positive half-cycle, and the transistor that is connected to the negative DC bus voltage will conduct to produce the negative half-cycle.

The timing for these transistors is much more critical, since they must be biased at just the right time to produce the six steps of each sine wave, and they must be synchronized with the biasing of the pairs for the other two phases so that all three phases will be produced in

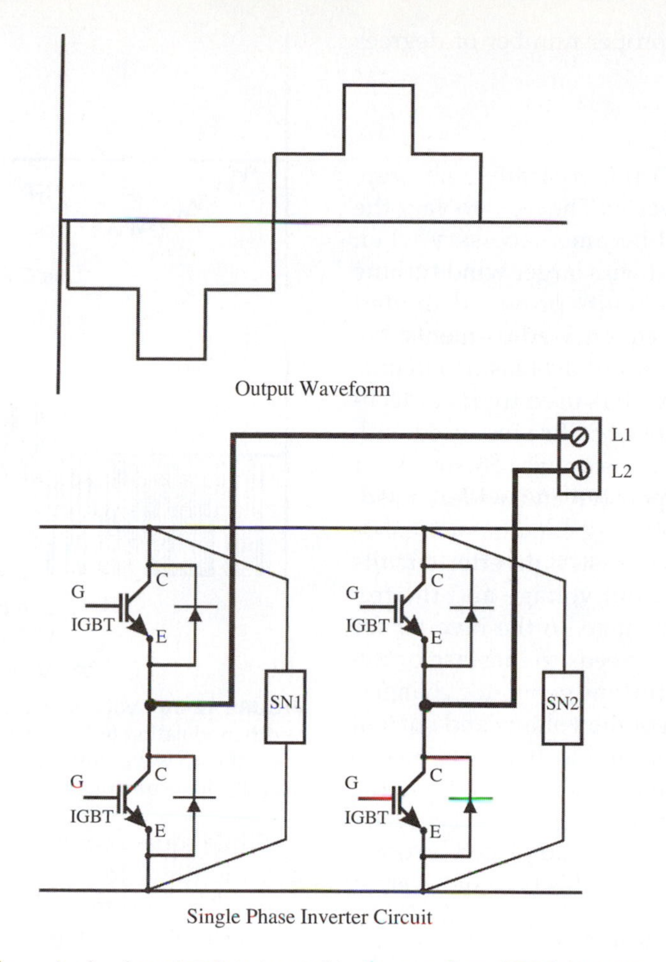

FIGURE 12-85 Electronic diagram for a single-phase insulated-gate bipolar transistor (IGBT) inverter with the output waveforms for the AC voltage.

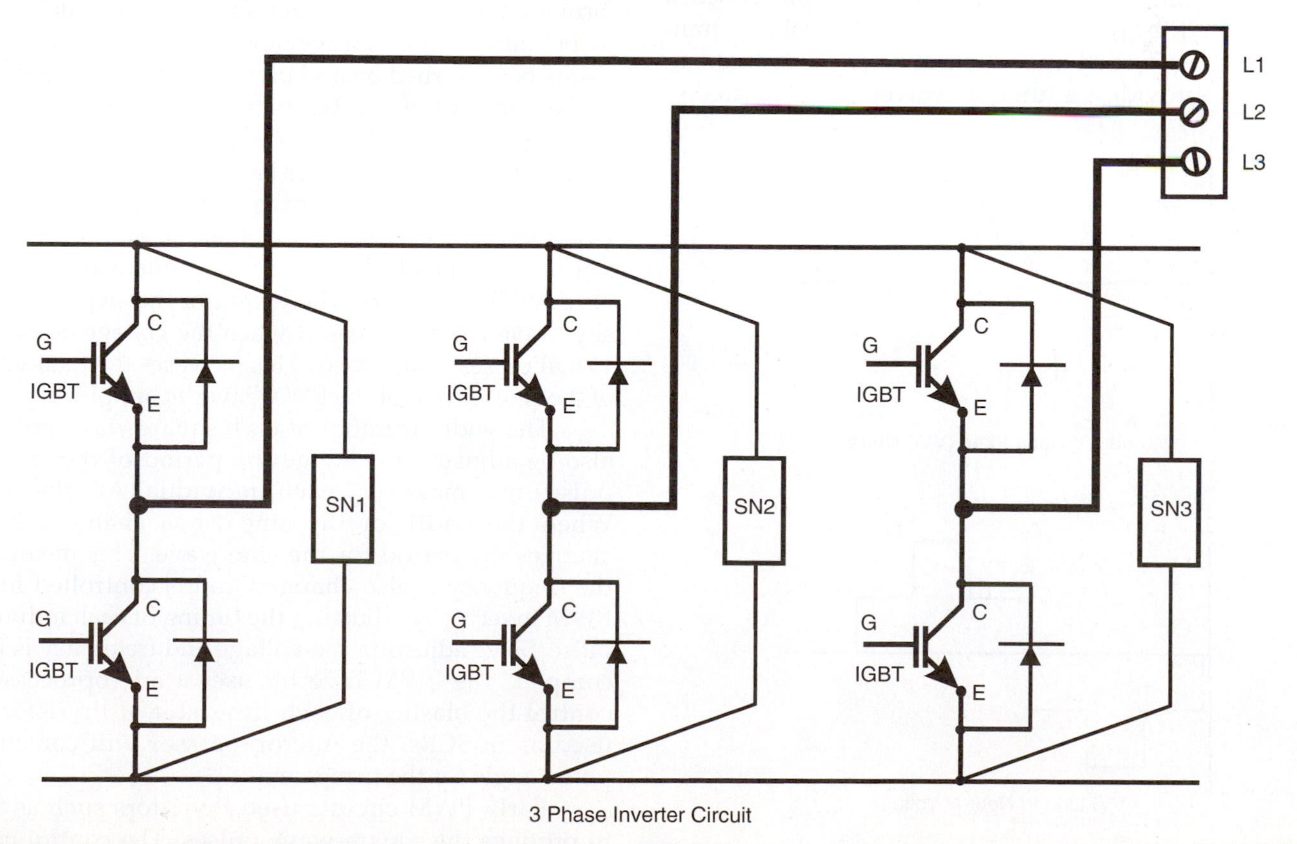

FIGURE 12-86 Electronic diagram for a three-phase inverter that uses six IGBTs.

the correct sequence with the proper number of degrees between each phase.

Variable-Voltage Inverters

A variable-voltage inverter (VVI) is basically a six-step, single-phase or three-phase inverter. The need to vary the amount of voltage to the load became necessary when these inverter circuits were used with larger wind turbine generators. Originally these circuits provided limited voltage and limited variable-frequency adjustments, because oscillators were used to control the biasing circuits. Also, many of the early VVI inverters used thyristor technology, which meant that groups of SCRs were used with chopper circuits to create the six-step waveform. After microprocessors became inexpensive and widely used, they were used to control the biasing circuits for transistor-type inverters to give these six-step inverter circuits the ability to adjust the amount of voltage and the frequency through a much wider range. In the inverter for the wind turbine, the frequency needs to increase or decrease as the shaft of the wind turbine generator changes.

Figure 12-87 is a diagram of the voltage and current waveforms for the VVI inverter. In this diagram you can see that the voltage is developed in six steps and that the resulting current looks like an AC sine wave. These are the waveforms that you would see if you placed an oscilloscope across any two terminals of this type of inverter.

Pulse-Width Modulation Inverters

Another method of providing variable-voltage and variable-frequency control for inverters is to use pulse-width modulation (PWM) control. This type of control uses transistors that are turned on and off at a variety of frequencies. This provides a unique waveform that makes multiple square wave cycles that are turned on and off at specific times to give the overall appearance of a sine wave. The outline of the waveform actually looks very similar to the six-step inverter signal. An example of this type of waveform is provided in Figure 12-88. In this diagram you can see that the overall appearance of the waveform is an AC sine wave. Each sine wave is actually made up of multiple square-wave pulses that are caused by transistors being turned on and off very rapidly. Since the bias of these transistors can be controlled, the amount of voltage for each square-wave pulse can be adjusted so that the entire group of square waves has the overall appearance of a sine wave. If you look at the voltage waveform for the PWM inverter, you will notice that the outline of the AC sine wave still looks like the six-step sine wave originally used in VVI inverters. The height of the steps of the AC sine wave is also increased when the voltage of the individual pulses is increased. This increases the total voltage of the sine wave that the PWM inverter supplies.

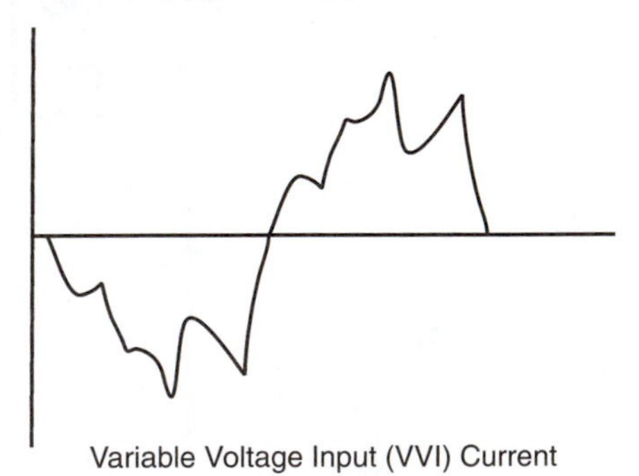

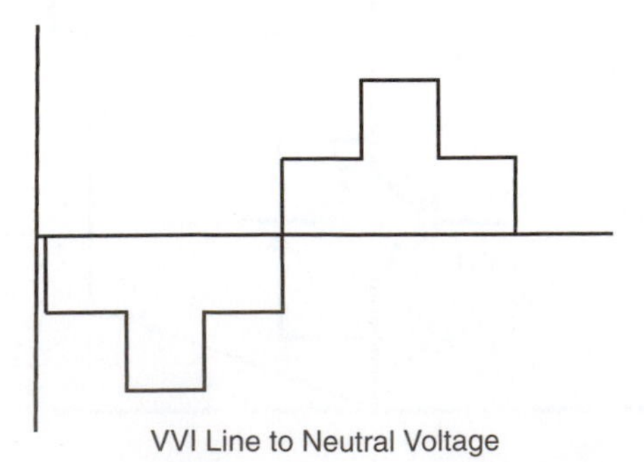

FIGURE 12-87 Voltage and current waveforms for the variable-voltage input (VVI) inverter.

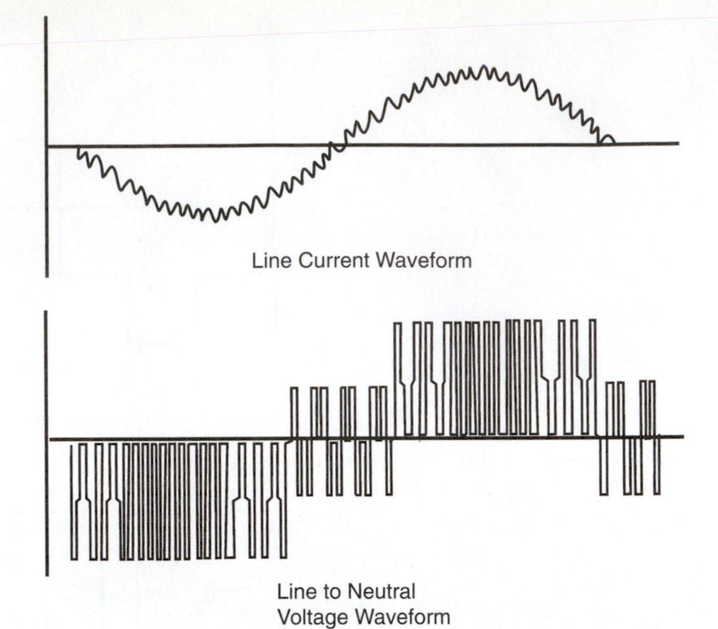

FIGURE 12-88 Voltage and current waveforms for the pulse-width modulation (PWM) inverter. Notice that the overall appearance of each waveform is an AC six-step sine wave and that it is actually made of a number of square-wave pulses.

The width (timing) of each square wave pulse can also be adjusted to change the period of the group of pulses that makes up each individual AC sine wave. When the width of the sine wave changes, it also changes the period for the sine wave. This means that the frequency is also changed and is controlled for the PWM inverter by adjusting the timing of each individual pulse. Since adjusting the voltage and frequency is fairly complex, the PWM inverter uses a microprocessor to control the biasing of each transistor. If thyristors are used as in SCRs, the microprocessor will control the phase angle for the firing circuit.

Early PWM circuits used thyristors such as SCRs to produce the square-wave pulses. The control circuit included triangular carrier waves to keep the circuit synchronized. This sawtooth waveform was sent to the oscillator circuit that controlled the firing angle for each thyristor. Today's PWM inverters use transistors, mainly because of their ability to be biased from zero to saturation and back to zero at much higher frequencies. Modern circuits more than likely use transistors for these circuits because they are now manufactured to handle larger currents, well in excess of 1500 A.

Current-Source Input Inverters

The current-source input (CSI) inverter produces a voltage waveform that looks more like an AC sine wave and a current waveform that looks similar to the original on/off square wave of the earliest inverters that cycled SCRs on and off in sequence. This type of inverter uses transistors to control the output voltage and current. The on-time and off-time of the transistor are adjusted to create a change in frequency for the inverter. The amplitude of each wave can also be adjusted to change the amount of voltage at the output. This means that the CSI inverter, like the previously discussed inverters, can adjust voltage and frequency for fixed-frequency applications or other applications that require variable voltage. Figure 12-89 shows the voltage and current waveforms for the CSI inverter.

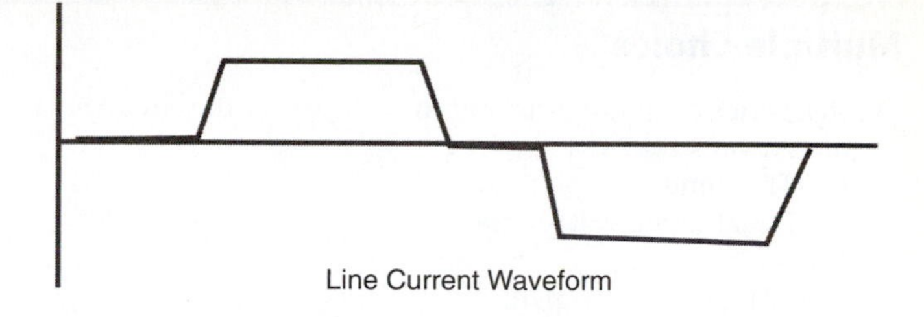

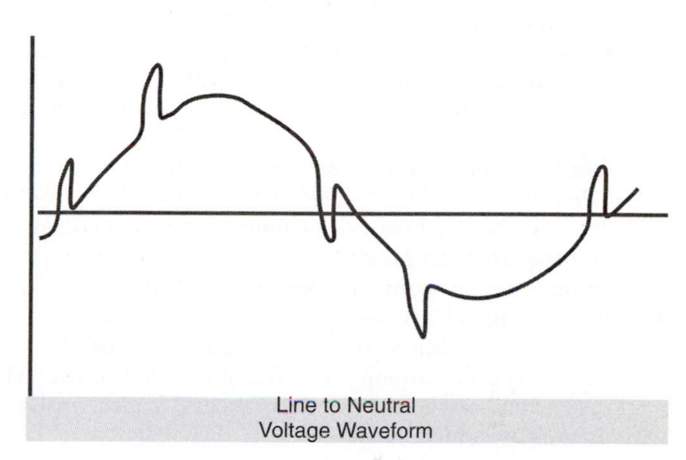

FIGURE 12-89 Voltage and current waveform for a current-source input (CSI) inverter.

Cycloconverters

A *cycloconverter* is a circuit designed to convert the frequency of AC voltage directly to another frequency of AC voltage without first converting the voltage to DC voltage. The history of this circuit dates back to the 1930s, when mercury arc rectifiers were used to control the frequency of railroad engines in Germany. The supply voltage for these original circuits was a fixed 50-Hz AC sine wave common in Europe. The train engines used low frequency (16.6 Hz), so their electric motors would turn slowly, creating a tremendous amount of torque. The input circuit for the cycloconverter uses a large transformer, and the output section uses thyristors to adjust the timing of the output stage, which allowed frequency to be changed.

Since modern electronic technology provides many ways to control voltage and frequency, the cycloconverter circuit is generally used in power transmission systems. In fact, many advantages are provided by converting the AC voltage to DC before the frequency is converted back to AC and adjusted for the output section. For example, when the AC input voltage is rectified to DC and filtered, all transient signals and voltage spikes are removed so that when the DC is converted back to AC, the output circuit is effectively isolated from the input.

Applications for Inverters

Inverters are seldom found as stand-alone circuits. You will normally find them used in conjunction with other circuits such as rectifiers and filter circuits in power supplies that provide a source for the DC voltage the inverter needs. You may also find the inverter as an integral part of the DC-to-DC converter circuitry in many types of DC power supplies. The major use of inverters in wind turbines today is to convert variable frequency from the generator to the fixed frequency required for AC use.

Questions

1. Explain why the voltage at each branch of a parallel circuit is the same as the supply voltage.
2. Explain the symbols M, k, m, and μ and provide an example of how each is used in units of measure.
3. Discuss the advantage of connecting switches in series with a load.
4. Explain the term *infinite resistance*.
5. List two things you should be aware of when making resistance measurements with an ohmmeter.
6. Explain why you should set the voltmeter or ammeter to the highest setting when you first make a measurement of an unknown voltage or current.
7. Which two materials are combined to make P-type and N-type materials?
8. Explain the operation of a diode (PN) junction and show the input AC waveform and the output DC waveform.
9. Identify the terminals of a transistor and explain its operation.
10. Explain the operation of a power electronic frequency converter (inverter).

Multiple Choice

1. In a series circuit, the current in each part of the circuit is
   a. Always zero
   b. The same
   c. Equal to the voltage
2. Milliamps are
   a. 1/1,000 of an ampere
   b. 1/1,000,000 of an ampere
   c. 1 million amperes
3. If a temperature control switch, high-pressure switch, and oil-level switch are connected in series with a hydraulic motor and the oil-level switch is opened because of low oil level, the motor will
   a. Still run, because two of the other switches are still closed
   b. Not be affected, because no other loads are connected to it in series
   c. Stop running, because current flow will be zero
4. Voltage in a parallel circuit
   a. Increases as additional resistors are added in parallel
   b. Decreases as additional resistors are added in parallel
   c. Stays the same across parallel branches as additional resistors are added in parallel
5. Resistance in a parallel circuit
   a. Increases as additional resistors are added in parallel
   b. Decreases as additional resistors are added in parallel
   c. May increase or decrease when resistors are added in parallel, depending on their size
6. An inverter
   a. Converts DC voltage to AC voltage and back to DC voltage
   b. Converts AC voltage to DC voltage and back to AC voltage
   c. Converts DC frequency to a new DC frequency
7. When one branch circuit of a multiple-branch parallel circuit develops an open, voltage in other branch circuits
   a. Decreases to zero
   b. Increases because fewer resistors are using up the voltage
   c. Stays the same as the supply voltage
8. When you are measuring AC current with a clamp-on ammeter, you should
   a. Open the claws of the meter and place them around the wire where you want to measure the current.
   b. Create an open in the circuit and place the meter probes so that the meter is in series with the load.
   c. Turn off all power to the circuit so that you don't get shocked.
9. A rectifier diode
   a. Converts DC voltage to AC voltage
   b. Converts AC voltage to DC voltage
   c. Can be used as a seven-segment display for numbers
10. When a voltage measurement is made, you should
    a. Place the meter probes across the load.
    b. Turn off all power to the circuit during the voltage reading so you don't get shocked.
    c. Create an open in the circuit and place the meter probes so that the meter is in series with the load.

GLOSSARY

AC voltage Voltages that are alternating in polarity, reversing positive and negative over time.

Acceleration The rate at which an object changes its velocity.

Active stall control The control that adjusts the blade pitch, with the goal of taking control of the blade and rotor and bringing them to a complete stop if necessary.

Air density A measure of how much mass is contained in a cubic foot of air (density = mass/volume). Air density is affected by the amount of moisture (humidity) in the air, the air pressure, and the temperature. Air becomes more dense when it is cooled down, when its pressure increases, and when there is more humidity in the air.

Air pressure The force exerted by air, whether compressed or unconfined, on any surface in contact with it.

Airfoil The turbine blade on a wind generator.

Alternating current (AC) A circuit in which the electrons travel in one direction and then change direction and travel in the other direction. This movement by the electrons can best be shown by its characteristic waveform.

Alternator An energy converter (generator) that converts mechanical (rotational) energy to AC current.

American National Standards Institute (ANSI) A U.S. organization that recommends standards for many products in various industries.

American Wind Energy Association (AWEA) The national trade association for the wind energy industry, which includes wind power project developers, equipment suppliers, services providers, parts manufacturers, utilities, researchers, and others involved in the wind industry. The AWEA has established some standards for wind turbine hardware performance.

Amplifier An electronic circuit that takes a small electrical signal and converts it into a larger electrical signal that can power a hydraulic valve or other electrical control device.

Amps (amperes) Unit for electrical current.

Analog control A signal that can be varied between 0 and 100% and causes the system to respond between 0 and 100%; if the signal is from a sensor, it can measure any variable value from 0 to 100%.

Angle of attack The angle at which the wind strikes the turbine.

Apparent power Power caused by multiplying voltage times current when the current is caused by inductive reactance, capacitive reactance, and resistance. The amount of apparent power will appear to be larger than the true power because the true power has a phase shift caused by capacitance or inductance. True power, by comparison, calculates only current caused by pure resistance. *Note:* Apparent power has the same value as true power when a circuit contains only resistance.

Armature The moving part of a magnetic component such as a relay, solenoid, motor, or generator.

Asynchronous AC generator An AC generator in which the rotational speed of the rotor is not proportional to the system frequency. This type of generator is actually an induction motor that has its shaft turned by the wind turbine blades, and it produces a large amount of voltage and current.

Automatic descent Equipment designed especially for an automatic and quick descent from the top of a wind turbine.

Axial force Direction forces that are applied to a shaft.

Ball bearing A friction-reducing bearing made up of a ring-shaped track containing freely revolving hard metal balls against which a rotating shaft or other part turns.

Barometric pressure The pressure created by the atmosphere pushing down on the earth. It is expressed in inches of mercury (in Hg) or in pounds per square inch (psi).

Bearing race The inner ring and outer ring of a ball bearing, generally made from hardened steel.

Bevel gear A type of gear in which the teeth are cut at an angle and are beveled. This type of gear allows the teeth to mesh and transfer power with less wear on the teeth, and these gears are quieter than other types of gears. The bevel gear can run at much higher speeds than other gears with less wear. Also known as a miter gear.

Betz's law Theory by Albert Betz that states that the best efficiency a wind turbine can achieve is approximately 59%.

Blade pitch The position or location of the turbine blade with respect to the wind turbine. The position or pitch of the blade can be adjusted on some wind turbines from 0 to 90 degrees.

Blade pitch control A control that is used to change the orientation of the blades to moderate the speed of the turbine.

Blade tip speed The speed at which the blade moves through the air as it rotates. This speed is measured at the very tip of the blade. The tip speed of the blade is much faster than the speed at the point where the blade is attached to the rotor.

Boundary-layer wind Wind that is closest to the surface of the earth and exists in the lowest part of the atmosphere. The behavior of boundary-layer wind is directly influenced by its contact with the earth's surface. Boundary-layer wind usually responds to changes in flow, velocity, temperature, and moisture near the earth surface within an hour or less.

British Wind Energy Association (BWEA) The trade and professional body for the wind power and marine renewable energy industries in the United Kingdom, and the UK's leading renewable energy trade association.

Brownout A reduction or cutback in electrical power, especially as a result of a shortage, a mechanical failure, or overuse by consumers. When the demand for electricity begins to outpace the supply on any given day, the voltage on the entire system will begin to become lower or droop.

Brushes Carbon conductors that maintain an electrical connection between stationary and moving parts of a motor, generator, or other type of rotating machine; also known as slip rings.

Bushing A bearing designed for sliding loads; also called a plain bearing.

Capacitive load A type of load that includes current flowing through capacitors which is used to correct power factor. Capacitive loads are also used in some switch-mode computer power supplies.

Carabineers Devices used to establish secure anchorage points and connections very quickly while a technician is climbing. They also allow technicians to tie themselves off safely in a working position when necessary.

Cartridge valves Hydraulic valves that can provide directional control, pressure control, or flow control. Cartridge valves are mounted in a manifold block rather than as standalone valves.

Check valves Valves that block fluid flow in one direction and force all fluid flow to move through the flow-control valve.

Closed-loop control Control that utilizes feedback to measure the actual system operating parameter being controlled.

Coil A part of a relay or solenoid that is made by tightly winding a long piece of wire into loops. When current flows through the wire, it creates a strong magnetic field in the coil that can be used to open and close contacts in a relay, or change the position of a solenoid valve from open to closed or vice versa. Coils of wire are also used in motor windings and transformers.

Commutator segments Bars of a commutator that are connected electrically to the two ends of one or more armature windings.

Contactor One or more sets of contacts that are opened or closed by a magnetic coil. The contactor is similar to a relay but normally is larger. By definition, contactor contacts are rated for more than 15 A.

Continuity test A test that measures the amount of resistance. The test is for two states, very low resistance that indicates a fuse is good or a switch is closed, or very high resistance that indicates a fuse or switch is open.

Cube-type wind turbine A wind turbine that is designed so that it has a cowling around the five wind turbine blades it uses to produce energy from the wind.

Current The flow of electrons.

DC voltage The electrical force (potential) that has polarity and causes the flow of electrons in one direction.

Department of Energy (DOE) Cabinet-level department of the U.S.government concerned with federal policies regarding energy. The DOE has a number of standards in place that cover some parts of the wind turbine, such as the electrical system and towers.

Diode A two-terminal, solid-state semiconductor that allows current flow in one direction.

Direct Dir current (DC) The flow of electricity in a circuit, which flows in only one direction.

Directional control A hydraulic control valve that changes the direction of fluid flow through a valve. This valve can change the direction of fluid flow to cause a cylinder to extend or retract.

Double helical gear Two helical gears of opposite hands on a common gear blank, with a space between the meeting of the two gears.

Doubly fed induction generator A type of rotating field AC generator that has DC voltage supplied to the rotor and AC voltage supplied to the stator. This type of machine is sometimes called a double-excited induction generator.

Downwind horizontal-axis wind turbine A downwind tower design in which the wind blows over the generator and then through the blades.

Downwind turbine A type of turbine with its rotor on the back or lee side of the turbine.

Drag The force that opposes the motion of the airfoil as it moves through the air.

Drive gear The gear where energy is put into the gear system; the driven gear is the gear that the drive gear moves.

Drive train The components of a wind turbine that consist of the turbine blades, rotor, low-speed shaft, gearbox, high-speed shaft, and generator.

Driven gear The gear that the drive gear moves in a gear system.

Electrical demand The maximum amount of electrical energy that is being consumed at a given time. It can be measured by the total amount of electricity that is needed to supply all consumers over a period of an hour, day, week, month, or year.

Electrical grid An interconnected network for delivering electricity from suppliers to consumers.

Electromagnet A magnet that is produced by passing current through a coil of wire.

Electromagnetic interference (EMI) Electrical interference in the radio-frequency range that is caused by magnetic induction. This interference interrupts radio and other communication transmissions.

Electronic inverter An electronic circuit that converts AC electricity output from a wind turbine that is at any frequency and converts it to DC voltage and then back to AC voltage at exactly 60 Hz. Some versions of the inverter are used to convert DC voltage output from a wind turbine generator to AC voltage at exactly 60 Hz. *See also* Power electronic frequency converter.

EPROM (erasable programmable read-only memory) An electronic memory chip in which a program is stored and that does not require a battery backup. It can only be erased using a special EPROM writer.

Error The difference between the setpoint and the feedback signal when compared in the summing junction in a closed-loop control system.

European Wind Turbine Certification (EWTC) The European Wind Turbine Certification (EWTC) is the certification body for wind turbines manufactured and/or installed in Europe. Their standards are specific for wind turbines and applications used in Europe. A similar certification body in the United Kingdom is the British Wind Energy Association (BWEA).

Evacuation The removal of persons or things from an endangered area.

Fail-safe system A secondary system that ensures continued operation even if the primary system fails.

Federal Aviation Administration (FAA) The division of the U.S. Department of Transportation that inspects and rates civilian aircraft and pilots, enforces the rules of air safety, and installs and maintains air navigation and traffic control facilities.

Feedback A signal from a sensor to the summing junction of a closed-loop system.

Feedback sensor A sensor that measures a process variable and sends the signal back to the summing junction. An example is a speed sensor.

Finite-element analysis (FEA) A type of computer program that uses large mathematical algorithms to test the complex design of the turbine blade in harvesting wind energy.

Flicker power A short-lived voltage variation in the electrical grid which might cause a load such as an incandescent light to flicker.

Flow control A hydraulic valve that controls the amount of flow through a hydraulic valve. This control allows for minor adjustments to be made in various parts of a system to balance the speed of its moving parts.

Flux lines Invisible magnetic lines of force that emanate from the south pole to the north pole of a magnet.

Force A unit of mass being accelerated. Force is calculated using the equation $F = ma$.

Frequency The number of periodic cycles per unit of time.

Furling Occurs when the blade's pitch is changed to a point where it begins to stall. It produces less torque to the shaft so the turbine blades will begin to slow down, just as if the wind speed became lower.

Fuse A device designed as a one-time protection against overcurrent or short-circuit current. The fusible link melts and causes an open circuit when its current rating is exceeded.

Fuse disconnect A special switch designed to provide a disconnect for electrical equipment as well as to provide a mounting for fuses. When the disconnect is in the open position, the fuses will be isolated from the source of power, so they can be removed and replaced or tested.

Gear backlash The small amount of space that remains after two teeth mesh, which is normal and necessary for the gears to mesh.

Gear ratio When a number of gears are put together in a transmission or gearbox, the number of teeth in each gear will create a ratio.

Gearbox The protective casing for a system of gears which helps modify the speed of blade rotation.

Generator A device for converting energy into electrical power.

Gigawatt A unit of power equivalent to 1000 megawatts or 1 million kilowatts.

Gin pole A tilting-type tower arrangement in which an additional pole is placed on the base of the main wind turbine tower pole at a 90-degree angle. When the main wind turbine pole is moved into the base, it will be located parallel to the ground and the right angle gin pole will be pointing upward. A winch or vehicle can be used to pull right-angle arm of the gin pole downward, which will cause the main wind turbine pole to move into an erect vertical position.

Gradient wind Wind that blows at a constant speed and flows parallel to curved isobars immediately above the earth's surface, where friction from irregularities such as mountains, trees, etc., cause changes in the flow.

Green technology A wide variety of technologies that involve developing energy that produces the minimal amount of pollution. Green technology today includes energy developed from wind, solar, bio fuels, geothermal sources, and other evolving technologies.

Grid An integrated network of electricity distribution, usually covering a large area.

Grid companies (Gridco's) Companies that manage the grid function, which is for interconnection and routing of electricity through hardware cables so that no area has a brownout or a blackout due to insufficient power.

Ground The negative side of a circuit.

Ground-fault protection A system that continually checks the current that a generator produces and ensures that it is all going into the grid system. This type of system will also detect irregular voltage or currents that may be fed back into the system.

Grounding A method of connecting metal in a wind turbine to a ground (rod) stake that is pushed into the earth to a depth of 8 ft. Grounding makes everything connected to the ground rod have the electrical potential of the earth.

Guy lattice tower A tower for wind turbines or cellular telephone tower that is constructed as an open structure of free-standing steel that is connected at tie points. The geometry of the tie points creates triangles which make the tower stronger.

Guy wire A tensioned cable designed to add stability to a structure. One end of the cable is attached to the structure, and the other is anchored to the ground at a distance from the structure's base.

Helical gear A type of gear that allows the driver and the driven shafts to meet parallel, like the spur gear, or at 90 degrees to each other.

Herringbone gear A gear that is similar to a double helical gear in that it consists of two helical gears, but it is different in that the two gears are cut so that the teeth are nearly touching. Herringbone gears are more expensive to manufacture than other types of gears, but the design of the teeth allows the load to be dispersed over twice the area, because two sets of teeth are used at the same time.

High-speed shaft The shaft in a horizontal-axis wind turbine that connects the generator to the gearbox. The high-speed shaft for a horizontal-axis wind turbine turns at approximately 1500 rpm.

Horizontal-axis wind turbines Wind turbines in which the axis of the rotor's rotation is parallel to the wind stream and the ground. Horizontal-axis wind turbines can be further classified into upwind and downwind turbines.

Hydraulic amplifier An electronic circuit that takes a small electrical signal and converts it into a larger electrical signal that powers a hydraulic valve.

Hydraulic control A control that uses hydraulic oil under pressure to extend and retract cylinders and cause hydraulic motor shafts to rotate. Hydraulic controls are also used to adjust the pitch of the turbine blades by rotating them at the point where they are attached to the rotors hub, and to rotate the complete nacelle to ensure that the blades are positioned into or out of the wind at different times.

Hydraulic pump A pump designed specifically to pump hydraulic fluid. Typically, the pump is a positive-displacement pump which moves a constant amount of fluid for every rotation of the pump shaft.

Hydraulic reservoir The tank of the hydraulic system that provides a place to store sufficient fluid for applications of the system.

Hydraulic servo valve A valve that allows the blade pitch positioning system and the yaw positioning system to control the wind turbine to adjust continually, to make the wind turbine as efficient as possible.

Image register A memory section of a programmable logic controller (PLC) that is designed specifically to keep track of the status (on or off) of each input and output. It is also called the image register or the image table.

Induced current Current created by electromagnetic induction. When current flows through a coil of wire and builds a magnetic field and then the current stops (alternating current passing through its zero point), the magnetic field collapses and the flux lines cut across a second coil and induce a current in the second coil.

Induction Creating current in a coil of wire by passing it through a magnetic field.

Induction generator A type of AC generator which generates alternating current.

Induction motor A motor designed to take advantage of the characteristics of the three-phase voltage it uses for power.

Inductive Load A load consisting of coils from transformers, motor coils, or solenoid coils.

Input module A module for a programmable logic controller (PLC) to which switches and other electrical devices are connected. The input module converts a higher-voltage signal from a switch to a lower-voltage signal that flows into the PLC. Each circuit in the input module typically has an indicator lamp that tells whether voltage is present at the input circuit.

Institute of Electrical and Electronics Engineers (IEEE) An engineering organization (pronounced I triple E) in the United States that develops, promotes, and reviews standards for the electronics, computer, and electric power industries.

Internal gear A gear having teeth cut on an inner cylindrical surface.

International Electrotechnical Commission (IEC) The world organization that prepares and publishes global standards for all electrical, electronic, and related technologies.

International Electrotechnical Commission (IEC) Wind Turbine Standards IEC standards that are set forth to demonstrate the power curve of the wind turbine, the annual energy performance curve, and the sound pressure levels. The strength and safety tests and the duration tests are administered as pass/fail.

International Organization for Standardization (ISO) The world's largest developer and publisher of international standards. The ISO enables a **consensus** to be reached on solutions that meet both the requirements of business and the broader needs of society.

Inverter An electronic circuit that changes DC voltage to AC voltage and provides a set frequency such as 60 Hz.

Journal The part of a bearing or shaft where the bearing rides.

Kilovolt-ampere (kVA) 1000 volt-amperes. A rating standard for transformers.

Kilowatt A unit of power, equal to 1000 watts.

Ladder diagram An industry standard for representing relay control logic. The name comes from the fact that the overall form of the diagram looks like a ladder. Also called a schematic diagram.

Ladder logic An industry standard for representing relay control logic. The name comes from the fact that the overall form of the diagram looks like a wooden ladder. It is also referred to as a ladder diagram.

Laminated steel rotor An AC induction motor that uses a laminated steel rotor. The core of the rotor is made of die-cast aluminum or copper in the shape of a cage. The caged rotor is sometimes called a squirrel-cage rotor because the overall shape of the rotor is the same shape as an exercise wheel for a hamster or squirrel.

Land breeze A breeze that blows from the land toward open water. Land breezes are more common in the fall and winter as the land begins to cool more than the water.

Lanyard Any of various small cords or ropes for securing or suspending something.

Lattice tower A three-legged tower that is put together in sections with the wind turbine mounted on its top section.

Leading edge The front area of a turbine blade surface that first comes into contact with the wind.

Lift A condition when air moves past the airfoil similar to an airplane wing and causes a low-pressure area on the top side of the blade, which allows it to rise.

Lift-to-drag ratio The ratio of the value of the lift to the value of the drag. Higher lift value and lower drag value provide higher lift-to-drag ratio. The higher the lift-to-drag ratio, the more efficient the turbine blade is at converting wind energy into shaft torque, which produces more electricity at the generator.

Light-emitting diode (LED) A two-lead solid-state PN junction device that produces a small amount of light when it is forward-biased.

Lightning arrestor A type of lightning protection system that absorbs the lightning strike and transfers the large amounts of high voltage and high current to the ground, where it is safely dissipated.

Load efficiency The rate at which the wind turbine blade can harvest wind energy and the ability of the generator to convert the torque to electricity. The load efficiency is measured in percent from 0 to 100%. The load efficiency is changed by adjusting the pitch of the blades to ensure that the maximum amount of wind energy is harvested and converted to electricity through the generator.

Low-speed shaft A shaft that makes turbine blades turn at somewhere between 10 and 20 rpm.

Low-voltage ride-through (LVRT) The safety system in place for when low-voltage occurs in the grid system to which a wind turbine is connected. The wind turbine must have the capability to ride through this condition. It does not matter if the wind turbine is causing the low voltage or if the low voltage exists in the grid to which the wind turbine is connected.

Magnet Material that has an attraction to iron or steel.

Magnetic coil A coil of wire that has electrical current passed through it and becomes a very strong electromagnet. Magnetic coils are used in solenoids and relays.

Mass The quantity of matter, determined by its weight.

Megawatt A unit of power, equal to 1 million watts.

Megger A specialized ohmmeter that uses approximately 600 to 2500 V to check cables, transformer windings, and motor windings for current leaking through their insulation.

Monopole tower A tower for a wind turbine that is a single pole assembled in sections and installed on site using a large crane. The towers are generally made of steel and bolted together.

Motor starter A large relay that has a single coil, multiple sets of contacts, and a set of overload heaters and contacts that are used to protect the motor against overcurrent. The current that is supplied to the motor flows through the contacts of the motor starter and also flows through the heater element of the overload. When the current becomes too high for the motor to handle safely, the heat that is produced by the heating elements is sufficient to cause the overload contacts to " trip" and open. The contacts of the overload are connected in series with the motor starter coil, and when the overloads trip and open the contacts, the current supplied to the coil is interrupted so that the motor starter drops out its main contacts. When the main contacts open, the motor stops running.

Nacelle An enclosure or housing for the generator, gearbox, and any other parts of the wind turbine that sits on top of the tower.

National Electrical Code (NEC) An electrical code written by the National Fire Protection Association (NFPA). Wind turbine electrical systems and wiring in the United States must conform to the NEC. This includes all wiring in panels, switches, and generator controls. The NEC must also be followed for correct grounding of all electrical components and circuits to ensure personnel safety and prevent electrical shock hazards.

National Renewable Energy Laboratory (NREL) Wind Turbine Certification The primary U.S. laboratory for renewable energy and energy efficiency research and development. NREL certification ensures that wind turbines of any particular type, defined by size, form, and use, are designed, manufactured, and have documentation to meet specific standards and other technical requirements.

Needle bearing A type of bearing that does not have an outside race but does have an inside race. This arrangement allows for the needles to be in contact with the shaft it is supporting. In this type of bearing the rollers are called needles because they have a much smaller diameter than a typical roller. The needle bearing has the ability to carry much larger radial loads at much higher speeds than other types of rolling-element bearings.

Net meter A meter that runs in one direction when the wind turbine is providing power to the grid and in the opposite direction when the application uses power from the grid rather than from the wind turbine.

Normally closed A condition of switch contacts or relay contacts in which they have low resistance. A normal condition is considered to exist when no power is applied to the circuit.

Normally open A condition of switch contacts or relay contacts in which they have high resistance. A normal condition is considered to exist when no power is applied to the circuit.

N-type material Semiconductor material in which a majority of carriers are electrons and have negative charge.

Occupational Safety and Health Administration (OSHA) The division of the U.S. Department of Labor that sets and enforces occupational health and safety rules.

Off-peak power Electrical power that is used during off-peak hours, such as late at night, to heat residential hot water and other electrical storage applications. The advantage of off-peak power is that it shifts electrical power usage to times when high-demand industrial and commercial loads are not consuming as much power as during peak times.

Offshore wind farm A wind turbine farm located at least 10 km (6 miles) away from land.

Open circuit A circuit that has a point in it with extremely high resistance, which stops the flow of electrical current. The open can be created intentionally with a switch or contacts, or it can be created for safety reasons by a fuse or circuit breaker. Other fault conditions such as broken wires or broken components can also cause the high-resistance open.

Open fuse A fuse that is bad because it has extremely high resistance and will not pass current. The open fuse is caused by a fault that causes high current which creates excessive heat through the fuse and causes its link to open.

Open loop A type of control loop in which the output signal is controlled directly by the operator; this type of loop does not use a feedback signal.

Open switch A switch that has high resistance between its contacts. When a switch is in the open condition, no current flows through it.

Overload A part of a motor starter that consists of a heater element and a set of normally closed contacts.

Output Power or force produced by a machine.

Output module An electronic circuit connected to a programmable logic controller (PLC). This module receives a low-voltage signal from the PLC and converts the signal to a higher-voltage signal that controls a motor starter coil, solenoid coil, or other electrical load.

Parallel circuit A circuit that has two or more loads connected so that each load receives the same voltage. If two or more switches are connected in parallel, one switch can be opened and current can continue to flow through the other switches if they are closed.

Passive pitch control A type of control that depends on the original design of the turbine blades to help slow the turbine down in the event that winds become too strong. It is also referred to as passive stall control.

Passive stall control A type of control that depends on the original design of the turbine blades to help slow the turbine down in the event that winds become too strong. It is also referred to as passive pitch control.

Peak electrical demand The largest amount of electrical energy that is needed during any hour at any time over a 24-hour period on any day of the week or month.

Peak performance When the output of the generator of a wind turbine is at or above its rated output.

Periodic maintenance Maintenance on a wind turbine consists of a series of visual inspections, small tests, and checks that are carried out on a schedule. The schedule for when the inspection should occur will be defined by the number of operating hours or by a calendar date, whichever comes first.

Permanent-magnet generator A low-speed generator that produces low-voltage AC that can be rectified and used for battery charging or that can be sent through an inverter and transformers to provide 60 Hz for immediate use or for grid-tied systems.

Pillow block A block designed to hold a bushing in place against all types of axial and radial loads.

Pilot pressure Pressure that can be created from the main system pressure through a pressure-reducing valve, or with a separate smaller hydraulic pump designed to create pilot pressure.

Planetary gears A set of gears or a gear system that consists of one very large elliptical outer gear that has one or more gears, called planet gears, which engage it on the outside. The planet gears revolve about a central gear located in the middle of the set, called a sun gear. The planetary gear set is used because the energy is dispersed over the planetary gears rather than being absorbed by a single gear. The name comes from the similarity to the way the planets rotate around the sun.

Power circuit The part of an electrical circuit that has motors or other power loads

Power curve A graph that shows the wind speed and the output power of a wind turbine over a range of wind speeds from zero to the maximum wind speed the wind turbine is designed for.

Power factor (PF) The ratio of true power (TP) to apparent power (AP). The formula for power factor is TP/AP, and since true power is always less than apparent power, the power factor is always less than 1.00. The closer the PF is to 1.00, the more efficient the circuit is and the less the losses are.

Power quality The condition of electrical power provided to a residence, industrial application, or the grid with regard to frequency and voltage level. Power quality standards require the voltage to remain at the proper frequency and voltage level to within a small percentage. Events such as voltage sags, impulses, harmonics, and phase imbalance are not tolerated, and the system causing these problems must either be corrected or disconnected.

Pressure control A hydraulic control that adjusts the pressure of a system by allowing a portion of the hydraulic fluid to return to the tank and relieve the pressure of the system.

Pressure gradient force One of the main forces acting on the air to make it move as wind. Generally, the pressure gradient force points from high-pressure zones toward low-pressure zones, which influences weather patterns to move from high-pressure areas toward low-pressure areas, and this movement or pressure differential is one of the things that makes the wind move. The amount of pressure difference will influence the speed the wind blows, causing the wind to blow harder when the pressure difference is greatest.

Prevailing winds Winds that blow predominantly from a single general direction over a particular point on the earth's surface.

Primary winding One of two transformer windings by which voltage is taken into the transformer. Supply voltage is connected to the primary transformer winding.

Process variable (PV) The variable that is being sensed by the process sensors, such as temperature, pressure, or flow.

Program mode A mode for a programmable logic controller (PLC) that does not execute the scan cycle.

Programmable logic controller (PLC) A solid-state control system that continually scans its user program. The controller has a user-programmable memory for storage of instructions to implement specific functions.

Project development plan A plan that includes things such as selecting the proper site for a wind turbine or wind turbine farm, identifying available wind resources and predicting energy output for various-sized wind turbines, identifying landowners and developing landowner agreements, identifying grid interconnection if necessary and utility companies that control the grid, identifying government agencies that control the country, state, county, or city regulations that will affect your selection.

Proportional valve A valve that allows the swash plate to be moved in very small increments from minimum to maximum, which causes the pump output flow to vary between 0 andf 100%.

P-type material Semiconductor material in which a majority of carriers are holes that have positive charge.

Public Utility Regulatory Policies Act (PURPA) A federal law passed in 1978 as part of the National Energy Act. This law requires utility companies to purchase power from independent providers, but the law does not determine the rate that should be paid for the power.

Rack A type of gear used to convert rotary motion into linear motion. When the shaft with the driver gear rotates, the teeth of the spur gear mesh with the teeth of the rack, which causes the rack to move in a linear motion. When the rotation of the drive shaft is reversed, the motion of the rack is also reduced.

Radial force A force that originates at the center of a bearing or shaft and moves outward, like the spoke of a wheel, to the outside of the bearing.

Reactive power Electrical power produced by current flow through a capacitor or inductor.

Rectifier A device (usually a diode) that conducts current in only one direction, thereby transforming alternating current (AC) to direct current (DC).

Relay An electrical device that consists of a single coil and one or more sets of normally open or normally closed contacts.

Rescue equipment Equipment used to rescue people in case of emergency.

Residential wind turbine A wind turbine system designed for personal use to offset the use of electricity from the power company. Most residential users consume between 1000 and 2000 kW per month, so the system may only be able to provide a portion of the total energy needed each month.

Resistance The opposition of a body or substance to current passing through it, resulting in a change of electrical energy into heat or another form of energy.

Resistive load An electrical load that has electrical resistance. When current passes through the resistance, it creates heat.

Return on investment (ROI) The amount of money or other things of value that becomes available after money is invested into a project. For example, if a wind project costs $100,000 to install and start up, the money that the investors receive from the energy the wind turbine produces over a specific period of time is the ROI.

Roller bearings A bearing made of solid steel rollers that look like small steel tubes. The roller bearing is made of an inner race, an outer race, and the rollers. A retainer ring ensures that the rollers stay on track inside the races.

Rotating magnetic field A magnetic field which changes direction at a constant angular rate.

Rotor A rotating part of an electrical or mechanical device that is connected to the generator by a shaft, or the rotating part of a wind turbine to which the blades are attached.

Rotor hub The hub or mechanical part of the wind turbine where the blades are attached. The rotor hub connects the blades to the low-speed shaft.

Run mode A mode for a programmable logic controller (PLC) in which the PLC actively scans inputs, solves logic, and turns outputs on or off based on the results of the logic circuits.

Safety brakes Brakes for a wind turbine that are mounted on a larger rotor and use discs that are activated by hydraulic pressure, which causes the discs to come into contact with the rotor to apply friction. The brakes are fail-safe in that they are always closed to stop the rotor, and they will automatically return to the closed position, where the friction will stop the rotor from turning, when they are not energized.

Safety harness A device of belts or restraints that hold a person or thing to prevent falling or injury.

Salt-water corrosion Gradual wearing away of material due to exposure to salt water.

SCADA (Supervisory Control And Data Acquisition) A data collection system that provides information such as wind velocity, wind direction, and the amount of electrical power that is being produced. When this data is gathered over a long period of time and stored in a computer data bank, it can be analyzed and very accurate projections can be developed.

Schematic (ladder) diagram A diagram that is designed to show the sequence of operation of an electrical control system.

Sea breeze A thermally produced breeze from a cool ocean surface onto adjoining warm land.

Secondary winding A transformer has two windings, a primary winding and a secondary winding. Voltage is applied to the primary winding, and voltage comes out the secondary winding. The amount of voltage at the secondary winding is proportional to the number of turns in the primary and secondary windings.

Self-excited generator A generator that requires field current to create armature voltage.

Sensor A device that detects or measures something and generates a corresponding electrical circuit to an input circuit or controller.

Separately excited generator A type of generator in which an outside voltage (called a battery voltage) is used to increase the current flow in the field, which in turn creates a stronger magnetic field. A separately excited generator can be used only where a second power source is available. The amount of voltage that is used to control the magnetic field is relatively small in comparison to the amount of voltage and current produced at the armature of the generator.

Series circuit A circuit that has only one path. Whenever there is an open anywhere in a series circuit, all current flow stops.

Series-parallel circuit A circuit that has components connected both in series and in parallel.

Service drop The wires that connect from a building to the main power lines.

Service platform A platform which allows individuals to work for extended periods at locations high above the ground. If any maintenance is required, the platform can be stopped at the correct height and the work can be completed. Service platforms make it safer and more comfortable for technicians to work on wind turbines at extreme heights.

Setpoint The setting for the maximum speed at which the blades can rotate.

Shut-down brakes On a wind turbine, the control system and the brake system operate together to shut down the brakes. When the wind speed is too low, the blades are stopped in what is called a shut-down. When the wind turbine is in a shut-down state, its brakes are set and the blades are locked so they cannot rotate.

Site preparation Preparation that includes preparing the concrete pad and its footings, as well as preparing all underground conduits for the electricity portion of the system, which must be located under the concrete pad. The exact location of the mounting bolts for the tower, and the location for electrical conduits and panels, must be precisely determined during this process.

Slip The difference between the rated speed and the actual speed of a motor. The more slip the motor has, the more torque it can create at its shaft so it can move larger loads.

Slip rings Mounted metal rings that are used so that current can be conducted through stationary brushes into or out of a rotating member.

Smart grid Update of the present-day grid in which information will flow in both directions on the grid. This data will be able to be recorded and analyzed so that predictions can be made about the times energy will be consumed in the future.

Spur gear A gear on which the teeth are cut straight and parallel with the shaft. The number of teeth in the drive gear and the number of teeth in the driven gear are selected to create a gear ratio that will cause a specific speed or force.

Squirrel-cage rotor An AC induction motor that uses a laminated steel rotor.

Stall A condition that occurs by increasing the angle at which the wind strikes the blades (angle of attack), reducing the induced drag (the drag associated with lift). This allows the blade to flex in higher winds and reduces the fatigue and wear associated with two-bladed turbines.

Stator The stationary part of a generator.

Step-down transformer A transformer in which the amount of secondary voltage is smaller than the primary voltage.

Step-up transformer A transformer in which the amount of secondary voltage is larger than the primary voltage.

Substation A small group of transformers located close to the location where wind turbines create electrical power. The transformers at these substations step up the voltage to a level so it is large enough to transmit to the place where it will be used.

Summing junction A point in a control algorithm where the setpoint is compared to the feedback signal (process-variable sensor).

Surge protection Protection designed to take any voltage that is above a safe value and route it directly to ground, where it can be dissipated harmlessly.

Switch An electrical device with one or more sets of contacts.

Synchronous AC generator A generator that produces AC voltage with a frequency proportional to the speed at which its field or armature is rotated by external means.

Teeter A back-and-forth motion that allows the blade of a wind turbine to flex rather than remain rigid, reducing wear and fatigue which may cause cracks to develop. Teetering is so-named because of its resemblance to the movement of a child's teeter-totter or see-saw.

Three-phase generator An AC generator that produces three distinct voltages that are 120 degrees apart. Each voltage is called a phase.

Three-phase voltage AC voltage that is supplied in three distinct circuits that are 120 degrees apart. Each voltage source is called a phase.

Tilting-type tower A tower designed with a tilting fixture that secures the pole at its base and allows the tower to be lifted into place until it is upright, after which it is secured with guy wires. When the wind turbine needs inspection or maintenance, the tower is lowered until the unit is on the ground, where it can be worked on safely and conveniently.

Tip speed The measured speed at the blade tip as it rotates through the air.

Torque The turning force applied to a shaft, tending to cause rotation. Torque is equal to the force applied multiplied by the radius through which it acts.

Trailing edge The part of a wind turbine blade that is last to contact the wind.

Transformer An electrical device with two coils (primary winding and secondary winding) that are located in close proximity. Voltage is applied to the primary winding and voltage is taken off the secondary winding.

Transistor A three-terminal electronic device that has an emitter, collector, and base.

Transmission companies (Transco's) Individual companies that manage the hardware and own the actual transmission lines, towers, transformers, and switchgear on the grid.

Troubleshooting Discovering and eliminating the cause of trouble in equipment.

Troubleshooting matrix A table that shows the sequence of operations in troubleshooting.

True power The voltage and current that flows only through resistive loads. Also called real power.

Turbine blades Flat objects connected to a center shaft that convert the push of the wind into a circular motion.

Turbulence Wind that does not blow in a straight line.

Type certification A type of certification used to ensure that wind turbines of any particular type (defined by size, form, and use) are designed, manufactured, and have documentation to meet specific standards and other technical requirements.

Underground transmission lines Underground lines that carry electric energy from one point to another in an electric power system.

Underwater cable Cable located under water that carries electric energy from one point to another in an electric power system.

Underwriters Laboratory (UL) An independent product-safety certification organization that tests products and writes standards for safety.

Uninterruptible power supply (UPS) A power supply that is backed up by one or more batteries and an inverter that converts DC voltage to AC voltage. An uninterruptible power supply takes in AC voltage and converts it to DC voltage which goes into the bank of batteries. When needed, the electricity is then converted back to AC electricity that has a frequency of 60 Hz.

Upwind horizontal-axis wind turbine A wind turbine in which the wind blows over the blades first and then over the generator and nacelle.

Upwind turbine A type of turbine in which the rotor faces the wind so that the wind blows over the blades and then over the generator nacelle.

Valve spool The movable part of a directional valve. The spool changes position and causes the flow through the directional valve to change.

Vane pump A hydraulic pump with vanes that slide inward and outward to adjust the amount of flow or volume the pump provides. The vanes are called a cartridge and can be changed out as a package.

Vertical-axis wind turbine A wind turbine that can harvest the wind regardless of the wind direction because its axis of rotation is perpendicular to the wind stream and the ground.

Volt-ampere (VA) A rating for a transformer found by multiplying the voltage times the current (amperes). This rating does not take into consideration any impedance or reactance losses, so it is an apparent power rating.

Voltage drop The voltage that occurs when current flows through a resistor or load in an electrical circuit. In a series circuit, all of the voltage drops across each resistor can be added, and their total will equal the supply voltage to the circuit.

Volts A unit of electrical force given to electrons in an electrical circuit.

Wind farm Two or more wind turbines at the same location.

Wind friction The friction between two layers or two currents of air that move in different directions or at different speeds. The friction that occurs at the boundary of these two currents is an indication of wind shear. Wind friction may cause winds to become turbulent or cause the direction of the wind to constantly change.

Wind map Information that provides precise locations to connect with wind data information. Other useful information can be provided in photographs or satellite images that will show all the necessary information about the location and the surrounding environment.

Wind turbine peak performance Any time the output of the wind turbine generator is at or above its rated output.

Wind turbulence A condition in which the wind does not blow in a straight line and does not blow continually at the same speed.

Wiring diagram An electrical diagram that shows the location of all the electrical components in a circuit and the wires that connect to them.

Worm gear A gear in which the teeth intersect at a 90-degree angle and which provides a high ratio of speed reduction in a small space. This type of gear is usually the quietest type of gear.

Wound rotor A rotor with wire coils.

Yaw The direction in which the nacelle is pointed with respect to where the blades are, compared to the direction in which the wind is blowing. Yaw control is a system designed to change the direction in which the wind turbine points and can point the wind turbine blades into the wind when wind speed is moderate or low, and move the nacelle so the wind turbine blades are not pointed into the wind when the wind speed is too high.

Yaw control The control that manipulates the rotation of a http://www.daviddarling.info/encyclopedia/H/AE_horizontal-axis_wind_turbine.html horizontal-axis wind turbine around its tower or vertical axis.

Yaw drive An electric motor that is used to rotate the yaw ring that changes the position of the nacelle yaw on a wind turbine.

Yaw mechanism A mechanism on a wind turbine that rotates the nacelle to control the direction in which the blades are pointing. The nacelle can be rotated so that the blades are directly into the wind when the turbine is trying to harvest the largest amounts of wind. It can also move the blades out of the wind when the winds are too strong and may damage the wind turbine. The yaw mechanism is a large ring that is mounted on the bottom of the nacelle.

ACRONYMS AND ABBREVIATIONS

AC	Alternating Current
ANSI	American National Standards Institute
AP	Apparent Power
ASSE	American Society of Safety Engineers
AWEA	American Wind Energy Association
BWEA	British Wind Energy Association
CEMF	Counter Electromotive Force
CRT	Cathode-Ray Tube
CSA	Canadian Standards Association
DC	Direct Current
DNH	Determination of No Hazard
DOE	Department of Energy
DOH	Determination of Hazard
EWI	Emergya Wind Technologies International
EWTC	European Wind Turbine Certificate
FAA	Federal Aviation Administration
FEA	Finite-Element Analysis
GW	Gigawatt
HAWT	Horizontal-Axis Wind Turbine
IEC	International Electrotechnical Commission
IEEE	Institute of Electrical and Electronics Engineers
IFR	Instrument Flight Rules
ISO	International Organization for Standardization
kVA	Kilovolt-Ampere
kW	Kilowatt
LED	Light-Emitting Diode
LVRT	Low-Voltage Ride-Through
mA	Milliampere
MW	Megawatt
NASA	National Aeronautics and Space Administration
NEC	National Electrical Code
NFPA	National Fire Protection Association
NREL	National Renewable Energy Laboratory
NSF	National Science Foundation,
OPEC	Organization of Petroleum Exporting Countries
OSHA	Occupational Safety and Health Administration
PEFC	Power Electronic Frequency Converter
PF	Power Factor
PLC	Programmable Logic Controller
PPE	Personal Protective Equipment
PSI	Pounds per Square Inch
PURPA	Public Utility Regulatory Policies Act
ROI	Return on Investment
RPM	Revolutions per Minute
SCADA	Supervisory Control and Data Acquisition
SWCC	Small Wind Certification Council
TP	True Power
UL	Underwriters Laboratories
USDA	United States Department of Agriculture
VA	Volt-Ampere
VAWT	Vertical-Axis Wind Turbine
VFR	Visual Flight Rules

INDEX

A

AC circuit
  impedance
    total opposition for inductive and resistive circuit calculating, 257
  power factor
    calculating in, 257
    change with inductors/capacitors, 258
  resistance and inductance in, 256
    inductive reactance, calculating in, 257
AC generator
  electromagnetic theory use, 122–123
  field, 123
  rotary-armature alternator, operation theory, 125
  sine wave, 123
  stationary-armature alternator, operation theory
    exciter current, 123
    programmable logic controller (PLC) use, 124
    separately excited machine, 124
  three-phase voltage waveform, 125
  voltage in stator rotor, 124–125
  wind turbine generator, 122
AC motors
  exploded view of, 119
  induction motor
    counter-EMF, 122
    EMF, 122
    operation, 121
    torque and slip use, 121–122
  laminated steel rotor, 120
  motor end plates, 120–121
  motor speeds at 60 Hz, 121
  squirrel-cage rotor for induction motor, 120
  stator, 119–120
  synchronous motor, 122
  three-phase AC voltage, characteristics
    rotating magnetic field, 118–119
  wound-rotor
    armature with two slip rings, 120
    operation, 121
Actuators types, 99
ADAMS. *See* Automatic Dynamic Analysis of Mechanical Systems (ADAMS)
AeroDyn subroutine library, 13
Air density, 66–67
Airfoil
  air moving, 46
  angle of attack, 46
  blade geometry, 54–55
  design of, 47
  drag coefficient for, 48
  furling, 47
  leading and trailing edge, 46
  lift, 46
  lift-to-drag ratio, 48
  number of blades, 55
  pitch and stall control, 47
  pitching, 47
  position of, 47
  stall-regulated concepts, 47
  tip speed, 48
Air pressure, 33
Alternating-current (AC), 110
  frequency, 111
  period of sine wave, 111
  sine wave moving through 360 degrees, 111
Alternator, 110
  DC voltage production use in, 128
American National Standards Institute (ANSI), 12–13, 191
American Society of Safety Engineers (ASSE), 191
American Wind Energy Association (AWEA), 12
Amperes, 239
Amplifier, 62–63
Analog-type voltmeter, 242
Anemometer and wind vane
  ultrasonic wind measurement instrument, 79–80
  wind direction indicator, 79
ANSI. *See* American National Standards Institute (ANSI)
Apparent power, 163
Arbor press, 143
Arc flash, 150, 157
Armature, 262
Asynchronous AC generators, 125
  controlling frequency of voltage from PEFC and IGBTs, 126
  slip in, 126
Automatic descent devices and block-and-tackle rescues, 193–194
Automatic Dynamic Analysis of Mechanical Systems (ADAMS), 13
AWEA. *See* American Wind Energy Association (AWEA)
Axial force, 139

B

Ball bearings
  bearing cage, 140
  bore of, 140
  double-row ball bearings, 140
  examples and structure of, 140
Bar magnet and flux lines, 112
Bearings
  axial force, 139
  journal in, 138–139
  maintaining of
    bearing and gear puller, 142–143
    bearing press, 143
  plain bearings (bushing)
    pillow block, 139
    split pillow block, 139–140
  radial force, 139
  rolling-element bearings
    ball bearing, 140
    bearing races in
    needle bearing, 141–142
    roller bearing, 140–141
  thrust bearings, 142
Betz's law, 65. *See also* Wind turbines
Bevel gear (miter gear), 132, 134–135
Blade pitch, 74
Blade service platform, 188–189
Brakes, 105
  drive train and yaw drive, 106
  dynamic braking, 107
  hydraulic brakes, 107
  hydraulic caliper brakes, 106
  hydraulic cylinder, 106
  mechanical parking brakes, 107
British Wind Energy Association (BWEA), 13
BWEA. *See* British Wind Energy Association (BWEA)

C

Cable installation work, 166
Canadian Standards Association (CSA), 12–13
Capacitive load, 163
Capacitors
  capacitive and resistive circuit
    impedance formula for, 256
    total opposition, calculation of, 256
  capacitive reactance, 255
    calculation, 256
  impedance, 255
  inductive reactance, 255
  operation of, 255
  resistance and capacitance in AC circuit, 255
Capacity margin, 151
Carabineers
  locking, 191
  nonlocking, 191
  self-locking, 191
Cartridge-type fuses, 268
Cartridge valve
  in hydraulic circuit, 100
  screw in manifold block, 100
Circuit breakers, 269. *See also* Fuses
  single-pole, two-pole and three-pole, 270
Clamp-on ammeter
  multiplier, 243
  transformer, 243
  used to measure current, 243
Climbing and safety equipment, 191
  carabineers
    locking, 191
    nonlocking, 191
    self-locking, 191
  inspection and maintenance, 192–193
  safety belt, 192
  safety harness, 192
  shock-absorbing lanyard, 192
Closed-loop pitch control
  amplifier, 62
  error, 62
  feedback sensor, 62

PLC and PID, 62–63
setpoint, 62
summing junction, 62
yaw control, 63
Closed switch, 249, 272
Coil, 245, 262–267, 277
magnet, magnetic field strength, 114
Commercial wind turbine systems, 204, 206
at marine sales and service facility, 205
railroad switching yard, 205
Compound generator
short and long shunt, 117
Computer control system for mid-sized wind turbine, 81
Conductors, 238
atomic structure for, 270
electrons flow in, 270
Contactor
difference between relay and, 264
NEMA ratings for, 264–265
Copper atom
electron flow in, 238–239
valence electron and current flow, 239
CSA. *See* Canadian Standards Association (CSA)
CSI inverter. *See* Current-source input (CSI) inverter
Cube-type wind turbine. *See* Towers for wind turbines
Current, 110, 238
large and small currents measuring, 243–244
meter probes placement, 244
Current-source input (CSI) inverter, 285
Cycloconverters, 285

D

Darrieus type VAWT, 6, 51, 184
for residential system or very large system, 7
Database of State Incentives for Renewables & Efficiency (DSIRE), 26
DC generators
amount of voltage and polarity, control voltage regulator, 117–118
armature, 114
wire cutting through flux lines of, 115
compound generator
short and long shunt, 117
DC shunt generator
armature shaft, 115
brushes and commutator, 115–116
end plate, 115
resistance, 115
stator, 115
field poles and flux lines of, 115
PEFC, 114–115
self-excited shunt generators, 117
flashing field, 116
separately excited shunt generators, 116
series generator, 117
DC voltage utility, 130
Delta-connected three-phase transformer, 259
Department of Energy (DOE) in United States, 11
Digital clamp-on ammeter, 243
Digital-type voltmeter, 242
Diode
with anode and cathode identified, 276
electronic symbol, 273
identifying terminals with ohmmeter, 276
PN junction for, 273
for rectification, 273
testing, 275–276
used in circuit to convert AC voltage to DC voltage, 273
Direct current (DC), 110
Direct-drive wind turbine, 146–147
Directional control devices, 96
cartridge valve, 100–101
check valves use in, 97
directional control valves, 97
directional, pressure and flow-control valves, 97, 101–102
pilot-operated hydraulic valves, 99–100
pressure-relief valve, 101
spool, cutaway view of, 98
three-position valve with four ports, 98
two-position valve
inner working, 98
with four ports, 97
valve spool, 97–98
Double-pole, single-throw (DPST) relay, 263
Double-row ball bearings, 140
Doubly fed induction generator, 127
Downwind turbines, 8–9, 50. *See also* Horizontal-axis wind turbines (HAWT)
DPST relay. *See* Double-pole, single-throw (DPST) relay
Drive trains
components mounting arrangement, 145
drive train compliance, 145–146
inspection through open nacelle, 146
vibration analysis, 146
DSIRE. *See* Database of State Incentives for Renewables & Efficiency (DSIRE)
Dual-element fuse, 268
Dynamic braking, 107

E

East Central Area Reliability Coordination Agreement (ECAR), 21
Electrical circuit
basic parts of, 240–241
closed switch, 249, 272
coil, 245, 262–267, 277
conductors, 238
copper wire use, 238–239
electrical current, 239–240
electron flow in, 238–239
fundamentals of
ladder diagram, 248
open and closed switch, 249
power source, switch and fan motor, 248
wiring diagram, 248
hole, 239
large and small currents measuring, 243–244
measurements with digital VOM meters, 244–245
milliamps measuring, 243
prefixes, symbols and exponents, used in, 248
resistance and continuity with digital VOM, measuring
continuity test, 245
high resistance in open fuse, 245
low resistance in, 245
resistance in, 240
steam, 239
switch controlling motor
manually operated, electrical diagram, 238
switch defined, 238
voltage, 238–239
measuring, 242
as water system, 241–242
wiring and ladder diagrams, 241
Electrical grid system, 20
for United States, 21
Electrical heating element, power consumption calculation, 251
Electricity
electrical current, 238
electrical power, 247
distribution from generating station to residential customer, 21
electron orbit, 238
elements, 238
nucleus, 238
Electromagnet
magnetic field, 113
reversing polarity, 114
magnetic strength, adding coils by, 114
Electromotive force (EMF), 122, 240. *See also* Voltage
Electronic inverters, use in wind turbines, 130, 162
EMF. *See* Electromotive force (EMF)
Enercon E-126, Europe, 176
Energy Research and Development Administration (ERDA), 38
Environmental and ecological assessments for wind power, 22
Erasable programmable read-only memory (EPROM), 81
ERDA. *See* Energy Research and Development Administration (ERDA)
European Wind Turbine Certificate (EWTC), 13
Europe, large offshore wind farms in
Lynn and Inner Dowsing Wind Farm, 212
Nysted Wind Farm, 213
EWTC. *See* European Wind Turbine Certificate (EWTC)

F

FAST code, 13. *See also* Standards and certification for wind turbines
Federal Aviation Administration (FAA), 180–181
tower height restrictions for
Determination of No Hazard (DNH), 180
Determination of No Hazard to Air Navigation (DNH), 180
electromagnetic interference and, 181

Federal Aviation Administration (*Contd.*)
Form 7460–1, filling of, 180
imaginary surfaces, issue of, 180
NPH, issue of, 180
operational impacts, issue of, 180
Finite-element analysis (FEA), 12, 66
Five-bladed wind turbines, 57, 74
Flicker power, 164
Florida Reliability Coordinating Council (FRCC), 21
Flow-control valves
with check valves, symbols for, 101
hydraulic
circuit diagram, 102
flow controls symbols, 101
FRCC. *See* Florida Reliability Coordinating Council (FRCC)
Frequency, 111
Full-wave rectifier, 273–274
Fuses, 267
circuit breakers and load centers, 269–270
disconnect panels, 269
plastic fuse pullers as safety, 269
sizes and shapes, 268
types of, 268

G

Gearboxes in wind turbines, 129–130
bearings, use of (*see* Bearings)
categories of gears, 132
differential gearbox, use of, 143–144
driven and driver gear, 131
gear meshing and backlash, 137
measurement of, 137–138
gear ratio, advantages of, 131–132
idler gear, 132–133
improvement in design of, 131
internal gears in, 131
location of, 130–131
need of, 130–131
operation of gears, 132
parts of, 144–145
planetary gears, 136
advantages and disadvantages, 137
specialized gears for wind turbine shafts, 137
troubleshooting gears, 138
types of gear in
bevel gear (miter gear), 132, 134–135
helical gears, 132–134
spur gears, 132–133
worm gear, 132, 135–136
use of, 131
Gear pumps
double-gear positive-displacement pump, 94
Generator
AC and DC electricity
current, 110
ground, 110
voltage, 110
Gigawatt, 2
Gin pole, 172
Global Positioning System (GPS) information, 201
Green technology
green energy payback calculations, 24–26
need for energy from, 22
skills needed for jobs
for maintaining and overhauling, 28–29
Grid
capacity margin for United States, 151
code requirements for, 159
definition of, 151–152
electrical power distribution system in US, 150
federal law to purchase power, 152
frequency and voltage control on, 162–163
grid sections in United States, 152
Eastern Interconnect, 151
Texas Interconnect, 151
Western Interconnect, 151
net meter, to measure electricity, 152
and rules and regulations on safety and power quality, 158–159
smart grid
blackout area, identification of, 153–154
brownout condition, 153
computer data transmission by high-voltage lines, 153
concept of, 152
creation and integration of, 154
information network in, 53
rate at low-peak times, 152–153
residential and commercial controllers, 153
transmission efficiency, increase in, 154
subsections of, 150–151
substation of, 151
voltage levels for long-distance transmission, 152
Grid companies (Gridco's), 152
Grid-tied wind turbines, 130
Ground-fault protection system, 195
Grounding, 160, 195
Guy wires, 170
for tilting-type tower, 173
tubular steel towers and, 171
use in lattice-type tower, 175–176

H

Half-wave rectifier, 273
Heater assembly, 266
Helical gears, 132
crossed helical gears, 134
double helical gear, 134
gear pitch, 133
herringbone gears, 134–135
thrust load, 134
Herringbone gear, 134
High-leg delta system, 260–261
Home-made wind turbine systems, 203–204
Horizontal-axis wind turbines (HAWT)
actuators, types, 99
anemometer and wind vane
ultrasonic wind measurement instrument, 79–80
wind direction indicator, 79
basic parts of, 71
gearbox and generator, 2
low-speed and high-speed shaft, 2
shaft and electrical generator, 53
turbine blades and rotor, 2, 53
brakes, 105
drive train and yaw drive, 106
dynamic braking, 107
hydraulic brakes, 107
hydraulic caliper brakes, 106
hydraulic cylinder, 106
mechanical parking brakes, 107
for commercial application, 53
computer control system for mid-sized wind turbine, 81
data acquisition and communications, 78–79
directional control devices, 96
cartridge valve, 100–101
check valves use in, 97
directional control valves, 97
directional, pressure and flow-control valves, 97
flow-control valves, 101–102
pilot-operated hydraulic valves, 99–100
pressure-relief valve, 101
spool, cutaway view of, 98
three-position valve with four ports, 98
two-position valve, inner working, 98
two-position valve with four ports, 97
valve spool, 97–98
directional valves, valve actuators and types of passageways, 98–99
hydraulic control systems
hydraulic actuators use, 89–90
hydraulic cylinders, 102–103
hydraulic motor, 103
hydraulic proportional control valves, 103
hydraulic servo control valve, 105
proportional directional and flow control, 104
proportional pressure control, 105
symbol for, 105
typical voltage and current signals, 104
hydraulic pump
AC motor and smaller pilot, 93
gear, 94
lobe, 95
vane, 94–95
variable-volume, 95–96
hydraulic solenoid valve, electrical control through
symbol for, 102
hydraulic systems
basic parts, 90
check valve in, 91
fluid dynamics, 92
hydraulic cylinder use, 90–91
pump and motor mounted on hydraulic reservoir, 90
reservoir, 92–93
terminology, 91–92
individual pitch control of blades, 76
latch coils use, 85
lever-actuated valve, 98
load efficiency, control of blades for, 76

magnetizing current in generator, controlling
PLC role, 89
nacelle and nacelle bedplate, 71
completely assembled nacelle, 72
number of blades
five-bladed wind turbine, 74
single-bladed wind turbine, 73
three-bladed wind turbines, 74
two-bladed wind turbines, 74
PLC
analog control, 86–87
basic parts, 82–83
blade torque, 89
EPROM, 81, 89
error, 88
feedback control, 87
image register for, 83–84
inputs and outputs modules, 85–86
ladder logic program in, 82
laptop/desktop computer, use, 83
nacelle, 76
nacelle directional control system, 88
open loop system, 88
operation, 81
for power control, 88–89
process variable, 87
program mode and run mode, 85
program storage and permanent memory, 89
program understanding, 85
reservoir with pump and valves, 92
rotor blade pitch adjustments and teetering, 74–75
run mode, scanning, 83
SCADA system, 80, 89
scan cycle for, 82
summing junction, 87–88
three-port directional valves, configurations for, 99
type dedicated controller used, 81
wind turbine control systems, 80
yaw control, 77–78
yaw drive brakes, 78
yaw mechanism, 77
rotational speed and rotor speed control at high-speed end, 75
stall control in small wind turbines, 76
rotor hubs and blade types, 72
blade assembly, 74
mounting plate for rotor, 73
three blades with, 7
upwind and downwind turbines, 8
Hydraulic brakes, 107
Hydraulic cylinders, 102–103
Hydraulic proportional control valves, 103
hydraulic servo control valve, 105
proportional directional and flow control, 104
proportional pressure control, 105
symbol for, 105
typical voltage and current signals, 104
Hydraulic pump
AC motor and smaller pilot pump, 93
gear pumps, 94
lobe pumps, 95
vane pump, 94–95
variable-volume pumps, 95–96
Hydraulic servo control valve, 105
Hydraulic systems
basic parts, 90
check valve in, 91
fluid dynamics, 92
hydraulic cylinder use, 90–91
pump and motor mounted on hydraulic reservoir, 90
reservoir, 92–93
terminology, 91–92
Hydrogen atom with electron flow, 238–239

I

IEA. *See* International Energy Agency (IEA)
IEC. *See* International Electrotechnical Committee (IEC)
IGBTs. *See* Insulated-gate bipolar transistors (IGBTs)
Image register, 83–84
Impedance, 255
total opposition for inductive and resistive circuit calculating, 257
Induction AC motor
counter-EMF, 122
EMF, 122
operation, 121
torque and slip use, 121–122
Induction generator. *See* Asynchronous AC generators
Inductive load, 163
Inductors, 255
inductive reactance, 256
Installation of wind turbine
steps in
amount of wind available identifying, 216
electrical cables and connections, 219–222
electrical connections to wind turbine electrical panel, 222–223
mounting nacelle to tower, 218–219
mounting of blades to rotor and nacelle, 217–218
permits and other documents, 217
site preparation, 216–217
testing for initial run, 223–224
Institute of Electrical and Electronics Engineers (IEEE), 159
Insulated-gate bipolar transistors (IGBTs), 126, 162, 282
Insulator
atomic structure for, 271
electrons flow in, 270
International Electrotechnical Committee (IEC), 13
wind energy standards, 14
wind turbine standards, 14
International Energy Agency (IEA), 13
International Standards Organization (ISO), 13–14
Interstate Renewable Energy Council (IREC), 26
Inverter, 4
applications for, 285
CSI inverter, 285
DC voltage to AC voltage, change, 282
PWM inverter, 284
single-phase, 282
six-step inverter, 282
three-phase, 282
VVI inverter, 284
IREC. *See* Interstate Renewable Energy Council (IREC)
Islanding, 164–165
ISO. *See* International Standards Organization (ISO)

J

Jobs in wind power industries
assembly technicians, 29–30
blade technicians, 30
educational requirements for, 29
electricians and electrical transmission technicians, 30
field service technicians, 29
planning and sales, 29
project manager and architectural jobs, 29
riggers and tower installers, 30
teachers and technical trainers, 29
websites, 30

K

Kilowatt, 2

L

Ladder diagrams
sequence of operation, 241
Laminated steel rotor, 120
Land breeze, 34–35
Large AC motor
hydraulic and smaller pilot pump, 93
Large horizontal-axis wind turbines
basic parts of, 3, 53–54
Large wind farm projects
Capricorn Ridge Wind Farm, 209
Fowler Ridge Wind Farm, 210
Horse Hollow Wind Energy Center, 210
Klondike Wind Farm, 208
Lone Star I and II Wind Farm, 208
Peetz Table Wind Energy Center, 209
Roscoe Wind Farm, 210
San Gorgonio Pass Wind Farm, 209
Sherbino Wind Farm, 210
Sweetwater Wind Farm, 209
Tehachapi Pass Wind Farm, 210
Lattice towers. *See also* Towers for wind turbines
assembly of, 174
drawback of, 173–174
electrical power cable, placing of, 174
guyed lattice tower, 175–176
lifting harness to wind turbine, securing of, 174–175
mounting wind turbine on, 174
nuts and washers, installing of, 175
structure of, 173
use of crane for lifting, 175
Light-emitting diode (LED), 276
Lightning arrestor, 194
Lightning protection system, 194–195
Load centers, 269–270. *See also* Fuses
Loads, 240
on electrical circuit, 163
Lobe pumps, 95
Locked-rotor amperage (LRA), 266–267
Low-voltage ride-through (LVRT), 163

LRA. *See* Locked-rotor amperage (LRA)
LVRT. *See* Low-voltage ride-through (LVRT)

M

MAAC. *See* Mid-Atlantic Area Council (MAAC)
Magnetic coil. *See* Electromagnet
Magnetic theory
  bar magnet and flux lines, 112
  coil magnet, magnetic field strength, 114
  domains/dipoles, 112
  electromagnet, 113
    magnetic field, reversing polarity, 114
    magnetic strength, adding coils by, 114
  magnesia, 111–112
  magnet, 111
    first and second laws of magnets, 112
  permanent magnet, 112
Maintenance of wind generation systems
  checklist for, 234
  electrical system, 234–235
  grease fittings on rotor and hub, 234
  hydraulic system, 235
  lubrication, 234
  oil in gearbox, 233
  safety harnesses and lanyards, 234
  of tower, 234
  visual inspection of blades, 234
Mechanical brakes, 107
Megawatts, 2
Megger test, 231–232
Mid-Atlantic Area Council (MAAC), 21
Mid-sized wind farm projects, 208
Milliamps meters measuring, 243
Motor starters
  basic parts of
    auxiliary contacts, 265
    control circuit, 265
    pushbutton contacts, 266
  commercial applications in, 265
  exploded view of, 267
  operation and overload, 266–267
  sizing of, 267
MSC software, 13
Multiplier, 243

N

Nacelle
  and bedplate, 71
    completely assembled nacelle, 72
  defined, 3
National Electrical Code (NEC), 13, 159, 190
National Electrical Manufacturers Association (NEMA), 264
National Fire Protection Association (NFPA), 13, 159
National Renewable Energy Laboratory (NREL), 11, 13, 159
National Science Foundation (NSF), 38
NEC. *See* National Electrical Code (NEC)
Needle bearings, 141–142
NEMA. *See* National Electrical Manufacturers Association (NEMA)
Net meter, 152, 161
NFPA. *See* National Fire Protection Association (NFPA)
Northeast Power Coordinating Council (NPCC), 21
Notice of Presumed Hazard (NPH), 180–181
NPCC. *See* Northeast Power Coordinating Council (NPCC)
NPN transistors
  base–collector junction
    testing for forward and reverse bias, 278
  electronic symbol and diagram for, 277
  operation of, 276–277
  troubleshooting, 277–278
NREL. *See* National Renewable Energy Laboratory (NREL)
NSF. *See* National Science Foundation (NSF)
N-Type material
  arsenic and silicon combining, 271

O

Occupational Safety and Health Administration (OSHA), 190
Offshore wind farms in Europe
  Lynn and Inner Dowsing Wind Farm, 212
  Nysted Wind Farm, 213
Ohio Department of Development's (ODOD)
  grant program for renewable energy projects, 28
Ohm's law
  for calculating current, 246
  for calculating resistance, 246
  electrical power, calculating, 247
  formulas, 246
    multiply line and divide-by line, 247
    Ohm's law wheel for remembering, 246–248
    pie, memory tool for, 247
Oil embargo, 6
Organization of Petroleum Exporting Countries (OPEC)
  amount of oil and, 6
OSHA. *See* Occupational Safety and Health Administration (OSHA)
Overhaul of wind turbine, 235
Overload device, 266

P

Parallel circuits, 251
  calculating power, formula, 254
  product over sum method, 253
  series-parallel circuit, 252
  voltage, resistance and current, 252
    calculation, 253
Passive pitch control, 63
Payback calculators, 25
Peak electrical demand
  backup power, 20
  brownout, 19
  shedding demand, 19–20
PEFC. *See* Power electronic frequency converter (PEFC)
Periodic maintenance, 233–235. *See also* Maintenance of wind generation systems
Permanent-magnet generator, 127–128
Personal protective equipment (PPE), 150
Pilot-operated hydraulic valves, 100
  four-port directional valves, common configurations for, 99
  hydraulic circuit, 99
  pilot pressure, 99
Plain bearings (bushing)
  pillow block, 139
  split pillow block, 139–140
Planetary gears, 136
Plug-in type relay, 262
PN junction material
  forward biasing, 272
  reverse biasing, 272–273
PNP transistors
  electronic symbol and diagram for, 277
  operation of, 276–277
  troubleshooting, 277–278
Power
  diode with anode and cathode identified, 276
  distribution system, 15–16
  factor
    calculating in AC circuit, 257
    correction, 163
  isolation, 161
  poles, 166
  quality, 162
  surge, 160
Power electronic frequency converter (PEFC), 54, 114–115, 126, 202
Pressure-relief valve
  hydraulic symbol, 101
Prevailing winds, 35
Programmable logic controller (PLC), 81
  analog control, 86–87
  basic parts, 82–83
  blade torque, 89
  EPROM, 81, 89
  error, 88
  feedback control, 87
  image register for, 83–84
  inputs and outputs modules
    electrical diagram of, 86
    instructions, 85
    latch coils use, 85
    processor, 85
    wiring input switches and output devices, 86
  ladder logic program in, 82
  laptop/desktop computer, use, 83
  nacelle, 76
    directional control system, 88
  open loop system, 88
  operation, 81
  for power control, 88–89
  process variable, 87
  program
    and run mode, 85
    storage and permanent memory, 89
    understanding, 85
  rotational speed and rotor speed control
    at high-speed end, 75
  rotor blade pitch adjustments and teetering, 74–75
  run mode, scanning, 83

SCADA system, 80, 89
scan cycle for, 82
summing junction, 87–88
three-port directional valves, configurations for, 99
type dedicated controller used, 81
wind turbine control systems, 80
yaw control, 77–78
yaw drive brakes, 78
yaw mechanism, 77
Project development plan, 201
Proportional, integral and derivative (PID), 62–63
P-Type material
aluminum and silicon combining, 271–272
Public Utility Regulatory Policies Act (PURPA), 1978, 152
Pulse-width modulation (PWM) inverters, 284–285
PWM inverters. *See* Pulse-width modulation (PWM) inverters

Q

Quietrevolution QR5 vertical-axis wind turbines, 52, 184

R

Ratchet mechanism, 266
Rectifier
half-wave and full-wave, 273
three-phase, 273–275
Red leg delta system. *See* High-leg delta system
Relay, theory and operation
coil and contacts, 261
ratings for, 262
difference between contactor and, 264
identifying by arrangement of contacts, 263
normally open and closed contacts, 262–264
plug-in type relay, 262
pull-in and hold-in current, 262
Residential wind turbine, 202
Resistance, 238
and inductance in AC circuit
inductive reactance, 256
phase shift for, 256
Resistive load, 163
Roller bearings
double-row roller bearing, 141
features of, 140–141
tapered roller bearing, 141
Rolling-element bearings
ball bearing, 140
bearing races in
needle bearing, 141–142
roller bearing, 140–141
Rotary-armature alternator, operation theory, 125
Rotational speed and rotor speed control
at high-speed end, 75
stall control in small wind turbines, 76
Rotor hubs and blade types, 72
blade assembly, 74
mounting plate for rotor, 73
Rural location
multiple larger megawatt wind turbines in, 3
wind turbines use in, 203

S

Safety brakes, 196–197
Safety issues in wind turbine. *See also* Towers for wind turbines
electrical safety and surge protection, 195
ground-fault systems, 195
grounding system, 195
lightning safety, 194–195
other protection circuits, 195
safety connection switches, 195–196
Savonius type VAWT, 6, 52
with additional wing, 49
advantages of, 7
SCADA system. *See* Supervisory Control and Data Acquisition (SCADA) system
Sea
breeze, 34
large megawatt wind turbines, 3
Self-excited shunt generators, 117
flashing field, 116
Semiconductors
atomic structure for, 271
Separately excited shunt generators, 116
SERC. *See* Southeastern Electric Reliability Council (SERC)
Series circuits
connecting switches in, 249
current and voltage by Ohm's Law, 250
electrical heating element, power consumption
large resistance load, 251
fuse, 249
fuse disconnect, 249–250
hydraulic pump controlling by, 249
loads adding in, 250
open fuse, 249
power consumption of each resistor, 251
solving for current in, 250
voltage drop across each resistor, calculation, 250–251
Series generator, 117
Series-parallel circuits
with hydraulic pump and lube pump, 254
power circuit, 254
used in wind turbines, 254
oil pressure switch and oil temperature switch, 255
Service drop, 166
Short circuit element, 268
Shut-down brakes, 197–198
Silicon-controlled rectifier (SCR)
anode–cathode junction, 278
controlling, 279
electronic symbol for, 278
operation of
in circuit controlling DC voltage and current to load, 279
conduction angle, 279
firing angle, 279
used to control DC motor, 279
P-type and N-type material in, 278
testing, 279–280
Single-bladed wind turbines, 73
advantage of, 57
applications in, 56–57
with one counterbalance, 57
Single-element fuse, 268
Single-phase full-voltage magnetic controllers
NEMA sizes for, 268
Single-phase inverter, 282
electronic diagram for, 283
Single-pole, single-throw (SPST) relay, 263
Single-source power system, 167
Site
assessment, 201
issues
environmental, 201
maintaining, 202
noise, 201–202
visual impact, 201
preparation
pad and mounting, 216–217
Six-diode bridge
for rectifying three-phase AC voltage to DC voltage, 274
Six-step inverter, 282
Small community wind farm projects, 207–208
Small residential-type horizontal-axis wind turbine, 53
Small Wind Certification Council (SWCC), 12
Smart grid
blackout area, identification of, 153–154
brownout condition, 153
computer data transmission by high-voltage lines, 153
concept of, 152
creation and integration, 154
information network in, 53
rate at low-peak times, 152–153
residential and commercial controllers, 153
transmission efficiency, increase in, 154
Southeastern Electric Reliability Council (SERC), 21
SPST relay. *See* Single-pole, single-throw (SPST) relay
Spur gears, 132–133
applications of, 133
gear and rack, 133
idler gear, 133
Squirrel-cage induction generator, reactive power generation by, 163
Squirrel-cage rotor for induction motor, 120
Stand-alone remote power sources, 22
Standards and certification for wind turbines
agencies for, 13
ANSI standards, 13
AWEA standards, 12
BWEA certification, 13
computer-aided design and modeling for safety testing, 12
EWTC certification, 13
IEC standards, 14
ISO standards, 14
NEC standards, 13
NREL certification, 11–12

Standards and certification (*Contd.*)
small wind turbine certification, 12
standards organizations in North America and Europe, 12–13
Stationary-armature alternator, operation theory
exciter current, 123
programmable logic controller (PLC) use, 124
separately excited machine, 124
Stinger leg delta system. *See* High-leg delta system
Summing junction, 87–88
Supervisory Control and Data Acquisition (SCADA) system, 25, 80, 89
Surge protection, 195
SWCC. *See* Small Wind Certification Council (SWCC)
Switches
controlling motor
manually operated, electrical diagram, 238
and connections for power distribution, 160
circuits to grid connection, switch for, 160
disconnect switches, use of, 161
grounding equipment, 160
protection system for grid faults, 161
safety disconnect switch, 160
surge-protection equipment, 160
voltage phase, monitoring of, 160–161
defined, 238
Synchronous AC generators, 126–127
Synchronous AC motor, 122

T

Thermistor, 280
Three-bladed wind turbines, 74
disadvantage of, 56
energy conversion in, 55
Three-phase contactors. *See also* Contactor
NEMA ratings for, 264
Three-phase inverter, 282
electronic diagram for, 283
Three-phase transformer
delta-connected, 259
high-leg delta system, 260–261
neutral terminal with
delta-and wye-connected transformers, 259–260
wye-connected, 258–259
Thrust bearings (washers), 142
Tilting-type tower, 171–172
Torque, 114
Towers for wind turbines, 170
birds and bird safety, 198
climbing towers to top
electrical service lift inside tower, use of, 187
inside stairs and ladders, 186–187
by outside ladders and lattice-type, 188–189
FAA tower height restrictions for
Determination of No Hazard (DNH), 180
Determination of No Hazard to Air Navigation (DNH), 180
electromagnetic interference and, 181
Form 7460–1, filling of, 180
imaginary surfaces, issue of, 180
NPH, issue of, 180
operational impacts, issue of, 180
foundations and concrete support
concrete pad or pier for support, 184–185
finished poured pad, for monopole tower, 185
metal reinforcement material in pad, use of, 185
monopole installation, on small concrete pad, 186
pad for lattice-type tower, 186
larger monopole towers, 176–178
lattice towers
assembly of, 174
drawback of, 173–174
electrical power cable, placing of, 174
guyed lattice tower, 175–176
lifting harness to wind turbine, securing of, 174–175
mounting wind turbine on, 174
nuts and washers, installing of, 175
structure of, 173
use of crane for lifting, 175
lifting equipment and cranes for
mid-size hydraulic crane, 179
small crane for lattice tower, 179–180
tall crane with lower-section boom, 179
maintenance of, 198–199
monopole towers, 176
near or over water, 181
deep-water installations, 182
location on seabed, 182
location over shallow water, 183
mid-depth installation, 182
problem in offshore installations, 183
technology for mounting of tower, 182–183
overspeed safety and overload controls for protection, 196
fail-safe systems, 198
hydraulic parking brake, 198
safety brake or tip brake, 196–197
service brake, 196
shut-down brakes, 197–198
safety procedures and regulations, 189–190 (*see also* Climbing and safety equipment)
anchorage equipment, 189
ANSI/ASSE Z359 Fall Protection Code, 191
automatic descent devices, 193
fall prevention and fall arrest equipment, 189
inspection and care of rescue equipment, 194
OSHA regulations for working safely, 190–191
quick-descent system, 193–194
rescue and evacuation strategies, 193
safe evacuation from nacelle, 193
safety during construction and installation of turbines, 190
safety exercises, conduction of, 194
U.S. and Canadian standards for safety, 190
steel tower, sections connection in, 178
types of
cylindrical pipe, for wind turbine applications, 171
lattice tower, 170
monopole towers, 170
self-supporting towers (cubes), 181
tilting-type tower, 171–173
tubular steel towers, for small wind turbines, 170–171
turbine mounted on light pole, 183–184
vertical-axis wind turbine, 184
Transformer, 154
kilovolt-amperes, as unit of power, 154
operation of
induced current, 155
induced voltage, 155–156
induction, principle of, 155
transformer windings and AC voltage, relationship between, 155
overview of, 154–155
primary winding, 154
secondary winding, 154
step-down transformer, 154, 156–157
step-up transformer, 154, 156
substations and power distribution in city, 157–158
troubleshooting application, 157
and turns ratio, 155
secondary voltage and secondary current, calculation of, 156
VA (volt-ampere) rating for, 157
voltages used in power distribution systems, 157
Transmission companies (Transco's), 152
Triac
electronic symbol for, 280
as solid-state relay, 280
testing, 281–282
used as switch to turn on voltage, 280
for variable voltage control, 281
Trip mechanism and overload contacts, 266
Troubleshooting in wind turbine, 225
chart for small wind turbine electrical problems, 228
checking, 225
data collection system for, 226
electrical problems
circuit and fuse, 230
clamp-on ammeter use, 232
coil or contacts of relays, 230
continuity test, 231
ohmmeter use, 231
power source, 230
testing components and wiring with Megger, 231–232
voltmeter use, 229–230
generation and transmission problems
blade control, 229
mechanical system, 228
sensor to determine, 229
shafts and gearbox, alignment of, 228
hydraulic problems, 232–233
keeping records and maintenance log, 226
mechanical and tower problems, 229
swapping parts, 225–226
understanding of, 226–227

using company resources
manuals, bulletins, safety alerts and help desks, 225
using tables and matrix, 227
warranty work, 226
True power, 163
Two-bladed wind turbines, 74
advantage of, 56
efficiency, 56
teetering
mounted on teeter hub, 56
Type certification, 11. *See also* Standards and certification for wind turbines
Type characteristic measurements, 11

U

Underwriters Laboratory (UL), 12–13
Uninterrupted power supply use, 15
United States
electrical demand in, 16–17
electrical energy production and capacity from wind turbines in, 207
electrical generation
capacity for, 17
by fuel and sector, 18
electrical grid system for, 21
electricity
commercial and industrial demand, 19–20
consumption by sector, 17
generation by fuel and sector, 17
residential demand, 20
total daily demand for, 18–19
total peak demand, 19
use in, 2
grid
capacity margin for, 151
electrical power distribution system in, 150
grid sections in, 152
Eastern Interconnect, 151
Texas Interconnect, 151
Western Interconnect, 151
investments in wind energy, 6
net internal demand, actual/planned capacity resources and capacity margins, 21
wind farms in, 206
offshore installations, 211–212
wind resources, map of, 43
Upwind turbines, 8, 50. *See also* Horizontal-axis wind turbines (HAWT)

V

Vane pump, 95
cartridge, 95
vanes of, 94–95
Variable-voltage inverter (VVI), 284
Variable-volume pumps
axial piston pump, 96
piston pump, 96
pump housing, 95–96
Vertical-axis wind turbines (VAWT), 49
advantages and disadvantages, 53
Darrieus and Savonius types, 6
Visual and landscape assessment, 202
Voltage, 110, 238
measurements and safety, 243
Volt-ampere reactance (VAR), 258
Voltmeter
analog-type voltmeter, 242
digital-type voltmeter, 242
probes placement of, 242–243
Volt-ohm-milliammeter (VOM), 243
digital VOM meters
measurement with, 244–245
resistance and continuity, measuring with, 245
RMS digital VOM, 244
VVI inverter. *See* Variable-voltage inverter (VVI)

W

Watts, 247
Wild leg delta system. *See* High-leg delta system
Wind
extreme wind speeds, 37
farms
advantages of, 206
large wind farm projects, 208–210
mid-sized wind farm projects, 208
small community wind farm projects, 207–208
substations, 167
gusting wind speeds, 37
nature of, 35
predicting wind speed, 37–38
speed, variations in
day/night, 36
seasonal and annual, 36
system, geographic considerations for, 35–36
turbulence, 36–37
in complex terrain, 39
NASA, smoke tests with wind turbine for, 38
in wakes on wind farms, 38–39
Wind power
for commercial application, 40–41
environmental and ecological assessments, 22
financial implications and return on investment, 23
future directions for, 23
growth in recent years, 6
history of, 4–6
industries, jobs in
assembly technicians, 29–30
blade technicians, 30
educational requirements for, 29
electricians and electrical transmission technicians, 30
field service technicians, 29
planning and sales, 29
project manager and architectural jobs, 29
riggers and tower installers, 30
teachers and technical trainers, 29
websites, 30
political implications for, 23
as source for electricity, 2
tax considerations for, 26
additional U.S. investment tax credit, 27
checking out latest offers, 28
examples of state programs, 28
terminology
acceleration, 33
air pressure, 33
amount of work, 33
barometric pressure, 33
boundary-layer wind, 34
force and mass, 33
gradient wind, 34
pressure, 33
pressure gradient force, 34
wind direction and velocity, 33
wind friction, 33–34
Wind turbines
in agricultural area, 203
angle of attack and blade pitch, 61
basic parts and electricity generation, 2–4
blade pitch control, 61
active stall control, 63–64
closed-loop pitch control, 62–63
energy source use, 63
individual blade pitch control, 64
optimum wind speed for, 62
passive pitch control, 63
calculating cost and income, factors in, 24
classification
HAWT, 7–9
by number of blades, 9
by output electricity, 9
by size, 9
VAWT, 6–7
codes for, 42
comparison of blade types, 57–58
constant rotational speed, operation, 60
output shaft, 61
controlling, 4
data for calculation of payback period, 25
determining power at site, 41–42
drive train in (*see* Drive trains)
electrical cabinet for, 223
electricity to grid from (*see* Grid)
electricity transmission limitations, 15–16
electronic circuits, use of, 164
energy converted, 64
Betz's law, 65
gearbox and generator, efficiency, 65
factors affecting power quality, 164
five-bladed turbines, 57
flicker power, control of, 164
generator, 122
grounding system for, 165
installed electrical generation capacity, 17
located at railroad switching yard, 41
low-voltage fault in grid system, 163
mounted on existing lighting poles, 40
need of gearbox in (*see* Gearboxes in wind turbines)
and net metering, 161–162
overhead feeder circuits, 166
peak performance
air density, 67
generator efficiency measuring, 68
laboratories testing, 67–68
power curve, 66–67
wind curve in, 67
PLC control, 4

Wind turbines (*Contd.*)
power
connection to grid, 160
curve data, 26
quality issues, 162
supply to building or residence from, 159–160
transmission line, located on, 40
for schools and colleges, 41
series-parallel circuits used in, 254–255
single-bladed, two-bladed and three-bladed turbines, 55–57
advantages and disadvantages, 58
and single-source power system, 167
site requirements, 42
specifications of, 9–11
standards and certification
agencies for, 13
ANSI standards, 13
AWEA standards, 12
BWEA certification, 13
computer-aided design and modeling for safety testing, 12
EWTC certification, 13
IEC standards, 14
ISO standards, 14
NEC standards, 13
NREL certification, 11–12
small wind turbine certification, 12
standards organizations in North America and Europe, 12–13
testing, 65
FEA, 66
SCADA, use, 66
tool for selecting site for locating, 42
towers used for (*see* Towers for wind turbines)
true power and reactive power, 163
turbulence, problems, 66
types of, 2
underground conduits from, 165–166
at variable wind speeds, 64
voltage regulation on, 162–163
ways to obtain income from, 22–24
in wind farm, 167
Wiring diagrams, 241
Worm gear
drawbacks of, 136
as speed reducer, 135
Wound-rotor AC motor
armature with two slip rings, 120
operation, 121
Wye-connected three-phase transformer, 258–259

Y

Yaw control, 4, 77–78
Yaw drive, 77
Yaw drive brakes, 78
Yaw mechanism, 8–9, 51, 76–77